Visual masking: an integrative approach

OXFORD PSYCHOLOGY SERIES

Visual masking:
an integrative approach

BRUNO G. BREITMEYER
Department of Psychology
University of Houston
Texas

OXFORD PSYCHOLOGY SERIES NO. 4

CLARENDON PRESS : OXFORD
OXFORD UNIVERSITY PRESS : NEW YORK
1984

Oxford University Press, Walton Street, Oxford OX2 6DP
London Glasgow New York Toronto
Delhi Bombay Calcutta Madras Karachi
Kuala Lumpur Singapore Hong Kong Tokyo
Nairobi Dar es Salaam Cape Town
Melbourne Auckland
and associates in
Beirut Berlin Ibadan Mexico City Nicosia

Oxford is a trade mark of Oxford University Press

Published in the United States
by Oxford University Press, New York

Breitmeyer, Bruno G.
  Visual masking.—(Oxford psychology series; no.4)
  1. Visual perception
  I. Title
  152.114   BF241

  ISBN 0–19–852105–7

Set by Promenade Graphics Ltd., Cheltenham, Glos.
Printed in Great Britain by
Butler and Tanner Ltd., Frome, Somerset

Preface

This book represents an accumulation and culmination of experimental, theoretical, and scholarly work carried out in the last 7–8 years. My interest in visual masking was inspired in the early 1970s by three confluent events: the completion of two separate dissertations on metacontrast by my colleagues, Drs Sybille Sukale–Wolf (now Dr Sybille Anbar) and Bruce Bridgeman, during our graduate studies at Stanford University; the subsequent seminar on visual masking conducted by Dr Leo Ganz of Stanford University; and my own concurrent work and interest in motion perception. In 1971, Dr Anbar and I had fancied and tentatively planned writing a book on spatiotemporal aspects of visual perception. For a variety of reasons, including geographical separation and vocational changes, our plan was never realized. Nevertheless, the desire eventually to write a book of this genre remained dormant in me for several years. During this period progressively greater amounts of relevant empirical and theoretical information coalesced around my own experiments and contemplations on visual masking. My suspended desire was stirred by an increasingly felt need for a systematization and integration of this vast amount of information and has found expression over the last two years in the writing and publication of the current monograph on visual masking.

The published literature on visual masking and related topics is extremely voluminous. To keep the length of the book within reasonable limits, I have chosen to restrict coverage to what, in my opinion, are the most germane and relevant empirical and theoretical findings. However, within these limits, the citation of published material is fairly extensive; for some, perhaps too extensive. Among its various functions, the book is intended to serve as a reference source for readers interested in pursuing in greater detail some of the discussed topics. The extensive references are a courtesy extended not only to these readers but also to the many researchers whose work has contributed in varying degrees to our understanding of visual masking. To those investigators whose work I did not incorporate and cite, I can give only two reasons: my witting choice of relevant topics and my unwitting ignorance of their work. In this connection, I also have deliberately omitted a separate chapter on visual coding, despite its suggested inclusion by one of the earlier anonymous reviewers of my initial, tentative chapter outline. Visual coding is a highly complex topic and itself would require at least a book-length manuscript to do its discussion and analysis justice. Several recent publications (C.S. Harris 1980; Uttal 1973, 1981) in fact are devoted entirely or in large part to discussions of visual coding, and the interested reader can go through these or other sources.

Since visual masking is itself a highly complex and technical phenomenon, the book focuses on it and what are considered to be more manageable associated topics. To facilitate the reader's comprehension of these topics, numerous illustrations have been chosen that hopefully enrich their descriptions. None the less, some readers may find that certain sections or entire chapters of the book are difficult to read. The responsibility for this difficulty lies mainly with my approach to the study of visual perception, which is more extensively grounded and discussed in Chapter 11.

Briefly, the approach to visual masking adopted in the current book is psychobiological. The attempt is to relate many of the known perceptuo-cognitive phenomena, typically relegated to psychological disciplines, to known neurophysiological and neuroanatomical findings. Moreover, both the psychological and related neural phenomena, in turn, are related to what are assumed to be their common function in extralaboratory, naturalistic visual behaviour.

The nature of the former relation of psychological to neural phenomena is not specified, except by the following exclusion. It should be made very plain here, at the outset of this book, that I am not proposing a neural reductionism of perceptual phenomena or conversion of psychological theories to neural data. Interpretation by readers of such an attempt usually is based on the unquestioned premise that a description of psychoneural relationships is by definition a form of such reductionism.

One may as well interpret the current approach as a form of 'elevationism' attempting to explain or predict neural findings in light of pyschological or behavioural ones rather than vice versa. However, neither 'ism' seems appropriate for an approach that primarily advocates conceptual integration of diverse perceptual and neural phenomena. Its aim is to give coherence both to psychological and to neural laboratory findings in terms of their common behavioural, functional, and ecological significance. As such, the approach is more of a heuristic device guiding a research programme than a formal theory relating neural and psychological findings via specific linking hypotheses.

The realization of this approach therefore rests on discussing perceptual, neural, and behavioural phenomena. On the one hand, for the perception psychologist not familiar with neurophysiology or neuroanatomy several chapters, especially Chapter 6, may prove to be difficult reading. On the other hand, the same difficulty may hold for the anatomist or physiologist not familiar with psychological terminology, theories, or findings discussed in other chapters. However, initial or temporary difficulties with certain topics should not discourage the reader, since they become clarified as the explanatory context of the book broadens.

The book is organised along three main themes. Chapter 1 outlines the *history* of visual masking. In Chapter 10 the study and theory of visual masking is placed in the context of phylogenic and ontogenic development and *purposive visual behaviour* executed under a naturalistic, extralaboratory environment. The eight intervening chapters concentrate on the experimental *methods* and *findings* and the inferred *mechanisms* and *processes* of visual masking as well as the *theories* that have been proposed to account for them. In particular, Chapters 2–4 review methods and findings of visual masking by light, visual response integration and persistence, and visual pattern masking. In Chapter 5, recent theories of lateral masking are reviewed. The readers thoroughly familiar with these findings and theories may skim these chapters; however, for the sake of continuity, more careful reading of these chapters is strongly suggested. Chapter 6 reviews recent neurophysiological and neuroanatomical findings on sustained and transient cells in the visual system and serves as a preliminary to the introduction, in Chapter 7, of a theory of visual masking, based on their assumed existence in human vision. Chapters 8–10 outline and discuss extensions of this theory to higher order processes in perception, abnormal vision, local and long-range masking effects, and the role of masking and related phenomena in dynamic visual behaviour characterized by saccadic changes of fixation or the maintenance of fixation when the observer or objects in the environment are moving. When going over Chapters 2–9, the reader should keep the following topics in particular focus as preparation for their integration with themes additionally introduced in Chapter 10:

1. The relation of (a) precortical *visual response persistence* to (b) *visual response integration* and, in turn, their relation to *visual masking*, particularly the *forward* variety (mask precedes target) (Chapters 2–4).

2. The *suppression* of precortical response persistence and, hence, of response integration by a retrochiasmic, centrally located form of visual inhibition (Chapter 6) indexed pyschophysically by (a) a spatially local, lateral masking phenomenon known as *metacontrast* (Chapters 4, 5, and 7) and (b) a long-range masking phenomenon known as the *jerk effect* (Chapter 9).

3. The existence of a variety of *target disinhibition* or *recovery* effects when a second mask stimulus is added to a typical target–mask sequence (Chapter 9).

4. The role of *selective attention* and its control and direction in visual masking (Chapter 8).

The reading of Chapter 11, the Epilogue, is optional. It presents no new findings or theories of visual masking. Its intent is to give a reasonable justification for the organization of topics on historical, methodological, and philosophical grounds and was included as much for my own as for the reader's benefit. My ranking as philosopher and historian of science is, at best, as an interested non-expert. Nevertheless, I ventured into these fields in order to place my metatheoretical approach to the study of visual masking into clearer perspective.

I hope that the approach adopted in the book will yield a comprehensive and integrative outlook on the study of visual masking. I take full responsibility for any failure in achieving this end. Moreover, it is my genuine hope that this book will inspire the readers interested in visual masking and related topics to experiment and think creatively in their own work. I wish them pleasant and productive reading.

Finally, I would like to express my gratitude to a number of individuals and organizations who contributed to the writing of the book. Thanks go to Mss Betty Baldwin, Jo Ann Johnston, Teresa McCuan, Heidi Muller, Carolyn Watkins, and Sabrina Reagan for typing portions of the original and final manuscript versions of the book, and to Eugene Gilden, Alysia Banta, Kevin Olive, and Frank Sergi for aiding in much of the library research; to an anonymous reviewer of the original manuscript for the patient reading and the extensive chapter-by-chapter, itemized suggestions for revising and improving the final manuscript; to Mary Williams, who, by volunteering much of her time to a critical and constructive proofreading of both versions of the manuscript, contributed significantly to an improved rendition of the monograph; to Marcus Boggs and Oxford University Press for facilitating its editorial review and publication; to the Center for Public Policy of the University of Houston for assuming the costs incurred by my inclusion of the many illustrations copied from published material and to the University of Houston Publications Committee for generously supporting a portion of the typing costs.

Houston, Texas   B. G. B.
November 1983

Contents

This book is dedicated to my parents,
Gustav and Julia Breitmeyer

1 A history of visual masking

1.1. Introduction

If we view science as a means of giving cognitive coherence to our observations or perceptions of the world, then the history of science in turn can be viewed as the study of the evolving contents and modes of such cognitive structuring. Typically, a science progresses when it confronts puzzles, problems, or inconsistencies that require solutions or when fortuitous discoveries are made. The resulting advances usually are piecemeal and methodical. By reinforcing and adding to the edifice of observations structured around extant theory and method, these advances indicate the more or less continuous development of what Kuhn (1962) has called 'normal science'. However, significant anomalies crop up occasionally, which remain deeply and inextricably rooted within a particular scientific system. Their persistent intractability signals an intellectual crisis, the resolution of which demands nothing less than a radical, profound reorganization not only of the prevailing and manifest theoretical and methodological framework but also of the oft-tacit presuppositions or metatheoretical foundations on which it is based. Relative to the normal-science time frame, such radical, saltatory resolutions of scientific anomalies comprise what Kuhn (1962) has termed 'paradigm-shifts'.

What follows from the above, briefly sketched definition of the scientific enterprise is that the history of a science consists not merely of a catalogue and chronicle of its theories, methods, and observations but also, and more importantly, of the *analysis* and *interpretation* of their attendant problematic situations (Popper 1972) and fundamental presuppositions (Collingwood 1940). Situational analysis and interpretation of this sort is akin to what Hempel (1966) has termed 'explication' or 'rational reconstruction' and to Collingwood's (1956) method of 'reenactment of past thought' (for further discussions and critiques of these historical methods, see Donagan (1966) and Skagestadt (1975)).

Besides attempting to render an adequate situational analysis of the particular science of visual masking, the historical approach adopted here also highlights numerous significant similarities between past and present presuppositions, theories, methods, and observations. This aspect of the approach is not premised on any *a priori* notion of history as recurrent or repetitive but rather on an *a posteriori* analysis drawing on relevant prior and current sources. Nor is the approach meant to yield the impression that in regard to visual perception nothing new is under the sun. On the contrary, the study of visual perception in the last century and in particular in the past three decades has been marked by vast accumulations of new findings and by significant conceptual and technical developments (see, for instance, Uttal's (1981) extensive review and critical analysis of recent and current developments in vision and visual perception). None the less, despite all of these advances, the claim that nothing as radical or profound as a 'Copernican Revolution' (Kuhn 1957) has occurred up to now in the science of visual perception or, specifically, visual masking, seems hardly disputable. In fact, the currently adopted choice of highlighting, whenever possible, significant similarities between past and present aspects of the study of visual masking was made to illustrate its piecemeal, continuous, and normal-science mode of development.

1.2. Background

Visual masking refers to the reduction of the visibility of one stimulus, called the target, by a spatiotemporally overlapping or contiguous second stimulus, called the mask. Historically, masking has played a leading role in the study of spatial and temporal properties of visual perception. In this role it remains of great importance in the present and promises to continue as such in the future. To understand the historical roots of our present and future interest in visual masking, let us review the scientific context within which masking developed not only as a methodological tool but also as a phenomenon deserving empirical and theoretical investigation *per se*.

Initially, as now, the use and study of visual masking were grounded in attempts to delineate the temporal stages and parameters of the perceptual process. These attempts involved the logical parsing and the experimental measure of, among others, the following temporal stages: the time for a stimulus to reach focused awareness (*Apperceptionzeit*), perception time (*Wahrnehmungszeit*), perceptual duration (*Wahrnehmungsdauer*), sensation time (*Empfindungszeit*), the rise and fall times of sensation, sensory persistence, retinal (transduction) latency, conduction velocity, cortical processing latency, and so on. Cattell (1885a, 1886) condensed these measures into a coarser parsing of four basic temporal parameters, each corresponding to one sensory–perceptual operation or stage:

1. The duration that a stimulus must be present in order that a sensation be excited (threshold level).

2. The duration of a stimulus required to produce the maximum sensory intensity (saturation level).

3. The time required for a stimulus to be changed into a nervous impulse (transduction latency).

4. The time taken up in the nerve and brain before the stimulus is seen (perceptual latency).

As Baade (1917b) pointed out, this 'microtomization (*Mikrotomierung*)' or temporal slicing of the perceptual process occurred in a context concerned with the complementary studies of (a) 'pure' sensations and (b) the initial stages in the microgenesis of the perceptual process. Perhaps it is best to quote Baade's characterization of this context:

. . . Out of the following consideration it will become clear what especially great significance the study of the initial stages of the perceptual process has for the investigation of sensations. The chemist seeks to produce pure chemical bonds; the physicist seeks to free the research-relevant processes from all superfluous and accidental side effects; the bacteriologist seeks to isolate his research objects in a 'pure culture' [or medium]. Should the psychologist not strive to observe, in an isolated state, a state of purity, or however one wishes to express it, those objects of psychology to which he grants the name and role of 'element', 'primitive form' or something similar?

Should one search for the above-mentioned isolated sensations, one would naturally expect their presence only in the first stages of the perceptual process; for it is, so to say, palpably clear that nothing of the later stages will be as simple as the initial ones (Baade 1917*b*, pp. 99–100; translation mine)

Within this microgenetic context, some of the earliest research on the initial stages and temporal parameters of perceptual processes was conducted in the latter half of the nineteenth century (Baxt 1871; Cattell 1885*a*, 1886; Erdmann and Dodge 1898; Exner 1868; Tigerstedt and Bergqvist 1883) and continued well through the initial decades of the twentieth century (Fröhlich 1923; Monjé 1927). Although these investigations were criticized throughout and even discredited on logical and methodological grounds (Cattell 1885*a*, 1886; Erismann 1935; McDougall 1904*b*; Rubin 1929; Wundt 1899, 1900), they left in their wake a host of interesting experimental techniques and empirical observations. Moreover, they raised theoretical and related methodological problems which, as yet, are not resolved (Gibson 1979; Neisser 1976; Shaw and Bransford 1977; Turvey 1977).

A thorough, extensive discussion of these problems is beyond the scope of this chapter. However, briefly, they revolved fundamentally around the following questions:

1. Is it desirable or possible to isolate perceptual elements (e.g. pure sensations)?

2. Correspondingly, is it desirable or possible to determine the primitive stages or mechanisms of the perceptual process in which the elements can be isolated?

3. As a methodological corollary, by employing brief and static, i.e. tachistoscopic, stimuli, is one not introducing laboratory artefacts into the perceptual process which bear little resemblance or relevance to more naturalistic, extralaboratory perception?

The present, in-vogue recognition of the scientific importance attached to these questions (Neisser 1976; Turvey 1977) is a recurrent theme in the study of perception. To underscore this recurrence let us quote verbatim the introductory paragraph of Ebbecke's (1920) article entitled *Über das Augenblicksehen* ('On Momentary Seeing'):

Typically our seeing process is one characterized by a roving view. As soon as one is prevented, through some *unnatural way*, from running one's eye over the objects in the visual field and, so to speak, probing them, all sorts of disturbances intrude into visual sensation. Under [prolonged] rigid fixation, visibility begins to blur in that brightness and hue differences disappear and afterimages appear. Conversely, when the eye catches only a brief glimpse of a visual object, the visual impression is rendered inaccurate or altered [relative to free viewing conditions] (p.13; translation and italics mine).

Despite this time-honoured recognition of the problems introduced by the attempt to delineate stages of the perceptual process and the corollary use of the tachistoscopic method, chronometric research on the perceptual process flourished then as it does now (e.g. Posner 1978). As will be seen in Chapters 10 and 11, the problem with the use of the tachistoscope and mental chronometry is not so much a methodological one *per se* as it is a more pervasive, paradigmatic one clothed or, so to say, masking as a methodological one.

Relegating to Chapters 10 and 11 further discussion of these methodological and theoretical issues, I shall turn to that subclass of tachistoscopic· techniques and observations relevant to this historical understanding of visual masking. I shall focus primarily on the following topics:

1. The types of masking termed *metacontrast* and *paracontrast*.

2. The relation of *stroboscopic motion* to metacontrast.

3. The type of masking termed *masking by light*.

4. The existence of response *persistence* and temporal *integration* in vision; and

5. The role of central, *cognitive* (non-sensory) processes involved in visual masking.

This choice of topics was not arbitrary; rather, they reflect the main thrust of past as well as present research in visual masking and cognate areas.

1.3. Metacontrast and paracontrast

By metacontrast I mean the reduction in the visibility of one briefly presented stimulus, the target, by a spatially adjacent and temporally succeeding, briefly presented second stimulus, the mask. As such metacontrast is a form of backward masking in so far as the masking stimulus exerts a retroactive effect on the target stimulus. By exchanging the temporal order of the above two sequential stimuli—that is, by designating the first stimulus as the mask and the second as the target to be masked—the conditions for paracontrast, a type of forward masking, are met.

The coining of the terms *metacontrast* and *paracontrast* and the first extensive investigation of these two masking effects are credited to Stigler

(1910, 1926). However, as noted in Alpern's (1952) historical review of metacontrast, evidence for these effects predated Stigler's 1910 work by about a decade or two. Moreover, the use and importance of metacontrast as a methodological tool in the study of the time course and elementary processes of visual perception, although explicitly acknowledged by Stigler (1926) and Piéron (1935) in the early twentieth century, was already apparent several decades earlier.

According to Stigler (1908, 1910), Exner, in 1868, was the first to employ the experimental technique which eventually developed to become known as metacontrast and paracontrast masking. Exner used the subjective comparison of two spatially bordering stimuli, one of which was flashed slightly before the other, in order to investigate the time course of light sensations produced by a brief, tachistoscopic light stimulus. Without going into the details of Exner's rationale and method, I shall briefly analyse the following crucial assumptions made by Exner to justify his method. Exner assumed that two objectively equal light stimuli, which fell, one briefly after the other, on immediately adjoining retinal areas, elicit two equal sensations. What was implied is that the two sensory effects produced by the brief stimuli are independent and do not interact spatially. Without such a *constancy assumption*, whether in fact true or not, Exner could not have employed the sensation produced by the second stimulus to monitor the time course of sensation produced by the spatially bordering first one. This working hypothesis, later also adopted in related investigations by Kunkel (1874) and Petrén (1893), may have been premised, additionally, on the then-prevailing notion of 'local signs' introduced in 1852 by Lotze in his *Medizinische Psychologie* (*Medical Psychology*). According to Lotze, local signs resulted from stimulation of spatially delimited 'sensory circles', each of which was connected via a separate, independently acting nerve fibre to its appropriate cortical area. Lotze's notion of local signs was adopted and adapted by many of his contemporaries, particularly by the noted nativist Hering and two of the most influential and pre-eminent empiricists, Wundt and Helmholtz. Furthermore, to produce independent local sensory activity, Exner (as well as Kunkel and Petrén) assumed that the use of very briefly flashed stimuli effectively eliminated reciprocal spatial contrast mechanisms. This assumption was refuted by Stigler's (1910, 1913, 1926) subsequent work on masking. Consequently, it should be clear why the former investigators, who saw their work primarily of significance to the study of the time course of visual perception, failed to consider the significance of their investigation in terms of spatial contrast phenomena. Moreover, the use of brief flashes, presumably to eliminate spatial contrast phenomena, implies that these investigators were aware of such contrast effects. In fact, Exner (1868) makes reference to the existence of edge or border contrast effects (Exner 1868; p. 615). As early as 1834, Müller, in his *Handbuch der*

Physiologie des Menschen (*Handbook of human physiology*) proposed that spatial contrast phenomena were based on mutual and reciprocal action between separate retinal areas, and several of Exner's contemporaries (Hering 1872, 1878; Hermann 1870; Mach 1865, 1866*a*, *b*, 1868) had by that time already published some major works on spatial contrast and the reciprocal dependence and interactions between adjacent retinal areas of stimulation. Therefore, it seems more credible that Exner, Kunkel, and Petrén simply did not consider spatial contrast phenomena as a significant influence in their studies rather than that they were entirely ignorant of prior and contemporaneous work on spatial contrast.

Be that as it may, it was not until Sherrington's (1897) work on reciprocal action in the retina and McDougall's (1904*a*, *b*) investigations of the sensory intensity of brief visual stimuli that spatiotemporal contrast phenomena were viewed as being of prime importance in the study of the time course of visual sensations. In this regard, Sherrington (1897) stated that 'the physiological result of applications of a stimulus to any point of a sensifacient surface is decided by not only the particular stimulus there and then incident but also by circumjacent and immediately antecendent retinal events in determining the final physiological or sensory effect produced by a given retinal point of stimulation.' Sherrington's (1897) definition of 'simultaneous contrast' and 'successive contrast' as *reciprocal* sensory relations or interactions across an interval of space and *time*, respectively, also implied the existence of and distinction between metacontrast (temporally backward masking) and paracontrast (temporally forward masking), although his study failed to draw this distinction either conceptually or experimentally.

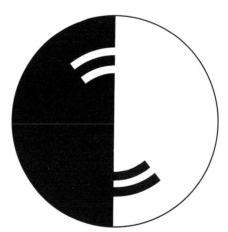

Fig. 1.1. One of the rotating discs used by Sherrington (1897) in his investigations of reciprocal, spatiotemporal interactions in human vision. (From Sherrington (1897).)

None the less, from Sherrington's (1897) and also McDougall's (1904*a*) investigations—both published prior to Stigler's work—one can infer that both investigators were aware of metacontrast and paracontrast effects, although neither named or isolated these effects as such. It was left for Stigler (1908, 1910, 1913, 1926) to do so, and for subsequent investigators to rediscover, reconceptualize, or elaborate on them (Alpern 1953; Baroncz 1911; Baumgardt and Segal 1942; Fry 1934, 1936; Fry and Bartley 1936; Piéron 1935). In the following section, some of the major findings, theories, and techniques will be reviewed.

1.3.1. Principal investigations: findings and theories

Sherrington's (1897) investigation was perhaps the first to indicate the existence of a metacontrast *effect*. One of the stimuli that Sherrington employed was a disk, as shown in Fig. 1.1, which could be-rotated, at varying angular speeds, in either clockwise or counterclockwise direction. When the disk spins in the clockwise direction the single white arc bracketed by the two black arcs shown at the bottom of the figure is followed in time by the two spatially adjacent double white arcs shown at the top of the figure. The percept is one of three bands which appear to flicker at rates dependent on the speed of the rotation. Sherrington found that as the disk spun at increasing speeds in the clockwise direction, flicker persisted longest in the middle band which corresponded to the single white ring. Conversely, when rotation reversed, flicker persisted longest in the bands produced by the double white rings. This result can be summarized as follows: when brief illumination of a given retinal area is followed shortly by brief illumination of spatially adjacent area, the critical flicker fusion (cff) in the former area is increased relative to that of the latter one. Since the interactions are assumed to be reciprocal in time, one can say alternatively that the cff in the latter area decreased relative to that of the former one. Under similar experimental arrangements (see Fig. 1.3(a)), Piéron (1935) reported essentially the same result. Now, these results are somewhat puzzling in the context of meta- and paracontrast. The fact that stimulation of a given area can decrease cff in a subsequently illuminated and spatially adjacent area is easily reconcilable with a forward, paracontrast masking effect. However, if metacontrast is viewed as a backward *suppressive* effect, why is it that cff is *increased* in the first of two illuminated, spatially adjacent areas?

To answer this, we must draw a distinction between flicker and brightness sensations; the two types of sensation may rely on different sensory mechanisms. In fact, what Piéron reported (and Sherrington failed to report) is that at intermediate rates of rotation the *brightness* visibility of the leading stimulus could be *suppressed* entirely by the succeeding spatially adjacent stimulus. However, as rotation rate increased, or alternatively as the stimulus onset asynchrony (SOA) between adjacent retinal areas decreased, the brightness of the leading stimulus became progressively greater; that is, the suppressive effect that the lagging stimulus had on the leading one decreased. Hence, by taking brightness as a sensory response index, Piéron did obtain a metacontrast suppression effect, although, like Sherrington, he reported a complementary facilitatory effect on cff. The empirical and theoretical work discussed in Chapters 6 and 7 should indicate why flicker sensitivity does not serve as an index of metacontrast suppression but rather as a correlated index of a complementary metacontrast facilitation. On the other hand, it will also become clear why decreases in flicker as well as brightness sensitivity may be two indices of paracontrast suppression. To anticipate in summary fashion, I shall make the following two claims:

1. The fast 'flicker-detectors', activated by the spatially adjacent, temporally lagging stimulus, suppress, and are also *reciprocally* suppressed by, the activity of slow pattern 'brightness (or contrast) detectors' generated by the leading stimulus.

2. Consequently, this in turn results in (a) an inhibition of the flicker-detectors responding to the lagging stimulus and (b) an inhibition of contrast-detectors and a simultaneous disinhibition of flicker-detectors responding to the leading stimulus.

Given the restriction that we index metacontrast (or paracontrast) by studying variations in brightness or spatial contrast, let us now turn to McDougall's (1904a) investigation which, to my knowledge, yielded the first clearly interpretable metacontrast effect. McDougall also used an experimental apparatus similar to that employed by Sherrington. Specifically, McDougall used an opaque disk, having two small apertures as shown in Fig. 1.2(a), which was rotated before a luminous background and visually fixated at its centre. Both apertures were continuously transilluminated when the disk rotated. However, when aperture *b* was covered and the disk was rotated at fairly slow speeds ($\leq$ 1 revolution per second), McDougall reported seeing a travelling pattern of alternate dark and light bands, known as Charpentier's bands, produced by aperture *a*. These bands, as shown in Fig. 1.2(b), in turn were trailed by a greyish band known as Bidwell's ghost or the Purkinje image (Brown 1965). The former bands are related to the temporal oscillations in the visibility of a clear positive after-image produced by a flashed stationary stimulus (see Chapter 3, Sections 3.5.1 and 3.5.2); the latter, greyish band is related to a more prolonged dull and ill-defined phase of the after-image. McDougall hypothesized that Bidwell's ghost represented the trailing end of the primary sensation (*Primärempfindung*) produced by the leading aperture *a*. Moreover, he reasoned that by again uncovering the temporally trailing aperture *b*, the initial stronger portion of the primary sensation produced by this aperture could suppress the weaker trailing end of the sensation,

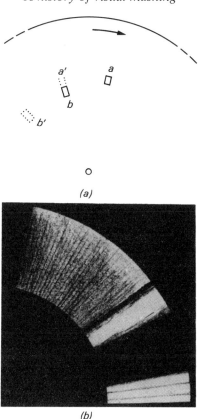

Fig. 1.2.(*a*) McDougall's (1904*a*) stimulus apparatus, consisting of an opaque rotating disc with two apertures (*a* and *b*) transilluminated by a light source behind the disc. (*b*) The appearance of the leading Charpentier bands and the trailing Bidwell's ghost or Purkinje image produced, for instance by aperture *a* while *b* is masked when the disc is rotated clockwise. (From McDougall (1904*a*).)

i.e. Bidwell's ghost, produced by the leading aperture *a*. In fact, when aperture *b* was uncovered, the trailing Bidwell's ghost of aperture *a* did disappear.

Note that McDougall's explanation of the 'backward' suppression of Bidwell's ghost rests on the following implicit assumptions: (i) although two physical stimuli are spatially and temporally segregated, their sensory responses interact not only via reciprocal lateral interaction but also via their temporal overlap, thus in effect simulating simultaneous brightness contrast at the physiological level; (ii) as a corollary, the sensory response of a brief stimulus must persist, in some form or other, beyond the momentary duration of the stimulus.

These explanatory assumptions were made explicit and elaborated in subsequent investigations of meta- and paracontrast conducted by Stigler

(1910, 1913, 1926). Based on Exner's (1868) notion of 'positive after-image', Stigler (1910) drew the following distinction between the initial and trailing portions of the primary sensation produced by a flashed stimulus. The (initial) part of the primary sensation (*Primärempfindung*) which was produced by the presence of the stimulus was designated the *homophotic* image; that (trailing) part which outlasts the stimulus was designated the *metaphotic* image and corresponded to the Exnerian positive after-image. As stated by Stigler, the metaphotic image was further distinguishable in that it, relative to the *homophotic* image, is particularly susceptible to masking by spatially adjacent stimuli. Although this masking was believed to tap the same mechanisms as simultaneous or homophotic contrast, Stigler named it *metaphotic contrast* or *metacontrast* in order to highlight the fact that the metaphotic image of a temporally leading stimulus was affected by (the homophotic image of) a succeeding stimulus.

As stated so far, this explanation, based on (i) a form of visual persistence and (ii) a mechanism related to simultaneous brightness contrast, seems reasonably and simply stated in terms of temporal integration of interactive sensory responses. In 1926 Stigler, however, complicates this theoretical explanation somewhat by explicitly acknowledging another possible mechanism which we can alternatively term the *overtake* or *delay* hypothesis. Stigler (1926) states that:

. . . metacontrast shows that it is possible for an excitation to be *overtaken* on its way from the retina to the central organs and masked via a contrast effect [produced] by a succeeding [spatially] neighboring stimulus.

Metacontrast shows further that the visual excitation is:

measurably *delayed* at one and probably several sites along its way from the periphery to the center [i.e. the brain] (pp.950–1, translation and italics mine).

Thus Stigler inadvertently or, more likely, deliberately raised two hypothetical explanations of metacontrast: one could conveniently be called the *integration–inhibition* hypothesis; the other, the *overtake–inhibition* hypothesis. The likeliest reason supporting Stigler's deliberate introduction of the latter and later hypothesis is his finding of dichoptic metacontrast in 1926 following his failure to find such metacontrast in 1910 (see below). Based on his 1910 results, Stigler proposed on the basis of Exner's (1898) similar proposal (see footnote at bottom of p. 14) that horizontal cells in the retina mediate the metacontrast effects between adjacent retinal areas. Consequently, in order to be suppressed the metaphotic image or the Exnerian positive after-image would also have to be localized at retinal, in particular at receptor levels of activity, a hypothesis which antedates a very recent explanation of visual persistence proposed by Sakitt (1975, 1976; see Chapter 3, Section 3.5.1). However Stigler's (1926) dichoptic metacontrast results could no longer be explained by the above-outlined scheme. As a result, the integration–inhibition hypothesis could only be saved if a *sui generis,* cortical source of visual persistence, or else, one which simply depends on transferred temporal properties of sensory activity arriving from more peripheral levels, existed. Except for Monjé's (1931) somewhat later work (see Section 1.5) not many experimental findings prior to or contemporaneous with Stigler's bore clearly on the issue of the locus of visual persistance. However, that the receptor signal must traverse one or more ganglia before it reaches the central visual areas of the brain was already an established anatomical fact (e.g. Minkowski 1913, 1920 *a, b*). It is likely that, being aware of this fact, Stigler (1926) opted for the more adequately supported overtake–inhibition hypothesis. At any rate, what was furthermore implied by this hypothesis and the dichoptic metacontrast results is that the brightness suppression of the leading stimulus by the following one must depend on central sources of inhibition. Stigler's two hypotheses, in some form or another, have been adopted in subsequent theories of metacontrast and masking, and today are expressed in abbreviated terms as the integration and interruption hypotheses, respectively (Kahneman 1968; Scheerer 1973).

Within the framework of these theoretical explanations, Stigler (1910, 1913, 1926) employed the use of two spatially adjacent light-on-dark semicircles or rectangles as target and mask, with the target temporally preceding the mask. On the basis of his studies, extending over roughly two decades, Stigler came to the following main empirical conclusions, most of which have been repeatedly corroborated in later and recent investigations:

1. Metacontrast can be obtained binocularly as well as monocularly (1910).

2. Although initially denied by Stigler (1910), later he found that metacontrast also could be obtained dichoptically (1926; see also Kolers and Rosner 1960; May, Grannis, and Porter 1980; Schiller and Smith 1968, Weisstein 1971; Werner 1940).

3. Metacontrast is not only a dichoptic effect, it is also an interhemispheric (transcallosal) effect (1926; see also McFadden and Gummerman 1973).

4. 'Metacontrast is exclusively dependent on the luminance of the two [adjacent] contrast fields and [is] essentially unaffected by color. It [i.e., metacontrast] is as evident when one semicircle is red, and the other green as when both are white' (Stigler, 1926; p. 967; but see Chapter 4, Section 4.3.5 for discussions of both corroborative and disconfirming results reported in more recent studies).

5. Metacontrast depends on the temporal onset asynchrony (SOA) of the two adjacent stimuli; or as Stigler (1910) stated: 'should the time difference between the onsets of the two luminous stimuli be increased beyond the temporal-resolution threshold . . . a difference value is finally attained . . . at which Field I [the leading stimulus or target] appears

darker than Field II [the succeeding stimulus or mask]' (p. 418) (a reworked version of this 'SOA Law' has recently been stated by Turvey (1973) to account for central pattern masking).

6. In foveal view, which was employed by Stigler (1910), metacontrast was eliminated or progressively attenuated when an increasingly wide, dark vertical strip separated the two successive and neighbouring stimuli, a result also reported more recently by Kolers and Rosner (1960).

7. Metacontrast depends on the direction of gaze (Stigler 1910). When the border between the test and mask field was viewed parafoveally rather than foveally or directly, metacontrast seemed to be stronger and more immune to spatial separation between the two neighbouring stimuli (Alpern 1953; Kolers and Rosner 1960; Merikle 1980; Saunders 1977; Stewart and Purcell 1970, 1974).

8. The first stimulus can reduce the apparent brightness of the succeeding one. This latter effect is known as *paracontrast* (Stigler 1926; see also Alpern 1953; Kolers and Rosner 1960; Weisstein 1972).

Before discussing later, relevant studies of metacontrast and masking, let us examine one more of Stigler's (1910) findings which in hindsight seems to be of relevance to his theoretical explanations of metacontrast. What Stigler found was that metacontrast obtains not only when the second stimulus is equal to or greater in intensity than the first but also when the second stimulus is substantially weaker than the first (see Stigler (1910); p. 394, Experiment 9; p. 399, Experiment 45). This raises the following set of problems. Stigler (1910), among others (e.g. Cattell 1885b), was aware that a relative increase in stimulus intensity decreased the effective response latency and persistence of a visual sensation, and vice versa. The puzzle this fact poses is that with a decrease in latency and persistence of the first stimulus relative to the second, how is it that (i) according to the integration–inhibition hypothesis, the slow and weaker sensory effects of the second stimulus integrate temporally with and inhibit the briefly persisting and stronger ones of the first; or (ii) according to the overtake–inhibition hypothesis, the second, weaker stimulus sensation overtakes and inhibits the faster, stronger one produced by the first stimulus? This theoretical puzzle, also inherent in a later, related theory (Ganz 1975; see Chapter 5 below), was not apparent to Stigler, his contemporaries, or his immediate successors, and as such did not attain central status in theories of metacontrast until much later (Alpern 1953).

Of additional, particular interest to the present historical review of theories and findings are the studies reported by Fry (1934), Piéron (1935) and Alpern (1953). Fry (1934), independently of Stigler's work, redis-covered the metacontrast effect, and his theoretical explanation is basically a composite of Stigler's two hypotheses. To explain the temporally backward influence in metacontrast, Fry stated: 'what seems to happen is that the response of the retina to the first stimulus is considerably *delayed* and *prolonged* and *overlaps in time* the response to the second [spatially adjacent] stimulus and is *inhibited* by it by some kind of interaction between retino-cortical pathways at synapses either at the retina, or at the basal ganglia, or at the cortex' (p. 706; italics mine). Within this theoretical framework, Fry (1934) extended Stigler's basic findings by obtaining the following important quantitative results: similar to Stigler's (1910) results, Fry (1934) found that parafoveal metacontrast could be obtained when a spatial (dark) gap up to 1.25 degrees was introduced between the target and the mask stimuli. Moreover, Fry found that the strength of metacontrast decreased with the size of the gap, indicating that metacontrast interactions are, relatively speaking, a local spatial phenomenon.

Piéron (1935) added the following technique to the basic metacontrast motif. As shown in Fig. 1.3, he employed not only the, by now, standard metacontrast paradigm (Fig. 1.3(a)) but an interesting variation in which several spatially staggered and adjacent stimuli could sequentially mask each other (Fig. 1.3(b)).

Figure 1.3(b) shows that as the disk rotates clockwise stimulus *d* is followed in time by adjacent stimulus *c*, which in turn is followed by adjacent stimulus *b*, and so on. Piéron found that at a given optimal rate of clockwise rotation, stimulus *c* suppressed the visibility of stimulus *d*; stimulus *b* in turn supressed stimulus *c*; and finally stimulus *a* suppressed stimulus *b*. In effect, only stimulus *a* was visible as such. What, in the context of the above theoretical explanations, is perhaps surprising and puzzling in this demonstration is not so much the staggered masking of stimuli *d*, *c*, and *b* but rather the fact that stimulus *b*'s inhibition of the visibility of stimulus *c* (or stimulus *a*'s inhibition of the visibility of stimulus

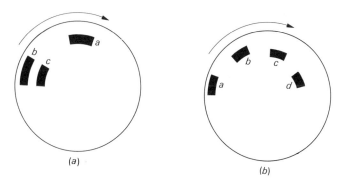

(a)

(b)

Fig. 1.3.(*a*) One of Piéron's (1935) modifications of McDougall's (1904*a*) stimulus apparatus. With Piéron's apparatus stimulation from one aperture, *a*, was followed by adjacent stimulation from two adjacent apertures, *b* and *c*, when the disc was rotated clockwise. (*b*) A second of Piéron's modifications showing the spatially staggered series of apertures giving rise to a sequential blanking or masking of stimulation arising from apertures *d*, *c*, and *b* when the disc is rotated clockwise. (After Piéron (1935).)

b) in turn failed so disinhibit the visibility of stimulus *d* (or of stimulus *c*). That is to say, if the sensory processes elicited by stimulus *b* inhibit the sensory processes which otherwise would have been elicited by stimulus *c*, would one not expect that the sensory processes elicited by stimulus *d* in turn not be inhibited since this inhibition relies on the presence of stimulus *c*'s sensory processes? As shown in Chapters 7 and 9, the resolution of the puzzle depends on a theoretical reconceptualization of metacontrast.

In his extensive studies of metacontrast, Alpern (1953) confirmed most of these prior findings. Of notable exception, however, was Alpern's (1953) failure to obtain metacontrast (i) when the stimuli fell within the foveal region; or (ii) when the stimulus intensities of the mask fell below photopic threshold levels. The failure to obtain foveal and dichoptic metacontrast is somewhat surprising in view of the fact that Stigler (1910) reported foveal metacontrast (Stigler's stimulus display subtended the central 1.5° of vision), and the fact that Stigler (1926) and, later, Werner (1940) reported dichoptic metacontrast. Be that as it may, Alpern (1953; see also Alpern 1965) on the basis of his results and analyses, tentatively concluded that metacontrast is a peripheral retinal phenomenon which is probably due to inhibitory interactions between fast cone and slow rod processes.*

To my knowledge, Alpern (1953) was the first investigator of metacontrast to publish the by now, typical U-shaped or type B (Kolers 1962) metacontrast function, although such a function also could have been inferred from Werner's (1935) less quantitative, more phenomenally oriented study of metacontrast (which in addition indicated that metacontrast is an orientation-specific masking effect). That is to say, Alpern's quantitative experimental methods demonstrated clearly that the optimal suppression exerted by a circumjacent mask on the brightness of a target occurred not when both simuli were presented simultaneously, but rather when the mask lagged the target by about 50 ms—this, despite stimuli of equal energy and, thus, of equal sensory latency and persistence. Moreover, Alpern (1953) was also the first investigator to note the theoretical significance of such type B functions obtained even when the mask energy was substantially lower than the target energy. This reintroduced the puzzle already raised in connection with prior discussions of some of Stigler's (1910) results. Given a second stimulus of equal (or greater) latency than the first, how is it, as Stigler (1910, 1926) and later investigators believed, that excitation of the second overtakes and

---

* In his initial metacontrast investigation Stigler (1910) also failed to obtain dichoptic metacontrast. He, like Alpern (1953), interpreted this result as indicating a peripheral, retinal locus of metacontrast interaction. Stigler's (1910) explanation differed from Alpern's (1953), however, in that the response integration hypothesis was maintained in the context of hypothetical inhibitory lateral interactions transmitted between adjacent receptors via horizontal cells (Exner 1898), whereas Alpern's (1953) theoretical explanation rests on the faster cone response being able to *overtake* and inhibit the slower rod response.

suppresses that of the first? Perhaps it was the inconsistency raised by this question of the prior explanations, based implicitly on equivalent response persistences or response lags *within the same types of sensory processes*, that led Alpern to posit an alternative explanation based on *differential response latencies* of cone and rod processes. As such, Alpern's (1953) explanation, although only partially correct, was in spirit the first to highlight what presently is called a dual-channel approach to metacontrast (Breitmeyer and Ganz 1976; Matin 1975; Weisstein, Ozog, and Szoc 1975)

One half century prior to Alpern's (1953) work McDougall (1904a) had already implicitly invoked the longer response latency of rod, relative to cone, processes in explanations of his lateral and backward masking effects. McDougall hypothesized that many, though not all, of the trailing Bidwell's ghost phenomena were due to rod processes, which have longer response latency than the cone processes responsible for the earlier (Charpentier) bands produced by a rotating illuminated window (see Fig. 1.2(b)). Further *implied* by McDougall's (1904a) investigation and made explicit by Alpern (1953) is the belief that cone mechanisms can inhibit rod mechanisms (and vice versa). In fact this issue of rod–cone interactions was brought into the limelight by Alpern's (1953) study and, as we shall see, continues to draw appreciable attention in the present (see Chapters 4 and 5).

1.3.2. Metacontrast and stroboscopic motion

The existence of stroboscopic motion had been demonstrated experimentally by Exner (1875, 1888) several decades prior to publication of Wertheimer's (1912) phenomenon-rich and epoch-making paper, *'Experimentelle Studien über das Sehen von Bewegung'* ('Experimental studies on seeing motion'). Although Wertheimer's study was primarily addressed to clarification of the sensory processes giving rise to (stroboscopic) motion perception, it nevertheless is of relevance to the study of masking and metacontrast. In fact, Wertheimer was aware of the potential relation between masking and stroboscopic motion. He reported, for instance, as Schulz (1908) had done before him, the masking or spatial displacement of one, usually the first, of the two stimuli used in a stroboscopic motion demonstration. This was particularly true for the type of motion called *phi*. Referring to the apparent rotation of (a) a flashed vertical toward (b) a subsequent flashed horizontal line, Wertheimer noted that:

in the extreme sense, the subject had no inkling that the vertical [first line] was actually exposed . . . In the course of the experiments several cases resulted in which one of the two exposed objects plainly was not seen, nor could it be imagined; and the subject judged that only one was exposed; in regard to the other, perceived one, φ motion was clearly apparent either coming from the first (a) or alternately approaching the second (b) locus [or stimulus], (Wertheimer 1912, p. 217, translation mine).*

Moreover, Wertheimer was aware of prior work reporting masking phenomena, and he cites Schumann's description of an apparent explosion, (*'Explodieren'*) or expansion of a stimulus at one location when followed by a masking stimulus. Wertheimer did not emphasize this aspect of his study; and, therefore, its importance and relation to metacontrast and masking seems to have been not clear to him.

What is of importance, however, is that like Exner (1875) and Anstis and Moulden (1970), Wertheimer demonstrated that stroboscopic motion could be obtained dichoptically. This as noted by Wertheimer, implicated the involvement of central or cortical mechanisms in stroboscopic motion perception. Although not explicitly stated by Wertheimer, we can infer that the masking phenomenon accompanying stroboscopic motion under monocular or binocular viewing also were obtained here and, therefore, to extend Wertheimer's conclusion, masking also partakes of a central or cortical component. Of additional importance is Wertheimer's finding that differences between the colours and shapes of the two successive stimuli did not eliminate stroboscopic motion. What one perceived during the stroboscopic motion when using such heterogeneous stimuli was a transformation (*Veränderung*) from one colour or pattern to the other, an observation recently replicated by Kolers and von Grünau (1976). It was as though the mechanisms producing motion sensations were more or less unaffected by those concerned with pattern or colour discrimination.

Finally, Wertheimer also noted, as did Kolers and von Grünau (1977) more recently, the importance of attention in stroboscopic motion (and, by inference, in masking). The degree and smoothness of stroboscopic motion depended on where one directed one's attention or gaze. Of course, here, changes in the direction of attention were confounded with changes in the retinal location of stimuli. As reported earlier by Exner (1888), the retinal periphery is particularly sensitive to stroboscopic motion—an important fact, since, as Stigler (1910) had shown, metacontrast also was stronger under indirect rather than direct viewing of the edge separating the test from the masking flash.

1.4. Masking by light

Masking by light is a form of visual masking in which a briefly flashed, uniformly illuminated field obscures the visibility of a prior or later flashed

---

* Wertheimer (1912, p. 226) also notes the disappearance of both vertical and horizontal lines when they were presented alternately in back-and-forth sequence. This sequential blanking is similar to the aforementioned, temporally staggered metacontrast effects reported by Piéron (1935). That is to say, the two stimuli were alternated at a temporal rate such that each was able to mask the brightness and contour aspects of the other. Yet each left unmasked the information giving rise to stroboscopic motion perception. This and other of Wertheimer's observations, to my knowledge, are the first clear demonstrations of a perceptual dissociation of brightness or contour perception and motion perception. Unfortunately, Wertheimer failed to explore further this psychophysical segregation of figural and motion aspects of a stimulus sequence; and to my knowledge such a segregation or distinction was not made again until Saucer (1954) mentioned it.

target stimulus. In the history of masking, backward masking by light (mask flash presented after target flash) had been employed extensively ever since Exner's (1868) pioneering work on the time course and fate of visual sensations elicited by the leading target stimulus (Baade 1917*a*, *b*; Baxt 1871; Cattell 1885*a*, 1886; Fröhlich 1923; Monjé 1927; Schumann 1899; Tigerstedt and Bergqvist 1883). By and large, the main empirical results of these studies can be summarized as follows: (i) the more intense the after-coming mask flash, the less the visibility of the prior target (e.g. another light flash, a letter, or a word (Baxt 1871; Cattell 1885*a*, 1886; Sperling 1965; see Fig. 2.5)); and (ii) the greater the temporal interval between prior target and succeeding mask, the weaker the masking magnitude (Cattell 1885*a*, 1886; Schumann 1899; Sperling 1965; see Fig. 2.5)—that is, with increases in temporal separation, backward masking by light is a monotonically decreasing or type A function (Kolers 1962). Most theoretical explanations of this masking effect offered then and now (e.g. Eriksen 1966) were straightforward in terms of the sensory persistence hypothesis in which it was assumed that the sensory response of the leading target stimulus persists in the form of a decaying positive after-image which can be suppressed by integrating with the response of an after-coming mask.

In these early studies, the mask was used as a tool to monitor the temporal processing stages, e.g. the response persistence, of the target stimulus. However, as noted later by Sperling (1964), one can conversely use the visibility of the target to monitor the time course of the sensory response elicited by the mask. The latter approach was successfully employed and exploited by Crawford (1940, 1947) to investigate the effects on a small test flash produced by the onset, duration, and offset of a larger, bright conditioning flash. Since the onset and offset of a conditioning flash produce sudden changes in the adaptation level of the visual system, related studies on such adaptation effects are of prime importance to the historical understanding of masking by light.

The study of the sensory effects of sudden changes—in particular, of sudden increases—in luminance has had a long history. McDougall (1904*b*) and Stigler (1908) cite the relevant work of Plateau, Exner, and Helmholtz in the first half and middle of the nineteenth century; and McDougall (1904*b*) himself investigated the currently well known transient overshoot and sudden decline in visual activity following an abrupt stimulus onset. These sudden overshoots and subsequent declines in brightness were well documented by Stainton (1928) and related not only to the Broca–Sulzer effect (Broca and Sulzer 1902) but also to masking effects such as metacontrast (Baumgardt and Segal 1942). Moreover, both light and dark adaptation were well known and extensively studied phenomena by the turn of the twentieth century (Blanchard 1918; Lohmann 1906; Piper 1903). However, it was not until several decades later that Crawford (1947)

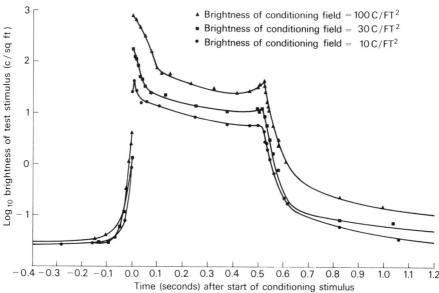

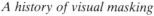

Fig. 1.4. The change in test field threshold (ordinate) as a function of the time of presentation of the test field relative to the onset of the conditioning field (abscissa). Negative and positive time values indicate that the test field was flashed, respectively, before and after the onset of the conditioning field. (From Crawford (1947).)

quantitatively investigated the sudden changes of visual sensitivity or, alternatively, of visual adaptation produced by the onset and offset of a uniform light flash. The method employed by Crawford was to measure the detection threshold of a small, 10-ms test flash spatially centred on a larger conditioning field flashed for a duration of about 500 ms. By varying the temporal interval between the onset of the test flash and the onset and offset of the conditioning flash, Crawford was able to monitor the changes in visual sensitivity produced by rapid light and dark adaptation, respectively.

The main and, by now, often replicated result is shown in Fig. 1.4. On the abscissa is given the time interval between the onset of the test flash and conditioning flash. Positive values indicate that the test flash was presented after onset of the conditioning flash. On the ordinate are given the test flash threshold values. Note several important results:

1. The greater the intensity of the conditioning or masking flash the greater the overall test flash threshold or, alternatively, the lower the overall visual sensitivity.

2. The rise in test flash threshold occurs up to 100 ms before the onset of the conditioning flash.

3. There are transient overshoots of the test flash threshold at and near the time of the onset and offset of the conditioning flash.

4. After offset, the immediately following sensory effects of the conditioning flash are prolonged on the order of 200–300 ms, before assuming the longer lasting phase of dark adaptation. The first and last of the results are merely replications of aforementioned findings on (a) the increase in mask effectiveness with intensity and (b) sensory response persistence. It is the second and third results which are of greater theoretical importance.

The problem posed to Crawford (1947) by the second result was to explain how visibility of a prior test flash is masked by a following conditioning flash onset. Crawford (1947) offered two alternate explanations which basically are similar to Stigler's (1910) overtake and temporal integration accounts of metacontrast. Crawford (1947) maintained that:

either the relatively strong conditioning stimulus overtakes the weaker test stimulus on its way from retina to brain and interferes with its transmission; or the process of perception of the test stimulus, including the receptive processes in the brain, takes an appreciable time of the order of 0.1 s, so that the impression of the second (large) stimulus within this time interferes with the perception of the first (p. 285).

Although Crawford (1947) made note of the suddenness of the threshold rise at onset and the presence of the smaller rise just prior to offset, he does not elaborate on these threshold rises or overshoots *per se*. In hindsight, however, it turns out that these overshoots play very important roles in theories of light and dark adaptation (Baker 1963; Wald 1961). At issue in these theories are the adequacies of photochemical and neural explanations of visual adaptation. As noted by Wald (1961), disproportionately large rises of visual threshold (e.g. the transient overshoots reported at onset or offset by Crawford (1947)) can accompany fairly little bleaching of photopigments. Some non-photochemical, most likely, neural process also seems to be required to explain these dramatic changes of visual sensitivity. In fact, as Wald (1961) points out, evidence for such a neural component can be inferred from extrapolation of data reported as early as 1918 by Blanchard. Similarly, extrapolation of the results of Lohmann's (1906) study suggests the presence of a neural component in light adaptation; and independent, yet related work by Bartley (1938) on brightness perception and Bernhard (1940) on electrophysiological correlates of light stimulation indicated that some of these neural components may be centrally or cortically located (Battersby, Oesterreich, and Sturr 1964). An earlier related view on the role of central mechanisms in backward masking by light flashes was already expressed by Cattell (1885a, 1886).

1.5. Sensory response persistence and temporal integration in vision

As noted above, one of Crawford's (1947) findings was that the sensory effects consequent to a conditioning or masking flash outlasted its offset by several hundred milliseconds. The fact that sensory responses elicited by a brief visual stimulus outlast its duration had been confirmed and reconfirmed many times since the late nineteenth century (Aubert 1865; Baxt 1871; Cattell 1885*a*, 1886; Charpentier 1890; Exner 1868; Fechner 1840*a*, *b*; Fröhlich 1921, 1922*a*, *b*, 1923, 1929; Helmholtz 1866; McDougall 1904*a*, *b*; Martius 1902; Monjé 1931; Müller 1834; Plateau 1834; Schumann 1899; D'Arcy 1773 (cited in Boynton 1972)). Exner (1868) described the brightness sensation following a brief light stimulus in terms of a relatively fast rise toward a peak value followed by a more gradual decline or decay. Exner (1868) designated that part of the primary sensation (*Primärempfindung*) which outlasted the stimulus as the 'positive after-image', and subsequently Monjé (1931) differentiated this primary sensation from the longer lasting secondary after-images. We have seen above that in some form or another, e.g. the metaphotic image, such visual persistence was an integral part of theoretical explanations of metacontrast (Fry 1934; Stigler 1910, 1926). In today's nomenclature, Exner's positive after-image or Stigler's metaphotic image could correspond to what is termed *iconic persistence* (Coltheart 1980; Neisser 1967). In fact, one of the major current conceptualizations of iconic persistence is that it is basically a type of after-image (Hochberg 1968, 1978) whose source, in particular, is thought to reside in retinal receptor activity (Sakitt 1976; Turvey 1977).

The issue of whether iconic persistence is based on peripheral (retinal) or central (cortical) processes, or perhaps both—an issue which is central to current theories of iconic persistence (see Coltheart 1980, and Long 1980 for reviews)—was already implied or even made explicit (Schumann 1899; cited in Baade 1917*b*, p. 123) in many of the investigations of the late nineteenth century cited previously. In order to address this issue historically, I shall introduce a recently made distinction between *previsible* or *neural* and *visible* or *phenomenal* persistence (Coltheart 1980; Turvey 1978). By previsible persistence we mean the physiological response persistence at peripheral sites which, though contributing to visible persistence, does not constitute it; by visible persistence we mean the persisting, phenomenally observable image, most likely associated with some central, cortical process, consequent to a brief stimulus flash. Such a distinction is not entirely without historical foundation, since it was at least implicit in the work of Baxt (1871), Cattell (1885*a*, 1886), Tigerstedt and Bergqvist (1883), Baade (1917*b*), and Monjé (1931).

Keeping in mind this distinction, let us look at the following methods and principal findings on response persistence in vision which were reported between the mid 1800s and the early decades of the twentieth century. To my knowledge, four main methods of investigating visual persistence were employed at or near the start of the twentieth century. One was a variation of the backward masking method introduced by Exner (1868). The rationale attending this method ran something as follows. If a prior brief stimulus is followed at some temporal interval by a bright masking flash, the latter will in effect disrupt the processing of the positive after-image produced by the former stimulus. Thus by determining the minimal value of that interval after the prior stimulus at which the after-coming flash no longer exerted its masking effect, one could conclude that sensory effects elicited by the prior stimulus must have persisted for at least that minimal duration (Baade 1917*b*; Baxt 1871). A lower bound of visual persistence could thus be established.

A second method, an adaptation of the one initially employed by D'Arcy (1773), was to exploit the presence of Bidwell's ghost and Charpentier bands produced by a rotating stimulus (see Fig. 1.2). Fröhlich (1921, 1922*a*, *b*, 1923, 1929) proposed the following rationale. By measuring the spatial extent of the bands or ghosts trailing a moving stimulus (and subtracting the time required for the entire stimulus to move across a given point), one can, from knowledge of the velocity of the stimulus, measure the duration of that part of the primary sensation which outlasts the presence of the stimulus at a given point.

A third method was to measure the cff of a light source. The cff is defined as the frequency at and above which the flickering stimulus appears *steady* rather than flickering. For example, Ferry (1892) reasoned that by measuring the temporal interval between successive, isochronal exposures of a flickering stimulus at which perceptual fusion just occurs, one has an estimate of retinal persistence. Finally, a fourth procedure relies on what is called the 'seeing-more-than-there-is' phenomenon (McCloskey and Watkins 1978) in which either a narrow, vertical slit-aperture is moved left or right in front of a much wider, stationary pattern display or alternatively the slit-aperture is stationary and the pattern display is moved behind the aperture. Consequently, at any moment, an observer has only a small portion of the otherwise occluded pattern in view. However, despite these momentary limited views, as the aperture or the display move, the observer perceives the entire pattern of the display. Since some of the partial views of the pattern occur later in the motion sequence than others, these separate views must be integrated over time to form a complete pattern percept, suggesting that at some level of the visual system the activity produced by earlier views persists until at least the onset of the activity produced by later views. This phenomenon was reported as early as the 1860s by, among others, Zöllner (1862), Helmholtz (1866), and Vierordt (1868). In addition to newer methods, variations of these four procedures are still employed currently to investigate response persistence in vision (Allport 1970; Erwin 1976; Haber and Nathanson 1968; Meyer and Maguire 1977; Parks 1965, 1968, 1970; Spencer 1969). Employing these methods, the early main findings reported in investigations of persistence is vision and replicated more recently (see Coltheart's 1980 review) can be summarized as follows:

1. As the intensity of a stimulus increases, response persistence decreases (Bowen, Pola, and Matin 1974; Exner 1868; Ferry 1892; Martius 1902; Monjé 1931).

2. As the photopic luminosity or physiological efficacy of a band-limited light source increases, persistence decreases (Ferry 1892).

3. As a corollary, variations in persistence are not attributable to variations in wavelength (colour) but rather to the covariations in photopic luminosity (Ferry 1892).

4. Persistence decreases as light adaptation level increases (Fröhlich 1923; Haber and Standing 1970; Schumann 1899; cited in Baade, 1917b, p. 111).

5. For durations smaller than the critical duration limiting temporal integration (Bloch's law; see McDougall 1904b), persistence decreases with increases in stimulus duration (Baroncz 1911; Bowen *et al.* 1974; Fröhlich 1923; Haber and Standing 1970; Martius 1902).

6. According to Exner (1868) persistence is greater foveally than extrafoveally (Breitmeyer and Halpern 1978) although subsequent work by Fröhlich (1923) indicated contrary findings.

7. Visual persistence, at least for stimuli of intermediate intensity, is longer (by a factor of $\sqrt{2}$, indicating central, binocular brightness summation between the two eyes) under monocular than under binocular vision (Monjé 1931).

Up to now, these results, except perhaps the last one, indicate the importance of peripheral sensory variables in determining persistence in vision. As such, they refer to what we have called previsible or neural persistence. The seventh or last finding pointed out the importance of binocular compared with monocular vision and suggested the possibility of central cortical processes involved in determining brightness and persist-ence. In fact, the work of Baxt (1871), Cattell (1885a, 1886), and Schumann (1899; cited in Baade 1917b) also pointed to the involvement of such central processes; although, in hindsight, we can say that it did so equivocally.

Baxt (1871) used a backward masking technique to study the temporal parameters of persistence in vision. When using three letters as a target stimulus presented for 12.9 ms followed at varying interstimulus intervals (ISIs) by a second uniform flash of light of 55 ms duration, Baxt found that the target escaped the backward mask's influence when the ISI was at least 57.9 ms. On the other hand, when, other things being equal, the same subjects were required to recognize a more complicated and hence difficult Lissajou figure, the critical ISI value increased to 195.6 ms. Apparently the difficulty of the task determined how long the sensory response must persist in order to recognize the stimulus correctly.

Unfortunately, a Lissajou figure is not only more complicated than alphabetic characters but also contains finer detail and more figural elaboration. Thus, the greater difficulty associated with the Lissajou figure may have tapped sensory rather than cognitive sources of difficulty. Baxt (1871) himself makes note of the fact that a stimulus containing small spatial differences, i.e. detail, requires a longer exposure duration or integration time in order to be recognized than do larger stimuli (Kahneman 1964; see Chapter 3, Section 3.3.2). Thus, the longer mask-escape ISI and hence the greater inferred response persistence of a Lissajou figure may be correlated with a concomitant increase in the required temporal integration time (Bowling and Lovegrove 1980, 1981; see Chapter 3, Section 3.3.2). Related findings, interpretable in terms of the above argument, were also reported by Cattell (1885*a*, 1886) in his comparison of backward masking effects on letters composed of the simpler Latin type as opposed to the more complicated and detailed Gothic type.

However, that cognitive difficulty *per se may* affect response persistence in vision was demonstrated by Schumann (1899) in the following statement regarding recognition performance on words of (equal type but of) varying difficulty:

When, for example, one [briefly] exposes a word which is difficult to recognize, the experimental subject experiences recognition problems when a mask follows the word by 0.2 sec; whereas with a more easily recognized word, the perceptual image seems already to have decayed completely (cited on p. 123 in Baade 1917*b*; translation mine).

Baade (1917*b*) stresses and elaborates this finding as follows:

the fact, that the afterimage of difficultly read words persists longer than that of easily read ones, points via this strong influence of central factors, to the fact that one must seek its seat at a central level (p. 123; translation mine),

a conclusion similarly and more recently arrived at by Erwin (1976).

What, if any, were some of the central factors one can infer from these and contemporaneous works? Of prime significance was the factor of effort or activity (*Tätigkeit*) and attention (*Aufmersamkeit*). We have seen how this latter factor, in the form of directed gaze, played a role in stroboscopic motion and metacontrast (Stigler 1910; Wertheimer 1912). Reference to attentional variables in the study of basic sensory processes had been made on numerous other occasions (Baade, 1917*a*, *b*; Exner 1888; Rubin 1929; Tigerstedt and Bergqvist 1883). Of particular significance is Baade's (1917*a*, *b*) distinction between the possible sensory and attentional consequences of an after-coming mask for prior stimulus visibility. Baade (1917*a*, *b*) notes that the sensory consequences of such a mask are often confounded with its ability to evoke and thus divert attention from the target or test stimulus. As such, Baade implicitly raised the possibility of non-sensory, cognitive masking mechanisms (Turvey and Michaels 1979), and hence undermined his own rationale of using backward masking to index visuosensory persistence and his conclusion that *such* persistence is influenced by central factors.

Be that as it may, the stage of visual processing at which such attentional masking might play its role was, to my knowledge, not made explicit in any of the above investigations. However, there is some implicit indication that it was conceived of as at least intervening between *conscious registration* of a stimulus and its *recognition* or *cognitive categorization* (Cattell 1885*a*, 1886; Tigerstedt and Bergqvist 1883). Tigerstedt and Bergqvist (1883) employed the following distinction initially made by Wundt in his treatise, *Psychologische Psychologie* (*Physiological Psychology*). Wundt partitioned the psychophysical process into three stages of which the first two were: (i) the entry of the visual impression into consciousness, otherwise designated *perception*; and (ii) the entry of the conscious impression into the focal point of awareness, otherwise designated *apperception*. The former is similar to Neisser's (1967) more recent notion of preattentive process; the latter is, of course, a focally attentive process. An alternative to Wundt's perception stage or Neisser's pre-attentive process is the notion of iconic memory or persistence, a level of (parallel) visual processing at which a literal visual and visible representation of the stimulus is given without its yet being further processed, identified, or cognitively categorized. The latter, focal recognition process presumably requires effort or attention. It is at this transitional stage from perception (iconic memory) to apperception (focal awareness or recognition) that, according to Tigerstedt and Bergqvist (1883), attention would exert its role. In today's masking terminology, attention is diverted or interrupted at this stage so that, in turn, the transfer of information from a precategorical, literal visual representation or icon of the stimulus to a higher, cognitive categorical representation is also interrupted (Michaels and Turvey 1979; see also Chapter 8, Section 8.3). In this regard, using a backward masking technique, Tigerstedt and Bergqvist (1883) determined that the rate at which information is transferred from the perception to the apperception stage averaged one stimulus item per 13.8 ms; a rate very similar to that determined by Sperling (1963), Scharf, Zamansky, and Brighthill (1966) and Scharf and Lefton (1970) in more recent times.

What is the fate of items not so transferred? Here, Cattell (1885*a*) made the following relevant observations on what he called 'the limits of consciousness':

In making these experiments I notice that the impressions [of briefly exposed letters, words, or sentences] crowd simultaneously into my consciousness, but beyond a certain number, leave traces too faint for me to grasp. Though unable to give the impression, I can often tell, if asked, whether a certain one was present or not. This is especially marked in the case of long sentences; I have a curious feeling of having known the sentence and having forgotten it. The traces of impressions beyond the limits of consciousness seem very similar to those left by my dreams (p. 312).

One can infer from Cattell's observation that the process of focally attending to and cognitively grasping these visual imprecisions or icons takes time and effort; and since these iconic impressions decay rather rapidly, not all of them can be transferred into focal awareness.

1.6. Summary

One of the major tasks undertaken by visual scientists in the latter half of the nineteenth century was the identification of (i) perceptual elements or building blocks; and (ii) the stage or stages at which these elemental components manifested themselves during perceptual microgenesis. Under various guises this programme has continued to the present, although its limitations were pointed out then as they still are now.

The initial emphasis of this programme was on establishing the time course of sensation and perception. In this context, the discovery that the duration of sensations could outlast that of brief stimuli producing them, gave visual persistence a central role. Determining its intensive and temporal properties was a major project in which two varieties of masking were employed. For instance, we noted that lateral masking (para- and metacontrast) was unwittingly employed by, among others, Exner to index sensory intensity over time, since he assumed the absence of spatial contrast effects. However, at about the same time, estimates of visual persistence also quite wittingly employed backward masking by light.

The former paradigm, with a shift of emphasis to *spatial* interactions, led to the early twentieth century studies of meta- and paracontrast by Stigler, although both lateral masking effects were already investigated several decades earlier, without naming them as such, by Sherrington and McDougall. The latter paradigm of masking-by-light, for one, highlights its disruptive effects on central cognitive processes such as attention and, what is now called, read-out of iconic information. Moreover, it took a significant turn in Crawford's investigation of early light and dark adaptation. Here the mask or conditioning flash was no longer used to index the persisting response to a brief prior stimulus; but rather, a brief test probe—flashed prior to, during, and after the mask—was employed to monitor the response of the visual system to the mask.

Besides the past anticipation and definition of many of the more global issues, problems and paradigms surrounding visual information processing as presently studied, the history of visual masking and persistence also shows that many of the then-reported specific findings, methods, and theories anticipated those found in the more recently published literature. Although much new of empirical and theoretical importance to the study of visual masking and related phenomena has been placed under the sun, its connection to the past, although not always acknowledged, is one of methodical and gradual theoretical refinement and elaboration. Consequently, the history of visual masking exemplifies the history of a scientific discipline in its normal-science phase.

2 Masking by light

The history of masking by light reviewed in Chapter 1 was marked by a transition from using a uniform masking flash to estimate the (lower bound of) visual persistence of a prior target stimulus to employing instead a target probe to monitor the visual response to the masking flash. Crawford's (1947) use of the latter procedure, foreshadowed by the earlier related work of McDougall (1904b) and Stainton (1928), provided a means of investigating abrupt as well as sustained changes of visual sensitivity produced by the sharp onset and offset of a prolonged mask or conditioning flash. Moreover, Crawford's (1947) explanation of these sensitivity changes indexed by a brief target probe essentially was a slightly modified version of Stigler's (1910, 1926) metacontrast account based on the temporal integration and overtake hypotheses. Incorporated in Stigler's (1910) integration hypothesis were the activity of two physiological process: (i) the response persistence of receptors; and (ii) the later neural interaction mediated by horizontal cells. Of course, the overtake hypothesis also implied neural interactions between target and mask responses at a post-receptor level. As we shall see in the following sections, theoretical explanations of maskings-by-light based on Crawford's (1947) pioneering work and on subsequent investigations rely alternately or jointly on receptor and post-receptor neural processes, as implied by Blanchard's (1918) and Lohmann's (1906) earlier studies of light adaptation. Before discussing these investigations, I shall first introduce some definitions and distinctions.

The term *masking by light* refers to the reduction in visibility of a brief test flash (TF) by a brief mask flash (MF) consisting of a spatially uniform field (Sperling 1964). Typically (but not always) the test stimulus is relatively small (e.g. a small circular spot) and is spatially superimposed on a much larger mask field (e.g. a large concentric field). Under these conditions the test and mask contours have little opportunity to interact, and, consequently, the mask's effect on the visibility of the test stimulus is produced by luminance changes devoid of confounding lateral contour interactions. Of course, when the spatial dimensions of the test and mask fields do not differ substantially (e.g. if the diameter difference between test and mask fields is less than 1 or 2°), the role of contour effects cannot be ruled out. We shall see below that the role of contour effects becomes important in sorting out peripheral and central determinants of masking by light.

Masking by light can be differentiated further according to the following methodological distinctions:

1. When the test flash itself consists of a spatially uniform stimulus (e.g. a circular spot) and the observer is required simply to detect the presence of the test stimulus, the requisite conditions for *masking of light by light* are met.

2. When the test stimulus consists of a pattern or form such as an alphabetic character, we apply the term *masking of pattern by light*. Here the observer is required not only to detect but also to identify, recognize, or discriminate the test stimulus.

2.1. Masking of light by light

I shall restrict my discussion in this section to masking of light by light, with emphasis on those masking effects in which few, if any, contour interactions between target and mask stimuli manifest themselves. The first and complete such masking function was reported by Crawford (1947; see Fig. 1.4) and subsequent investigations have corroborated and substantially elaborated on his principal findings (Baker 1953, 1955, 1963, 1973; Battersby *et al.* 1964; Battersby and Wagman 1959, 1962; Boynton 1958, 1972; Glass and Sternheim 1973; Matsumura 1976*a*, *b*, 1977; Sperling 1965).

Sperling (1964) has outlined the rationale of studies of masking of light by light as follows. Variations in the test threshold as a function of the temporal interval between the test stimulus and the significantly stronger, suprathreshold mask stimulus monitor the physiological or sensory effect of the mask stimulus over time. By noting the magnitude variations of the time course of the visual system's response to the mask and relating them to systematic variations of experimental procedure, one can infer the source, within the visual system, of response variations.

Crawford (1947) used a mask with a duration of roughly 500 ms. Consequently, Crawford was able to monitor *step response functions* or masking effects not only at onset (on-response) but also at offset (off-response) of the mask stimulus. For substantially shorter stimuli whose durations are less than about 40 ms, no off-effects are observed (Ikeda and Boynton 1965). Here one can monitor an *on impulse response* produced by the pulsed mask. The masking effects produced by either of these two types of mask transients (step or impulse) are analysed and reviewed by Sperling (1965), Kahnemann (1968), and Boynton (1972). Moreover, Sperling (1965) offers a theoretical explanation tying the impulse to the step response. The interested reader is encouraged to go to these sources for a thorough introduction and review of the theoretical, methodological, and empirical issues involved in masking of light by light.

Without extensively reviewing all of these issues in detail, I shall focus on the following topics. (i) the effects of variations of the temporal wave form of the mask; (ii) the effects of varying the energy of the mask; (iii) wavelength effects in masking by light; (iv) the contribution of photo-chemical and neural components to masking by light; (v) the contribution of peripheral (retinal) and central (post-retinal) neural components to masking of light by light; and (vi) as a follow-up to the last two topics, a review of related anatomical and physiological work.

2.1.1. Effects of the temporal waveform of the mask

The temporal waveform of the mask means its rise-time (onset), duration, and fall-time (offset) characteristics. The duration of the mask stimulus can vary from what is called the 'prolonged' as, for example, Crawford's (1947) or Battersby and Wagman's (1959) use of a 500 ms or longer duration to the 'instantaneous' as, for example the short durations, less than 20 ms, used by Boynton and Siegfried (1926), Battersby and Wagman (1962), and

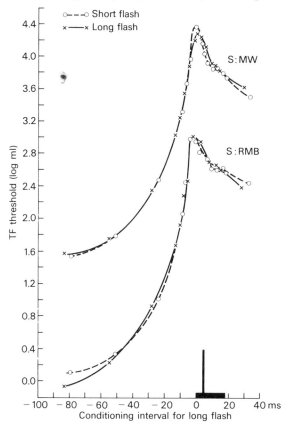

Fig. 2.1. Psychophysical estimates of on-responses for two human observers measured at two conditioning (mask) flash durations of 2.5 and 16.7 ms. The two masks were equated for apparent brightness. The curves for observer MW have been raised one log unit to avoid overlapping with observer RMB's results. (From Boynton and Siegfried (1962).)

Sperling (1965). The results obtained when varying mask duration can be summarized as follows: at short durations, provided that the brightness or the time integrated energies of the masks are equated, the physiological effects or on-responses produced by the masks are equivalent (Battersby and Wagman 1962; Boynton and Siegfried 1962). A clear illustration of this result is shown in Fig. 2.1, which shows the masking functions obtained by Boynton and Siegfried (1962) when two mask or conditioning flashes of 16.7 ms and 2.5 ms durations, respectively, were equated for brightness. The functions show the change of the test flash threshold at negative (test before mask onset) and positive (test after mask onset) conditioning intervals and were taken as indices of the visual system's equal on-responses to brief mask pulses of equal brightness.

The relationship between masking magnitude and mask duration can be extended to durations which exceed the critical value at or below which temporal integration of mask-stimulus energy (or time–intensity reciprocity) holds. Sperling (1965) argued that the first 50 or 60 ms interval of a mask presentation can account for a major portion of the variations in the peak threshold elevation or masking effect. When using the peak masking magnitude as criterion, Sperling's (1965) reanalysis of Battersby and Wagman's (1959) results showed that the threshold elevation or masking effect is a predictable funcion of the energy of the mask integrated over the first 60 ms of its presentation. In particular, the reanalysis, shown in Fig. 2.2, indicated a clear linear relationship between the log of the time-integrated test threshold energy and the log of the time-integrated energy in the first 60 ms of the masking stimulus. Slight deviations from this

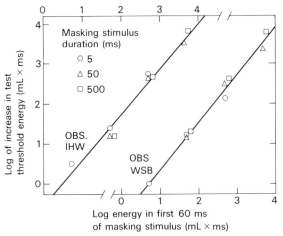

Fig. 2.2. The effect of mask stimulus luminance and duration on the peak test threshold. The abscissa (upper one for observer IHW; lower one for observer WSB) are graduated in log units of the mask luminance integrated over the initial 60 ms of the masks presentation. (From Sperling (1965).)

relationship as shown in Fig. 2.2 are correlated with the existence of a high mask intensity and longer mask durations (Sperling 1965). The Broca–Sulzer effect, characterized by an increase of brightness as flash duration increases up to about 50–100 ms, followed by a decrease at greater durations (Alpern 1963; Boynton 1961; Broca and Sulzer 1902), may be involved in yielding these similar, optimal durations.

The general conclusion one can draw from these results is that it is the temporal transient, i.e. the sudden incremental step or impulse of

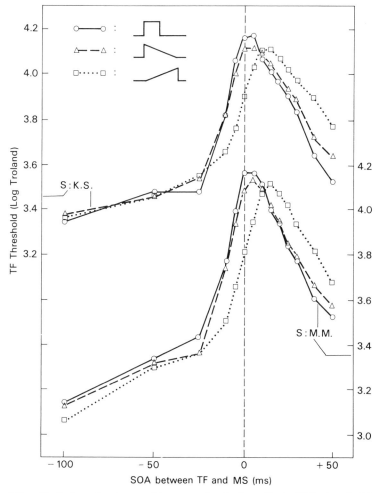

Fig. 2.3. Variations in the masking magnitude (test flash (TF) threshold) as a function of stimulus onset asynchrony (SOA) between the test and mask flashes for three temporal waveforms of the mask (see inset and text). Left ordinate corresponds to observer KS's results; right ordinate, to observer MM's results. (From Matsumura (1976*a*).)

luminance introduced by the mask, which determines peak masking magnitude. Recently Matsumura (1976*a*, *b*; 1977) investigated the effects of variations of the rise and fall times of a prolonged or pulsed mask stimulus on variations of the masking magnitude function. In one study, Matsumura (1976*a*) varied the temporal waveform of a pulsed mask (see inset of Fig. 2.3). A pulse of 15 ms duration and an energy of 2512 trolands (td) was used, superimposed on a background of 628 td. Two sawtooth pulses of 30 ms duration were also used, one characterized by an instantaneous rise time to a peak energy of 2500 td and a 30 ms decay time; the other by a 30 ms rise time and an instantaneous decay time. The time-integrated energies of all pulses were equal.

The masking results obtained with these three mask pulses are shown for two subjects in Fig. 2.3. The results can be summarized as follows:

1. Masking functions corresponding to the three temporal waveforms of the mask were nearly identical in shape.

2. They also were characterized by roughly equal peak masking magnitudes.

3. The temporal position of the peaks differed, occurring equally early at mask onset for the rectangular and negative (i.e. left-to-right decreasing) sawtooth pulse and latest (10–15 ms after mask onset) for the positive sawtooth pulse, characterized by a 30 ms rise-time and instantaneous fall time.

Consequently, even at these short durations perfect temporal integration of mask energy, which would be indicated by identical superimposed masking functions, does not occur—a finding concurring with a conclusion arrived at by Sperling (1965) in his analysis of the effects of mask duration.

In another experiment Matsumura (1976*b*) investigated the effects of variation in the rise and decay times of a prolonged mask stimulus on the masking magnitude. The prolonged mask stimulus, a 2516 td luminance increment on a 624 td background, was characterized by linear rise or fall times of 0, 50, 100, and 200 ms. Representative effects of variations of rise and decay times can be seen in Fig. 2.4(a) and 2.4(b), respectively. The following conclusions can be drawn from the effects of rise time variations. As rise time increases the threshold overshoot or peak at onset (originally reported by Crawford (1947)) attenuates; and there also is indication that it is delayed. Moreover, the shape of the peak or overshoot varies with rise time; it is sharper for shorter, more abrupt rise times than for longer, more gradual ones.

For variations in the decay time, the following general results are noteworthy:

1. In contrast to variations in rise time, no dramatic changes in the magnitude of the masking peak at offset (again originally reported by Crawford (1947)) occurred.

2. As with rise times, the peak masking effect seemed to be delayed.

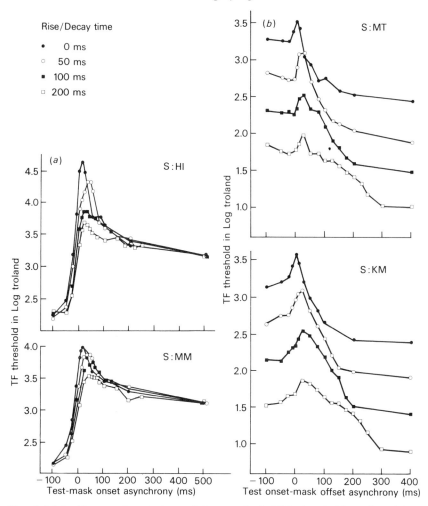

Fig. 2.4.(*a*) Temporal variations of the test flash (TF) threshold produced by a masking stimulus whose onset (increment) rise time varied from 0 to 200 ms. (*b*) Temporal variations of the test threshold produced by a masking stimulus whose offset (decrement) decay time varied from 0 to 200 ms. Here, for ease of inspection, the results for different mask decay times are staggered vertically by 0.5 log units. (From Matsumura (1976*b*).)

3. The peak masking effect seemed to broaden temporally as decay times increased.

4. The third result is correlated with a more gradual or slower decline in the masking effect after offset.

These results, as well as those indexing the effects of variations of rise time, again point to the importance of temporal transients in determining masking magnitude.

2.1.2. *Effects of mask and background intensity*

We have seen in our historical review of masking by light that decreases in the intensity of the mask stimulus produce an overall decrease in masking magnitude. Moreover, the threshold peaks at onset and particularly at offset of a prolonged mask stimulus are substantially attenuated as its intensity decreases.

A related finding, employing a pulsed rather than prolonged mask presentation, has been reported by Sperling (1965). In Fig. 2.5 are

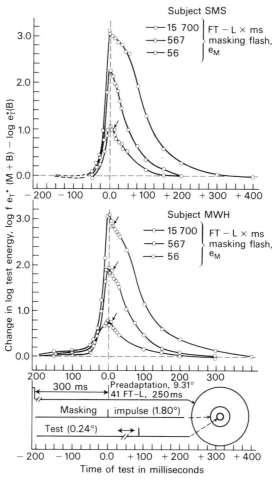

Fig. 2.5. Temporal variations of the detection threshold of a small test spot produced by larger concentric masking pulses at three time-integrated energy levels as indicated. The masking functions can be considered estimates of the visual system's on impulse responses to the three different mask pulses. (From Sperling (1965).)

displayed mask impulse responses (on-responses) as a function of mask intensity. The results support the following conclusions:

1. Impulse masking magnitude generally decreases as mask intensity decreases.

2. Forward masking is stronger and can be obtained over longer intervals than backward masking.

3. The duration of the forward masking effect outlasts the mask duration and decreases as mask intensity decreases. At least as indexed by changes in test threshold, this result indicates that the physiological effects of the mask seem to persist noticeably longer at higher mask intensities. (This finding will become relevant again in our discussion of temporal integration and persistence in the following chapter.)

4. The impulse response function is characterized by secondary peaks (see arrows), a result resembling, on a shorter time scale, oscillations in the apparent brightness following a brief flash (Stainton 1928).

The effects of mask intensity variations on the on-response of the visual system were also investigated as a function of the luminance of the background on which the mask stimulus is flashed (Boynton and Kandel 1957; Onley and Boynton 1962; Sperling 1965). What, at first glance, seems manifest is that the background *per se* does not affect the response to the mask stimulus. This is illustrated in Fig. 2.6(a) and (b) taken from Sperling (1965) which show the variations of the log of the peak test threshold, as a function of the log energy of an impulse mask, for backgrounds varying from a dark one to one of 41 ft-L. In Fig. 2.6(a) are shown variations of log-test flash threshold as a function of log mask flash energy without factoring out the effects of background intensity. Note that as background intensity progressively increases by about 2.0 log unit, the test threshold, at least at lower mask energies, where mask variations are relatively ineffective, progressively increases by about 1.6–1.8 log unit depending on the observer. Thus, at the low, ineffective mask energies, the relation between log threshold energy and background intensity roughly follows a modified Weber's Law. In Fig. 2.6(b) are shown the same results with the effects of background factored out. Here, particularly at higher mask energies and irrespective of the background intensity used by Sperling, variations in log test threshold energy can be accounted for almost entirely by linear covariations of log mask energy. The relation between test threshold and mask energy could again be expressed as a modified Weber's Law. In summary, as a function of background and mask energy, variations in test threshold could be expressed additively by separate modified Weber's Laws when the effects of background and mask energy were treated separately. The following chapter shows that visual persistence of a briefly flashed stimulus also is effected separately and systematically by variations of background and stimulus energy.

2.1.3. Effects of wavelength

Chromatic or wavelength effects on masking magnitude also were reported by Sperling (1965). Sperling used masks and test stimuli each of which could be either green or red independent of the colour of the other.

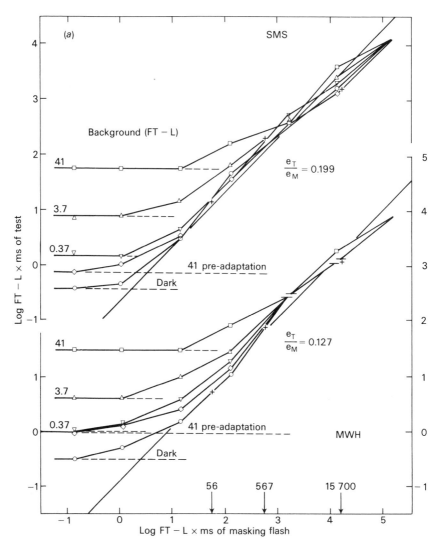

Fig. 2.6.(*a*)   Test threshold as a function of mask-impulse energy and background luminance. Points with the same background luminance are connected. The 41 preadaptation curve refers to a 250-ms field of 41ft-L terminated 300 ms before the masking impulse. (*b*) Test threshold as a function of mask-impulse energy with the effects of the background luminance on test threshold factored out. (From Sperling (1965).)

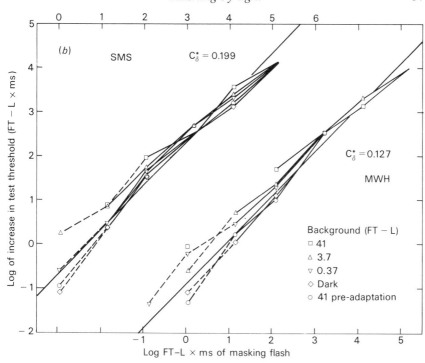

Sperling's (1965) results showed that the masking magnitude was greatest at or near mask onset and depended on the wavelength of the mask and test stimuli. Generally, green masks (Wratten filter No. 74) were stronger than red ones (Wratten filter No. 70), and green test stimuli were masked more than red ones. However, since Sperling (1965) did not equate the different chromatic sources for brightness or luminosity, one cannot separate out the effects of luminance and hue transients on test threshold.

An important variation of the basic Crawford paradigm was introduced by Glass and Sternheim (1973). What they measured psychophysically were the effects of an abrupt substitution of hues, equated for brightness, on the threshold for lights of varying wavelengths. The work of Boynton and Siegfried (1962; see also Onley and Boynton, 1962) showed that achromatic mask stimuli of equal apparent brightness yield roughly equal masking effects. If chromatic or wavelength transients were to have no effect, i.e. if brightness transients were the sole determining factor, one would not expect masking effects at or near the time of a transition of hues equated for brightness. However, what Glass and Sternheim reported, as shown in Fig. 2.7(a) and (b), are dramatic masking effects produced at and near the time of the hue transition when hues varying significantly in wavelength were employed. Fig. 2.7(a) shows the temporal masking function for a test stimulus, having a wavelength of 469 nm, as a function of

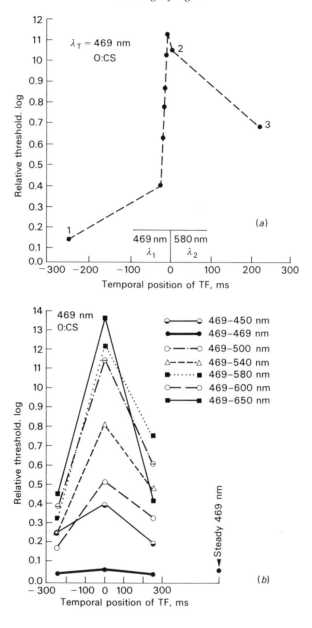

Fig. 2.7.(*a*)   The time course of changes of the test flash (TF) threshold when a background of 580 nm is suddenly substituted for a brightness matched background of 469 nm. The TF is at a wavelength of 469 nm as indicated. (*b*) TF thresholds at three temporal intervals relative to the time that a 469-nm wavelength of the conditioning field is suddenly replaced by another wavelength as indicated. (From Glass and Sternheim (1973).)

the time preceding and following the abrupt hue substitution of brightness equated fields from a wavelength of 469 nm to one of 580 nm. Note that just as with achromatic brightness transients there is (i) a rise in threshold that starts several tens of milliseconds prior to the transition; (ii) a peak masking effect at or near the instant of hue transition followed by (iii) a decline. Moreover, Fig. 2.7(b) illustrates that the magnitude of this effect depends on the wavelengths of the test and mask stimuli. Generally one can conclude that the larger the wavelength difference between test and mask stimuli the stronger the masking effect produced by transient hue substitution. This result is of particular importance since it becomes relevant again in subsequent discussions of empirical and theoretical aspects of metacontrast (see Chapters 4 and 7).

2.1.4. The contributions of photochemical and neural components

One way of viewing masking of light by light is an indicator of the effects of early light and dark adaptation produced by the onset and offset of a mask stimulus, respectively. In fact, Crawford's (1947) study was conducted in the framework of empirically indexing transient adaptation effects. The theoretical importance of the study of masking by light to our understanding of the initial stages of light and dark adaptation has been emphasized by Boynton (1958), Baker (1963), and Boynton and Miller (1963). Moreover, the practical importance of studying transient adaptation to our understanding visual performance under conditions of peripheral glare as, for example, during night-time driving has been noted, among others, by Schouten and Ornstein (1939), Fry and Alpern (1953), and Lie (1981).

As indicated by Wald (1961; see also Chapter 1, Section 1.4), the separate effects of photochemical and neural components of visual adaptation were implied by work conducted as early as 1918 by Blanchard and by Lohmann's (1906) still earlier investigation of light adaptation. Subsequent work and analyses carried out by Boynton and Kandel (1957), Boynton (1958), and Baker (1955, 1963) have investigated the existence and extent of these two components in determining transient light and dark adaptation.

As an instructive and illustrative case, let us first look at the investigation and results reported by Boynton and Kandel (1957). As shown in Fig. 2.8(a), three stimuli were employed: (i) a preadapting stimulus field, 13° in diameter, which, though held constant for any series of trials, could be varied in luminance from 0.00015 to 3000 mL across an experimental series; (ii) a concentric conditioning or mask stimulus, 11° in diameter, which was fixed at an intensity of 38 mL; and (iii) a foveal test spot, 1° in diameter, whose luminance could be adjusted to determine threshold values. As shown in Fig. 2.8(b), two types of trials were run. One series of trials was designated 'on' trials. Here the stimulus sequence was as follows (see top panel of Fig. 2.8(b)):

1. Prior to any series of trials the preadapting stimulus was exposed for at least 5 min at a fixed luminance level.

2. Following a warning signal the preadapting stimulus was extinguished.

3. 280 ms after this offset the onset of the mask stimulus occurred.

4. Relative to this mask onset the conditioning or masking interval separating the onset of a 40 ms test flash varied from -200 ms (test onset prior to mask onset) to $+300$ ms (test onset after mask onset).

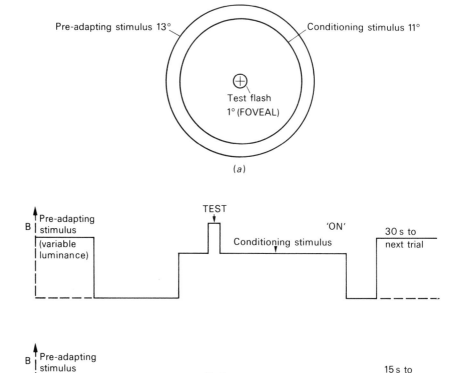

Fig. 2.8.(*a*)   The three stimuli used by Boynton and Kandel. The plus inside the test flash represents a cross-hair reticule used for fixation. (*b*) The temporal order of stimuli used by Boynton and Kandel under the 'on' and 'off' experimental conditions. (From Boynton and Kandel (1957).)

5. The conditioning or mask stimulus was extinguished 560 ms after its onset.

6. 160 ms thereafter the preadapting stimulus reappeared for 30 s prior to the next 'on' trial. For 'off' trials (see bottom panel of Fig. 2.8(b)) the procedure was basically the same with the exception that the mask or conditioning stimulus was not presented and that the intertrial interval was 15 rather than 30 s.

The rationale employed by Boynton and Kandel (1957) was as follows. It was assumed that the initial 5-min exposure of the preadapting field and its 30 (or 15) s reappearances between trials of a given session produced a temporally averaged and stable bleaching or photochemical adaptation level. By varying the luminance level between experimental sessions, the photochemical adaptation level could also be varied. Furthermore, for the 'off' trials, it was assumed that the test threshold was determined only by the temporally stable photochemical adaptation level. For 'on' trials it was assumed that the test threshold was affected separately and additively by two components: (i) the temporally stable photochemical bleaching level produced by the preadapting field; and (ii) the temporary neural, on-response produced by the conditioning or mask flash. Given these assumptions, it follows that one can factor out the 'pure' masking or neural on-response by simply subtracting the threshold values obtained in the 'off' condition from those obtained in the 'on' condition.

Based on this rationale, Boynton and Kandel's (1957) results are shown in Fig. 2.9. What is shown are log test thresholds, at a fixed conditioning interval of 100 ms, as a function of log preadapting luminance. The former effect, B_{mp}, presumably is regulated by the transient neural effects of the mask stimulus onset additively superimposed on the stable photochemical effects of the preadapting stimulus; the latter, B_p, presumably monitors only the stable photochemical adaptation effects. Consequently the curve labelled B_m, which is the function plotting the threshold difference between the 'on' and the 'off' condition, presumably monitors the pure neural on-response produced by the mask onset.

Two conclusions can be drawn from these results. For one, the depletion of photochemicals via bleaching (B_p) increases as preadapting luminance increases and can account for a major portion of threshold increments, particularly at higher preadapting luminances. Conversely, the neural on-response (B_m) produced by the onset of a mask of constant (38 mL) luminance decreases as preadapting luminance increases.

This last result seems to be at odds with Sperling's (1965) finding, shown in Fig. 2.6, which indicated (i) that the mask effectiveness does not depend crucially on the luminance level of a uniform background on which the mask is superimposed and (ii) that test threshold varies directly and additively as background and mask intensities increase. However, several aspects distinguished Sperling's methods from Boynton and Kandel's. For

Masking by light

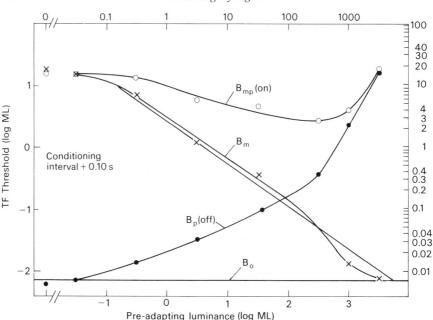

Fig. 2.9. Test flash (TF) thresholds as a function of the preadapting stimulus luminance for the 'off' and 'on' conditions at a conditioning interval of 100 ms. The curve labelled B_{mp} resulting from the 'on' condition is considered to be determined by a combination of photochemical bleaching (caused by the preadapting stimulus) and neural masking (caused by the conditioning or mask stimulus). The curve labelled B_p which results from the 'off' condition is thought to be determined by photochemical bleaching alone. The curve labelled B_m represents the threshold that would have been obtained when the effects of photochemical bleaching are factored out from the B_{mp} curve; B_m therefore represents the theoretical neural masking effect produced by the 'on' condition. B_0 is the absolute cone threshold. The B_m curve is the difference between the B_{mp} and B_p curves relative to a B_0 baseline. (From Boynton and Kandel (1957).)

one, the background field used by Sperling was presented continuously throughout the mask and/or test flash intervals, whereas Boynton and Kandel's preadapting field was extinguished for the 1-s trial during which the mask and/or test stimuli were presented. Secondly, Sperling's backgrounds ranged from dark to 41 ft-L (about 44 mL): the preadapting luminance employed by Boynton and Kandel ranged from 0.0015 to a much higher value of 3000 mL. Thirdly, Sperling's mask intensity varied over several log units, whereas the mask intensity employed by Boynton and Kandel was fixed at 38 mL. That is, whereas Sperling used a small range of background intensities and a large range of mask intensities, Boynton and Kandel conversely used a fixed masking intensity and large range of background intensities. Careful inspection of Sperling's results (see Fig. 2.6) reveals that there is a systematic variation (increase) of about 0.5–1.0 log unit among the test thresholds produced by a constant mask-energy flash at varying background luminances (increasing from 0.37 to 41 ft-L). As shown in Fig. 2.9, for the same, roughly 2.0 log unit, range of preadapting luminances used by Boynton and Kandel (0.4 to 44.1 mL), the systematic variations of test threshold along the function labelled B_p also range over approximately a 1.0 log unit interval. Had Sperling used much higher background field luminances, his obtained results might have shown an even greater systematic spread (e.g. 2–3 log units) around his best-fitting linear functions. However, this would not violate Sperling's conclusion that the background and mask exert separate effects on the test threshold.

Be that as it may, one can nevertheless ask whether Boynton and Kandel's assumption that the preadapting stimulus contributed only to the photochemical or bleaching component of light adaptation was entirely warranted. Recall that the test flash was presented anywhere from 200 ms before to 300 ms after the onset of the mask. Since mask onset occurred 280 ms after preadapting field offset, the test fell from 80 to 580 ms after the extinction of the preadapting field. The results shown in Fig. 2.9 were obtained at a conditioning interval of 100 ms; that is, 380 ms after the offset of the preadapting field. It is entirely possible that within that short time interval, the fast neural component of dark adaptation or what has been called the *off-response* can still exert an effect on test threshold (see Fig. 2.8(b) and Crawford's (1947) results displayed in Fig. 1.4). This may be especially applicable at high preadapting field intensities when response persistence and positive after-images are highly durable (Kriegman and Biedermann 1980; Sperling 1965). Consequently Boynton and Kandel's (1957) results obtained under the 'off' condition (Curve B_p in Fig. 2.9) may contain a minor contamination of a transient neural off-effect produced by the offset of the preadapting field. Of course, such a contamination if present may also manifest itself in the 'on' condition (Curve B_{mp} of Fig. 2.9). In particular, if the presumably contaminating, transient, neural off-effect were to combine by linearly adding to the neural on-response of the conditioning or mask stimulus, the B_m curve would still monitor a pure neural on-response. However, any non-linearity (Boynton 1961) such as sub- or superadditivity would result in under- and overestimates of the neural on-response. None the less, despite the possibility of such minor non-linear contamination, Boynton and Kandel's (1957) results stand as an ingenuous demonstration of the separable photochemical and neural components in early light adaptation.

Baker (1963) performed a more extensive analysis of photochemical and neural components of transient adaptation by including early light as well as early dark adaptation. Baker arrived at the following conclusions on early light and dark adaptation which are schematized in Fig. 2.10(a) and

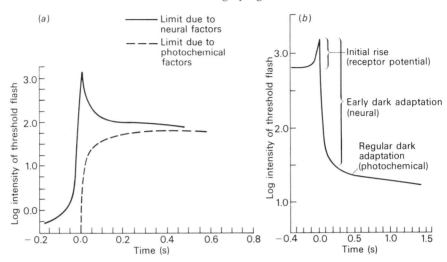

Fig. 2.10.(*a*) The theoretical contribution to early light adaptation (at mask stimulus onset) of neural factors (solid curve) and photochemical factors (dashed curve). (*b*) The theoretical contribution of receptor, neural, and photochemical factors to early dark adaptation. (From Baker (1963).)

(b). Fig. 2.10(a) shows the hypothetical contributions of neural and photochemical components in early light adaptation. Note that the neural component produces the transient threshold peak or overshoot at mask onset, whereas the photochemical component yields a rapidly, monotonically increasing function stabilizing about 300 ms after mask onset. Fig. 2.10(b) shows the hypothetical events contributing to early dark adaptation. They consist of an early threshold rise beginning about 100 ms prior to mask offset and attributed to transient changes of the receptor potential, followed by a rapid threshold decline attributed to a post-receptor neural component, which in turn is followed by a gradual threshold decline attributed to photochemical factors such as the regeneration of bleached photopigments.

Several lines of evidence, more fully discussed in the following three sections, for the most part support this breakdown of early light and dark adaptation. One line of evidence concerns the effects of the size of the mask field, relative to the test stimulus, on masking magnitude (Baker 1973; Battersby and Wagman 1962; Markoff and Sturr 1971; Matthews 1971; Teller, Matthews, Phillips, and Alexander 1971); another line derives from the comparison of masking under monoptic and dichoptic viewing (Battersby *et al.* 1964; Battersby and Wagman 1962; Markoff and Sturr 1971); and a third line of evidence derives from neuroanatomical and neurophysiological investigations of early light and dark adaptation.

2.1.5. The effects of stimulus size on masking magnitude

Crawford (1940) investigated the effect of varying the diameter of a steady (non-transient) conditioning field (CF) on the threshold of a flashed test stimulus (TS) of fixed, 0.5°, diameter. What Crawford found was that TS threshold varies as an inverted U-shaped function of CF diameter. As CF diameter increased from a value below that of the TS diameter, TS threshold initially increased, reached a maximum when CF diameter was equal to or slightly greater than the TS diameter, and then decreased again. Using a slightly altered method and prolonged, steady test stimuli rather than brief transient ones, Westheimer (1965, 1967, 1970) replicated and interpreted this finding as indicating the existence of excitation pools (Rushton 1963; Rushton and Westheimer 1962), which produce a desensitization between adjacent or nearby retinal areas of stimulation and a sensitization between retinal areas separated by a greater distance. In particular, as the diameter of the CF initially increases, it covers

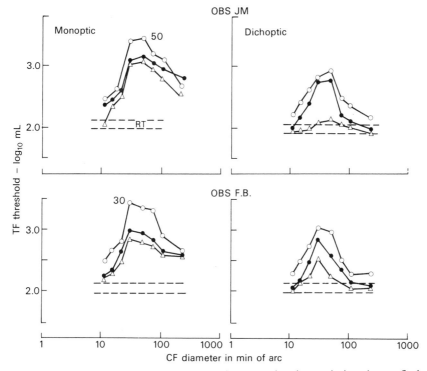

Fig. 2.11. Westheimer functions for two observers showing variations in test flash (TF) threshold as a function of conditioning (mask) flash (CF) diameter under transient (open and filled circles) and steady (open triangles) conditioning stimulus presentation. Under transient conditions the test flash was presented at the onset of a 50-ms (filled circles) or 200-ms (open circles) conditioning flash. (From Markoff and Sturr (1971).)

progressively more of the excitation pool, raising the background activity of that pool, in turn necessitating a higher TS threshold intensity, until, at a given diameter, a maximum TS threshold is attained. As CF diameter increases further, the CF invades more and more of the antagonistic lateral inhibitory pool; this, in effect, reduces the background activity of the excitation pool which is monitored by a concomitant decrease in TS threshold intensity. For the sake of brevity, we shall refer to the inverted U-shaped function defining the relationship between TS threshold and CF diameter as the *Westheimer function*.

As noted, Westheimer functions also can be obtained when the conditioning or mask field is flashed transiently. Markoff and Sturr (1971) measured Westheimer functions obtained under transient and steady CF presentations. The TS was 3.5′ in diameter and was presented for 5 ms, synchrononously with the on-set of a 200 ms or 50 ms CF or during a steady CF. The TS and CF were presented either monoptically (to the same eye) or dichoptically (to separate eyes). Moreover, their location on the retina was varied from the fovea to the 15° periphery, and the level of CF luminance could be either scotopic (0.01 mL) or photopic (1.0 mL). Results obtained at a retinal eccentricity of 5° and at a photopic CF intensity are shown in Fig. 2.11. Note the generally inverted U-shaped Westheimer functions peaking at the same CF diameters (50′ and 30′ for J.M. and F.B., respectively) obtained under transient CF presentation (open and filled circles) and under steady CF stimulation (open triangles). Furthermore, the overall masking effect seems to be larger under transient than under steady CF presentation; and relative to monoptic viewing, dichoptic viewing yields similar transient Westheimer functions but attenuated steady Westheimer functions. The last finding held for all retinal loci except the fovea where monoptic and dichoptic steady Westheimer functions were of about the same shape and magnitude. Other relevant results (not shown in Fig. 2.11) reported by Markoff and Sturr can be summarized as follows: (i) at a given retinal location the scotopic Westheimer functions peaked at greater CF diameters than did the photopic functions (see also Westheimer 1970); (ii) as retinal eccentricity increased, the peak of the Westheimer functions occurred at progressively higher CF diameters. These two generalizations applied for transient and steady CF conditions, and under monoptic and dichoptic viewing.

Several conclusions can be drawn from these results:

1. At a given retinal location (other than the fovea) the cone excitation pool is smaller than the rod excitation pool—a result consistent with Westheimer's (1970) findings.

2. As retinal eccentricity increases, the excitation pool diameters both for cones and for rods increase.

3. Based on comparison of the results of dichoptic and monoptic viewing in non-foveal areas, the effects of the excitation and lateral

inhibitory pools during a steady CF seem to manifest themselves largely at early, perhaps retinal levels of processing, although some post-retinal processes may also be used.

Moreover, the importance of rod and cone excitation pools in determining the shape of the Crawford-type masking function is demonstrated by the fact that the transient overshoots at on- and offset of a prolonged mask flash are obtained only when the diameter of the mask exceeds that of the rod excitation pool at scotopic luminance levels (Teller *et al.* 1971) or that of the cone excitation pool at photopic luminance levels (Matthews 1971). This indicates that the activity of single receptors must summate at a later neural level at or beyond which potent transient responses to on- or offset of a light flash are generated; and contradicts Baker's (1963) explanation of the transient overshoot at CF offset on the sole basis of transient receptor-potential changes.

2.1.6. Monoptic and dichoptic contour interactions

The involvement of later neural processes is given further support and clarification in the following analyses of psychophysical results implicating the role of contour effects in determining masking magnitude as monitored by variations in the test threshold (Baker 1963; Battersby *et al.* 1964; Battersby and Wagman 1962; Bouman 1955; Brussel, Adkins and Stober, 1977; Hood 1973). As CF diameter increases relative to a smaller TS diameter, two correlated processes may be used in determining masking magnitude. One may be the interaction between excitation and lateral inhibition pools as monitored by steady or transient Westheimer functions; the other process may involve the decreasing role of contour interactions between the TS and CF. These latter contour interactions were investigated under monoptic and dichoptic viewing conditions by Battersby and Wagman (1962) and Weisstein (1971) and their existence was reconfirmed under strictly dichoptic viewing by Battersby *et al.* (1964).

Battersby and Wagman (1962) used the following variation of the method employed by Crawford (1947). Thresholds for a 40' diameter TS were measured 7° in the peripheral visual field at CF diameters of 40', 1°20', 2°, and 4°40'. The 100 mL CF was presented for 500 or 1500 ms and the TS was presented for 5 ms at variable temporal intervals relative to CF onset. Viewing could be either monoptic or dichoptic (Battersby and Wagman used the terms *monocular* and *binocular*, respectively). Typical results are shown in Fig. 2.12.

Note several important aspects of the results:

1. The greatest amount of masking occurs under monoptic viewing; this is particularly evident at smaller CF diameters where contour interactions between TS and CF can occur. From this one can conclude that most of the masking magnitude can be accounted for by monocular, prechiasmal

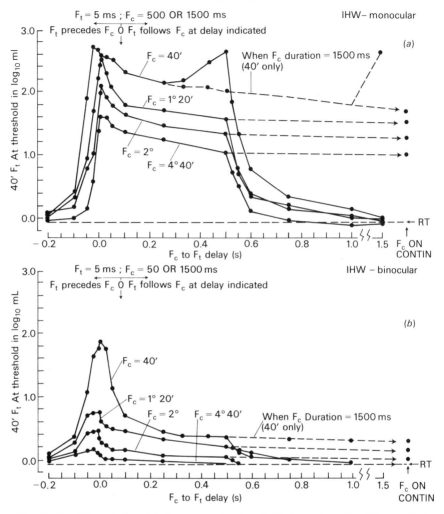

Fig. 2.12. The relationship between the threshold intensity of a 40' test flash (ordinate) and the temporal interval between the conditioning (mask) flash and test flash (abscissa) under monocular observation (*a*) and dichoptic (binocular) observation (*b*). (From Battersby and Wagman (1962).)

mechanisms not only sensitive to the luminance of the CF but also manifesting TS–CF contour interactions.

2. Compared with monoptic viewing, the transient threshold peak at CF onset is attenuated substantially under dichoptic viewing as CF diameter increases. The fact that one does obtain the transient threshold peak at all under dichoptic viewing implicates the role of postchiasmal, perhaps cortical mechanisms. Moreover, since the transient peak at onset decreases

substantially under dichoptic viewing as CF diameter increases, one can infer that these central mechanisms entail lateral contour interactions—an inference subsequently confirmed by Battersby *et al.* (1964).

3. For all CF diameters, the rapid decline of the TS threshold subsequent to CF offset seems to be largely or entirely a monoptic effect. From these particular results we can conclude that the rapid phase of dark adaptation is not determined by postchiasmal, contour-interactive processes but rather by more peripheral, perhaps retinal processes sensitive to purely temporal luminance changes.

4. The rise in TS threshold just prior to CF offset was found only under monoptic viewing when TS and CF diameters were equal; that is, when contour interactions are maximal.

This last result suggests that part of this threshold rise can be accounted for by prechiasmal contour interactions (in addition, see results reported by Breitmeyer and Kersey (1981) discussed in Chapter 4, Section 4.4.3, for evidence supporting the additional role of similar postchiasmal contour-interactions). However, pure peripheral luminance effects most likely are also involved since the threshold rise prior to CF offset also is evident under relatively high CF luminances and relatively large CF diameters (Crawford 1947; see Fig. 1.4), which effectively eliminate the role of contour interactions. In fact, Baker (1973) ascribes the threshold rise prior to CF offset *entirely* to transient changes in the retinal receptor potentials and therefore exclusive of stimulus area or contour effects. However, this interpretation is difficult to reconcile with those of Battersby and Wagman (1962) (see result (1) above), Battersby *et al.* (1964; Fig. 3, p. 1184), Matthews (1971) and Teller *et al.* (1971), all of which indicate the use of post-receptor activity in determining the transient overshoots at on- and offset of the mask flash. Consequently, a more adequate explanation involves receptor effects as well as post-receptor, pre- and postchiasmal contour effects, each of which may contribute a variable extent to the entire masking magnitude depending on variations of TS and CF area, TS-CF contour separations and CF luminance.

2.1.7. Neurophysiological and neuroanatomical studies of early light and dark adaptation

In recent years, significant advances have been made in our understanding of the sites and processes involved in luminance masking and light and dark adaptation. These conceptual advances have resulted from work on the vertebrate visual system (Barlow, Fitzhugh, and Kuffler 1957*a*, *b*; Boynton and Triedman 1953; Boynton and Whitten 1970; Brown, Wanatable, and Murakami 1965; Coenen and Eijkman 1972; Dowling 1963, 1967; Enroth-Cugell and Shapley 1973*a*; Fehmi, Adkins, and Lindsley 1969; Gordon and Graham 1973; Jakiela and Enroth-Cugell 1976; Salinger and Lindsley 1973; Schiller 1968; Werblin 1971; Werblin and Dowling 1969) as well as that of the invertebrate system (Felsten and Wasserman 1978, 1979, 1980; Riggs 1940; Riggs and Graham 1940). On the basis of most of this research, one could conclude that most or all of visual adaptation to light or dark in the absence of significant contour effects occurs at the retinal level, using not only photoreceptor processes but also neural processes (Werblin 1971, 1974; Werblin and Copenhagen 1974).

As a point of departure for my analysis of the neurophysiology and neuroanatomy of the masking of light by light I shall summarize the main points of Baker's (1963) analysis of the typical Crawford masking result (see Fig. 1.4):

1. Early light adaptation (at mask onset) is characterized by a sudden threshold increase which overshoots the terminal level, followed by a rapid decrease in threshold back to the terminal level. Baker argues that this transient light adaptation is not attributable to photochemical factors but rather depends on a combination of peripheral (retinal) and central (cortical) neural processes. I have outlined above how and under what stimulus and viewing conditions these two processes may contribute to early light and dark adaptation.

2. The dark adaptation transient is characterised by (i) a rapid threshold rise about 100 ms prior to mask offset; and (ii) a subsequent rapid threshold drop followed by the slower phase of dark adaptation. Baker (1963) attributes the former rapid threshold rise exclusively (and incorrectly) to transient changes in receptor potential and the latter rapid threshold decline to post-receptor neural processes.

Let us now look at the neurophysiological and neuroanatomical evidence concerning these conclusions. The early light-adaptation transient (devoid of contour effects discussed above) is evident in the photoreceptor response of *Limulus* (Felston and Wasserman 1978, 1979), on–off ganglion cells of frogs (Gordon and Graham 1973), cat retinal ganglion cells (Enroth-Cugell and Shapley 1973a), optic tract (Coenen and Eijkman 1972; Salinger and Lindsley 1973) and lateral geniculate nucleus (LGN) cells (Coenen and Eijkman 1972). Moreover, it seems to correlate significantly with changes of retinal sensitivity as monitored by a fixed criterion amplitude of the human ERG, in particular, the amplitude of the b-wave component (Boynton and Triedman 1953). Since most of the ERG amplitude variations are thought to originate somewhere between the receptor and bipolar cell level (Riggs and Wooten 1972) one can place some of the mechanisms of early light adaptation at the outer plexiform layer of the retina—a conclusion consistent with Werblin's (1971) finding that most of the adaptive processes to light (or dark) in the vertebrate retina appear to be completed at the bipolar cell level.

The early dark-adaptation transient can be monitored in frog on–off ganglion cells (Gordon and Graham 1973); and there is some evidence that at least the early threshold rise prior to mask offset can be monitored by the response of single optic nerve fibers (emanating from a single photoreceptor) of *Limulus* (see Fig. 5 in Baker 1963), although subsequent work on *Limulus* by Felsten and Wasserman (1978, 1979) failed to find any indication of this early threshold rise at the photoreceptor level. This discrepancy is of some theoretical importance since, as noted above, Baker (1963, 1973) attributes the early threshold rise prior to mask offset solely to transient changes in the photoreceptor potential.

Of course, the comparison of neurophysiological data derived from animal preparations with that derived from human psychophysics assumes that the responses of single cells, photoreceptors, and so on in non-vertebrate and vertebrate species correspond closely to psychophysical behavior. This assumption may not be entirely warranted for several reasons. For one, the human visual *system* is much more complex than, for instance, the photoreceptor of *Limulus* or the ganglion cell of a frog or cat. Secondly, although one can successfully relate electrophysiological findings to psychophysical findings of masking by light (e.g. Fehmi *et al.* 1969), one must be cautious not to convert *an* electrophysiological finding into *the* theory of masking (Uttal 1971, 1981). Merely because an electrophysiological result obtained from the study of a single receptor neurone at a given level of visual processing correlates well with the psychophysics of masking does not warrant the conclusion that *the* site or mechanism of visual masking by light has been isolated. Indeed, much of masking of light by light can be related to peripheral, retinal processes; however, dark and light adaptation are known also to affect the response of neurones at higher levels of visual processing (Nunokawa 1973; Sasaki, Saito, Bear, and Ervin 1971; Virsu, Lee, and Creutzfeldt 1977). Without more extensive investigations of the effects of early light and dark adaptation on LGN or cortical cells, one lacks information of major importance in formulating a comprehensive description of the psychophysics or psychophysiology of masking of light by light. After all, the psychophysical masking functions obtained from human observers monitors a system response of which the single-cell response is a mere and miniscule index. None the less, drawing cautious correspondences between psychophysical and neurophysiological and neuroanatomical results can be a fruitful enterprise—a fact that should become increasingly apparent to the reader as the reading of this and the following chapters progresses.

2.2. Masking of pattern by light

Up to now we have considered masking of light by light as a form of early light and dark adaptation. An alternative way of looking at masking by light is to consider it as a form of contrast reduction-by-luminance summation, a hypothesis which has been advocated most consistently by Eriksen (1966) and Thompson (1966). This hypothesis states that due to temporal resolution limits of the visual system, imposed by either response persistence or responce integration, the luminances of the target and mask stimuli effectively combine to produce a contrast reduction of the target. Strictly speaking, it is not the luminances of the two stimuli which combine (except in the case when their presentations overlap in time) but rather the sensory responses evoked by them. Moreover, the combination of response is not additive but seems to be a non-linear summation (Boynton 1961).

The test stimulus need not be an incremental luminance flash as typically employed in studies of early light and dark adaptation. It usually is a pattern or form of either positive or negative contrast value (i.e. light on dark or dark on light, respectively). At any rate, what is assumed by the luminance summation hypothesis is that the sensory activity of the target is obscured by the sensory activity of the mask whenever these activities persist and overlap in time or when they are integrated during a brief temporal interval (irrespective of whether or not they overlap temporally). Another way of viewing this reduction in the perceived contrast of the target is to consider masking as a way of reducing the signal-to-noise ratio within sensory channels that carry both the target and mask activity. Such a conceptualization is akin to Barlow's (1957) explanation of incremental threshold performance. If we designate the target-generated activity as the 'signal' and the mask-generated activity as the 'noise', we see that by increasing the mask effectiveness, we concurrently decrease the signal-to-noise ratio.

Two types of masking of pattern by light which have been investigated and compared extensively are forward and backward masking (the mask stimulus temporally precedes and follows the target stimulus, respectively). The general conclusions which can be derived from the results of these masking studies are the following:

1. Both forward and backward masking functions are Type A or monotonic functions (Kolers 1962). That is to say, masking is strongest when the target and mask stimuli are presented simultaneously and decreases progressively as the forward or backward interval of the mask relative to the target increases (Boynton 1969; Boynton and Miller 1963; Donchin 1967; Eriksen and Hoffman 1963; Eriksen and Lappin 1964; Fehmi *et al.* 1969; Holtzworth and Doherty 1971, 1974; Schiller 1965*b*, 1966; Schiller and Smith 1965; Turvey 1973). This would be consistent with an explanation based on temporal response persistence or one based on temporal response integration. We shall evaluate these two possibilities in the next chapter dealing specifically with the topics of temporal persistence and integration.

2. Although several investigators (Eriksen and Hoffman 1963; Eriksen and Lappin 1964; Mowbray and Durr 1964) report that monocular (or binocular) backward masking is as effective as forward masking; others

(Braddick 1973; Schiller and Smith 1965; Smith and Schiller 1966) find that forward masking is more effective than backward masking. This is a theoretically important discrepancy, since, as the next chapter shows, symmetry between forward and backward masking is consistent with one version of the temporal response integration hypothesis; whereas asymmetry between these types of masking by light are consistent with a temporal response persistence hypothesis.

3. Masking of a pattern by light is spatial frequency-specific. Green (1981*b*), using a brief, 30-ms sine-wave grating as target and a more prolonged, 700-ms uniform field as a mask, recently reported the following findings. The typical transient mask overshoots found at on- and offset of the mask (Crawford 1947), although obtained with test gratings of low, 1.0 c/deg., spatial frequency was eliminated at a higher, 7.8 c/deg., spatial frequency. Based on these results, Green (1981*b*) concluded that visual mechanisms sensitive to on- and off-luminance transients are characterized by a selectivity for low spatial frequencies, whereas high spatial frequency detectors do not respond well to such luminance transients. These findings also are discussed in the context of a particular theory of visual masking introduced in Chapter 7.

4. Like masking of light by light, masking of pattern by light cannot be obtained dichoptically when contour or edge interactions between mask and target stimuli are eliminated (Braddick 1973; Mowbray and Durr 1964, Schiller 1965*b*; Schiller and Wiener 1963; Smith and Schiller 1966). Despite the fact that cortical processes are most surely involved when contoured patterns such as alphabetic characters or grating orientation must be recognized by the observer (see, for example, Smith and Schiller (1966)), the forward or backward masking effect of a uniform, contourless, light flash none the less seems to manifest itself at precortical, perhaps retinal, levels. Indeed, much of contrast reduction-by-luminance summation may occur at the same peripheral levels as does early light and dark adaptation. Consequently a peripherally degraded target-contrast representation is transmitted to a central cortical site where, depending on the effective contrast or signal-to-noise ratio, the pattern is or is not identified.

Several electrophysiological results seem to concur with some of the above findings and conclusions on masking of pattern by light. For instance, Lindsley and his associates (Fehmi *et al.* 1969; Lindsley *et al.* 1967) have compared psychophysical and electrophysiological measures of backward masking by light in monkeys. They found psychophysical measures of the masking effect to be a monotonically decreasing Type A function of target-to-mask temporal interval. Correspondingly, averaged evoked potentials recorded from the optic tract, lateral geniculate nucleus and visual cortex revealed that the potentials evoked by the test stimulus were progressively less obscured by the mask-evoked potentials as the tarket-to-mask temporal interval increased. Since this was found at the

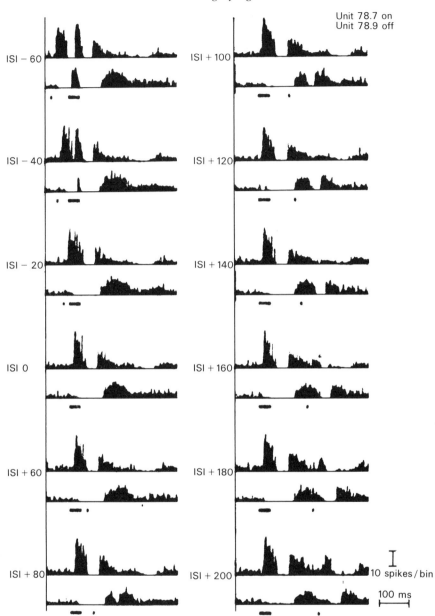

Fig. 2.13. Post-stimulus time histograms of an optic tract on-unit (upper row of each pair) and an optic tract off-unit (lower row of each pair). Masking flash of 40-ms duration is indicated by the black bar below the histograms. The temporal position of the 2-ms target relative to the mask flash, also given by the corresponding ISI values next to the histograms, is indicated by the black dot. Negative ISIs correspond to backward masking; positive ISIs to forward masking. (From Coenen and Eijkman (1972).)

earliest optic tract recording sites, Lindsley and his collaborators concluded that masking occurs as early as the retinal level of processing.

Coenen and Eijkman (1972) investigated electrophysiological measures of forward and backward masking by light in the optic-tract and lateral geniculate nucleus of cats. The reponse to the target and mask as a function of forward and backward masking interval are shown in an optic-tract on-unit and an optic-tract off-unit in Fig. 2.13.

Note that at negative target-to-mask interstimulus (ISI) values when the mask follows the target, the target response is increasingly obscured in both units as the ISI value approaches 0 ms. Similarly, at the positive ISI values when the mask precedes the target, the obscuring of the target again is highest at ISI values near 0 ms and decreases as ISI increases. Related results on masking by light were obtained from electrophysiological recordings of cells in cat lateral geniculate nucleus reported by Schiller

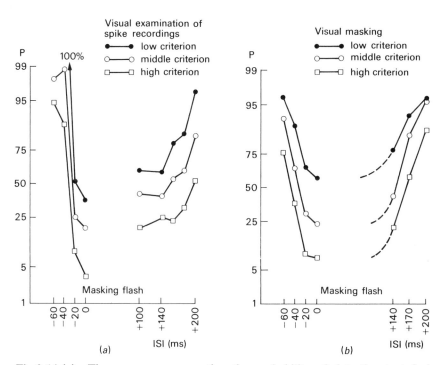

Fig.2.14.(*a*) Three curves representing the probability of detecting test flash responses at three criterion levels by means of visual inspection of single traces of spike recordings taken from a geniculate on-centre neurone. (*b*) Three psychometric curves of three human observers showing the probability of the detection of a test flash at three confidence levels as a function of forward and backward masking interval (positive and negative ISIs). The human observers who yielded the data shown in (*b*) were the same as those inspecting the single traces of spike recordings in (*a*). (From Coenen and Eijkman (1972).)

(1968). Moreover, what is evident from scrutinizing Fig. 2.13 and more manifest in Fig. 2.14 is that backward and forward masking effects, although of Type A, seem to be asymmetric. Fig. 2.14(a) shows the forward and backward masking functions (positive and negative ISIs, respectively) derived at three criterion levels from inspection of post stimulus time histograms of an on-centre neurone in the lateral geniculate nucleus. Fig. 2.14(b) shows the averaged forward and backward masking functions obtained at three confidence levels of target detection by three experienced human observers. Both neural and psychophysical functions, despite procedural and interspecies differences, yield more prolonged masking under forward masking than under backward masking.

It should be noted, however, that the human observers were asked to detect a test flash and not to identify a pattern. Consequently their function monitors what we have termed masking of light by light, which according to Kahneman (1967a) yields asymmetric forward and backward masking effects. Kahneman (1967a) maintains that symmetric functions are obtained when, for instance, one masks an acuity target or other contoured target by light (Eriksen and Hoffman 1963). This suggests the following explanation of the discrepancies between forward and backward masking results noted in the second of the four conclusions listed above. Recall from Chapter 1 that Baxt (1871), on the basis of his study of backward masking of pattern by light, came to the conclusion that stimuli such as spatial acuity patterns containing contour detail are characterized by a longer integration time than are simple luminance increment stimuli. Similar results have been reported more recently by Kahneman (1964) and Kahneman and Norman (1964). Moreover, as will be shown later (Chapter 7, Section 7.1), details of spatial contour and high-resolution information are processed at a longer latency than coarser stimuli (e.g. the interior of a luminance increment stimulus). Both of these temporal reponse properties characterizing channels transmitting spatial detail information of the target could contribute to an effectively longer backward masking interval relative to that obtained with spatially coarser target stimuli. As a corollary, one would expect the forward and backward masking functions to approach symmetry when a patterned test stimulus requiring fine visual resolution is masked by light, and to become asymmetric when the test stimulus consists of a uniform luminance increment or requires coarser visual resolution (see Fig. 3.5). Implicit in this explanation are the deployment of different response criteria and the assumption that the extent of forward masking is not as greatly altered when using a test stimulus consisting of a contoured pattern rather than a uniform luminance increment. This assumption seems a reasonable one since the same preceding masking flash will evoke the same persistent response irrespective of whether the following target is a pattern or a luminance increment.

2.3. Summary

The change of the visibility of a test stimulus flashed on a spatially uniform luminance-mask flash is known as masking by light. When the test stimulus itself consists of a spatially uniform luminance increment, masking of light by light prevails; when the test stimulus consists of a pattern (e.g. a letter, Landolt C, and so on, produced by either a luminance increment or decrement), masking of pattern by light prevails.

Masking of light by light can be obtained either with a brief mask flash or a more prolonged one. In the former case the change of test flash threshold as a function of test-to-mask onset asynchrony can monitor the visual systems on-response. In the latter case, one can similarly monitor the visual system's on- , maintained, and off-responses, or, in alternative terms, its transient and maintained light adaptation as well as its transient dark adaptation.

The on-response to a brief (less than about 50 ms) mask flash is usually characterized by a steep test threshold rise beginning about 100–150 ms prior to the mask onset, reaching a maximum at or near mask onset and declining more gradually after mask offset. The shape and magnitude of the on-response depends on the brief mask's energy (luminance × duration) and its temporal waveform (Sections 2.1.1 and 2.1.2). More rapid mask onsets produce more rapidly rising test thresholds. Higher energy masks produce greater threshold maxima and more prolonged test-threshold elevations after mask offset. This particular result is relevant to the following chapter's topic of visual response persistence.

The on-, maintained, and off-responses to a prolonged mask flash usually can be characterized by a sudden rise of test-threshold beginning when the test flash precedes the mask onset by about 100–150 ms, reaching a transient maximum at or near the mask onset, followed by a slower decline to a maintained threshold level, which in turn is followed by a sudden relatively gradual pre-offset of the test-threshold, followed by a decline at mask offset preceding a more gradual decline to an asymptotic threshold level.

The presence and magnitude of these transient threshold overshoots at on- and offset of the mask as well as the maintained threshold during the sustained portion of the mask depends on: (i) the temporal waveform of the mask (Section 2.1.1); (ii) the intensity of the mask (Section 2.1.2) and of the background (Section 2.1.4); (iii) wavelength properties of target and mask flashes (Section 2.1.3); (iv) the diameter of the mask flash (Section 2.1.5); and (v) the presence of contour interactions between test and mask.

From these dependencies, as well as related neurophysiological studies of early light and dark adaptation (Section 2.1.7), one can conclude that the maintained mask component and threshold overshoots at mask onset and offset depend not only on retinal receptor properties but to varying extents also on neural properties such as rod and cone excitation pools likely located at post-receptor retinal levels and more centrally mediated spatial contour interactions (Sections 2.1.5 and 2.1.6). The latter interactions implicate the role of contour-specific mechanisms which may also affect lateral masking such as meta- and paracontrast (discussed more fully in Chapters 4, 5, and 7).

Masking of pattern by a brief light flash can be thought of as a form of target-contrast reduction by luminance summation with the mask flash. As such, it yields a Type A forward and backward masking effect. A tentative explanation of observed symmetries and asymmetries between monocular (or binocular) forward and backward Type A functions, based on the spatial frequency content of the target and the criterion content (high or low spatial resolution) demanded of the task, was proposed.

In this regard, it was shown that the masking of pattern by a prolonged mask flash depends on the spatial frequency content of the target. Transient overshoots at on- and offset of the mask are produced only when low spatial-frequency targets are employed. Higher spatial-frequency targets eliminate the overshoots and yield only maintained masking functions. A unifying explanatory basis for the dual dependence of transient overshoots on the spatial properties or size of the target and the diameter or size of the mask flash (see point (iv) above) will be offered in Chapter 7 (Section 7.4.2, explanations 1 and 2).

3 Visual integration and persistence

The previous chapter showed that the visual responses produced by a uniform-luminance mask flash and a spatially superimposed test flash could interact at several levels of visual processing. Photochemical and neural processes occurring at the receptor level as well as post-receptor neural processes at subcortical and cortical levels were shown to be involved to varying degrees. In addition, the extent of response interaction at these various levels of processing tapped by the mask and target was found to depend on stimulus parameters such as intensity, size, retinal locus, contour proximity, and so on. What these results imply is that an adequate theory of masking by light cannot rely solely on positing one level of stimulus processing; on the contrary, it must invoke several different levels. Furthermore, the response to a brief or a prolonged, uniform, mask flash was shown to persist progressively longer beyond its duration or offset, respectively, as its energy is increased (see Fig. 1.4 and 2.5).

Whereas this last finding can be considered as one of several indices of the inertial properties generally characterizing visual stimulus processing, the prior findings additionally suggest that the source and locus of the inertia may reside at several levels of visual processing. Moreover, depending on the nature of the visual stimulus and task, the inertial aspect of the visual system can be indexed not only by a measurable response persistence but also by an equally measurable response latency and response integration. Although all three response characteristics are correlated, in the present chapter I shall discuss only the latter two, response persistence and integration. *Temporal persistence* means the tendency or the ability of the visual response to persist beyond the duration of a brief stimulus or the offset of a more prolonged one. *Temporal integration* means the tendency of two temporally separated visual stimuli to be treated as one at the level of the physiological or sensory response.

Generally speaking, visual integration is correlated with a limited temporal resolution. If two visual stimuli follow one another at sufficiently large temporal intervals, the two stimuli can be perceptually differentiated. However, if the temporal interval separating the successive stimuli is equal to or less than some critical value, they no longer are differentiated perceptually; rather, at the visual response level they are integrated or treated as one stimulus. Flicker provides a good example of visual differentiation and integration. At low flicker frequencies one can easily perceive the temporal modulation of luminance. However, as flicker frequency increases, and, therefore, the flicker period or cycle-time decreases, the perception of temporal luminance modulation becomes progressively more difficult until, at or above some critical flicker frequency (cff), one no longer perceives flicker but instead sees a temporally uniform or steady field. In Section 1.5 it was noted that, as early as 1892, Ferry used the cff as a measure of visual persistence. In later parts of this chapter I shall show how visual persistence is empirically related to one source of visual integration. For the moment, however, I shall review and distinguish among several possible sources of visual integration.

3.1. Visual integration

Visual integration prevails whenever two stimuli occur sufficiently close in time so that they cannot be perceptually segregated. From a psychological standpoint the two stimuli are treated as occurring simultaneously or as one. In the past, two separate sources of integration have been proposed. One source rests on the existence of a central cortical intermittency which measurably reflects the temporal integration and flow of information processing activities in the central nervous system; the other rests on the existence of response integration within peripheral sensory channels. These two explanations are not mutually exclusive (Kristofferson 1967*a*, *b*); one can have temporal integration in peripheral sensory channels as well as in central information processing channels. The problem is to devise experimental methods which allow both sources to be studied separately and not to confuse one source with the other.

3.1.1. Central intermittency as a source of temporal integration

Evidence supporting the hypothesis of central intermittency as a source of temporal integration derives from a variety of studies (Harter 1967; Kristofferson 1967*a*, *b*; Latour 1967; Pöppel 1970; Stroud 1956; Treisman 1963; White 1963). One way of viewing central intermittency is in terms of a free-running, endogenous, cortical oscillation or scanning mechanism driven by an 'internal clock' (Wiener 1948; Stroud 1956; White 1963). The period of the oscillation determines the scan duration or the duration of the 'psychological moment' (Stroud 1956). During this period separate data inputs into central information-processing channels are treated essentially as a single input. In an extreme form of the 'internal clock' hypothesis, such a source of central intermittency is thought to be totally independent of the time of occurrence of an external signal (Harter 1967; Stroud 1956).

Although Kristofferson (1967*a*) reports findings consistent with an endogenous, free-running scanning mechanism independent of stimulus input, Harter and White (1967), Latour (1967), Allport (1968), Efron and Lee (1971), and Pöppel (1970), in contrast report results which indicate that the 'internal clock' is synchronized with external stimuli. As noted by Pöppel (1970), the external control of central intermittency could arise in at least two likely ways. For one, the external stimulus could itself elicit or produce the temporal oscillations of the cortical response; and, *per force*, responses to all subsequent stimuli, depending on their temporal position relative to the first one, would fall within a predictable cycle or processing period (Efron 1967) and phase relative to the central oscillations. An alternative way of exerting external control is realized if the external stimulus resets rather than initiates the central intermittency. That is to say, an endogenous cortical oscillator simply has its phase reset by the occurrence of an external stimulus event (Latour 1967). Consequently, the responses to all subsequent stimuli are synchronized relative to this initial resetting of phase.

The general implications of central intermittency for visual masking should be quite apparent. Given that there is a central scanning mechanism whose scan rate is determined by the frequency of central oscillations (either intrinsically or externally generated or controlled), two stimuli falling within one scan cycle or one psychological moment (Stroud 1956), are treated as one, and therefore might mask each other. The temporal order of the two stimuli should not be a significant factor, i.e., backward and forward masking functions should be symmetric. As a simplistic analogy, one can think of a frame of film exposed to two briefly illuminated and spatially uncorrelated scenes falling within a fixed exposure duration (e.g. 1/30 s). The end result is a double exposure in which the image of *either* scene masks or adds noise to the image of the other scene. Extensions and details of this implication for models of visual masking are discussed later. For the time being, I shall discuss a form of temporal integration which seems to be located in peripheral, afferent sensory processes.

3.1.2. Temporal summation in afferent visual channels

Boynton (1961) has pointed out that one likely basis for perceptual time quantization (in addition to the central intermittency discussed above) can be sought in the temporal response properties of afferent visual pathways carrying information about external stimuli. Temporal summation is such a characteristic, and its existence in the visual pathway can be inferred from a number of psychophysical and neurophysiological studies (Alpern and Faris 1956; Barlow 1958; Bartlett 1965; Baumgardt 1972; Bloch 1885; Graham 1965; Hood and Grover 1974; Levick and Zacks 1970; Roufs 1972). A fundamental empirical regularity expressing the existence of temporal summation in vision is known as Bloch's law (1885). It is expressed by the following formula: $I \cdot t = c, t \leqslant \tau$; where I is flash intensity, t is flash duration, c is a constant time-integrated flash energy, and τ is the upper limit of the flash durations for which this empirical law holds. What this law states is that a constant visual effect—for example, a threshold (Barlow 1958; Roufs 1972) or a fixed suprathreshold brightness sensation (Servière *et al.* 1977; White and Rinalducci 1981)—is directly related to the

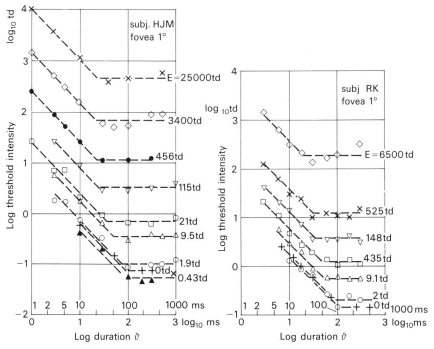

Fig. 3.1. Log threshold intensity as a function of log duration of the test flash. Results for different curves and symbols correspond to different background intensities, as indicated next to curves. Note decrease of log τ, the duration at and below which Bloch's law holds, as background intensity increases. (From Roufs (1972).)

product of flash intensity and duration, i.e. to the time-integrated energy of the flash, provided that the flash duration does not exceed its critical value. Another way of stating this is that the visual system is able to summate temporally flash intensity over a critical duration.

Figure 3.1 shows a typical example of psychophysical threshold performance confirming Bloch's law. Along the ordinate is plotted log threshold intensity; along the abscissa, log duration of the test flash. The values next to the empirically obtained data functions refer to the intensity of the background illumination against which thresholds were measured. Note the following aspects of the results. At any background intensity, the function relating log threshold intensity to log duration is bilinear. Up to a given critical duration, τ, log threshold intensity decreases linearly with a slope of -1.0 as log duration increases. That is to say, for $t \leqslant \tau$; $\log I = -\log t + k$ (k, a constant); or alternatively $I \cdot t = c$ ($c = 10^k$). For $t > \tau$, threshold intensity does not vary with t. Note also that as background intensity increases, the critical duration for which Bloch's law holds decreases (Barlow 1958; Graham and Kemp 1938; Roufs 1972; Sperling and Joliffe 1965). The specific results of Fig. 3.1 show that the duration over which temporal summation at threshold occurs decreases from about 100 ms to 50 ms as background intensity increases from 0.43 td to above 6500 td. Related results have been obtained in a number of neurophysiological studies of temporal summation in the visual system (Alpern and Faris 1956; Hood and Grover 1974; Levick and Zacks 1970). In particular, measuring the response of the vertebrate (frog) cone receptor, Hood and Grover (1974) obtained time–intensity reciprocity, which showed an increase in the temporal summation not only as background intensity decreased but also as flash intensity decreased.

Another variable which affects the time–intensity reciprocity expressed by Bloch's law is wavelength. Sperling and Joliffe (1965) reported a longer critical duration at threshold for detecting a 450-nm foveal target than for a 650-nm target. This result, however, held only for their 45′-diameter target but not the 4.5′-diameter one, indicating that by changing spatial summation one can change temporal summation (Owens 1972). In related studies, Krauskopf and Mollon (1971) found that the critical duration for short-wavelength flashes (430 nm) is substantially longer than that for long-wavelength flashes (500 or 600 nm), a result replicated by King-Smith and Carden (1976). Moreover, like previous investigators using achromatic stimuli (Barlow 1958; Roufs 1972), Krauskopf and Mollon found a decrease in critical duration at all wavelengths as the intensity of the backgrounds (either 500 or 600 nm) increased.

These findings are also relevant to suprathreshold tasks in which brightness sensations are elicited by brief flashes of variable intensity and duration. Here one would expect that, up to some critical duration determined by the background intensity, a constant suprathreshold brightness can be elicited by a constant time-integrated flash energy. Figure 3.2, taken from Servière, *et al.* (1977) confirms this expectation. The curves, shown in a log–log coordinate system in Fig. 3.2 portray not only the results of these investigators but also others' related findings. Analogous to the threshold data shown in Fig. 3.1, what these results show is that in order to obtain a constant suprathreshold brightness sensation corresponding to a given reference luminance, one can multiplicatively trade off flash intensity for flash duration up to some critical duration value. Moreover, as with the threshold data of Fig. 3.1, the critical duration for which one obtains time–intensity reciprocity or temporal summation increases as reference luminance decreases. This is evident from inspection of curves *a–d* in Fig. 3.2, which show suprathreshold temporal summation curves at progressively lower background intensities (see also Kahneman 1965).

Alternatively, we would expect that, up to some critical duration, perceived brightness of a suprathreshold flash of fixed intensity increases linearly with log duration. Additionally, one would expect that the critical

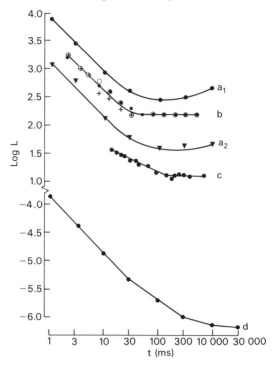

Fig. 3.2. The inverse dependence of physical luminance (log L) of a stimulus on log duration required to produce a constant subjective brightness produced by a fixed reference luminance. Curves labelled a_1 at top through to d at bottom correspond, respectively, to reference luminances of 640, 160, 64, 24, and 10^{-6} cd/m^2. (From Servière, Miceli, and Galifret (1977).)

duration increases as the intensity of the suprathreshold flash decreases. Past and recent psychophysical results reported by Stainton (1928), Kahneman (1965), and White, Irvin, and Williams (1980) tend to confirm these expectations; however, in these studies the sampling of flash durations was relatively coarse, and a more credible confirmation would require a more extensive, fine-grained analysis of temporal integration of suprathreshold flashes of a given intensity.

Hood and Grover's (1974) aforementioned findings are also relevant for the following reasons. The fact that temporal integration in afferent visual pathways can be monitored as early as the photoreceptor level suggests that a free-running 'internal clock' or central intermittency cannot be the only source of temporal integration. This is underscored by the additional fact that temporal integration at the afferent sensory level is a function of external stimulus parameters such as background or stimulus intensity. The studies of afferent vision discussed up to now have used the existence of time-intensity reciprocity, as expressed by Bloch's law, to define temporal summation. What, if any, implications might these findings have for a two-transient paradigm such as that used in forward or backward masking?

A simple extrapolation of the above findings could run along the following line of reasoning. Assuming that temporal integration, at a fixed background and stimulus intensity, extends over a given duration, one might predict that the visual response is indifferent to the way the stimulus is distributed over this duration. For example, let the temporal summation duration be 100 ms. If temporal summation occurs irrespective of the distribution of a stimulus over that duration, then a 50-ms flash of light at intensity *I* should elicit the same response as two 25-ms flashes of light at intensity *I* whose onsets are separated by an interval greater than 0 ms but less than or equal to 75 ms. This would follow from the fact that the two 25-ms flashes occur within the critical 100-ms duration over which summation occurs.

However, empirical results, using subthreshold summation as an index of double flash interactions, where the two flashes are spatially coincident, disconfirm this seemingly reasonable and simple extension of afferent temporal summation. What in fact happens is a bit more complicated. Ikeda (1965) and subsequently Rashbass (1970) have investigated temporal summation at threshold for double flashes of light. In particular, Ikeda (1965) used, among other conditions, double incremental flashes presented against either a 61.2 td background or a 328 td background. The duration of either of the two flashes was 12.5 ms. Ikeda measured the detection threshold of the double pulses as a function of the temporal interval, *t*, between their onsets. Initially, however, he determined *T*, the detection threshold of a single flash. By definition *T* is equal to the isolated thresholds, T_1 and T_2, of Flash 1 (F_1) and Flash 2 (F_2), respectively, i.e. $T_1 = T_2 = T$. Flash-threshold intensity levels were normalized relative to this reference threshold, which itself had a normalized value of 1.0 ($= T/T$). If we let $S_{1,t} = (T_1/T)$ and $S_{2,t} = (T_2/T)$ be the normalized threshold intensities of F_1, followed at an onset interval, *t*, by F_2, we can define an empirical summation index, σ_t, by the following formula:

$$\sigma_t = 0.30 - \log (S_{1,t} + S_{2,t})$$

At *t* = 0 ms (onset simultaneity), where the response summation for F_1 and F_2 is optimal, $S_{1,0} + S_{2,0}$ by definition, must equal 1.0; that is:

$$S_{1,0} + S_{2,0} = [aT_1/T] + [(1 - a)T_2/T] = \frac{[T_2 + a(T_1 - T_2)]}{T} = \frac{T_2}{T} = 1.0$$

where $0 \leq a \leq 1$, so that $\sigma_0 = 0.30$. However, for *t* > 0 ms, $[aT_{1,t} + (1 - a)T_{2,t}] > T$; since the empirically determined response summation for F_1 and F_2 decreases from its optimum as *t* increases. Therefore, since $\log (S_{1,t} + S_{2,t}) > 0$, $\sigma_t < 0.30$. In particular, for *t* >> 0 and exceeding the maximal summation interval, F_1 and F_2 ideally must each be equal to their isolated thresholds, T_1 and T_2, in order for *both* to be individually and independently detected. Thus, here $S_{1,t}$ and $S_{2,t}$ must both equal 1.0 so that $S_{1,t} + S_{2,t} = 2$ and $\sigma_{t>>0} = 0.03 - \log 2 = 0$. However, in a threshold detection task such as Ikeda's it is sufficient to detect either one of the stimuli above. Assuming that at $t >> 0$, each flash provides an independent threshold detection event, then, due to probability summation across these two events, both $S_{1,t}$ and $S_{2,t}$ are generally less than 1.0. In Ikeda's particular experiment the sum, $S_{1,t} + S_{2,t}$, was approximately equal to 1.58 so that $\sigma_{t>>0} = 0.30 - \log 1.58$ (≈ 0.10) corresponds to the probability summation index. Thus, when $0.30 \geq \sigma_t > 0.10$, one can infer total to partial *physiological response summation* for the two flashes; when $\sigma_t = 0.10$, one can infer their independent *probability summation*; and, finally, when $\sigma_t < 0.10$, one can infer *physiological response* cancellation or *inhibition*.

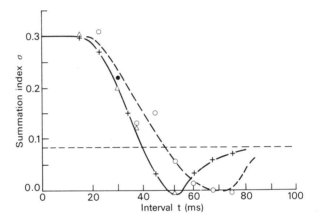

Fig. 3.3. The summation index, σ, as a function of temporal interval between two subthreshold flashes at two levels of background illumination (crosses, 328 td; open circles, 61.2 td). (From Ikeda (1965).)

A typical result, using double incremental flashes, is shown in Fig. 3.3. On the ordinate is shown the summation index, σ; on the abscissa is shown the temporal interval, t, separating the onsets of the two flashes. Note that for values of t less than or equal to about 25 ms, temporal summation is complete; as t increases to about 40 or 50 ms, the summation index progressively decreases; and then it falls below 0.10 as t increases even further until it again reaches a final value of 0.10. As noted, the drop of the summation index below 0.10 indicates that inhibition (negative summation) exists between the responses to the two flashes. It should also be noted that temporal summation is extended over a larger interval as background intensity decreases, a result consistent with the single flash studies reported above.

In contrast, however, to single-flash temporal summation, double-flash temporal summation is complete over a surprisingly small flash onset asynchrony of about 20 ms whereas partial summation extends to about 40–60 ms depending on background intensity, beyond which one even obtained a subtractive, inhibitory effect. We see then that the visual system is not indifferent to the distribution of flash energy over time even if the double pulses fall within the duration of complete temporal summation as measured with the single flash paradigm (Bloch's law). One probable source of this incommensurability is the fact that double pulses introduce additional transients relative to a single flash.

The further implication is that when one invokes temporal integration as an explanation for forward and backward masking (e.g. Eriksen 1966) one must be careful to refer to the proper index of temporal summation. Casual reference to Bloch's law, derived from single flash threshold or brightness judgements, may not be as appropriate as reference to results obtained with double flashes whose interactions can yield not only complete summation at small flash onset asynchronies but, depending on the range of flash onset asynchronies, also show merely partial summation or even inhibitory interactions.

3.2. Visual persistence

Chapter 1 showed that response persistence in vision has been investigated and, thus, known to exist for at least the past two centuries. The past two decades and, in particular, the last 7 years have been marked by an especially concentrated interest and effort in the study of visual persistence. Recent articles by Dick (1974), Coltheart (1980), and Long (1980) extensively review most of the extant methods, findings, and theories of visual persistence; and the reader is encouraged to read these sources in addition to the following review, which, although overlapping substantially with these other reviews, presents some novel and complementary analyses.

Coltheart (1980) distinguished among the following types of visual persistence. By *neural persistence* he means that neural activity evoked by the visual stimulus may continue after stimulus offset. *Visible persistence* refers to the fact that the visual stimulus may continue to be visible for some time after its offset. Finally, *informational persistence* refers to the finding that information about visual characteristics of the stimulus may continue to be available to an observer for some time after offset of the stimulus.

This information *may* persist even when visible persistence has already terminated. Coltheart (1980) identifies this latter type of informational persistence with *iconic memory* as operationally defined by the post-exposure, partial report methods introduced by Sperling (1960) and Averbach and Coriell (1961) and distinguishable from neural or visible persistence (see also Phillips 1974); although Long and Beaton (1982) recently reported results at odds with Coltheart's distinction between iconic memory and visible or neural persistence. Nevertheless, in drawing the distinction, Coltheart has given a clearer operational and, therefore, testable definition to the concept of iconic memory, which up to now, ever since its initial introduction and definition by Neisser (1967), has been facilely confused with either neural or visible persistence. Despite this new distinction and definition, in our discussion we shall concentrate on visible and neural persistence and, in keeping with past convention, refer to them as forms of *iconic* persistence. *Informational* persistence is discussed only when deemed relevant. A more complete analysis and discussion of informational persistence is given by Adelson and Jonides (1980), Coltheart (1980), Long (1980), and Long and Beaton (1982). The interested reader is urged to go to these sources.

3.2.1. Visible persistence

Experimental techniques Several experimental techniques have been developed in recent years to measure visible persistence. Some of these techniques are given extensive and critical discussion in Coltheart's (1980) and Long's (1980) recent excellent reviews. I shall briefly mention these techniques and outline the rationale behind their use. Coltheart (1980) mentions the following techniques: temporal synchrony judgements, reaction time measurements, stroboscopic illumination of a moving stimulus, the moving-slit technique, judgements of phenomenal continuity, temporal integration of form parts, and stereoscopic persistence. I shall not discuss the last type of persistence other than to raise the point later in the chapter that stereoscopic persistence indicates the existence of a central source of persistence at or beyond the locus of binocular combination of monocular responses.

Sperling (1967) introduced the technique of temporal synchrony judgements, and it was subsequently taken up, among others, by Efron (1970*a*, *b*, *c*), Haber and Standing (1970), Bowen *et al.* (1974), Bowen (1981), Long and Gildea (1981), and Sakitt and Long (1979*a*). The rationale behind the technique is to, for instance, synchronize the apparent disappearance of a visual stimulus after its offset with another event. In the intramodal method of synchrony judgement the other event could be the perceived onset of a brief visual probe (Efron 1970*a*); in the cross-modal method the other event could be the perceived onset of an auditory stimulus such as a tone or click (Haber and Standing 1970). At any rate, by measuring the interval between the physical offset of a visual stimulus and the onset of the intra- or cross-modal probe, one can obtain an estimate of the post-offset, visible persistence of the stimulus.

Measurements of reaction time (RT) to brief flashes also can be employed to measure visual persistence. This technique has been employed by Pease and Sticht (1965), Briggs and Kinsbourne (1972), and Breitmeyer, Levi, and Harwerth (1981*a*). The rationale behind this method assumes that, if the visual response persists for some time after offset, there will be a measurable delay between physical and apparent offset which presumably determines part of the RT. However, since there is a minimal, irreducible motor component to any RT (Williamson *et al.* 1978; Woodworth 1938), offset RT does not give an entirely accurate measure of persistence. Ideally, for very brief stimuli (e.g. less than 10 ms) one would want to subtract onset RT from offset RT to obtain an estimate of the entire duration of the perception. However, if the duration of the brief stimulus is longer (e.g. greater than 10 ms), one would want also to subtract the stimulus duration from the offset–onset RT difference. This value would then be an estimate of the post-offset visible persistence. This technique rests on the following assumptions, each of which could in turn be put to an empirical test. First, under ideal conditions one would want the irreducible motor components contributing to onset and offset RTs to have equal durations; under less ideal conditions, the motor component at onset ought to be shorter than that at offset. Secondly, the constant characterizing the sensory rise time at onset ought to be less than or equal to the constant characterizing the sensory decay time of visible persistence. If either or both of these two assumptions are invalid, which sometimes may be the case (Lewis, Dunlap, and Matteson 1972; Pease and Sticht 1965), one will obtain inaccurate or possibly even negative persistence estimates; a result which leads to the curious conclusion that the psychophysical observer was prescient of the stimulus offset. If one is interested in making only ordinal or interval comparisons between post-offset persistences, comparisons of offset RTs are sufficient. Assuming again that the offset RTs are characterized by an irreducible motor component of fixed duration, ordinal variations of offset RT are positively correlated with ordinal variations of visible persistence. Moreover, differences between offset RTs can also be used to estimate interval differences of visible persistence.

Stroboscopic illumination of a moving stimulus also has been used extensively to measure visual persistence (Allport 1968, 1970; Dixon and Hammond 1972; Efron and Lee 1971; Wade 1974). The discussion of the method employed by Allport (1968, 1970) is particularly instructive. Allport painted luminous lines on a cathode ray tube (CRT) at varying scan rates so that each line was present for a variable but brief interval. As the line was being painted at regular repeated intervals, the CRT was simultaneously rotated. This, in effect, produced a sequential series of luminous radial lines imaged on the retina. By controlling separately the CRT scan rate and rotation rate, Allport could paint radii at desirable temporal and spatial (angular) intervals on the retina. Since the radii are painted sequentially, the observer's estimate of the number of radii continuously visible for any combination of scan and rotation rate can in turn be used to estimate visible persistence. For example, if radii are painted at a rate of one every 20 ms and the observer sees six radii rotating as a unit at any given time, then the visible persistence is at least 100 ms in duration.

The moving-slit technique, already mentioned in our historical review of Chapter 1, provides another means of measuring visible persistence. In this technique an opaque mask with, say, a vertical, slit-like aperture is moved in front of a stimulus. At any given instant of time only that part of the stimulus directly behind the aperture is exposed to the visual system. Yet, if the aperture moves over the stimulus at a sufficiently high speed, the stimulus is seen in its entirety, an effect known as the 'seeing-more-than-there-is' phenomenon. In order for the stimulus to be visible in its entirety, the initially exposed portion of the stimulus must persist visibly at least until the finally exposed portion becomes visible. The slowest slit-speed or scanning rate (if the slit is repeatedly moved over the stimulus) at which the stimulus is visible in its entirety can thus be used to obtain an estimate of the maximal duration of visible persistence. An alternative procedure is to keep the slit aperture stationary and move the stimulus behind it.

Anstis and Atkinson (1967) have reviewed some of the earlier uses of these techniques in the study of pattern vision. Based on their own results, these investigators argued that the visible effect produced by the techniques occurs only if different portions of the stimulus are presented to different retinal locations, i.e. if 'retinal painting' occurs; a condition which would prevail, for example, if the observer kept his or her eyes stationary as the slit moved over the stimulus or else moved the eyes in the direction of the stimulus sliding behind a stationary aperture. Haber and Nathanson (1968) concurred with this conclusion, which, however, seems to contradict the results obtained in Park's (1965) earlier and McCloskey and Watkins' (1978) later related investigations. In Park's (1965) study, the stimulus moved behind a stationary slit rather than the slit moving in front of the stationary stimulus. According to Park (1965, 1968), the observers were able to recognize the stimulus (for instance, an outline drawing of a camel) without retinal painting. However, McCloskey and Watkins (1978) argued that Parks (1965, 1968) very likely did not control adequately for the observers' inadvertent use of eye movements. Hence, if, as noted, the eyes tracked in the direction of the stimulus moving behind that stationary slit, visible persistence may have been produced by retinal painting rather than some post-retinal process. However, in their own investigation, McCloskey and Watkins (1978) effectively eliminated eye movements and replicated Park's (1965, 1968) findings. This replication, in addition to other empirically supported arguments and related results (e.g. Parks 1970) noted by these investigators, suggests that although retinal painting may be one process that is sufficient but not necessary for establishing visible persistence, other post-retinal processes, as indicated by their own and Park's (1970) results, are also used.

Judgements of phenomenal continuity also have been used to measure visible persistence. This technique basically is a modification of the cff technique employed by Ferry (1892) to measure visual integration. It was reintroduced in various forms by Haber and Standing (1969), Kulikowski and Tolhurst (1973), Meyer and his collaborators (Meyer 1977; Meyer, Lawson and Cohen 1975; Meyer and Maguire 1977), Corfield, Frosdick, and Campbell (1978), and Breitmeyer and co-workers (Breitmeyer and Halpern 1978; Breitmeyer *et al.* 1981*a*). Here the method employed by Meyer and Maguire (1977) serves as a good instructive example. In their experimental task, Meyer and Maguire had their observers adjust the off-time or the interval between offset and onset of continuously cycled, 50-ms presentations of a stimulus (in Meyer and Maguire's study, the stimulus was a vertical grating) until the threshold for perceived temporal continuity of the stimulus was reached. The maximal off-time at which perceptual continuity could still be obtained was taken as an estimate of visible persistence. (In this and similar experiments, perceived continuity refers to the pattern components of the grating, not the flicker component. In other words, although the pattern may be seen as continuously present, on–off flicker is still nevertheless visible).

Temporal integration of form parts is another method for measuring visible persistence. This method is related to the phenomenal continuity technique discussed in the above paragraph; however, here two spatially and temporally separated stimuli are presented only once in sequence. The observer's task is to make a perceptual judgement of the spatiotemporal composite. This technique has been employed by a number of investigators (DiLollo 1977*a*, *b*, 1980; DiLollo and Woods 1981; Eriksen and Collins 1967, 1968; Hogben and DiLollo 1974; Ikeda and Uchikawa 1978; Kinnucan and Friden 1981; Long and Sakitt 1980*a*, *b*; Pollack 1973; Rohrbaugh and Eriksen 1975; Sakitt and Long 1978). The rationale of this technique rests on the following argument similar to that adopted by Eriksen and Collins (1967). Suppose we have two (or more) stimulus displays, each of which separately contains an unidentifiable pattern; but which, when spatially juxtaposed, produce a recognizable figure. If the two stimuli are briefly and sequentially flashed and the first one visibly persists for some brief duration, their composite will be recognizable at short interstimulus intervals (ISIs) but not at long ones in which case two separate meaningless patterns are seen. The largest ISI at which an observer can still recognize the composite (above chance level) can be taken as an index of visible persistence of the first stimulus component.

A technique for measuring visible persistence, which was not discussed by Coltheart (1980), is monocular forward pattern masking. As already noted, this method, implicitly used in the studies of masking-by-light (e.g.

Crawford 1947; Sperling 1965) discussed in Chapter 2, was employed by
Breitmeyer and Halpern (1978) in their study of visual persistence. The
rationale for this method is straightforward. If the first pattern stimulus in a
two-stimulus sequence is designated as the mask, its sensory activity and
visibility will persist for some time after offset. During this time of post
offset persistence, the first stimulus effectively masks or decreases the
visibility of the second stimulus. The largest ISI between the first, mask
stimulus and the second, target stimulus at which one obtains forward
pattern masking can be used as an estimate of the visible persistence
produced by the first stimulus. Section 10.3.1. shows that such a definition
has particular relevance in relating the roles of masking and visual
persistence to naturalistic, dynamic viewing. For the moment, however, it
should be remembered that this method of measuring visible persistence is
easily confounded with measures of (previsible) neural persistence since
much of forward masking is a form of luminance-dependent masking
occurring at peripheral levels of visual processing merely providing input to
central levels of pattern recognition and persistence (Michaels and Turvey
1979; see Section 8.3, and Fig. 8.8(b); Sperling 1965; Turvey 1973; see
Section 2.2 and Fig. 2.5). This caution is reinforced by the fact that the
magnitude and temporal extent of forward masking generally is substantially attenuated under dichoptic viewing (Battersby and Wagman 1962;
Greenspoon and Eriksen 1968; Smith and Schiller 1966; Turvey 1973),
indicating that the extended response or effect of the mask during
monocular (or binocular) viewing incorporates primarily peripheral
processes. Thus, since neural persistence as a source of peripheral masking
could feasibly outlast a central source of visible persistence, monocular
forward masking may yield an overestimate of such visible persistence.

Another index of visual persistence, not mentioned by Coltheart (1980)
but implicitly used in Allport's (1968, 1970) studies, is apparent stroboscopic motion, a variant of what more generally may be regarded as event
perceptions. That motion and event perception persists visually has been
demonstrated by several investigators (Demkiw and Michaels 1976;
Treisman, Russell, and Green, 1975; Wilson 1981). In particular, the fact
that a temporal delay must exist between the first and second stimulus in
order to produce optimal apparent stoboscopic motion means that the
neural traces of the first stimulus must persist for some time to integrate
with those of the second one (Kahneman and Wolman 1970). Since
stroboscopic motion can be obtained with interocular presentations (Anstis
and Moulden 1970; Kahneman 1967*b*; Shipley, Kenney, and King 1945;
Verhoeff 1940; Wertheimer 1912), the use of cortical as well as peripheral
mechanisms is implicated. This seems to be particularly plausible since in
higher mammals direction-selective motion detectors are found at no
earlier stage of the retinogeniculostriate pathway than the striate cortex
(Hubel and Wiesel 1962, 1968).

A modification of this method has been used by Julesz and Chiarucci (1973). They presented a variable number, N, of successive frames of uncorrelated random dot patterns, but cycled this series of N frames many times. When presented at a rate of 8 frames/s or higher, this repeated frame sequence appeared as dynamic noise; however, at lower frame rates and, thus, longer cycle durations, observers would still perceive some periodicity of stroboscopic motion. Observers' judgements of seeing periodic motion decayed exponentially, with an average time constant of about 540 ms, as cycle duration increased. By this method, estimates of visual persistence are much longer than those obtained with other techniques.

Experimental findings Although these different experimental techniques yield correspondingly different and, at times, inconsistent estimates of visible persistence, two highly replicable and consistent effects, obtained under moderate levels of background illumination, are what Coltheart (1980) has called the *inverse duration effect* and the *inverse intensity effect*. Exceptions to one or both of these effects have been noted by Sakitt (1976; Sakitt and Long 1978, 1979a) and Long and Sakitt (1980a, b; 1981). However, as will be seen later (Section 3.5.1), these discrepancies can be explained by using extremes of intensity of either the stimulus and background (e.g. see Long and Beaton, 1982) as well as non-uniform response criteria adopted among the various studies of visual persistence.

The inverse duration effect refers to the finding that as stimulus duration increases, its visible persistence after stimulus offset decreases; the inverse intensity effect similarly refers to the finding that as stimulus intensity increases, visible persistence also decreases. Unfortunately, Coltheart's meaning of the inverse intensity effect, as stated above, seems to be restricted to the stimulus *per se* rather than to the entire stimulus display including the background. If one differentiates between stimulus intensity and background intensity, then the inverse intensity effect can apply both to the stimulus and to the background. Consequently, as shown below, we can speak separately of an inverse stimulus-intensity effect, an inverse background-intensity effect, or of their interaction.

The inverse duration, stimulus-intensity, and background-intensity effects have been confirmed repeatedly under a variety of experimental techniques (Allport 1968, 1970; Bowen *et al.* 1974; DiLollo 1977a, b 1980; Dixon and Hammond 1972; Efron and Lee 1971; Haber and Standing 1969, 1970). As illustrative cases we shall consider the particular investigations of Haber and Standing (1970) and Bowen *et al.* (1974).

Haber and Standing (1970) used a variation of the temporal synchrony judgement technique described above. Using this technique their observers yielded visible persistence estimates at variable durations of the stimulus exposure (consisting of 3 × 3 array of black-on-white letters) and under a variety of pre- and post-exposure conditions (Fig. 3.4). In

condition 1 the pre-exposure and post-exposure fields were the same
luminance (5 mL) as the stimulus field; in condition 2, only the
pre-exposure field was illuminated, the post-exposure field was dark; in
condition 3 the arrangement was reversed relative to condition 3 (the
pre-exposure field was dark and the post-exposure field was illuminated);
finally in condition 4 both pre- and post-exposure fields were dark. Of
special relevance to our discussion are comparisons of condition 1 and 4,
for they inform us about the effects of variable background intensities
against which a stimulus of variable duration is flashed. From a
methodological viewpoint conditions 2 and 3 are somewhat complicated.
In condition 2 there is a sudden transition from a prolonged light
background to a prolonged dark one; in condition 3 there is a sudden
transition from a dark background to a light one. In these cases, one
introduces Crawford-type masking effects (either early dark adaptation
[off-transients] or else early light adaptation [on-transients]) which may
interfere with either the persistence effects or with the temporal synchrony
judgements.

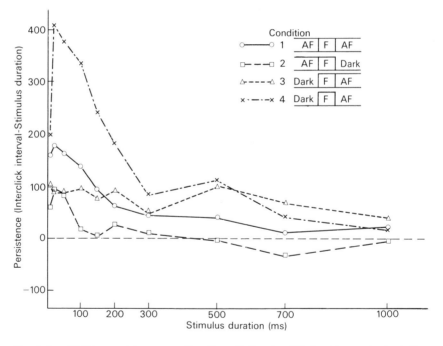

Fig. 3.4.   Visible persistence as a function of stimulus (F) duration and adaptation
field (AF) combinations, as shown in inset. (From Haber and Standing (1970).)

Inspection and comparison of the results of conditions 1 and 4 clearly
indicate the following. First, for both conditions, as stimulus duration

increases, visible persistence after stimulus offset decreases. This is true for all durations except the briefest, and this may indicate the existence of a minimal threshold duration which must be exceeded before the inverse duration effect holds. Second, for all durations, a stimulus flashed against a bright background (condition 1) visibly persists for a shorter interval than a stimulus flashed against a dark background (condition 4). These two results illustrate the existence of the inverse duration effect and inverse background-intensity effect.

Whereas Haber and Standing (1970) maintained a constant stimulus intensity but varied background intensity, Bowen *et al.* (1974) conversely maintained a constant (dark) background intensity but varied stimulus intensity as well as duration. The latter investigators also used a variation of the temporal synchrony judgement technique. Their main findings can be summarized as follows:

1. For any stimulus intensity, visible persistence varied inversely with duration.

2. For any stimulus duration, visible persistence decreased as stimulus intensity increased.

3. Up to a critical duration of about 100 ms, visible persistence was found to be a constant function of a constant time-integrated energy of the stimulus; beyond the critical duration of 100 ms visible persistence decreased (see White and Rinalducci, 1981, for related results). Another way of stating these results is that Bloch's law holds for visible persistence. This indicates the existence of a close relationship between temporal integration and temporal persistence in vision, a relationship which we shall analyse further in the following section.

For now, however, let us add the following caveat to the inverse stimulus intensity effect. The effect breaks down when the stimuli are very intense relative to the background on which they are flashed (Coltheart 1980). Presumably, these very intense stimuli produce strong and long-lasting after-images which, according to Coltheart (1980) and contrary to Sakitt (1976; Sakitt and Long 1978) are not to be identified with visible persistence obtained with moderately intense stimuli at some moderate level of background intensity. This distinction between long-lasting after images and primary visible persistence was already implied in Exner's (1868) and Monjé's (1931) work on the time course of a visible sensation. Strong and increasingly long-lasting after-image effects could also play a role in forward masking when it is used as an index of persistence. The discussion of Sperling's (1965) investigation of masking in Chapter 2 (see Fig. 2.5) showed that as mask intensity increases from 56 to 15 700 ft-L relative to a 41-L preadaptation field, the strength and duration of forward masking also increased. Here, the threshold-elevating effects of the ever-brighter masks persist in the form of ever-stronger and longer-lasting after images which, according to Barlow and Sparrock (1964) behave like 'equivalent backgrounds'; that is, backgrounds of light which appear as bright as the after image. Consequently, when measuring visible persistence via forward masking or, for that matter, any other technique, the experimenter should be careful to restrict the stimulus intensities to a lower range so as not to confound after-image persistence during dark-adaptation with visible persistence measured at a constant prevailing light-adaptation level. I shall defer further discussion of the distinctions and similarities between after-image and visible persistence to a later point (Section 3.5.1) in the present chapter and in Section 10.3.

3.3. Relationships between temporal integration and persistence

I shall now discuss two relationships existing between visible persistence and temporal integration in afferent visual channels. One relationship is conceptual and methodological and has some bearing on the interpretation of psychophysical data relevant to the previously discussed explanations of visual integration based on the alternative hypotheses of central intermittency, central scanning, or the psychological moment, presumably controlled by an internal clock or pacemaker. The other relationship is empirical and demonstrates that temporal persistence and integration in the afferent visual pathway may share the same mechanism.

3.3.1. Temporal persistence as a basis for temporal integration

Up to now I have discussed temporal integration either in terms of central intermittency or the psychological moment and the correlated notion of an internal clock; or in terms of time–intensity reciprocity as expressed by Bloch's law. Some of the psychophysical phenomena which have been interpreted as supporting the existence of an internal control scanning mechanism (White 1963) need serious reconsideration in light of related findings on visible persistence (Allport 1968, 1970; Efron and Lee 1971) which indicate that if there is a psychological moment, it is not determined by a free-running, central oscillator or scanning mechanism but rather is under the same external stimulus control as is visible persistence. For instance, in Allport's (1968) above-mentioned study using the method of stroboscopically and successively illuminated radial lines, observers perceived a constant number of visible radial lines rotating or travelling as a unit around the centre of the CRT display. A similar result was reported by Efron and Lee (1971). Allport (1968) took this as evidence for the existence of a continuously travelling psychological moment, rather than a temporally discrete or quantized one generated at a fixed rate by an internal clock. Moreover, both Allport (1968) and Efron and Lee (1971) showed that the duration of this psychological moment varied inversely with the intensity of illumination. As noted by Efron and Lee (1971), this undercuts the heuristic value of the psychological moment hypothesis; in short, these results do not support the existence of an internally generated, stimulus-independent scanning mechanism as envisaged by Wiener (1948), Stroud (1956), or White (1963).

This is not to say that such a mechanism does not exist; but only that many psychophysical studies which purport to have studied or indexed the existence of such a mechanism have instead merely tapped sensory persistence—in particular, visible persistence, the duration of which, as a function of stimulus parameters, co-varies directly with the duration of the travelling psychological moment. Consequently, by applying Occam's dictum explanations of visual integration based on the existence of a free-running, central oscillator are deftly eliminated; tentatively, an explanation based on visible persistence is entirely adequate. We shall see later (Section 3.6) that a related problem exists when attempts are made to explain visual masking and integration in terms either of the psychological moment hypothesis or of visual persistence hypothesis.

For now let us view temporal integration as a natural consequence of temporal persistence. Suppose two equally intense, brief, 10-ms stimuli are presented at an ISI of 50 ms. Suppose further that both stimuli generate a response of the same latency and that each stimulus generates a visual response which persists for 150 ms. Then, at the physiological or sensory response level, the two stimulus representations temporally overlap or coincide for 100 ms. At this level, they are *per force* temporally integrated or combined. In view of this it is not surprising that the two stimuli phenomenally share a temporal or spatiotemporal integrity. An explanation invoking a psychological moment or a central free-running scanning mechanism would be entirely superfluous; in particular because—as will become evident later in this chapter—response persistence, like integration, can be monitored as early as the photoreceptor level.

3.3.2. Temporal summation as a co-variate of temporal persistence

Based on the above analyses, I do not wish to be misinterpreted as stating that visible persistence determines visual integration. In fact, all I want to stress is the high correlation between these two temporal indices of the visual response. It was noted above (Section 3.2.2.) that Bowen *et al.* (1974) had established the existence of an empirical relationship between visible persistence and Bloch's law. In their study, up to a critical interval, a constant duration of visible persistence was obtained for a constant time-integrated stimulus energy. A related series of studies has been reported by Bowling and co-workers (Bowling and Lovegrove 1980, 1981; Bowling *et al.* 1979). This series of studies is based on a conjunction of two types of observations, one having to do with the study of Bloch's law, and the other with visible persistence as a function of size or spatial frequency.

Chapter 1 showed that Baxt (1871) over a century ago had found that a longer visual integration time is required to detect a detailed Lissajou figure than a coarser spatial pattern. More recently, this phenomenon has been replicated and extensively studied by Kahneman and co-workers (Kahneman 1964, 1966; Kahneman, Norman, and Kubovy 1967; Kahneman *et al.* 1967) in their comparison of time–intensity reciprocity under two tasks: one requiring subjective brightness estimates, the other requiring form identification or resolution of a Landolt C. The critical duration for which Bloch's law held in the form identification or resolution task was significantly greater than that obtained in the brightness estimate task.

Using flashed sinusoidal luminance gratings as stimuli, Breitmeyer and Ganz (1977) and Legge (1978) recently reported the related findings showing that, for a modified form of Bloch's law, time–contrast reciprocity at threshold holds for increasing durations as the spatial frequency of the grating increases. The particular results of Breitmeyer and Ganz's (1977) study are shown in Fig. 3.5. The ordinate shows the logarithm of contrast at threshold, the abscissa the logarithm of the duration of the flashed gratings. Time–contrast reciprocity functions were plotted, as shown, for gratings having a spatial frequency of 0.5, 2.8, or 16.0 cycles per degree (c/deg.) Note that up to a critical duration, increasing from about 50 ms at a spatial frequency of 0.5 c/deg to about 200 ms at a spatial frequency of 16 c/deg, log threshold contrast decreases linearly with log duration. The slopes of the functions are all equal to -0.7, corresponding to a modified Bloch's law for which duration, D, and threshold contrast, C, relate as follows: $C \cdot D^{0.7} = k$; k is a constant.

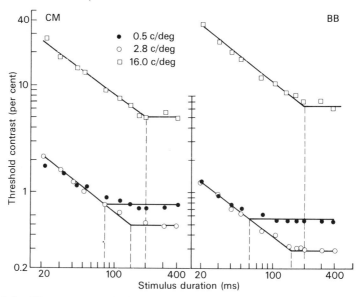

Fig. 3.5. Time–contrast reciprocity as measured by threshold contrast for sinusoidal gratings of 0.5, 2.8, and 16.0 c/deg as a function of the duration of the flashed gratings. (From Breitmeyer and Ganz (1977).)

This modified form of Bloch's law has been replicated by Legge (1978). Moreover, whereas the threshold contrast obtained at presentations longer than the respective critical durations remained constant in Breitmeyer and Ganz's (1977) study (see Fig. 3.5), Legge reported that log threshold contrast tended to further decrease with log duration; however, it decreased at a slower rate or at a still lower slope value. This discrepancy may simply be a curve-fitting artefact, since Breitmeyer and Ganz (1977) used durations ranging only from 10 to 400 ms, whereas Legge (1978) used a much longer range which would give a stronger indication of a second, more gradually declining branch of the time–contrast reciprocity function.

Recently several investigators (Bowling *et al.* 1979; Breitmeyer and Halpern 1978; Breitmeyer *et al.* 1981*a*; Corfield *et al.* 1978; Meyer and Maguire 1977), using the phenomenal continuity technique, showed that visible persistence also increases as spatial frequency of a grating stimulus giving rise to the persistence increases. Moreover, Bowling *et al.* (1979; Fig. 4) showed a clear positive correlation between spatial-frequency dependent increases of the critical durations obtained by Legge (1978) and their own spatial-frequency dependent increases of visible persistence. Bowling and Lovegrove (1980, 1981) subsequently capitalized on this relationship and studied it more extensively. A particular result which typifies their finding is illustrated in Fig. 3.6. It shows variations in visible persistence (along the ordinate) as a function of stimulus duration (along the abscissa) for three spatial frequencies. Note that the relationship between duration of visible persistence and grating is bilinear, similar to the time–contrast reciprocity findings reported by Breitmeyer and Ganz (1977) and Legge (1978) (see Fig. 3.5). For each spatial frequency visible persistence decreases as stimulus duration increases up to a critical stimulus duration of 70, 90, and 150 ms for the 1, 4, and 12 c/deg gratings, respectively. Beyond these respective critical durations, persistence continues to decline with stimulus duration but at a much slower rate, much like the gradual decline characterizing time–contrast threshold reciprocity reported by Legge (1978). Although Legge (1978) argued that the gradual decline of the time–contrast reciprocity function results from probability summation, Bowling and Lovegrove (1981) report that the size of the slope of the gradual decline defining their persistence results depends on grating orientation and therefore implicates a central, physiological process, whereas the earlier steeper decline reflects peripheral processes.

A comparison of the data in Fig. 3.5 and 3.6 shows that variations in temporal integration of gratings can predict variations in their visible persistence and vice versa. Since temporal integration varies inversely with wavelength (King-Smith and Carden 1976; Krauskopf and Mollon 1971; Sperling and Joliffe 1965; see Section 3.1.2), one would also expect wavelength-dependent variations in temporal persistence. Meyer

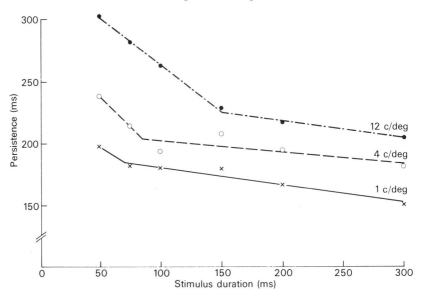

Fig. 3.6. Visible persistence for sinusoidal gratings of 1, 4, and 12 c/deg as a function of the duration of the flashed gratings. (From Bowling and Lovegrove (1980).)

(1977) obtained differences in visual persistence between green and red flashed gratings or uniform green and red flashes. Although not statistically significant, under monocular viewing conditions, the visible persistence obtained with red (Wratten 26) stimuli was consistently greater than that obtained with green (Wratten 53) stimuli. Bowen (1981) recently obtained similar non-significant, yet indicative, results for chromatic (530 nm and 570 nm) increment flashes, using the temporal-simultaneity-judgement method to estimate post-offset persistence. Moreover, Bowen (1981) also showed that, at any wavelength, stimuli comprised of brief hue substitutions against brightness-equated backgrounds (no luminance increments) produced longer persistence than chromatic luminance increments. In view of the wavelength-dependent results discussed in Sections 3.1.2, one would expect to obtain a greater and possibly significant difference in visible persistence when, for instance, comparing blue, short-wavelength flashes with red, long-wavelength ones. At any rate, the above spatial frequency- and wavelength-dependent results and considerations indicate the existence of common underlying properties or mechanisms which are responsible for the co-variations of temporal integration and persistence. The nature of these mechanisms is not clearly specifiable at this time; however, from what follows, it should be apparent that one such mechanism may exist at relatively early levels of visual processing, perhaps as early as the photoreceptor level.

3.4. Neural persistence

I shall proceed on the assumption that neural persistence is a necessary but not sufficient condition for visible persistence. Neural persistence very likely occurs anywhere along the visual pathway between the initial, previsible photoreceptor responses to whatever central, brain processes are used in the conscious (visible) registration of the stimulus. In other words, we assume that only higher levels of cerebral processes consciously register the occurrence of a stimulus event, whereas more peripheral, in particular, subcortical processes in the retinogeniculocortical pathway, although registering the event, do so at a previsible or preconscious level. At the photoreceptor level, response persistence is a well-documented fact. Both cones and rods in a variety of vertebrate retinae continue to respond for some time after stimulus offset (Fain and Dowling 1973; Whitten and Brown 1973*a, b, c*; Hood and Grover 1974). However, the post-offset cone signal decays more rapidly relative to that of rods (Fain and Dowling 1973; Normann and Werblin 1974; Whitten and Brown 1973*a*). This is shown in Fig. 3.7, taken from Whitten and Brown (1973*a*). The cone and rod receptor potentials were obtained by applying equal quantum stimuli at 560 and 508 nm, respectively, for a duration of 400 ms. Note that the cone receptor response is characterized by both a faster rise and decay time than the rod response.

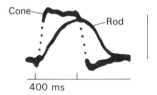

Fig. 3.7. Cone and rod receptor potentials elicited by two 400-ms, equal-quantum flashes at 560 and 508 nm, respectively. (From Whitten and Brown (1973*a*).)

In addition, the duration of the post-offset persistence of the photoreceptors is a function of stimulus and/or background intensity. Whitten and Brown (1973*a*) varied intensity of a foveal stimulus from 10^1 to 10^5 lm/m^2. They did obtain cone response variations as a function of flash intensity, some of which they attributed to stray light artefacts at the very high intensities. However, at stimulus offset, the response decay seemed to be crisper or faster at higher intensities (see Fig. 5 of Whitten and Brown 1973*a*), a finding also evident in Fain and Dowling's (1973) study. Moreover, Whitten and Brown (1973*a*) noted that the off *latency* of the rod response to a prolonged (400 ms) stimulus becomes much longer at low stimulus intensities, despite the fact that decay rates of rod responses are not much affected by variations of scotopic luminance. This brings up the following important point. Although rise and decay times may not vary over a range of stimulus intensities, variations in offset latency can effectively curtail or extend the response of the receptor beyond the offset of a stimulus flash. In other words, post-offset persistence can be a function of (i) offset response latency; or (ii) response decay rate; or (iii) both. Therefore, since the decay rate of the rod response varies very little, if any, with scotopic intensity but offsets latency decreases as intensity increases, the overall post-offset persistence in rods also decreases with intensity.

In a mixed response when both cone and rod signals are evident in the same response record, one has, at stimulus offset, an initial, short-duration cone response followed by a longer persisting rod response. Whitten and Brown (1973b, Fig. 9) showed that as intensity increases from low scotopic to high photopic levels the later, longer persisting rod response is progressively attenuated—a fact which Whitten and Brown (1973b) attribute to either rod saturation or the inhibition of the rod response, via horizontal cell input (Kolb 1974; Steinberg 1969; Steinberg and Schmidt 1970), by the increasingly stronger cone response.

Moreover, in addition to stimulus intensity, the level of dark or light adaptation also affects the receptor response persistence. Whitten and Brown (1973c), for instance, showed that for the cone response, persistence is longest at dark adapted levels and becomes progressively shorter until it is eliminated at higher light adaptation levels. A similar result has been reported by Hood and Grover (1974). This general finding is illustrated in Fig. 3.8, where the numbers next to the cone response traces indicate the relative levels of light adaptation. In summary, one can say that post-offset, receptor response persistence, be it determined by offset response latency or decay rate, generally decreases as stimulus intensity or background intensity increases. These inverse stimulus-intensity and background-intensity effects on neural persistence at the photoreceptor level bear a marked resemblance to the same effects on visible persistence discussed in the last section.

Evidence for the existence of neural persistence beyond the receptor level derives from a number of neurophysiological investigations of cells in the afferent visual pathway (Baker, Sanseverino, Lamarre, and Poggio 1969; Cleland, Levick, and Sanderson 1973; Enroth-Cugell and Shapley 1973a; Hammond 1975; Levick and Zacks 1970; De Monasterio 1978a). For instance, the total response duration produced in a sustained neurone by a brief, say, 2 ms flash can be in the order of 150 ms (Cleland *et al.* 1973; see Section 6.1 and Fig. 6.5). Moreover, for stimuli of longer duration, say, 500 ms, the post-offset response persistence depends on the level of background illumination. Figure 3.9 adapted from Hammond's (1975) investigation of cat retinal ganglion cells, illustrates response persistence to a prolonged stimulus in a sustained on-centre cell at scotopic and photopic background levels. Note that under both background intensities, the

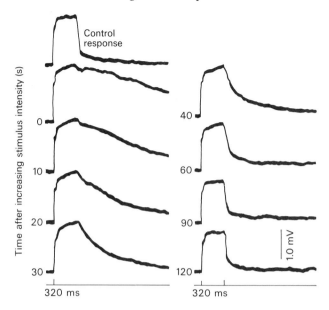

Fig. 3.8. Monkey cone receptor potentials produced by a 320-ms flash at increasing light adaptation levels. Control response is elicited when light adaptation is stabilized at its highest level. Note that the decay at offset becomes more rapid at higher adaptation levels. (From Whitten and Brown (1973c).)

response persists briefly beyond the offset of the stimulus; moreover, the post-offset response persistence is of longer duration at the lower, scotopic level than at the higher photopic level. This, as careful inspection shows, is due to two components. One is the shift to a greater offset latency, prior to the decay onset, at the scotopic relative to photopic background level; and the other is a corresponding decrease of the rate of response decay. We noted above that these two components also characterize the post-offset persistence at the receptor level. This correspondence suggests that the intensity-produced variation of response persistence at post-receptor levels of processing in the afferent visual pathway is determined partially by similar changes at earlier photoreceptor levels.

However, it raises an important issue concerning the source or sources of visible persistence. No doubt, in order to be visible a stimulus must produce cortical excitation. However, from this we cannot conclude that the properties of visible persistence may be determined only at the cortex. For instance, the duration of visible persistence may, as Sakitt (1976) argues, be determined largely by neural persistence at early, peripheral sites of processing, e.g. at the level of the photoreceptors. If that were the case, subsequent levels of neural processing would not significantly affect, but merely reflect, the durations of visible persistence. On the other hand, as indicated in Section 3.5.2. and later in Chapter 10, there may also be

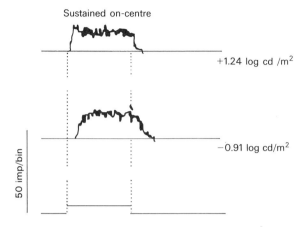

Fig. 3.9. Response of a sustained on-centre ganglion cell to a 500-ms on-annulus falling in the centre of its receptive field at photopic (upper panel) and scotopic (lower panel) background intensites. (From Hammond (1975).)

central, cortical sources of visible persistence in addition to earlier peripheral sources.

3.5. Peripheral and central sources of visible persistence

3.5.1. Peripheral (photoreceptor sources)

Based on her investigations of iconic memory or what I have termed visible persistence, Sakitt (1975, 1976) proposed the controversial claim that visible persistence has its source at peripheral levels of visual processing, in particular, at the rod photoreceptor level. Sakitt and her co-worker, Long (Long 1979, 1980; Long and Sakitt 1980*a*, *b*, 1981; Sakitt and Long 1978, 1979*a*, *b*) have extended these original investigations and, as a result of intervening work by Banks and Barber (1977, 1980) and Adelson (1978, 1979) have had to include also the cone response to give an adequate account of experimental findings on visible persistence (see also Monjé (1927)). None the less, Sakitt (and Long) maintain that '*the information about the icon* [and visible persistence, (Long and Sakitt 1981)] *is stored primarily in the photoreceptors in the retina*' (Sakitt 1976; p. 267). This is a strong and, as stated, a controversial claim heavily contested by other investigators claiming to have tapped central, cortical sources of visible persistence. However, before discussing these latter findings, let us look more closely at Sakitt's photoreceptor hypothesis of visible persistence.

In all of the studies reported by Sakitt and Long cited above, two critical experimental procedures were employed. First, the subjects were dark adapted; and, second, relative to a low intensity background the test stimuli consisted of bright flashes which could vary over several log units of luminance relative to the background or to each other. For instance, in Long and Sakitt's (1980a) study stimuli varying from 0.318 cd/m^2 to 318 cd/m^2 were flashed against a scotopically adapted background of roughly 0.06 cd/m^2. The other studies used similar but at times less extreme differences between scotopic background and more intense stimulus intensities.

Using these experimental methods Sakitt and Long generally found the following results: (i) visible persistence increased with stimulus intensity; and (ii) visible persistence increased with stimulus duration. This was particularly evident for rod vision where Sakitt (1976) also found a positive co-variation of persistence and scotopic luminosity. These results directly conflict with the above-discussed inverse intensity and duration effects on visible persistence (see also Ferry 1892; and Section 1.5). However, I also noted that the inverse intensity and duration effects may break down when extremely bright stimuli relative to a darker background are employed (Coltheart 1980). This may be due to the production of extremely strong after images, which, as shown above, can also extend the duration of forward masking produced by a bright stimulus flashed against a darker background (Crawford 1947; Sperling 1965; see also Figs. 1.4 and 2.5).

Moreover, Sakitt and Long often used relatively large stimuli (in particular, Sakitt (1976) used a 10° × 7° array of letters) which *per force*, under dark-adapted conditions, biases the longer persisting rod response. However, long persistence may also be found when the foveal cone response is dominant. This, on the basis of Whitten and Brown's (1973c) results (see Fig. 3.6), would be particularly evident when the fovea is dark adapted. In fact, employing a centrally fixated array composed of black alphanumeric characters on white surround subtending an angle of roughly 2°, Kriegman and Biederman (1980) found that when this array was briefly and intensely illuminated while the subject was in a dark-adapted state, a positive after image was produced which could last up to 5 s. Moreover, the strength and/or duration of the positive after image, as monitored by the observers' ability to correctly identify the alphanumeric characters, was directly related to the intensity of illumination, a result which is not found under light adapted states of vision (Coltheart 1980; Sperling 1963).

This latter finding that visible persistence does not vary appreciably at light adapted levels is also consistent with Whitten and Brown's (1973c) finding (see the traces labelled 60, 90, and 120 in Fig. 3.8) that neural persistence at the cone receptor level also does not change appreciably at higher light adaptation levels. The fact that cone responses can persist as long as rod responses given proper stimulus luminance and adaptation levels (here compare Kriegman and Biederman's (1980) results, implicating cone vision, to Sakitt's (1976) results implicating rod vision; in particular see Sakitt's Figs. 12–15) somewhat undercuts Sakitt and Long's (1979a) and Long's (1979) distinction between a short-duration cone persistence and a long-duration rod persistence.

The problem raised by all of the above results, particularly by the existence of durable rod and cone positive after images is not so much one of fact but rather one of conceptual interpretation. In essence, we are here concerned with legitimizing strong, long-lasting positive after images as forms of visible persistence. Is it proper, as suggested above and earlier in Exner's (1868) and Monjé's (1931) own investigations of the time course of visual sensations, to dichotomize between a long secondary after image and a shorter primary visible persistence? After all, both are more or less literal representations of the stimulus pattern in the form of positive after images whose time course may simply differ. One opinion is that it is proper to make this dichotomy when we make it on the basis of operational rather than ontological criteria.

Recall that at a dark adapted background level extreme increases in stimulus intensity or increases in duration of very bright stimuli produce increasingly strong and durable positive after images; that is to say, one obtains a direct intensity and duration effect. At progressively higher levels of light adaptation and for stimuli whose intensity does not deviate extremely from the background intensity, increases of stimulus intensity or duration curtails visible persistence. Here the inverse intensity and duration effect holds. These results are not necessarily contradictory, nor do they point to an essential dichotomy, particularly if one compares these results with correspondingly similar ones regarding photoreceptor persistence (Fain and Dowling 1973; Hood and Grover 1974; Normann and Werblin 1974; Whitten and Brown 1973a, b, c). What seems to be the case is that instead of supporting an essential, ontological dichotomy or contradiction, these results indicate a continuity of response from one extreme of relatively low background intensity and high stimulus intensity to another extreme of relatively high background intensity and moderate or low stimulus intensities. The fact that one obtains different durations of visible persistence under these very different stimulus conditions is not surprising when one also observes the same trends at the photoreceptor level, an interpretation consistent with more recent findings reported by Long and Gildea (1981), Long and Sakitt (1981), and Long and Beaton (1982). This should not be taken to imply that photoreceptor persistence is *the* source of visible persistence. Rather it suggests that photoreceptor persistence is a source of neural persistence; and, hence, one possible precursor of visible persistence. There are cogent reasons for believing that central, cortical sources of visible persistence also exist, and below I shall review the evidence pointing to the existence of these sources.

3.5.2. *Central cortical source*

The evidence pointing to more central, post-receptor sources of visible persistence rests on the following rationales. For instance, if, under binocular viewing, binocular brightness summation (Blake and Fox 1973) of equal monocular background intensities, leads to different, in particular, shorter estimates of visual persistence than those obtained under monocular viewing, as shown by Monjé (1931; and see Chapter I, Section 1.5), the existence of a central source can be inferred, which at least partially determines the duration of visual persistence. Moreover, if stimulus variables such as spatial frequency, size, or orientation—to which photoreceptors *per se* do not respond selectively but to which cells at subsequent, in particular, cortical levels do respond selectively (see Section 6.1) —affect the duration of visible persistence, then again central levels of processing are implicated in determining visible persistence. Another rationale exploits the use of random-dot stereograms developed by Julesz (1971). The stereograms consist of two random-dot matrices devoid of pattern information such that when one matrix is presented to the left eye and the other to the right eye, a stereoscopically-generated pattern is visible. Thus, if, using these stereogram displays, one can demonstrate the existence of stereoscopic or 'Cyclopean' visible persistence which outlasts or is independent of persistence in either monocular channel, one has effectively demonstrated a cortical source of visible persistence.

Coltheart (1980) carefully reviews Engel's (1970) use of the phenomenal continuity technique (see Section 3.2.1) to investigate stereoscopic and monocular persistence, and the reader is urged to go to these two sources for a more extensive discussion and review. In summary form, however, we can state Engel's (1970) particular results as follows: whereas monocular persistence lasted for at most 80 ms, stereoscopic persistence could last as long as 300 ms. That is to say, when the individual left-eye or right-eye, random-dot matrix was flashed intermittently at varying rates the maximal allowable blank interval at which monocular phenomenal continuity was perceived was about 80 ms; however, when both eyes were intermittently and synchronously stimulated by their respective random-dot patterns, the resultant stereoscopic pattern appeared to be phenomenally continuous for blank intervals up to 300 ms. A consequence of this four-fold shorter monocular persistence of random-dot matrices as noted by Ross and Hogben (1974) is that if the temporal interval between the left-eye and right-eye matrices is larger than about 80 ms, the monocular response persistence to the leading, say, left-eye matrix will have decayed by the time the response to the later, right-eye matrix is generated. Consequently, the two separate monocular inputs can no longer enter the binocular, stereoscopic level of processing as a unit and thus no stereopsis should result. Ross and Hogben (1974) employed a method of generating random-dot stereogram patterns in which two corresponding, continuous trains of random-dot matrix frames were presented so that any individual left-eye and right-eye frames, paired to produce stereopsis, were separated by temporal intervals ranging from 0 to above 100 ms. The results showed that stereopsis broke down at intervals exceeding about 70 ms, a value closely approximating Engel's (1970) 80-ms estimate of monocular, random-dot matrix persistence.

Moreover, van der Meer (1976), also employing a phenomenal continuity technique, showed that these stereoscopic or 'Cyclopean' sources of persistence are differentially affected by previously established disparity-specific (uncrossed or crossed disparity) adaptation effects (Blakemore and Hague 1972; Blakemore and Julesz 1971; Long and Over 1973; Mitchell and Baker 1973). In particular, van der Meer (1976) showed that after adaptation to a crossed-disparity Cyclopean pattern, visible persistence decreased for a similar pattern of crossed disparity but increased for one of opposite uncrossed disparity. Conversely, after adaptation to an uncrossed-disparity Cyclopean pattern, visible persistence decreased for a similar pattern of uncrossed disparity, but increased for one of opposite crossed disparity. Van der Meer assumed that these results reflected inhibition and disinhibition, respectively, among the antagonistically organized pools of adapted crossed (or uncrossed) and unadapted uncrossed (or crossed) pools of disparity detectors. As will be seen below, a similar explanation may hold for related spatial frequency- and orientation-specific adaptation effects on visible persistence.

Here, it should be cautioned that not every method of measuring stereoscopic persistence yielded positive results. As stated, Engel (1970) happened to employ the technique of phenomenal continuity (see Section 3.2.1.) of an intermittently presented, Cyclopean or stereoscopic pattern. However, other investigations, employing similar random-dot stereograms, conducted by Fox (personal communication) and co-workers (Fox and Lehmkuhle 1977; Lehmkuhle and Fox 1980) are equivocal at best. In particular, Fox and Lehmkuhle (1977) investigated *informational* persistence of Cyclopean patterns generated by the dynamic random-dot stereogram technique (Breitmeyer, Julesz, and Kropfl 1975; Julesz, Breitmeyer, and Kropfl 1976) which due to the rapid presentation of sequentially uncorrelated random-dot stereogram pairs, minimizes the role of monocular mechanisms in forming the Cyclopean image. Fox and Lehmkuhle employed the comparison of the post-exposure, partial-report technique to the post-exposure, whole-report technique, a comparison originally devised by Sperling (1960), to get an estimate of *informational* persistence. Fifteen Cyclopean letters arrayed in three rows by five columns were briefly presented to the observer. After offset of the Cyclopean array one of three tones—high, medium, or low—presented at varying intervals after offset signalled the subject to report the top, middle, or bottom row of the array, respectively. Fox and Lehmkuhle found no post-offset partial report advantage, as compared with the whole report, when using a Cyclopean array. However, when they used a standard monocularly visible array of letters, they did obtain a relative partial-report superiority. This finding, based on the post-exposure partial-report cueing method, on the one hand tentatively indicates that *informational* persistence is not obtainable with stereoscopic or Cyclopean stimuli, and may also indicate a clear difference between visible persistence and iconic memory or, equivalently, *informational* persistence as defined by Coltheart (1980). On the other hand, since Cyclopean visible persistence is four-fold longer than monocular visible persistence, one might expect the difference between whole and partial reports obtained under monocular or binocular viewing with standard stimuli to be attenuated when Cyclopean stimuli are employed.

These tentative conclusions are underscored by a more recent investigation of stereoscopic para- and metacontrast reported by Lehmkuhle and Fox (1980). They, like Engel (1970), obtained results consistent with long-lasting stereoscopic persistence. In this study a target consisting of a Cyclopean Landolt C was generated, whose gap could be located at the left, right, top, or bottom. Gap discrimination was correct at the 80 per cent performance level when the Landolt C was presented alone. However, when the Cyclopean Landolt C was surrounded by a brief Cyclopean masking annulus in the same depth plane as the Landolt C, gap-detection performance dropped significantly. In particular, both forward and backward Type A or monotonic masking functions were obtained, with a greatest decrement (a 40 per cent drop) in gap-detection performance at target–mask onset synchrony. Moreover, the forward masking effect, extending over 300 ms, was more pronounced than the backward masking effect—a finding (see Section 2.2) also often reported in many masking studies using standard, monocularly visible stimuli (Schiller and Smith 1965; Sperling 1965; Smith and Schiller 1966; Braddick 1973). One way to explain these results, specifically the forward masking result, is that the stereoscopic activity generated by the annular mask at or beyond the locus of binocular combination persisted for at least 300 ms and consequently interfered with or masked the visibility of the following Cyclopean Landolt C over that interval of time. This highly plausible explanation, besides being consistent with the existence of stereoscopic persistence, also illustrates the close correspondence between the 300 ms duration of stereoscopic persistence obtained in the masking study of Lehmkuhle and Fox (1980) and the maximal allowable stereoscopic blank duration reported by Engel (1970) in his study employing the phenomenal continuity technique. These few studies support the existence of a Cyclopean source of stereoscopic persistence, which, though occasioned by separate monocular inputs, none the less is independent of monocular sources of persistence. The theoretical importance of this Cyclopean source is manifest, since its existence would give strong support to a central, cortical locus of visible persistence besides more peripheral ones.

Let us now turn to the other rationale employed to demonstrate cortical sources of visible persistence. As mentioned, this rationale depends on tapping what are believed to be cortical processes sensitive to, among other stimulus dimensions, orientation and size or spatial frequency. If variations along these stimulus dimensions produce variations of visible persistence, neural processes at cortical sites are implicated in producing or controlling visible persistence.

Meyer and his collaborators (Meyer, Jackson, and Yang 1979; Meyer *et al.* 1975; Meyer and Maguire 1977) were the first to show the existence of orientation and spatial-frequency specific effects on visible persistence. For instance, using the phenomenal continuity technique, Meyer and Maguire (1977) and subsequently several other investigators (Bowling, Lovegrove, and Mapperson 1979; Badcock and Lovegrove 1981; Bowling and Lovegrove 1980, 1981; Breitmeyer and Halpern 1978; Corfield *et al.* 1978; DiLollo and Woods 1981; Lovegrove, Bowling, and Gannon 1981; Lovegrove, Heddle, and Slaghuis 1980; Meyer *et al.* 1979) have shown that visible persistence of a grating stimulus increases with spatial frequency.

These spatial frequency-dependent effects might be taken as evidence for a cortical component in visible persistence. However, Long and Gildea (1981) and Long and Sakitt (1981) argue that these effects on visible persistence are due to an artefact produced by probability summation. In Meyer and Maguire's (1977) study and also the other ones cited above, the stimulus display size was kept constant. As a result, the number of bars visible to the observer varies directly with spatial frequency, and consequently prolonged grating visibility after offset could simply be due to a correlated probability-summation effect. Long and Sakitt (1981) tested this alternative hypothesis by varying the horizontal dimension of a display of vertical gratings inversely in relation to the spatial frequency so that the number of visible bars stayed the same. They found, consistent with the probability summation hypothesis, that under these conditions visible persistence did not vary with spatial frequency. Moreover, varying the vertical dimension by the same spatial extent as the horizontal dimension, but leaving the latter fixed, failed to produce the probability-summation effect, indicating that it is the number of bars rather than the total area of the grating display which determines probability summation. These findings seem to undermine the cortical-locus hypothesis of visible persistence proposed by Meyer and Maguire (1977).

In addition to the probability-summation effect, Long and Sakitt (1981) found that the obtained duration of the visible persistence of grating depends on the method and response indicator employed. Using the phenomenal continuity technique employed by Meyer and Maguire (1977), Long and Sakitt (1981) found as expected: (i) that for spatial frequencies ranging from 1.0 to 14.5 c/deg, visible persistence increased with spatial frequency; and (ii) that, again as expected, visible persistence at all spatial frequencies decreased as the space-averaged luminance of the grating and the background on which it was repetitively flashed increased. However, using the temporal-synchrony-judgement technique, in which the perceived onset of a brief and small visual probe was to be matched to the moment that *all* traces of the grating had just disappeared, Long and Sakitt (1981) found, contrary to the results obtained with the phenomenal continuity method, that persistence of gratings *decreased* with increases of spatial frequency and decreases of their space-averaged luminance. This result supports an after-image explanation of visible persistence, since prior studies of after-image duration as a function of grating frequency (Brindley 1962; Corwin, Volpe, and Tyler 1976) reported related findings. Ever since Purkinje's (1819) work, in the early 1800s, on immediate after effects of brief visual stimulation—followed in the next 100 years by several major related investigations (Aubert 1865; Fechner 1840*a*, *b*; Helmholtz 1866; Müller 1834; Plateau 1834), culminating in Fröhlich's (1921, 1922*a*, *b*, 1929) studies in the early 1900s—it has been known that after images of brief flashes often produce a damped oscillation of alternating positive and negative phases. As intensity of the flash increases, the duration of this after-image sequence also increases (Alpern and Barr 1962; Feinbloom 1938). In addition, recently Corwin *et al.* (1976) reported that the oscillating after images are particularly evident at lower spatial frequencies or larger stimulus sizes, which, by the way, also produce longer-duration after images than do stimuli of smaller size (Nagamata 1954).

Moreover, Büttner, Grüsser, and Schwanz (1975) have provided physiological evidence showing that these oscillations are related to responses found no earlier than the ganglion-cell level of the retina. Certainly they are not manifest at the photoreceptor, horizontal- or bipolar-cell level (Büttner *et al.* 1975; Grüsser 1960; Steinberg 1969; Steinberg and Schmidt 1970; Werblin 1971). Although it is likely that cone and rod receptors, respectively, may be involved in producing fast and slowly decaying after images (Grüsser 1960) as argued by Sakitt and Long (1978, 1979*a*), the fact the receptor potential does not oscillate indicates that at least one of the properties, namely, oscillations, of visible persistence defined as a form of after image (Sakitt 1976) is located at post-receptor, neural levels. This, of course, raises the possibility that other properties, such as total duration, also are at least partially determined by such post-receptor, neural processes.

In fact, several other investigations (Bowling and Lovegrove 1981; Lovegrove *et al.* 1981; Meyer *et al.* 1975, 1979) point more positively to a later, cortical source of visible persistence. For instance, Bowling and Lovegrove (1981) and Lovegrove *et al.* (1981) report that, for a fixed display size and spatial frequency, the visibility of left-oblique or right-oblique gratings persists longer after grating offset than the visibility of horizontal or vertical gratings. This orientation-specific effect general-ized over a number of spatial frequencies is shown in Fig. 3.10, taken from Bowling and Lovegrove (1981). Photoreceptors are not sensitive to stimulus orientation and we know, ever since the pioneering work of Hubel and Wiesel (1962, 1968, 1977) that the visual cortex contains orientation-selective cells (see also Maffei and Campbell 1970). Consequently, these orientation-specific effects on visible persistence are best explained in terms of a cortical rather than peripheral locus of visible persistence. This conclusion is not meant to exclude the role of peripheral sites in contributing to visible persistence. In fact, Bowling and Lovegrove (1981) based on their empirically established relationship between orientation-specific temporal integration and persistence (Bowling and Lovegrove 1980, 1981; see Section 3.3.2), argue that, besides a later cortical component of persistence and integration, a peripheral component or source also exists.

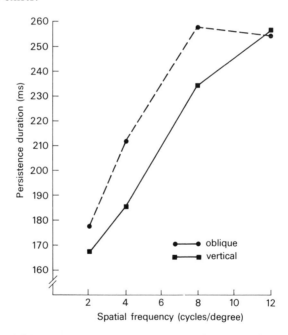

Fig. 3.10. Visible persistence as a function of the spatial frequency and orientation (vertical and oblique) of sinusoidal gratings. (From Bowling and Lovegrove (1981).)

Further evidence for a cortical component derives from orientation-specific and spatial frequency-specific adaptation effects on visible persist-ence of gratings (Meyer *et al.* 1975, 1979; Meyer and Maguire 1981). Recall from above that related, cortical, disparity-specific adaptation effects on

Cyclopean persistence have been reported (van der Meer 1976). For instance, Meyer *et al.* (1975) found that the visible persistence of a test grating (6.5 c/deg) decreased following adaptation to a grating of the same orientation and increased following adaptation to a grating at an orientation orthogonal to that of the test grating. These results are consistent with neurophysiological evidence of adaptation of orientation-specific cells in the visual cortex (Maffei, Fiorentini, and Bisti 1973; Creutzfeldt 1972; Creutzfeldt and Heggelund 1975) and with the intra-cortical interactions among cells or columns of cells selectively responsive to stimulus orientation. A reduction in visible persistence when test and adapting orientation match may be due to inhibition, produced by the adapting grating, within a cortical column of cells responsive to that orientation or between neighbouring columns responsive to similar orientations (Blakemore and Tobin 1972; Creutzfeldt 1972; Magnussen and Kurtenbach 1980). An increase in visible persistence when test and adapting orientations are orthogonal may be due to post-adaptation disinhibition (or facilitation) among columns selectively responsive to orthogonal or nearly orthogonal test grating orientations (Blakemore and Tobin 1972; Creutzfeldt and Heggelund 1975). This latter facilitation produced by orthogonal gratings is consistent with psychophysical studies of masking of one grating by another (Gilinsky 1967) and on related studies using the stabilized-retinal image technique (Schmidt, Cosgrove and Brown 1972). A mask grating facilitates rather than suppresses the visibility of an orthogonal test grating.

A related, spatial frequency-selective adaptation effect on visible persistence—particularly for briefly flashed gratings of intermediate to high (>5 c/deg) spatial frequencies (Meyer and Maguire 1981)—has been reported by Meyer *et al.* (1979). Moreover, the fact that this effect on visible persistence, as well as the orientation-specific one discussed above, transferred interocularly is consistent with the similar interocular transfer of orientation- and spatial frequency-selective adaptation effects reported originally by Blakemore and Campbell (1969). These and similar selective adapting effects, since they transfer interocularly, even when the adapted eye is pressure blinded and thus is prevented from transferring retinal activity to the visual cortex (Barlow and Brindley 1963), indicate a cortical locus of visible persistence and are difficult to reconcile with a view that restricts the source and locus of visible persistence to peripheral, receptor processes.

Additional evidence from studies on visible persistence and eye movements, to be more thoroughly discussed in Chapter 10, underscores the existence of a cortical source of visible persistence which is organized spatially or spatiotopically (Breitmeyer, Kropfl, and Julesz, 1982; Jonides, Irwin, and Yantis 1982; Ritter 1976; White 1976) rather than one entirely organized retinotopically and peripherally as believed by several other investigators (Doerflein and Dick 1978; Hochberg 1968, 1978; Sakitt 1976; Turvey 1977).

3.6. The temporal persistence-summation and psychological-moment hypotheses in masking and temporal integration of pattern

Section 3.3 showed that temporal persistence bears important empirical and conceptual relations to temporal integration. Based on empirical evidence a direct correlation between visible persistence and temporal summation in afferent visual pathways as expressed by Bloch's law was noted, as were several psychophysical phenomena that purport to demonstrate that the existence of a psychological moment can be as easily and more parsimoniously explained by persistence; therefore, for these phenomena the notion of the psychological moment is conceptually superfluous.

Masking of one stimulus by a spatially overlapping and temporally preceding or following second stimulus is an additional phenomenon which poses problems for either the temporal summation or the psychological moment hypothesis. The temporal summation hypothesis of masking is based on Bloch's law discussed in Section 3.1.2. There I stated that a single stimulus—via an inverse intensity–duration relationship—obeys Bloch's law of summation, but that double pulses do not; despite their falling within the critical duration, τ, at and below which the law is valid. Both forward and backward masking employ a double-flash technique. One might, therefore, expect that although certain stimulus parameters, e.g. background intensity, predictably affect temporal summation and masking intervals in the same *general* way, the presence of additional temporal transients in the double-flash stimulus presentation facilitates temporal interaction between, in addition to, summation of the two flashes. Moreover, even if the responses of two temporally separate flashes were simply to summate within the critical interval, τ, their temporal order would not be a critical factor; hence, one would expect symmetrical forward and backward masking functions. As noted in Section 2.2, under monocular or binocular viewing several investigators have reported such functions (Eriksen and Hoffman 1963; Eriksen and Lappin 1964; Mowbray and Durr 1964); yet others reported asymmetrical functions, with forward masking typically being more prolonged than backward masking (Braddick 1973; Coenen and Eijkman 1972; Schiller and Smith 1965; Smith and Schiller 1966; Sperling 1965).* Moreover, these differences, also noted in

* Although under monocular or binocular viewing forward and backward masking functions can vary from symmetry to asymmetry, under dichoptic viewing conditions backward masking by either light (see Fig. 2.12), noise, or patterns (Section 4.5) generally tends to be stronger than forward masking. This suggests that in the former viewing conditions the often-reported, asymmetrical, stronger forward masking effect is due to response persistence in precortical, afferent pathways. Section 8.3 shows that, despite this

Section 2.2, may be more readily explained and accommodated by predictable variations of response *persistence* of low as opposed to high spatial-frequency target information.

Rather than τ specifying Bloch's law of temporal summation in peripheral visual pathways, for the sake of argument, it can equally well characterize the duration of an internally or externally generated psychological moment. Here one again would predict that the temporal order of two integrated flashes falling within the internal, τ, would not matter. In terms of masking, this temporal-order indifference of the psychological moment to double flashes in turn also implies symmetrical forward and backward masking functions. The existence of asymmetrical functions, however, militates equally against the psychological-moment hypothesis as it does against the temporal-summation (Bloch's law) hypothesis.

In addition to masking by spatially overlapping stimuli, several investigators (DiLollo and Wilson 1978; Eriksen and Collins 1967, 1968) have tried to assess the possible roles of persistence and the psychological moment in temporal form integration of non-overlapping visual patterns. With the spatially non-overlapping and complementary stimulus displays used by these investigators, temporal integration can aid in pattern recognition. For instance, Eriksen and Collins (1967, 1968) put the persistence and psychological-moment hypotheses to a test in studies of temporal form integration in which two randomly appearing displays were to be integrated in order to recognize a capital letter. On the basis of these studies, neither hypothesis could be excluded; the data were consistent in part with either hypothesis. However, recently DiLollo and Wilson (1978) also investigated temporal integration of form; and whereas their results positively and clearly indicated the role of persistence in temporal form integration, they were not consistent with psychological-moment hypothesis of temporal integration.

A variation of a method adopted by Hogben and DiLollo (1974) provided DiLollo and Wilson (1978) with the following procedure and rationale. The visual display consisted of a briefly flashed array of 25 dots arranged in a 5 × 5, square matrix. One dot, whose coordinate position in the matrix was chosen at random from trial to trial, was omitted from the display. The observer's task was to indicate the location of the missing dot within the matrix. The twenty-four dots were sequentially displayed in the following manner, as schematized in Fig. 3.11. Six dots, picked at random, were flashed at a duration varying from 20 to 200 ms (flash A in Fig. 3.11). Another six dots, picked at random from the remaining 18, were presented

general asymmetry of dichoptic pattern masking, under chosen conditions one *can* nevertheless obtain an index of symmetric masking effects at central, cortical levels of (iconic) persistence and integration. Moreover, Chapter 7 (Section 7.4.2, Explanations 5 and 6) offers an explanation for the empirically obtained stronger, more prolonged backward masking functions typically found with interocular presentations of the target and mask stimuli.

for 10 ms, as designated by flash B in Fig. 3.11, and were timed relative to flash A so that flash B terminated 10-ms before flash A terminated. The remaining 12 dots were presented in a third 10-ms flash, designated flash C in Fig. 3.11. Note that the temporal intervals between flash C and flashes A and B were 10 and 20 ms, respectively. According to the psychological-moment hypothesis, the information in flashes A and C has a greater probability of falling within the same psychological moment than does the information from flashes B and C. This follows from the fact that the probability of two stimuli falling into the same psychological moment of fixed duration is inversely proportional to the temporal interval or ISI

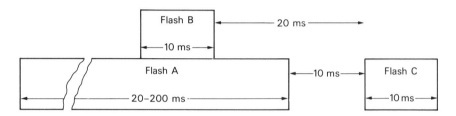

Fig. 3.11. Temporal sequence of flash presentations (A, B, and C) used by DiLollo and Wilson. (From DiLollo and Wilson (1978).)

between these two stimuli. Consequently, the psychological moment hypothesis predicts that more dot positions from flash B ought to be misidentified as blank positions than dot positions from flash A.

The persistence hypothesis makes just the opposite prediction, for the following reasons. I noted earlier that visible persistence is inversely related to flash duration. Accordingly, for progressively longer durations of flash A progressively shorter and weaker persistence would be expected, whereas the visible persistence produced by the brief flash B ought to be constant and relatively long. Consequently, despite the shorter ISI between flashes A and C, the persistence hypothesis predicts that, as flash A duration increases, progressively more dot positions from flash A ought to be misidentified as blank positions.

The results, as shown in Fig. 3.12, clearly are consistent with the persistence hypothesis. At no duration of flash A (except the 20 and 40-ms duration for subject LDC) are more dot positions of flash B misidentified as blank positions than dot positions of flash A, a result which is inconsistent with the psychological-moment hypothesis. These and related results (Allport 1968; Efron and Lee 1971) suggest that visible persistence is of much greater heuristic value in explaining pyschophysical indices of not only masking but also form integration than is the psychological-moment hypothesis.

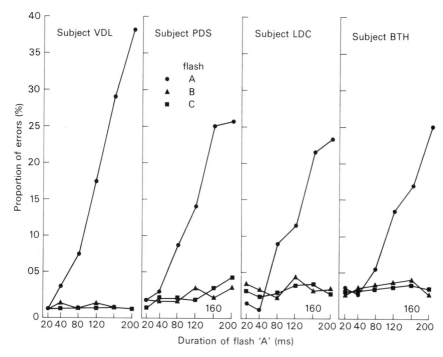

Fig. 3.12. The proportion of errors made by incorrectly identifying as missing a dot which was actually plotted in flashes A, B, and C (see Fig. 3.11), at each of six durations of flash A. The *a priori* probability of incorrectly identifying a dot in flash C (twelve dots) was twice that of flashes A and B (six dots each). To aid comparison among the three curves in each graph, the proportion of errors in flash C has been halved. (From DiLollo and Wilson (1978).)

3.7. Summary

Visual integration and persistence are two related psychophysical phenomena. Integration can be regarded in several ways:

1. In terms of the psychological-moment hypothesis according to which an internal source of intermittency (clock) controls the temporal interval over which two incoming sensory events are treated as one.

2. In terms of temporal response summation characterizing sensory mechanisms according to Bloch's law.

3. In terms of the post-offset persistence of a sensory response to the prior of the two sequential stimuli.

A distinction was made between two forms of visual persistence. Neural persistence refers to the post-offset response to a flashed stimulus of visual neurones along the afferent pathway; visible persistence refers to the phenomenal continuation of a percept beyond the duration of a stimulus producing the percept. Since neural persistence co-varies positively with visible persistence as a function of stimulus variables—e.g. intensity or duration—it most likely contributes to the stimulus-dependent variation of visible persistence.

Several methods have been developed to investigate visual persistence. Among these are (i) temporal synchrony judgements of the apparent on- and offset of a stimulus; (ii) reaction time measurements to on- and offset; (iii) judgements of the number of simultaneously apparent lines produced by stroboscopic illumination of a single moving line; (iv) judgements of phenomenal pattern continuity of repeated flashes of a stimulus; (v) temporal integration of complementary form parts across an intervening blank interval; (vi) the moving-slit technique; (vii) forward pattern masking; and (viii) stroboscopic motion.

Despite the variety of methods used to measure visible persistence, two highly replicable effects, the inverse duration effect and inverse intensity effect, can be obtained at moderate luminance levels. The inverse duration effect refers to the finding that post-offset persistence is inversely proportional to the duration of a stimulus. The inverse intensity effect refers to the finding that post-offset persistence is inversely related to either the background or the stimulus intensity. Both effects break down when relative to a low or moderate background intensity much more intense stimuli produce long-lasting positive after images generated by either cone or rod activity during the post-stimulus recovery of sensitivity.

It was also demonstrated that visible persistence is related to temporal summation expressed by Bloch's law. The inverse-duration effect seems to hold mainly for the temporal interval over which Bloch's law holds. Beyond that interval, stimulus duration either no longer or only weakly affects visible persistence.

Neural summation and post-stimulus response persistence can be demonstrated in visual units along the entire retinogeniculostriate pathway. Moreover, neural persistence co-varies predictably with visible persistence as stimulus parameters—for example, intensity, background luminance level, and duration are varied. Two sources of neural persistence at post-stimulus offset were identified: offset-response latency and the rate of the response decay back to its prior maintained level.

Current theories of visible persistence attribute its properties either to peripheral receptor activity or to central, cortical levels of activity. Evidence for the former peripheral source is based on exploiting known psychophysical properties of rod and cone vision which correlate strongly with neurophysiological measures of receptor activity, and an explicit assumption that visible persistence is tied to retinal coordinates, much like an after image. Evidence for the latter, central source of visible persistence rests, for one, on the demonstration of differences between monocular and stereoscopic or Cyclopean persistence studied with the use of random-dot stereograms and, secondly, on spatial-frequency- and orientation-specific adaptation effects on visible persistence. Moreover, in Chapter 10, we shall see that a form of visible persistence may exist which is tied to spatial (environmental) rather than retinal coordinates, and hence cannot be of peripheral origin.

An evaluation of the psychophysics of visual persistence indicates that it provides a more powerful and parsimonious explanation of the role of visual integration in stimulus recognition and in stimulus masking than does either the psychological-moment hypothesis or the summation sypothesis based on Bloch's law. Visual integration, according to the persistence hypothesis, results from the post-offset response to a first stimulus persisting for some time and thereby integrating with the initial post-onset response component of a following stimulus.

4 Masking by pattern: methods and findings

4.1. Introduction and definitions

The last two chapters discussed masking by light, temporal persistence, and temporal integration. I mentioned that the increase of the magnitude and duration of forward masking, produced by a brief light flash or the offset of a more prolonged flash as its energy increased, can be interpreted as one index of peripheral persistence. Evidence for additional, central sources of visual persistence was also reviewed and discussed. Moreover, in so far as temporal persistence was found to provide a basis for temporal integration of successive pattern components, Section 3.6 showed, and this chapter will show further, how such post-offset persistence can also be a source of masking by *integration* of not only homogeneous light flashes but also of a contoured or patterned flash which spatially overlaps with the mask. Besides reviewing this latter type of pattern masking based on visual pattern persistence, I shall also discuss meta- and paracontrast, two forms of lateral pattern masking in which the mask and target patterns do not overlap spatially and which serve to *interrupt* or *suppress*, rather than promote, the post-offset persistence and processing of a stimulus. The significance of these two basic (integration and suppression) masking mechanisms for processing visual information under dynamic viewing will be clarified in Chapter 10. For now, however, I shall review the methods and findings of masking by pattern.

Unlike masking by light, discussed in Chapter 2, pattern masking refers to experimental paradigms in which both the target and the mask, rather than consisting of spatially homogeneous flashes of light, consist of spatially patterned forms and contours. Consequently, by varying the temporal interval between the target and mask, pattern masking can be used to investigate the microgenesis of pattern-contrast and contours in human vision (Werner 1935).

As illustrated in Fig. 4.1, several variants of pattern masking exist. Fig. 4.1(a) shows typical target–mask stimulus combination used in the study of paracontrast and metacontrast as defined in Chapter 1. The target and mask need not consist of a disk and annulus as shown; equally useful would be a rectangle serving as a target and two spatially adjacent rectangles serving as mask or any other set of stimuli which, without overlapping, preserve spatial contiguity between the target and mask contours. When the mask temporally precedes the target *paracontrast* masking is obtained; when the temporal target–mask sequence is reversed, *metacontrast*

Fig. 4.1. Examples of target and mask stimuli typically used in (*a*) paracontrast and metacontrast, (*b*) pattern masking by noise, and (*c*) pattern masking by structure. (From Breitmeyer and Ganz, (1976)).)

masking prevails. Paracontrast and metacontrast consequently are specific cases of the more general forward and backward pattern masking techniques.

Fig. 4.1(b) illustrates a target–mask combination which is used in a pattern-masking procedure termed *masking by noise* (Kinsbourne and Warrington 1962*a*). Here the contrast elements and contours of the random-dot mask, although spatially overlapping those of the target, are designed to bear little, if any, structural relationship to the target contours. However, when the overlapping contours of the mask are designed so that they structurally resemble the contours of the target in terms of its orientation, curvature, angularity, or some other figural characteristic, we have, as shown in Fig. 4.1(c), a masking technique called *masking by structure* (Breitmeyer and Ganz 1976). On the basis of contour-contiguity, it should be obvious that metacontrast and paracontrast are two clear and special cases of structure masking.

These pattern-masking procedures can generate four fundamental types of masking effects schematized in Fig. 4.2. Using Koler's (1962) terminology, Fig. 4.2(a) shows idealized (i) monotonic, type A forward, and (ii) monotonic type A backward masking functions, corresponding to the conventionally assigned negative and positive stimulus onset asynchronies (SOAs) or target–mask onset asynchronies (mask precedes and mask follows target), respectively. Analogously, Fig. 4.2(b) illustrates idealized (i) non-monotonic, type B forward masking; and (ii) non-monotonic, type B backward masking functions. The term 'type B' is interchangeably used with the term 'U-shaped'.

The methods, findings, and theories relevant to these pattern-masking paradigms have been reviewed extensively elsewhere (Breitmeyer and Ganz 1976; Fox 1978; Kahneman 1968; Lefton 1973; Scheerer 1973; Weisstein 1972). I shall provide an additional review and analysis of methods and findings in the present chapter. Although some relevant theoretical topics also will be mentioned and discussed, a more extensive

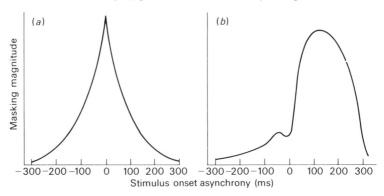

Fig. 4.2. Schematic representation of (*a*) type A and (*b*) type B forward and backward masking magnitude as a function of stimulus onset asynchrony (SOA) temporally separating the target and mask. Negative SOAs: mask precedes target; positive SOAs: mask follows target. (From Breitmeyer and Ganz (1976).)

review and analysis of these topics is deferred until Chapter 5, which specifically concentrates on theories and models of metacontrast.

4.2. Paracontrast

The number and types of stimulus dimensions which potentially may affect the magnitude and form of the forward, paracontrast function, up to now, has not been investigated sufficiently. Although one may at best infer from a recent result reported by Breitmeyer, Rudd and Dunn (1981*b*), that the range of target–mask spatial separations for obtaining paracontrast is more limited relative to the corresponding range for obtaining metacontrast, systematic research on the effects of stimulus size, retinal location, target–mask spatial separation, and their interactions remains to be done. Nevertheless, reports of several stimulus and response variables affecting the paracontrast function already have been published. For instance, paracontrast can yield either type B or type A masking functions (Alpern 1953; Kolers and Rosner 1960; Lefton and Newman 1976; Pulos, Raymond, and Makous 1980; Weisstein 1972). The type of paracontrast function one obtains depends on the experimental task required of the observer. When suppression of brightness or of spatial contrast is used as an indicator of the paracontrast masking effect, type B paracontrast functions are obtained (Growney, Weisstein, and Cox 1977; Kolers and Rosner 1960; Weisstein 1972). Here target brightness suppression is maximal when the mask leads the target by about 30–70 ms. Moreover, the magnitude of the effect decreases as target–mask spatial separation increases (Growney *et al.* 1977; Kolers and Rosner 1960) and increases as the mask energy increases relative to the target energy (Alpern 1953; Weisstein 1972). The type B effect can also be obtained dichoptically

(Foster and Mason 1977; Kolers and Rosner 1960). The latter finding shows that the type B paracontrast can be obtained when the target and mask responses interact at or beyond the level of binocular combination, indicating that cortical mechanisms may be used.

A type A paracontrast effect has been reported by Lefton and Newman (1976). In this study brightness or contrast suppression was not used as an indicator of masking; instead a target detection task was employed. At any target–mask SOA, duration thresholds for target detection were measured. Longer threshold durations indicate greater masking effects. Masking was greatest near target–mask synchrony and declined monotonically as the temporal interval separating the leading mask from the lagging target increased.

This dependence on the type of paracontrast effect on task variables points out the importance of the role of criterion content in visual masking (Kahneman 1968). Criterion content refers to the stimulus dimension along which an observer makes his or her perceptual judgement about the target. Since brightness judgements and detection thresholds may tap different stimulus dimensions, it should not be surprising that different paracontrast functions are obtained with correspondingly different experimental tasks. In fact, these differences between task parameters and corresponding criterion contents also are manifest in other masking paradigms.

4.3. Metacontrast

In metacontrast masking, one can also obtain either type A or type B functions; and again the type and magnitude of function one obtains depends on the experimental task and stimulus parameters. I shall review the experimental findings on metacontrast in the context of the following general variables: task parameters and criterion content, stimulus (target, mask, and background) parameters, and viewing condition.

4.3.1. Task parameters and criterion content

The fact that task parameters and criterion content can affect performance in visual masking is well documented (Bernstein, Fisicaro, and Fox 1976; Bernstein, Proctor, Proctor, and Schurman 1973; Haber 1969; Hernandez and Lefton 1977; Petry 1978; Stober, Brussel, and Komoda 1978; Ventura 1980). The dependence of metacontrast on task parameters and criterion content is most aptly illustrated when one compares the types of metacontrast functions obtained in a simple detection or target-reaction time (RT) task with those obtained in tasks requiring judgements of brightness or contrast and contour discriminations.

Type B metacontrast functions are generally obtained with suppression of the target's brightness or contrast (Alpern 1953; Flaherty and Matteson

1971; Growney and Weisstein 1972; Weisstein 1972; Weisstein, Jurkens, and Onderisin 1970), of its contour or contour detail (Breitmeyer 1978*a*; Breitmeyer, Love, and Wepman 1974; Burchard and Lawson 1973; Stober *et al.* 1978; Werner 1935; Westheimer and Hauske 1975) or of its figural identity (Averbach and Coriell 1961; Weisstein and Haber 1965; Weisstein *et al.* 1970). Type B functions also can be obtained when a choice RT task is employed in which the subject must respond as fast as possible to one of several possible targets on the basis of its discriminable figural properties (Eriksen and Eriksen 1972; May and Grannis, unpublished observations). Here choice RT to a target can reach a maximum at some intermediate positive SOA temporally separating the target from the mask. This is not too surprising since suppression of target contrast, contour or shape, upon which the choice reactions are based, also reaches its maximum at intermediate SOA values. In these studies, the target stimuli are typically characterized by an appearance of lower contrast, 'fuzzy' contours and consequent distorted shapes.

However, when using simple detection as the response criterion, one can obtain very different metacontrast functions.* For instance, in their simple RT studies of metacontrast, Fehrer and Raab (1962) and Fehrer and Biederman (1962) showed that choice RT to the target does not vary as a function of SOA; that is, neither a type A nor type B effect was obtained. This finding has been replicated in several other subsequent investigations of metacontrast (Harrison and Fox 1966; Bernstein, Amundson, and Schurman 1973; May and Grannis, unpublished observations). In a related study, Schiller and Smith (1966) showed that choice RT to, or forced choice detection of, a target disk displayed at one of two possible locations and followed by two simultaneously flashed rings—each of which could surround the target disk at each of its possible locations—did not change as a function of SOA. The above results indicate that the observers were able to detect the presence and location of a target on the basis of information (criterion content) which, unlike information about brightness, contour, or figure, is immune to the effects of the mask. When looking at a metacontrast display, although the target contrast and contours appear· to be entirely absent, one none the less can detect a form of 'explosive' or

* Cox and Dember (1972) obtained type B metacontrast functions using a detection task; hence their results are an exception to the general finding that no masking is obtained with a detection criterion. The discrepancy may be explained on the basis of employing different criterion contents. Cox and Dember (1972) used black-on-white target and mask stimuli and subjects were required to detect the target. One way to obtain a type B function in a detection task is to require subjects to detect the presence of a *black* target rather than just any target presence. In the former case the criterion content would be brightness or contrast; in the latter case, possibly apparent movement. Since brightness or contrast reversal can be obtained in backward masking by light and metacontrast (Barry and Dick 1962; Brussell, Stober, and Favreau 1978; Purcell and Dember 1968) at optimal SOAs, the location of the black target will appear even brighter than the background. Hence, if a contrast criterion is used, the target will not be detected as black.

split stroboscopic motion, which may provide the criterion content for the simple detection task.

It should be mentioned, however, that, when using their RT or detection criterion, Schiller and Smith (1966) obtained an absence of masking only when the target and mask were of equal energies (43 ft-L). When the target energy was lowered to 4.3 and 0.43 ft-L relative to the constant, 43 ft-L mask energy, type A functions were obtained in both target-RT and detection tasks—findings also reported in similar studies of detection in metacontrast where the target duration or time-integrated energy was substantially lower than that of the mask (Lefton and Griffin 1976; Lefton and Newman 1976). This, of course, points out that the type of metacontrast function one obtains depends not only on task parameters but also on variables affecting either or both of the target and mask stimuli.

4.3.2. Effects of stimulus intensity and contrast

Variations of the intensity or contrast of either the target or the mask stimulus can have measurable effects on the shape or magnitude of the metacontrast masking function. One of the prime determining variables is the target:mask (T/M) energy ratio. In her review of metacontrast, Weisstein (1972) noted the following general empirical relationships depending on T/M energy ratio: (i) when the ratio was greater than or equal to unity (target energy greater than or equal to mask energy), type B U-shaped backward masking functions were obtained; (ii) for ratios less than unity (target energy less than mask energy), the shape of the metacontrast function tends to shift from a type B to a monotonic, type A, function as the ratio becomes progressively smaller. This finding has been reported in several other investigations of metacontrast and backward masking (Breitmeyer 1978b; Fehrer and Smith 1962; Kolers 1962; Spencer and Shuntich 1970; Stewart and Purcell 1974).

In particular, Breitmeyer (1978b) investigated the relationship between the magnitude and shape of the metacontrast function, as indexed by contrast suppression of the target, and the T/M energy ratio as indexed by the T/M duration ratio. The target duration was fixed at 16 ms; the duration of the mask could vary in 0.3 log unit steps from 1 ms to 32 ms. The T/M duration or energy ratios thus ranged from 16.0 to 0.5. The results are displayed in Fig 4.3. When the mask duration was only 1 ms, little if any metacontrast masking was obtained. A noticeable type B function, peaking at an SOA of 56 ms, emerged at a mask duration of 2 ms and increased progressively in magnitude up to a mask duration of 8 ms. At further increments of mask duration, the masking magnitude at and beyond an SOA value of 56 ms did not change; however, what did change drastically was the masking magnitude at lower SOA values. Here the masking effect increased progressively as mask duration increased beyond 8 ms; and, consequently, a concomitant change in shape of the metacontrast function from type B to type A occurred. As an extension of this change from type B to type A metacontrast functions, one can offer the following plausible and testable hypothesis. Since type B paracontrast magnitude increases when T/M energy ratio decreases, as suggested by Weisstein's (1972) results, the corresponding shift from type B to type A metacontrast may reflect the decline portion of paracontrast magnitude at non-optimal, i.e. positive, SOAs (see also Fig. 5.2, panel 5).

Moreover, from a methodological viewpoint, this finding has important implications for explanations of backward masking and metacontrast. Since, as illustrated above, type A backward masking effects can obscure type B ones when the T/M energy ratio is substantially less than unity, experiments failing to obtain type B effects, without specifying the T/M energy ratio, cannot be used unequivocally as evidence against possible underlying mechanisms responsible for generating type B effects (Eriksen, Becker, and Hoffman 1970). As a rule of thumb, one can state that a type A metacontrast function obtained with a T/M energy ratio greater than unity and with a brightness or contour discrimination task would pose a problem for theories of type B metacontrast. However, to my knowledge no study has ever reported such a result.

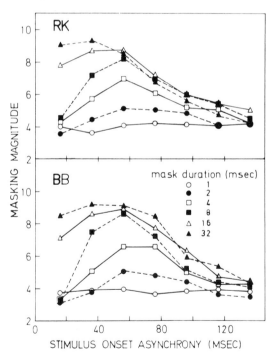

Fig. 4.3. Metacontrast masking magnitude as a function of stimulus onset asynchrony and mask durations as indicated. Target duration was fixed at 16 ms. (From Breitmeyer (1978b).)

In the above studies, despite variations of T/M energy ratio, the adaptation level determined by the level of background intensity was generally constant. Purcell, Stewart, and Brunner (1974) investigated not only the effects of T/M energy ratio but also the effects of background intensity. They measured metacontrast masking effects when the target and mask were flashed either against an unilluminated background or one at a luminance of 40 ft-L. U-shaped masking functions were obtained at each level of background intensity and at each T/M energy ratio. However, relative to the high background intensity the peak of the U-shaped metacontrast functions tended to shift to lower SOA values at the low background intensity. Similar results have been reported by Stewart and Purcell (1974) under transient dark adaptation interposed between target and mask flashes. A similar shift toward lower SOA values of peak type B metacontrast occurs when the intensity of the target and mask are both varied from high to low values. In particular, for stimulus energies ranging from 3000 down to 3.6 ft-L, Alpern (1953) found not only that the magnitude of metacontrast decreased but also that that the SOA at which peak metacontrast occurred shifted from 125 ms to approximately 75 ms.

Most studies of metacontrast usually have employed target and mask stimuli having a contrast of the same sign. That is, both stimuli were either black-on-white, or white-on-black surrounds. Using a detection mask, Sherrick, Keating, and Dember (1974) recently reported metacontrast suppression of both black and white targets by either black or white masks. However, they used a target duration of 15 ms, a mask duration of 100 ms and an ISI of 0 ms (mask onset at target offset). Consequently, only the single SOA value of 15 ms was sampled and no extensive inferences can be drawn about the shape or magnitude of the metacontrast functions.

Recently, Breitmeyer (1978c) extended the study of metacontrast with black-on-white, or white-on-black surrounds. Using a detection task, the mask duration or energy equal to that of the target so as to allow possible type B metacontrast effects to manifest themselves. Moreover, Breitmeyer, instead of employing a simple detection task, used a form or contour discrimination task, which, as noted above, is more likely to yield type B effects. Disk-like targets were employed and could be either black or white on medium grey background. Spatially surrounding rings served as masks and, again, could be either black or white on grey. The general findings can be summarized as follows: (i) U-shaped metacontrast functions were obtained for any combination of target and mask contrasts; (ii) the magnitude of the masking effect tended to be greater when the target and mask contrasts had the same sign. These findings indicate that although the magnitude of type B contour suppression in metacontrast is affected by target–mask contrast relations, the activation of interactive mechanisms producing these contour-related type B effects *per se* is largely indifferent to these same relations.

4.3.3. Spatial variables: stimulus orientation, size, separation, and location
Werner (1935) in one of his experiments (No. 18) investigated the effect of
the orientation of internal contours of a target on metacontrast. The target
consisted of either a vertically oriented grating or else a horizontally
oriented grating. The metacontrast mask consisted of adjacent black,
vertically oriented bars flanking the target area. Werner reported that
although metacontrast suppression of the vertical target grating was
obtained on about 80 per cent of the trials, the horizontal target grating
always remained clearly visible. In other words, the magnitude of
metacontrast suppression of the target visibility is orientation-specific and
depends on the similarity between the target's and mask's orientation.

The magnitude of type B metacontrast also can be affected by a number
of spatial variables, most notably the stimulus size, the spatial separation
between target and mask and the retinal locus of stimulation. Stimulus
size, in particular mask size, has been found to have an effect on
metacontrast magnitude, but the direction of empirical effects has not been
consistent from one study to another. Although some investigators have
reported decreases of metacontrast magnitude as the width of the
surrounding mask increases (Schiller and Greenfield 1969; Sturr and
Frumkes 1968; Sturr, Frumkes, and Veneruso 1965), others have reported
the opposite effect of increases of mask width on metacontrast magnitude
(Growney and Weisstein 1972; Kao and Dember 1973; Matteson 1969). In
fact, several of Growney and Weisstein's (1972) results indicate that
masking magnitude may be a U-shaped function of mask width. For stimuli
centred 1.0° from the fovea, Growney and Weisstein (1972) reported an
initial rapid increase of metacontrast magnitude as mask width increased
from 1' to about 10', followed by what seemed to be a more gradual
decline in mask effectiveness as the mask width increased even further.
This particular trend, by the way, may be an index of the antagonistic
processes of spatial summation and spatial (lateral) inhibition found with
variation of stimulus size as measured by the Westheimer function (Teller
1971; Teller *et al.* 1971; see Section 2.1.5). Given this nonomonotonic
relationship between mask width and metacontrast magnitude in addition
to the fact that the effects of target or mask size interact with the effects of
retinal locus (Bridgeman and Leff 1979), it should not be surprising that
the variation of mask width can yield different and even contradictory
effects on metacontrast magnitude.

What seems to be needed is a careful empirical investigation of
orthogonal variations of target size, mask size, and retinal locus. A study
which closely approximates these conditions is one reported by Bridgeman
and Leff (1979) who showed that large foveal brightness metacontrast
masking effects can be obtained only with relatively small targets (0.25°
diameter disks) and masks. As the target and mask dimensions increased,
foveal metacontrast magnitude decreased substantially whereas parafoveal

and peripheral metacontrast magnitude remained robust and either did not change noticeably or increased somewhat. In a related study, using a contour discrimination mask, Lyon, Matteson, and Marx (1981) also showed strong metacontrast effects in the fovea. In fact, masking magnitude was found to be a function of the difficulty of contour discrimination and retinal locus. Foveal metacontrast was as strong as parafoveal metacontrast when a finer contour detail had to be discriminated at the fovea than in the parafovea (see also, Breitmeyer 1978*a*; Westheimer and Hauske 1975). However, for discrimination of equal contour detail parafoveal metacontrast was stronger than foveal metacontrast.

The finding that metacontrast masking of brightness and figural identity more generally increases in magnitude as the eccentricity of the retinal locus of stimulation increases has been replicated often (Alpern 1953; Kolers and Rosner 1960; Lefton 1970; Merikle 1980; Saunders 1977; Stewart and Purcell 1970, 1974; Stoper and Banffy 1977). Moreover, the fact that several earlier investigations reported little if any foveal metacontrast (Alpern 1953; Kolers and Rosner 1960; Toch 1956) can, in light of Bridgeman and Leff's (1979) analyses, be attributed to the use of excessively large, and thus suboptimal, foveal targets (e.g. in Kolers and Rosner's (1960) study the smallest foveal target disk was 0.42° in diameter; the largest was 1.67° in diameter).

Target–mask spatial separation also is known to affect the magnitude and form of type B metacontrast functions. Generally, as the spatial separation of the target and mask increases, the metacontrast magnitude decreases (Alpern 1953; Breitmeyer and Horman 1981; Breitmeyer *et al.* 1981*b*; Growney *et al.* 1977; Kolers 1962; Kolers and Rosner 1960; Levine, Didner, and Tobenkin 1967; Weisstein and Growney 1969). Moreover, the peak of the U-shaped metacontrast function, besides decreasing in magnitude, at times also tends to shift towards shorter SOA values (Alpern 1953; Growney *et al.* 1977) as the spatial separation between the target and mask increases, although in other studies (Kolers and Rosner 1969; Weisstein and Growney 1969) no such trend occurred.

Perhaps of greater significance is the fact that, like stimulus size, target–mask spatial separation interacts with retinal locus of stimulation. For instance, Kolers and Rosner (1960), like Stigler (1910) in his pioneering work, noted that very small contour separations between the target and mask drastically reduced or even eliminated foveal metacontrast. In contrast to this finding, for non-foveal stimulus loci robust type B metacontrast functions can be obtained at target–mask spatial separations as large as 2° (Alpern 1953; Breitmeyer *et al.* 1981*b*; Growney *et al.* 1977).

Since the above studies indicate that the effects of variations of stimulus size, target–mask spatial separation and retinal locus on metacontrast magnitude mutually interact, it would be desirable in future investigations of these spatial variables to study them extensively and in detail using the same experimental design, method, and observers. Carefully measured results from such an intraexperimental comparison of these spatial factors may clear up some of the ambiguities and contradictions found when comparing the extant interexperimental results.

4.3.4. Viewing conditions: monoptic, dichoptic, and cyclopean

Although there is little controversy about the existence of monoptic metacontrast, the results of investigations of dichoptic metacontrast are somewhat equivocal. We noted in Chapter 1 that although Stigler's (1910) original investigation of metacontrast failed to show dichoptic masking effects, his later investigation (Stigler 1926) yielded dichoptic metacontrast. Although some subsequent studies (Alpern 1953) also reported an absence of dichoptic metacontrast, a majority of later studies did report dichoptically obtained type B metacontrast effects (Breitmeyer and Kersey 1981; Kolers and Rosner 1960; May, Grannis and Porter 1980; Schiller and Smith 1968; Weisstein 1971; Werner 1940). These discrepancies may in part be explainable in terms of interexperimental differences of stimulus size or location alluded to above. For instance, one may fail to obtain dichoptic, type B metacontrast if a relatively large foveal target is used and the mask contours, due to inaccurate interocular alignment, are separated from the target's contours (Bridgeman and Leff 1978; Kolers and Rosner 1960; Szoc 1973).

The above investigations reporting the existence of dichoptic metacontrast, and in particular, the fact that dichoptic metacontrast can be as strong or even stronger than monoptic metacontrast (Schiller and Smith 1968; Weisstein 1971) indicate (i) that the site of target–mask interactions responsible for generating type B metacontrast effects must be at or beyond the level of binocular combination of the respective monocular inputs; and (ii) that binocular rivalry between target and mask may be an additional source of target suppression under dichoptic as compared with monoptic viewing (Schiller and Smith 1968).

On the face of these results and interpretations, one may want to infer that type B metacontrast interactions also occur in Cyclopean or stereo-space when one uses random-dot stereograms to generate the target and adjacent mask patterns. That this is not the case has been shown by Vernoy's (1976), Lehmkuhle and Fox's (1980), and my own* (unpub-

---

* While I was a research associate during the year of 1973–74 with Dr Bela Julesz at Bell Laboratories, Murray Hill, New Jersey, I conducted an extensive set of experiments on backward lateral masking using Cyclopean stimuli consisting of a central vertical rectangle flanked by two vertical mask rectangles. Orthogonally varying such parameters as (i) lateral spatial separation between target and mask stimuli, (ii) size of target and mask stimuli, (iii) disparity (crossed or uncrossed) of both the target and mask stimuli relative to the Cyclopean background plane, and (iv) disparity or depth of the Cyclopean target relative to the mask, uniformly yielded type A backward masking functions—results entirely in line with those of the two experiments published by Vernoy (1976) and Lehmkuhle and Fox (1980).

lished) investigations of Cyclopean metacontrast, in all of which only type A backward masking effects were obtained. Consequently, a seemingly necessary condition for obtaining type B metacontrast is that one uses standard target and mask stimuli, each one of which is potentially capable of setting up its own non-Cyclopean, monocular contrast- and contour-forming operations in the visual pathway, although eventually they may also contribute to stereo-effects (Szoc 1973). These combined findings of dichoptic type B metacontrast when using standard stimuli, and the failure to obtain such metacontrast when using Cyclopean stimuli, indicate that binocular, cortical mechanisms sensitive to luminance contrast are used in generating type B metacontrast. Mechanisms sensitive to mere Cyclopean depth-contrast (devoid of luminance contrast) are not used in type B metacontrast, although they are in type A backward masking. Consequently, mechanisms giving rise to type B metacontrast, although located at the level of binocular combination probably are not located at as late a level of cortical processing as that used in generating Cyclopean contours (dependent not on spatial luminance transients but rather on abrupt spatial transitions produced by binocular image disparity or depth differences).

4.3.5. Wavelength variables: chromatic and rod–cone interactions

In the past three decades one of the most extensive and fruitful applications of the metacontrast masking method has been to psychophysical investigations of interactions (or their lack) between chromatic and rod–cone mechanisms. Much of the impetus to this line of research has been given by Alpern and his collaborators (Alpern 1965; Alpern and Rushton 1965; Alpern, Rushton, and Torii 1970*a*, *b*, *c*, *d*) whose results and conclusions have inspired further investigations to be discussed below.

To set the stage for the upcoming discussion of chromatic (cone–cone) and rod–cone interactions, let us digress briefly and first present in abstracted form the rationale and technique developed by Stiles (1939, 1949, 1959) for psychophysically isolating wavelength-specific or what are presently known as π mechanisms.* With the two-colour threshold technique originally employed by Stiles, the threshold of a small, briefly flashed test stimulus of a given wavelength, λ, is measured against a larger monochromatic adapting field of different and variable wavelengths, μ. A prerequisite assumption is that the underlying chromatic or π mechanisms behave independently at threshold. As a consequence, for a given intensity, B_μ of the adapting background at wavelength μ, the differential threshold intensity, T_λ, for the detection of the test flash at wavelength λ is determined by whichever mechanism has the highest sensitivity in the presence of B_μ. Let us follow up the consequences of this procedure under the assumption that either one or else more than one π mechanism is used.

---

* For a more extensive and illustrated account of Stile's rationale and of accompanying experimental results see Marriott (1962).

Case 1: Assuming the use of a single mechanism, T_λ will vary with B_μ in the following manner (see Fig. 4.4(a)). When B_μ is close to zero, T_λ will assume a constant value, namely, absolute threshold, irrespective of what wavelength, μ, is assumed by the adapting background. However, as B_μ increases, one eventually attains a background intensity value at and above which Weber's law holds (i.e., $T_\lambda/B_\mu = c$, a constant) and the function relating log T_λ to log B_μ is linear with a slope of 1.0. If we plot log B_μ on the abscissa and log T_λ on the ordinate, and if we, say, employ four values of μ,

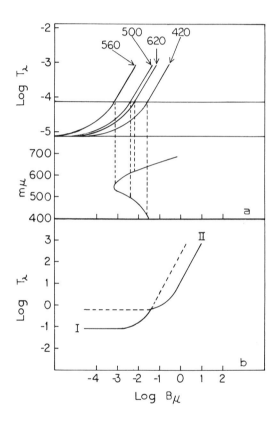

Fig. 4.4.(*a*)   The effect of varying background wavelength, μ, on the test-threshold (T_λ)/background (B_μ) intensity relation for a single—in this case—π_4 or 'green' mechanism. Upper curves show the test-threshold/background intensity functions for background wavelength, $\mu = 420, 500, 560$, and 620 mμ, and for test wavelength $\lambda = 500$ mμ. Lower curve relates the values of B_μ, which raises the test threshold by 1.0 log unit above the π_4 mechanism's absolute threshold level. (*b*) Hypothetical test-threshold/ background intensity function yielded when two mechanisms, I and II, are used. Here mechanism I is more sensitive to a test flash of wavelength λ over log B_μ (background-intensity) varying from -4 to -1.4, above which mechanism II is more sensitive to the same test flash.

e.g. $\mu_1 > \mu_2 \ldots > \mu_4$ then we can generate four such empirical functions, which will converge to the *same* asymptotic threshold value of T_λ for B_μ at or near zero; but when B_μ is sufficiently above zero so that Weber's law holds, these functions will be displaced laterally relative to each other by some fixed interval along the B_μ axis. If we now choose a fixed value (e.g. 1.0 log unit above absolute threshold) along the ordinate, log T_λ, and find the corresponding values of log B_μ which produce this constant value of log T_λ for each of background wavelengths, μ_1 through to μ_4, we can determine the wavelength-dependent response or the spectral sensitivity of the underlying single mechanism by plotting and smoothly extrapolating from log $B_{\mu1}$ to log $B_{\mu4}$ as a function of μ_1 through μ_4. A similar rationale applies when one maintains a constant log background intensity, log B, and wavelength, μ, and varies, instead, the test-flash wavelength, λ. Variations of log T_λ as a function of λ can then be used to construct the desired spectral sensitivity curve of the underlying single π mechanism.

Case 2: When two or more mechanisms, I and II, are used, the following rationale applies (see Fig. 4.4(*b*). Assume that I is more sensitive to the test-flash wavelength, λ, presented against background wavelength, μ, than is II. Consequently, the absolute detection threshold, for B_μ at or near zero, is determined only by I. Suppose further that the wavelength, μ, of the adapting background decreases the sensitivity of the I mechanism at a faster rate than of the II mechanism. Then, as B_μ increases further, at and above some critical value of B_μ, the II mechanism will be more sensitive than the I mechanism and, therefore, will determine the difference threshold. Consequently, over the entire tested range of $B_\mu > 0$, the threshold function is monotically increasing and continuous except for the discontinuity at that value of B_μ where a transition is made from the I to the II mechanism. In other words, the resulting threshold function will look similar to the extrafoveal dark-adaptation curve which is characterized by a discontinuity in sensitivity recovery known as the rod–cone or α-break. For more than two π mechanisms, the above rationale is simply applied iteratively.

Having now set the stage, I shall briefly review Alpern's (1965) rationale for the use of metacontrast in studying chromatic (cone–cone) and rod–cone interactions. On the basis of the above-cited work performed by Stiles (1939, 1949, 1959) on the increment threshold as a function of test flash and background wavelengths and by duCroz and Rushton (1963) on cone dark-adaptation, it was concluded that the π_0 rod mechanism and the π_1 (blue), π_4 (green), and π_5 (red) cone mechanisms by and large behaved independently of each other. For either the increment-threshold technique or the dark-adaptation technique, it was found that for each mechanism the test flash threshold was lowered in proportion to that mechanism's sensitivity to the test flash wavelength, λ, and raised in proportion to its sensitivity to the background wavelength, μ—a result one would predict from the independence of the four mechanisms. Furthermore, Alpern (1953) in his original metacontrast investigation found that the brightness of a test flash which activated only rod mechanisms could be optimally suppressed at an SOA of 50 ms by a flanking mask flash which activated both rod and cone mechanisms. Alpern (1965) entertained the possibility that the latency differences between rods and cones was equal to the optimal SOA for test flash suppression (i.e. 50 ms) and consequently attempted investigating the separate contributions of cone and rod mechanisms in suppressing the test flash brightness.

Using a fixed target–mask SOA of 50 ms, Alpern (1965) found that the visibility of a test flash activating only π_0 or rod mechanisms was suppressed only when the surrounding, after-coming mask flash also activated the π_0 mechanism but not when it activated any of the other cone π mechanisms. On the basis of this result it was concluded that rod and cone mechanisms do not interact in metacontrast. Using a similar rationale and technique Alpern and Rushton (1965) also found independence of the three different cone mechanisms in metacontrast. Both this apparent specificity of cone mechanisms and lack of rod–cone interactions alluded to above has been found in subsequent studies using metacontrast techniques (Alpern *et al.* 1970*a, c, d*) and is consistent with results reported in other studies of increment thresholds measured against flashed or steady backgrounds (Hallett 1969; McKee and Westheimer 1970; Westheimer 1970).

Besides apparently ruling out interactions among the π mechanisms, these results, as a corollary, also rule out the existence of latency differences between rod and cone responses (McDougall 1904*a*) or between different types of cone mechanisms (Mollon and Krauskopf 1973) as an explanation of type B metacontrast functions which peak at some positive SOA value, for example, the 50-ms value found and employed by Alpern (1953, 1965). However, we shall see below that, since the time they were reported, these negative results have been contradicted and superseded by the results of other, later investigations of interactions among π mechanisms.

Despite the earlier failure to obtain psychophysical indices of interactions among π mechanisms, several more recent studies using a variety of investigative techniques do report, for instance, rod–cone interactions (Barris and Frumkes 1978; Blick and MacLeod 1978; Buck, Peeples, and Makous 1979; Foster 1977; Frumkes, Sekuler, Barris, Reiss, and Chalupa 1973; Frumkes and Temme 1977; Latch and Lennie 1977; Makous and Peeples 1979; Sandberg, Berson, and Effron 1981; Temme and Frumkes 1977; von Grünau 1976). These psychophysical findings also are consistent with the results of several neurophysiological studies which indicate rod–cone signal convergence and interaction including post-receptor levels of neural processing (Andrews and Hammond 1970; Enroth-Cugell, Hertz,

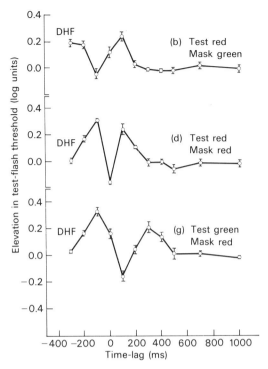

Fig. 4.5. Masking magnitude (test-flash threshold elevation) as function of SOA and the dominant wavelengths of the test and surrounding mask stimuli. In panels *b* and *d*, results are shown when the red test stimulus activated cone mechanisms; the green mask activated rod mechanisms. Panel *g* shows results when the dominant wavelength of the target and mask activated rod and cone mechanisms, respectively. (From Foster (1976).)

and Lennie 1977; Fain 1975; Gouras and Link 1966; Hammond 1968, 1971, 1972; Rodieck and Rushton 1976).

Specifically regarding metacontrast, more recent investigations have found evidence for both rod–cone interactions as well as interactions among different cone mechanisms (Foster 1976; 1978; 1979; Foster and Mason 1977; Yellott and Wandell 1976). Foster (1976; Foster and Mason 1977) reports rod–cone interactions in metacontrast when the test and mask stimuli were presented either monoptically or dichoptically. The last result implicates cortical processing and is consistent with Hammond's (1971, 1972) finding of interaction of rod and cone generated responses in lateral geniculate and cortical visual cells. In Foster's (1976) study, a short-wavelength, green stimulus and a long-wavelength, red stimulus were chosen as either mask or target. The short-wavelength green stimulus isolated the rod mechanism whereas the long-wavelength red stimulus isolated the cone mechanisms. Fig. 4.5 shows the monoptic masking results. Note that both type B paracontrast and metacontrast functions were obtained with this combination of wavelengths. Furthermore, it should be noted that when the test and mask flashes were green and red, respectively (panel g), optimal metacontrast was obtained at an SOA value about 200 ms higher than that obtained when the two flashes were red and green (panel b), respectively. Moreover, the reverse temporal shift seems to occur for paracontrast. Foster (1976) interpreted this shift of optimal SOA in terms of the latency difference between the slower (green-activated) rod and faster (red-activated) cone responses. This indicates, contrary to what might be inferred on the basis of Alpern's (1965) results, that rod–cone interactions and latency differences do play a role in determining the shape of metacontrast masking functions.

In light of Foster's (1976) findings, Alpern's (1965) failure to obtain evidence supporting rod–cone interactions in metacontrast no longer seems so puzzling. Note that in panel g of Fig. 4.5, the green test flash isolated the π_0 or rod mechanism whereas the red mask flash isolates the π_5 or red cone mechanism. This target–mask wavelength relation essentially replicates that of Alpern (1965). Furthermore, note that in this condition optimal metacontrast is obtained at an SOA of 300 ms, far exceeding the fixed 50-ms SOA value employed by Alpern (1965). In fact, if one takes the solid line connecting individual data points as an approximation of masking magnitude, at this shorter SOA value one finds, as did Alpern (1965), close to no metacontrast effect. Moreover, the same argument may hold for the reported failure of finding metacontrast interactions between cone π mechanisms (Alpern and Rushton 1965). However, both conjectures require clearer empirical demonstrations; in particular, ones which are based on a finer sample of SOA values than the rather coarse 100-ms sample interval used by Foster (1976).

Furthermore, the very fact that one can obtain type B metacontrast (paracontrast) functions when the test (mask) and mask (test) flashes activate faster cone and slower rod mechanisms, respectively (panels b and g of Fig. 4.5), indicates that latency differences between more central neural mechanisms must also be used in determining the metacontrast function—a conclusion consistent with the existence of dichoptic metacontrast (paracontrast) functions between rod and cone mechanisms (Foster and Mason 1977). As a matter of fact, to obtain type B metacontrast (paracontrast) when the target (mask) and mask (target) stimuli activate cone and rod mechanisms, respectively, the latency differences between the more central neural mechanisms must be larger and override the oppositely signed latency differences between rod and cone responses. Such an interpretation would dovetail neatly with that of Yellott and Wandell (1976), who on the basis of their monoptic and dichoptic comparisons of chromatic properties of metacontrast arrived at the conclusion that there are two types of metacontrast effects, one which originates in the retina and is receptor specific and the other which originates centrally and, unlike peripheral π mechanisms, is governed by spectral sensitivities that have been sharpened by opponent-process operations.

Moreover, Yellott and Wandell (1976) arrived at the following pertinent observation:

> If a 10 cd/m^2 10 msec red test flash (e.g. a 1° × 3° bar) is followed after about 70 ms by an equally intense 10 ms metacontrast mask (e.g. a pair of flanking bars) of the same color, its brightness is so dramatically reduced that naive observers will normally report that they have not seen any test flash at all. If the red mask is then replaced with a green one of the same luminance, the red test flash is greatly restored in visibility and seems to be hardly masked at all . . .
>
> . . . [These results] indicate that for the brightness reduction phenomenon ordinarily thought of as metacontrast, a mask's effectiveness depends on its subjective color similarity to the test flash . . . (p. 1279).

A similar conclusion was drawn by Holland (1963), Bevan, Jonides, and Collyer (1970) and Simon (1974) in their investigation of visual masking. They also found that the metacontrast effect was weakened as colour differences between the target and mask stimuli were introduced. On the other hand, we noted (Section 1.3) that Stigler (1910, 1913, 1926) found metacontrast to be indifferent to colour differences between the target and mask stimuli.

Based on these contrary findings one cannot maintain that colour similarity is either a necessary or else an unnecessary condition for obtaining metacontrast masking. Subsequent, more recent, investigations have shown that colour effects in metacontrast are much more complex. In the Yellott and Wandell (1976) experiment, for instance, the target and mask flashes were presented as luminance increments as well as hue substitutions against a uniform background. One therefore can charac-terize the two stimuli in terms of their luminance transients or their chromatic transients or both—one transient was confounded with the other. To unconfound these variables, Bowen, Pokorny, and Cacciato (1977) investigated the effects of pure chromatic transients and chromatic plus luminance transients on metacontrast. The target and flanking mask each consisted of a 620 nm, reddish orange flash presented on an achromatic background. In condition 1, the intensity of the stimuli were such that either stimulus introduced both a luminance transient and a chromatic transient relative to the background. In condition 2, the intensity of the two stimuli were set to match the brightness of the background; hence, only chromatic transients relative to the background were used. In condition 3, the target consisted only of a chromatic transient whereas the mask consisted of a luminance plus chromatic transient, and in condition 4 the reverse arrangement was employed. Bowen *et al.* (1977) found that type B metacontrast was obtained only in conditions 1 and 3 in which the mask produced both a brightness and a chromatic transient; in the other two conditions in which the mask produced only chromatic transients relative to the achromatic background no metacontrast effects were obtained. On the basis of these results one can arrive at the following conclusion: despite colour similarity between the target and mask stimuli, metacontrast is not obtained in the absence of brightness transients produced by the mask. As a corollary one could conclude as did Bowen *et al.* (1977) that the presence of brightness transients in the mask is a necessary condition for obtaining type B metacontrast.

However, in view of more recent findings reported by Reeves (1981), even this conclusion has had to be altered. Reeves capitalized on the finding reported by Glass and Sternheim (1973; see Section 2.1.3) that transient substitutions of hues equated for brightness can produce Crawford-type masking by light at and near the instant of hue substitution, provided that the hues are sufficiently far apart in wavelength. When the hues were the same or similar in wavelength, Glass and Sternheim (1973) reported little if any masking. Reeves (1981) used a variety of brightness-equated hue combinations for the target and mask stimuli and the background on which they were exposed. The results he obtained with these combinations of hues are shown in Fig. 4.6.

The upper two panels show metacontrast functions obtained for yellow (Y), white (W), blue (B), and red (R) test stimuli substituted against a brightness-equated green (G) background field. (Reeves (1981) used the term 'test' to refer to *both* the target *and* mask). Note that as the target and mask hue deviates progressively more from the background hue, progressively stronger type B metacontrast is obtained. The same general conclusion holds from inspection of the lower two panels, in which hue of the target and mask is held a constant red and the background field hues are varied. Thus, the failure of Bowen *et al.* (1977) to obtain metacontrast with their combination of orange (620 nm) test stimuli on a white, achromatic surround may have been due to the fact that the coloured test and neutral background hues were not sufficiently distant in chromatic space.

On the basis of these results, we can conclude that type B metacontrast can be obtained by pure chromatic transients provided that the wavelength differences between the target and mask are sufficiently large. This conclusion is given additional support by Foster's (1978, 1979) recent findings that type B metacontrast functions can be obtained when the test and mask flash selectively activate π_1 (blue-sensitive) and π_5 (red-sensitive) mechanisms, respectively. The π_1 and π_5 mechanisms were activated by stimuli with dominant wavelengths at 421 and 664 nm, respectively—sufficiently different in wavelength composition to produce type B metacontrast (Reeves, 1981). In addition, Foster (1979) obtained the following intriguing result. When the π_1 (blue) mechanism was masked by

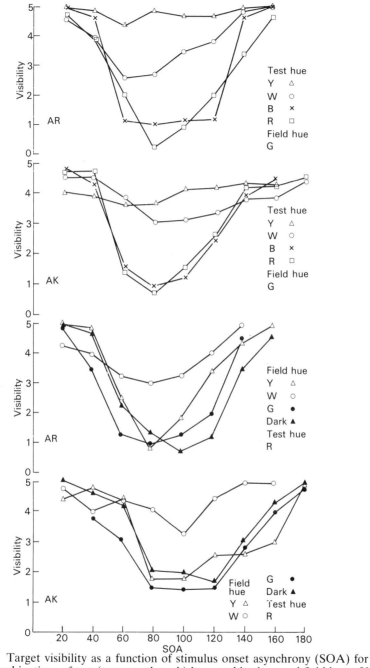

Fig. 4.6. Target visibility as a function of stimulus onset asynchrony (SOA) for several combinations of test (target and mask) hues and background-field hues. Y = Yellow, W = white, B = blue, R = red, and G = green. The lower the target visibility, the greater the metacontrast masking. (From Reeves (1981).)

the π_5 (red) mechanism, the optimal metacontrast masking effect was obtained at an SOA of 100 ms; however, when the π_5 mechanism activated by the target was masked by a similar π_5 mechanism, the peak masking effect occurred at an SOA of 50 ms. This relative shift in SOA at which peak masking occurs would be consistent with the existence of longer response latencies of short-wavelength, π_1 mechanisms and relatively shorter response latencies of long-wavelength, π_5 mechanisms (Mollon and Krauskopf 1973).

The above, seemingly complex effects of wavelength on metacontrast can now be summarized as follows:

1. Rod–cone interactions, contrary to Alpern's (1965) original study, exist in metacontrast (Foster 1976; Foster and Mason 1977).

2. The latency differences between rods and cones can play a role in determining the SOA at which peak, type B metacontrast masking occurs (Foster 1976).

3. Target and mask stimuli of the same or similar wavelength composition can produce type B metacontrast provided that at least the mask stimulus produces a brightness transient (Bowen *et al.* 1977).

4. Even in the absence of brightness transients, type B metacontrast can be produced by pure hue transients provided that the wavelength composition of the target and flanking mask is sufficiently different from that of the background (Reeves, 1981).

5. Result (4), as well as those of Foster (1978, 1979), show, contrary to Alpern and Rushton's (1965) earlier study, that different cone π mechanisms can interact among themselves to produce type B metacontrast.

6. The latency differences between cone mechanisms play a role in determining the SOA at which optimal, type B metacontrast occurs (Foster 1979).

7. There may be a metacontrast effect specific to retinally located, receptor processes and another metacontrast effect, more centrally located and showing spectral sensitivities which have been sharpened and influenced by opponent processes (Reeves, 1981; Yellott and Wandell 1976).

4.4. Variations of the metacontrast theme

The standard metaconstrast technique entails the presentation of a contrast target followed by a spatially surrounding mask. A typical example is the use of a target disk and a surrounding annulus as a mask (see Fig. 4.1(a)). However, other stimulus-presentation techniques not strictly adhering to this procedure can also yield metacontrast effects. Below are listed the major variations of the metacontrast method, and their associated empirical findings are also discussed.

4.4.1. Stroboscopic motion and masking

In Chapter 1, it was noted that Wertheimer (1912), in his extensive study of stroboscopic motion produced by two brief, spatially separated and sequentially flashed stimuli, had already alluded to masking effects which, in hindsight, can be attributed to metacontrast. More recent investigations have confirmed and elaborated on this relationship between stroboscopic motion and metacontrast. Several investigators, among them Toch (1956), Kolers and Rosner (1960), and Fehrer and Raab (1962), noted that the spatiotemporal stimulus conditions which produce metacontrast also are conducive to perceiving stroboscopic motion. In particular, Fehrer (1966) varied target–mask similarity, which is known to be directly related to the degree of beta (or optimal stroboscopic) movement (Kolers and von Grünau 1976; Neff 1936). She found that decreasing the target–mask similarity produced decrements in the magnitude of metacontrast and on that basis inferred a direct relationship between metacontrast and beta motion. Other observable similarities between stroboscopic motion and metacontrast also exist. For instance, both can be obtained dichoptically (see Section 4.3.4 for the metacontrast discussion, and Wertheimer (1912), Verhoeff (1940), Shipley *et al.* (1945),and Anstis and Moulden (1970) for evidence of dichoptic stroboscopic motion), and metacontrast contour suppression as well as a variety of stroboscopic motion known as *phi* motion (pure motion sensation not dependent on figural aspects) can be obtained with stimuli having oppositely signed contrasts (Anstis 1970; Breitmeyer 1978c).

Despite all of these, as well as Fehrer's (1966) and subsequent investigations showing a positive relation between metacontrast and stroboscopic motion (Breitmeyer, Battaglia, and Weber 1976; Breitmeyer *et al.* 1974; Didner and Sperling 1980; Fisicaro, Bernstein, and Narkiewicz 1977; Kahneman 1967b; Thorson, Lange, and Biederman-Thorson 1969), several others have shown that the relationship is neither direct nor simple (Breitmeyer and Horman 1981; Stoper and Banffy 1977; Weisstein and Growney 1969).

Kahneman (1967b) was the first to attempt a theoretical explanation of metacontrast in terms of stroboscopic motion. In his study, it was found that for stimuli spatially separated by approximately 19′, both the magnitudes of metacontrast and of apparent stroboscopic motion varied along the same or highly similar U-shaped functions of SOA. Kahneman (1967b) argued that metacontrast is simply an anomalous or perceptually impossible type of stroboscopic motion. For example, if a square target stimulus is followed by two flanking squares, one to the left, the other to the right, the target visibility is suppressed by the cognitive-perceptual system which cannot effectively accommodate both leftward and rightward motions simultaneously.

This simple and appealing cognitive formulation, however, does not square with several findings. First, Wertheimer (1912) had noted that the perceptual system can accommodate the simultaneous perception of multidirectional stroboscopic motion, a finding replicated by Stoper and Banffy (1977). That is, there is nothing impossible or anomalous about such percepts. Moreover, even when one employs a two-stimulus display to generate clearly perceivable beta motion, one can obtain type B suppression of contour detail or brightness of the leading stimulus by the lagging one (Breitmeyer *et al.* 1976; Breitmeyer and Horman 1981). Consequently, one need not invoke impossible stroboscopic motion, even if it existed, to explain metacontrast.

In addition, on the basis of several investigations (Breitmeyer and Horman 1981; Stoper and Banffy 1977; Weisstein and Growney 1969) one can conclude that stroboscopic motion is neither a sufficient nor a necessary condition for obtaining metacontrast brightness or contrast suppression. Kahneman (1967*b*) used a spatial separation of about 19′ between the first and second stimulus to generate his highly similar metacontrast and stroboscopic motion functions. However, Weisstein and Growney (1969) and Breitmeyer and Horman (1981) employed a wider range of spatial separations in their replication of Kahneman's (1967*b*) investigation. Their findings can be summarized as follows: co-occurrence of metacontrast and stroboscopic motion prevails only at small spatial separations. That is, whereas strong type B beta-motion functions could be obtained at all spatial separations, metacontrast magnitude decreased progressively as spatial separation was increased; at the largest spatial separation (in excess of 3°) little if any metacontrast suppression of brightness was obtained. These results demonstrate that the activation of mechanisms producing stroboscopic motion, particularly at larger spatial separations between stimuli, is not a sufficient condition for producing metacontrast brightness suppression, a conclusion also drawn on a different basis by Stoper and Banffy (1977) in their study on the relation of split apparent motion and metacontrast. Using a bar target and two flanking bars as masks, Stoper and Banffy (1977) also showed that under some conditions metacontrast can be obtained *without* simultaneously producing split or bidirectional stroboscopic motion, thus demonstrating that stroboscopic motion is not a necessary condition for obtaining metacontrast. They conclude that the mechanisms producing metacontrast brightness supression and stroboscopic motion by and large behave independently although they can interact under some favourable stimulus conditions.

However, it should be mentioned that when discrimination of contour detail, rather than brightness or contrast rating, is employed as a response criterion, one can obtain strong type B metacontrast suppression of the first of two stimuli, employed in a stroboscopic motion sequence, at spatial separations at which little if any contrast suppression of the first stimulus occurs (Breitmeyer *et al.* 1974, 1976). This may explain Eriksen and Colegate's (1970) finding showing that no masking of coarser form identification occurs during stroboscopic motion of letters flashed 3.5° apart. The high contrast, black-on-white letters, A, H, and V each had a stroke width of 0.19°. Although contour detail of the first of the two letters may have been suppressed during optimal stroboscopic motion (obtained at SOAs from 60 to 120 ms), its contrast would not be suppressed appreciably (Breitmeyer and Horman 1981; Weisstein and Growney 1969). Thus, despite a loss of sharp contour the first letter should still be identifiable on the basis of its coarser features. One need only defocus one's eye to note that the letters, A, V, and H are quite discriminable despite the loss of sharp contour.

To produce stroboscopic motion, typically two spatially separated and similar stimuli, e.g. two disks or bars, are presented in temporal sequence. Under these conditions one can produce a perception of beta motion (Wertheimer 1912) in which a single stimulus form is seen as moving from the location of the first stimulus to that of the second. An intriguing result reported by Kolers (1963) revealed that although the shape or form of the first of two separated disks can be masked by a surrounding metacontrast ring, it nevertheless can contribute to the perception of stroboscopic motion obtained when the second disk follows the first at optimal SOAs. This result, besides being confirmed, has recently been extended in an

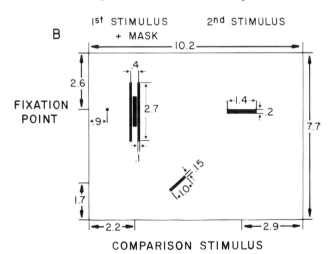

Fig. 4.7. The stimulus display used to generate apparent motion (1st and 2nd stimuli) and masking, here shown on the left as a pair of flanking vertical bars. The mask could also appear flanking the 2nd, horizontal stimulus. A comparison stimulus of desirable orientation could be placed below the left or 1st stimulus, below the right or 2nd stimulus or at positions 1/4, 1/2, or 3/4 of the way between the 1st and 2nd stimuli. (From von Grünau (1981).)

interesting series of experiments by von Grünau (1978, 1979, 1981). Von Grünau (1978) found that stroboscopic motion could be obtained if the form information of either one of two spatially separated disks used to induce (beta) motion was masked by a metacontrast ring. Subsequently he found that no stroboscopic motion is perceived if the form information in both disks is masked by metacontrast rings (von Grünau 1979). From this he concluded that form information is necessary for perception of motion. Below I shall show that this conclusion, like the one claiming that stroboscopic motion is a necessary accompaniment of metacontrast, may hold under some favourable stimulus conditions but need not hold generally. That is to say, one can produce totally formless stroboscopic motion: the variety of motion termed phi movement by Wertheimer (1912).

For the time being let us look more closely at another of von Grünau's (1981) simple, yet ingenious studies. In this study, von Grünau investigated the origin of pattern information of an apparently moving object produced by a two-stimulus sequence. The two motion-inducing stimuli, as shown in Fig. 4.7, consisted of a vertical bar and a rightward displaced horizontal bar. Two flanking mask bars could be presented with either the first, vertical stimulus, as shown, or with the second horizontal stimulus. What von Grünau (1981) found was that the orientation of the perceived moving object was determined by the orientation of whichever of the two stimuli was left unmasked.

Figure 4.8 shows results of an experiment in which the apparent orientation of the perceived moving bar was judged at several locations in its apparent movement from the first, vertical-bar location to the second horizontal-bar location. A comparison stimulus, as shown in Fig. 4.7, whose orientation was adjustable, was used to indicate the apparent orientation of the moving bar at locations either coincident with the left or right stimuli or 1/4, 1/2, or 3/4 of the spatial interval between them. Apparent orientations were judged in the absence of a set of mask bars flanking the first, vertical bar and in its presence. Note from Fig. 4.8 that without a metacontrast mask the perceived bar orientation changed continuously and linearly from vertical to horizontal during the path of perceived motion from left to right, results consistent with similar phenomenal descriptions made by Wertheimer (1912) and with pattern transformations, occurring during stroboscopic motion, reported by Kolers and von Grünau (1976). However, when the first stimulus bar was masked, the perceived orientation of a moving bar was available only at the later, rightward displaced stages of the apparent motion path; moreover, the apparent orientation of the bar at locations where no bar was physically present was judged to be that of the second, unmasked horizontal bar. Complementary results were obtained when the second, horizontal bar was masked instead of the first, vertical one.

From these results and the existence of pure phi motion (Wertheimer 1912; and see below) one can conclude that pattern or form information can but need not contribute to the perception of stroboscopic motion. There may be, as indicated by prior investigations of stroboscopic motion (Anstis 1970; Pantle and Picciano 1976), two systems generating strobosco- pic motion, one pattern or form sensitive, the other insensitive to pattern or form. These may very well correspond to beta and phi motion, respectively. In this regard, Anstis (1970) claims that the perception of phi (formless) movement depends only on brightness or contrast information but not on form.

The existence of stroboscopic motion, e.g. split apparent motion, in some metacontrast situations (Stoper and Banffy 1977), as well as the presence of stroboscopic motion even when the form information of one or the other of two stimuli used to produce the motion percept is suppressed by a flanking metacontrast stimulus, indicate that some of the information of the masked stimulus is immune to the masking effects and can contribute to a perception of motion and thereby to the detection of the (mere, formless) presence of a target stimulus (Fehrer and Raab 1962; Schiller and Smith 1966; see also Section 4.3.1).

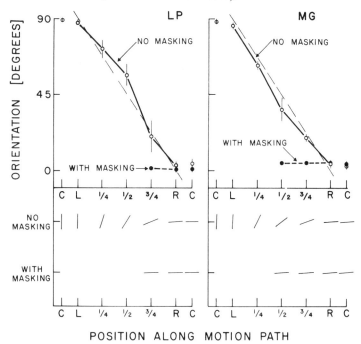

Fig. 4.8. The perceived orientation of a stroboscopically moving bar, at positions along the motion path as indicated by the abscissa, obtained without (open circles, solid line) and with (closed circles, dashed line) metacontrast masking of the first, vertical bar. (From von Grünau (1981).)

4.4.2. Sequential blanking

Another type of masking that is more directly relatable to metacontrast is what Mayzner and co-workers have called sequential blanking (Mayzner and Tresselt 1970; Newark and Mayzner 1973; Tresselt, Mayzner, Schoenberg, and Waxman 1970). In my opinion, these sequential blanking effects are a variety of sequential metacontrast, although the exact sequence-order and variables such as meaningfulness of the stimuli can affect performance (Mayzner and Tresselt 1970; Mewhort, Hearty, and Powell 1978; Newark and Mayzner 1973). The discussion of Piéron's (1935) work in Chapter 1 showed that an *optimal* temporal sequence of spatially adjacent stimuli can successively produce masking of any prior stimulus in the spatiotemporal sequence by its immediately following and spatially adjacent stimulus; only the last stimulus is immune to brightness or form suppression. At suboptimal temporal sequences, either slower or faster ones (i.e., at larger and smaller SOAs separating successive stimuli in the sequence), less sequential masking was observed.

Consequently, since metacontrast is typically optimal at intermediate SOAs of 50–100 ms, it would be reasonable to expect that sequential blanking is also optimal at successive SOAs of about 50–100ms. In an experiment reported by Hearty and Mewhort (1975) such a result was obtained. These investigators used a sequentially presented array of eight letters. The letters were each presented for 5 ms and were presented sequentially at varying ISIs either from left to right or else from right to left. That is to say, the temporal order reflected the left-to-right or else right-to-left spatial order. ISIs of 0, 50, 100, and 200 ms were used, corresponding to SOAs of 5, 55, 105, and 205 ms. Hearty and Mewhort (1975) found that optimal masking of the identity (and therefore also of the location) of the letter E embedded randomly in the spatial eight-letter array was obtained at an SOA of 55 ms. Thus, like meta-contrast using a two-stimulus sequence, multistimulus sequential blanking also was a type B, U-shaped function of SOA. Based on these results, one can reasonably conclude that sequential blanking partakes of a form of sequential metacontrast as originally demonstrated by Piéron (1935).

Moreover, in my own unpublished observations of the sequential blanking effect, when, for instance, employing an array of five adjacent vertical bars presented temporally in sequence from left to right, one can obtain not only sequential brightness and form suppression of the first four bars at an optimal SOA, but also one obtains a perception of rapid formless motion (phi motion) proceeding from the first to the fifth and hence through the fourth bar position. Moreover, as already reported by Wertheimer (1912), by alternately flashing two adjacent bars at optimal rates, I was able to produce pure, formless phi motion for some time, after which pattern or form, in line with the findings of Mewhort *et al.*(1978), was increasingly observable. These results suggest, contrary to von Grünau's (1978) claim, that form *per se* is not necessary to obtain stroboscopic motion. One could, of course, argue that, in the former five-bar sequence, the form of the final fifth bar is not masked and, therefore, is necessary for perceiving motion throughout the first four bar positions. Here, one could use an adaptation of von Grünau's (1978, 1979, 1981) procedure and mask the final bar by two flanking ones to see if one none the less can perceive formless, phi movement.

4.4.3. Metacontrast produced by stimulus offset

In section 4.3.2 it was noted that relative to a constant grey background the contour visibility of targets consisting of both luminance increments (white-on-grey) and decrements (black-on-grey) could be suppressed by an aftercoming flanking mask consisting of either a luminance increment or decrement (Breitmeyer 1978c). This indicates that either an on-transient or an off-transient produced by the mask can inhibit the contour processing of both incremental and decremental luminance targets. Moreover, it suggests the possibility that the offset of a mask may inhibit visibility of a target, a posibility previously entertained by Turvey, Michaels, and Kewley-Port (1974) in their study of pattern masking.

Breitmeyer and Kersey (1981) tested this possibility using the following experimental design. A typical disk–ring metacontrast display was used. The target and mask stimuli were black-on-white and were presented dichoptically. Under one condition, call the left-eye condition, the target was presented to the left eye, the mask to the right eye; in the other, right-eye condition the target and mask were presented to right and left eyes, respectively. The timing of the target and mask presentations deviated from the standard metacontrast procedure as follows. A trial consisted of a 2000-ms presentation of the mask ring, followed by a 50-ms presentation of the target disk, at one of the following mask–target onset asynchronies: 1750, 1810, 1830, 1850, 1870, 1890, 1910, 1930, or 1950 ms. In effect, besides eliminating the transient at the onset of the ring as an effective masking variable, the target *onset*-mask *offset* asynchronies varied from 50 to 250 ms. On any trial the subject was required to rate the apparent contrast of the target by matching it to one of 11 grey comparison stimuli whose contrast varied in 0.1 log unit steps from 0.095 (white) to 0.95 (black). The field containing the 11 comparison stimuli was presented for 2500 ms after the mask–target sequence. Lower contrast ratings corresponded to greater masking. The results of the experiment are shown in Fig. 4.9, and are plotted separately for the left-eye (open circles) and right-eye (closed circles) conditions. The average of the left- and right-eye conditions is also plotted (half-closed circles). For both subjects the contrast rating of the target varied as a U-shaped function of target

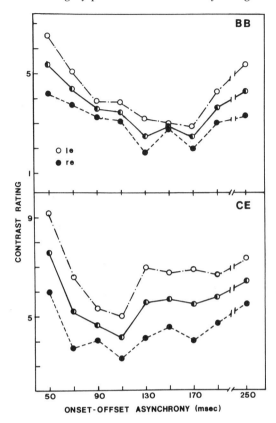

Fig. 4.9. Contrast ratings, in dichoptic metacontrast, of the target as a function of target onset-mask offset asynchrony and the eye to which the target was presented (open circles, left eye; closed circles, right eye) and averaged across eyes (half-closed–half-open circles). (From Breitmeyer and Kersey (1981).)

onset-mask offset asynchrony. The left-eye condition yielded less masking than the right-eye condition. Both subjects were left-eye dominant, therefore this difference may be related to eye dominance which controls the magnitude of either or both of the target and mask responses. Similar eye-dominance effects in metacontrast have been reported by Michaels and Turvey (1979).

These findings suggest that stimulus offset can be used as an effective metacontrast mask. We shall see subsequently (Section 7.4.2, Explanation 26) how this result may explain the fact that visible persistence is shorter for lower spatial frequency gratings than for higher frequency ones and for longer, relative to shorter, stimulus durations (see Sections 3.3.2 and 3.5.2 above).

4.4.4. The single-transient masking paradigm

In typical metacontrast studies both the target and the mask are presented briefly and transiently. Consequently these studies are examples of what Matin (1975) has called the two-transient masking paradigm. However, it is questionable whether or not the two-transient paradigm is a necessary condition for obtaining masking. Recently, Breitmeyer and Rudd (1981) investigated the possibility of obtaining suppression of the visibility of a target having a sustained or prolonged duration by a brief, transient-duration mask. The design of the experiment entailed a 10-s presentation of a black-on-white bar flanked symmetrically at spatial separations of 0.3°, 0.6°, 1.2°, or 2.4° by two mask bars flashed for 50 ms at a target–mask onset asynchrony of 2 s. At such a large SOA (2000 ms) one would not expect onset–onset response interactions between the target and the mask typically entailed in the production of type B metacontrast. If any suppression were to occur, it would have to be of the sustained neural response to the prolonged target by the transient response to the brief mask. The suppression of the target's visibility was monitored as follows. If the onset of the flanking masks produced a phenomenal disappearance of the target, the subject pressed a button as soon as he or she noticed the disappearance. This started a millisecond clock counter which ran until the subject indicated the reappearance of the target by releasing the button. In this way several measurements were obtained from which an average suppression duration was calculated. The results of Breitmeyer and Rudd's (1981) study are shown for two subjects in Fig. 4.10. The target–mask display, as indicated by the different symbols of the figure, could be centred at one of three eccentricities relative to the fovea: 1.7°, 6.0°, and 10.3°. For both subjects no suppression of target visibility was obtained at any target–mask spatial separation at the smallest eccentricity. The duration of suppression of the target increased progressively at higher eccentricities (at greater distances from the fovea) and at smaller target–mask spatial separations. Both trends are consistent with similar variations in the magnitude of the typical metacontrast function obtained when viewing eccentricity or target–mask spatial separation are systematically varied (see Section 4.3.3). It thus appears that the two-transient paradigm is not a necessary condition for obtaining masking effects, but rather that a single transient elicited by the mask is sufficient.

4.5. Pattern masking by noise and structure masks

Generally noise masks yield strong type A forward and backward masking effects under monoptic viewing conditions (Kinsbourne and Warrington 1962a, b; Schiller 1966; Schiller and Smith 1965). In the monoptic condition, the forward masking effects generally are stronger and extend over a longer temporal interval than backward masking effects (Kins-

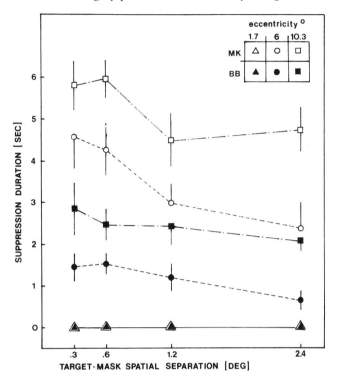

Fig. 4.10. The duration, in seconds, over which the visibility of the sustained target bar was suppressed after presentation of a 50 ms set of flanking mask bars as a function of target–mask spatial separation and viewing eccentricity. Vertical bars attached to data points indicate 1 SD. (From Breitmeyer and Rudd (1981).)

bourne and Warrington 1962*b*, Scharf and Lefton 1970; Schiller 1966; Schiller and Smith 1965; Smith and Schiller 1966; but see Section 3.6 for exceptions). Moreover, noise masking tends to be stronger at higher mask intensities and when the relative spatial overlap of the noise mask with the target increases (Schiller 1966). When the target and mask stimuli are presented dichoptically, one still obtains type A forward and backward masking effects; however, compared with the backward effect, the forward effect is relatively weak and extends over a smaller temporal interval (Greenspoon and Eriksen 1968; Smith and Schiller 1966; Turvey 1973). Moreover, in dichoptic backward noise masking, Turvey (1973) reported that the intensity of the mask does not seem to be an important masking parameter, although Monahan and Steronko (1977) recently reported that target and mask intensity both are important parameters. It should be mentioned, however, that Monahan and Steronko (1977) selected their subjects so that equal masking results were obtained with the target stimulus (a letter) presented to either eye. This procedure presumably eliminated eye-dominance differences between the left and right eyes which indicate, in line with the results reported by Breitmeyer and Kersey (1981; see Section 4.4.3) and Michaels and Turvey (1979), that eye dominance may play an important role in dichoptic pattern masking.

Masking by structure differs procedurally from masking by noise in that the mask is structurally related to the target pattern (compare Fig. 4.1(b) and (c)); the mask shares with the target figural features like orientation or curvature (Gilinsky 1967, 1968; Houlihan and Sekuler 1968; Sekuler 1965; Uttal 1973), thus producing spatial proximity of similar contours. Under these conditions, particularly at target to mask (T/M) energy ratios less than one, strong type A forward and backward masking effects can be obtained monoptically and dichoptically (Gilinsky 1967, 1968; Greenspoon and Eriksen 1968; Scharf and Lefton 1970; Sekuler 1965; Taylor and Chabot 1978). However, whereas monoptic or dichoptic forward masking effects generally are of the type A, monoptic backward structure masking effects also can be of type B (Bachmann and Allik 1976; Herrick 1974; Weisstein 1971) especially when the T/M energy ratio is at or above unity (Hellige, Walsh, Lawrence, and Prasse 1979; Michaels and Turvey 1973, 1979; Purcell and Stewart 1970; Spencer and Shuntich 1970; Turvey 1973) or when the stimuli are presented extrafoveally (Purcell, Stewart, Davis, Huntermark, Robbins, Rowland, and Salley 1975). Under dichoptic, foveal or extrafoveal, viewing, type B backward masking functions can be obtained at T/M ratios below as well as above unity (Michaels and Turvey 1979), indicating that under monoptic, especially foveal, viewing, the transition from type B to type A functions as T/M ratio decreases from unity is due to common integration of mask and target luminance or contrast information at peripheral levels (Hellige, Walsh, Lawrence, and Cox 1977; Turvey 1973). For instance, as with metacontrast (see Fig. 4.3), Hellige *et al.* (1979) and Michaels and Turvey (1979) found that the backward structure-masking function tended to shift from a type B to a type A function as the T/M energy ratio changed from 2/1 to 1/2. This shift, as shown in Fig. 4.11, taken from Hellige *et al.* (1979), was more pronounced for masks whose figural features were different from those of the target. The similarity between type B backward masking functions, produced, particularly, when the overlapping target and mask share many figural features (contour proximity) and when standard non-overlapping metacontrast stimuli are used, suggests the hypothesis that both type B effects result from interactions among correspondingly similar underlying mechanisms selectively sensitive to the same figural or contour information.

Moreover, transient overshoots at on- and offsets of a prolonged structure mask, similar to those obtained with a uniform light flash (Sections 2.1 and 2.2) also have been reported (Matin 1974*a*; Mitov, Vassilev, and Manahilov 1981). Mitov *et al.* (1981) used the following adaptation of Green's (1981*b*, see Section 2.2) study of the masking of gratings by a prolonged, uniform flash of light. In their study, Mitov *et al.* employed sinusoidal gratings as masks (500 ms) as well as targets (20 ms). As long as the spatial frequencies of both target and mask gratings were

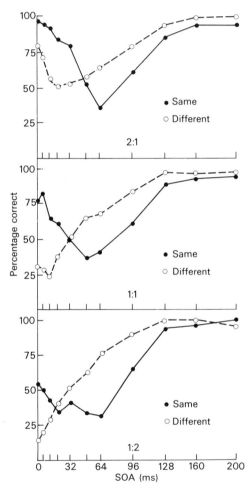

Fig. 4.11. Percentage of correct target recognition as a function of stimulus onset asynchrony (SOA). The results shown with solid circles and lines show masking functions obtained when the target and mask had the same or very similar features; the results shown with open circles and dashed lines show masking functions when the target and mask were composed of different features. Upper, middle, and lower panels show results at target:mask energy ratios of 2/1, 1/1, and 1/2, respectively. (From Hellige, Walsh, Lawrence, and Prasse (1979).)

below 6c/deg (but not necessarily equal to each other), pronounced transient masking overshoots were obtained at and near the on- and offsets of the mask flash. However, when the target and mask consisted of an 18c/deg and 6c/deg grating, respectively, only weak, attenuated transient overshoots were obtained. Moreover, when the latter target and mask spatial frequencies were reversed, no transient overshoots occurred; only a sustained masking effect (on the 6c/deg test grating for the duration of the 18c/deg mask grating) was produced. The former findings of transient overshoots when low spatial frequency masks were employed may be related to the results of Matthews (1971) and Teller *et al.* (1971) discussed in Section 2.1.5, showing that transient overshoots at on- and offsets of the mask occur only if the diameter of a uniform-light mask exceeds that of the spatial excitation pools of rod- and cone-driven processes.

Backward pattern and noise masking also have differential effects on determining the location as compared with the identity of a target. We already discussed such a difference between metacontrast suppression of target location and target contrast or identity information (see Section 4.3.1). Psychophysical work by several investigators (Finkel 1973; Finkel and Smythe, 1973; Mewhort and Campbell 1978; Schiller 1965*a*) indicates that there are separate, post-offset representations of stimulus-location and identity information which, furthermore, are processed by functionally or anatomically distinct systems (Dick and Dick 1969; see also below: Explanation 22, Section 7.4.2; and Sections 8.3.3, 10.2, and 10.3). Based on these results Smythe and Finkel (1974) showed that a backward noise mask suppresses target-identity formation over a longer temporal interval than it does target-location information. This indicates that post-offset location information persists for a shorter interval and is processed sooner than identity information. My own unpublished experimental findings confirm this finding. In my experiment, a random-dot pattern (Julesz 1971) served as a mask; and four capital letters randomly chosen from a set of 12 and arranged randomly at any four of 12 clock positions around an imaginary circle concentric with the fixation point, served as targets. Target and mask were flashed for 15 and 50 ms, respectively, with the target–mask SOA ranging, in equal 25-ms steps, from 15 to 315 ms. The proportion of errors made by observers in reporting location or else identity information (the two report-conditions were tested in separate subexperiments) was taken as a measure of masking magnitude. The results showed that both location and identity information were masked equally well at an SOA of 15 ms. However, although both functions declined in type A fashion as a function of mask-onset delay, the location-masking function declined at a significantly faster rate that the identity-masking function. Whereas no location masking was obtained at SOAs beyond 90 ms, identity masking was obtained up to an SOA of 240 ms.

4.6. Summary

Pattern masking can be differentiated operationally according to several procedures. When the target and mask patterns share common structural features and overlap spatially, *masking by structure* prevails. A mask consisting of random elements with no obvious structural relation to features of the target pattern is employed in *masking by noise*. Non-overlapping, spatially adjacent, mask and target patterns are used to investigate (forward) *paracontrast* and (backward) *metacontrast* forms of lateral masking.

In para- and metacontrast the type of function one obtains depends on the response criterion adopted by an observer and on several stimulus variables. Paracontrast is typically of type B when brightness ratings or matches are used and type A when simple detection is employed. With either criterion, the magnitude decreases as mask intensity decreases and as spatial separation between the target and mask stimuli increases. Type B paracontrast can also be obtained dichoptically, implicating mechanisms at or beyond binocular levels of visual processing.

Metacontrast yields type B functions when brightness, contrast, or contour provide the criterion content. Simple detection or reaction time tend to yield no masking—indicating that some target information is immune to the suppressing effects of the mask—or type A masking if the target/mask (T/M) energy ratio is significantly less than unity. Without corresponding shifts in criterion content, type B metacontrast shifts to a type A function as T/M energy decreases from a value above 1.0 to values below 1.0. For T/M energy ratios greater than or equal to 1.0, the SOA at which optimal type B metacontrast occurs shifts to lower values as either stimulus or background luminance decreases. Type B metacontrast suppression of contour is indifferent to the sign of contrast of the target relative to that of the mask; moreover, metacontrast can be produced by the offset of a prolonged mask, as well as by the typically employed briefly flashed mask. Besides these contrast and intensity variables, spatial variables also affect metacontrast. The magnitude of metacontrast decreases as the orientation- or contour-difference between target and mask increases and as the spatial separation between target and mask stimuli increases. Generally its magnitude also is weakest at the fovea and increases with retinal eccentricity. Stimulus size is an important variable which interacts with retinal location; smaller and larger stimuli are more effective foveally, and extrafoveally, respectively.

Interocular, dichoptic presentation of target and mask stimuli yield type B metacontrast which can be of greater magnitude than monocular presentations, indicating, for one, interaction of target and mask responses at or beyond the level of binocular combination and, for another, the possible additional contribution of binocular rivalry. At stereoscopic or

Cyclopean levels of visual processing, which bypass monocular pattern processing, metacontrast yields only type A functions, indicating that the mechanism responsible for type B metacontrast requires standard stimuli composed of luminance contrast or chromatic differences relative to a uniform background and resides at prestereoptic binocular levels of processing.

Although initial studies failed to yield evidence for metacontrast when target and mask stimuli selectively activated different wavelength or π mechanisms, more recent studies have found that both isolated rod–cone interactions as well as interactions between different isolated cone mechanisms can be obtained. Moreover, transients produced by stimuli comprised of hue substitution against brightness fields also are capable of producing type B metacontrast. Masking phenomena associated with stroboscopic motion, sequential blanking, and single transient stimulation can also be related to metacontrast masking effects.

Monocular pattern masking by noise or structure is typically a type A effect for target:mask (T/M) energy ratios less than 1.0. This holds for forward as well as backward masking, although the forward masking effect usually is effective over a larger time interval than backward masking. When the mask energy or contrast is low relative to that of target (T/M>1.0), a backward foveal type B function can result with monocular masking, particularly with masking by structure. Moreover, type B effects are also obtainable extrafoveally with equal energy stimuli. This shows that the type B effect obeys similar target-energy and retinal-location relationships as does metacontrast. Dichoptic masking by noise usually also produces type A effects; however, in contrast to the monoptic effects, here backward masking is usually stronger than forward masking. Furthermore, a backward, type B, dichoptic structure masking effect also can be obtained for equal-energy target and mask stimuli, indicating that this type B effect, like the metacontrast effect, includes central, cortical levels of processing. Additionally, the attenuation of monocular type A effects with relatively weak masks or under dichoptic viewing shows that it is due to energy-dependent integration of target and mask pattern information in common peripheral pattern processing channels.

A briefly flashed target pattern informs the visual system not only of its figural or structural aspects but also of its location in the visual field. Masking by noise and structure affects location and figural identification differentially. The temporal interval over which one can obtain masking of target location information is significantly shorter than the interval over which one can obtain masking of pattern information. This result suggests that channels signalling the location of a stimulus integrate information over a shorter interval of time than do channels carrying pattern information.

5 Models and mechanisms of lateral masking: a selective review

The previous three chapters reviewed most of the significant findings, and, when appropriate, briefly mentioned some of the theories of masking by light, visual persistence, and masking by pattern. This chapter deals in greater detail with models of lateral masking, i.e. paracontrast and metacontrast; and it will become apparent how the topics discussed in the prior chapters bear importantly on our evaluation of mechanisms proposed to explain lateral masking. This evaluation is intentionally selective in that only the most salient features are discussed. The empirical findings on lateral masking on the basis of which these evaluations potentially could be made are extremely extensive. A limited set of these findings was reviewed in Chapter 4, and some of these limited yet important results provide the empirical context in which the proposed models and mechanisms of lateral masking will be reviewed and compared in this chapter.

Ever since paracontrast, metacontrast, and cognate phenomena were initially discovered toward the turn of the century, investigators have speculated about their possible underlying mechanisms. For instance, in Chapter 1 it was noted that McDougall (1904a) proposed at least one possible mechanism for the suppression of Bidwell's ghost or the Purkinje image. In this particular experiment Charpentier bands and the trailing Bidwell's ghost were produced by a rotating transilluminated aperture which presumably isolated rod responses. When a second spatially displaced and temporally lagging aperture which activated cones was added to the display, the visibility of Bidwell's ghost normally produced by the leading aperture was suppressed. McDougall reasoned that the faster cone response to the temporally lagging aperture suppressed the slower rod activity and thus the Bidwell's ghost generated by the temporally leading aperture. McDougall (1904a) further noted that although cone–rod interactions may provide one possible mechanism of suppressing Bidwell's ghost another one based on cone–cone interactions, most likely also exists; for he demonstrated suppression of these images when both apertures presumably isolated the same type of cone activity.

Moreover, based on Exner's (1898) prior conjecture, Stigler (1910, 1926) in his pioneering work on lateral masking proposed that horizontal cells in the retina provided the basis for lateral inhibitory interactions between the neural responses (homophotic and metaphotic images) of the spatially adjacent stimuli typically used in his para- and metacontrast investigations. It should be noted that in his first series of investigations,

Stigler (1910) failed to demonstrate dichoptic metacontrast effects, and this failure may have motivated his placing the interactive mechanisms at a peripheral, retinal level. However, in a subsequent investigation (Stigler 1926) he demonstrated the existence of dichoptic metacontrast. This result, although not ruling out the interactions mediated by horizontal cells as a mechanism underlying metacontrast, certainly pointed to the existence of additional lateral interactions mediated at post-retinal levels of visual processing and to two additional properties, namely, synaptic delay and the associated overtake of the target response by that of the mask, respectively. A version of this overtake hypothesis was also adopted later by Crawford (1947) to explain target masking prior to the conditioning or mask-flash onset. Moreover, in line with Stigler's (1926) assumption of post-retinal interactions, Fry (1934) pointed out the possibility that in metacontrast the lateral effects of a temporally lagging flash on a leading, spatially adjacent one may be mediated by processes along the entire retinogeniculocortical pathway.

In the last three decades several additional models and mechanisms of para- and metacontrast have been proposed. A few of these mechanisms either explicitly or implicitly share features with those proposed by McDougall (1904a) and Stigler (1910). Others were already mentioned in Chapter 4, in the context of the empirical review of para- and metacontrast findings and methods. The specific models and mechanisms to be reviewed can be classified according to four general types:

1. Models based on fast inhibition and slow excitation.

2. Spatiotemporal sequence models.

3. Two-process models (based on interactive response persistence or on temporal integration and segregation).

4. Distributed network models.

These types are not mutually exclusive; for instance, Weisstein's (1968, 1972; Weisstein *et al.* 1975) model of lateral masking could be classified both under type 1 and type 4 models, since it is based on a neural network incorporating what in my opinion seem to be the more telling fast inhibitory and slow excitatory processes. Therefore, the typology proposed here, although somewhat arbitrary, is based on what may be regarded the most distinctive characteristics of a model or mechanism.

5.1. Fast inhibition and slow excitation

One of the most salient features of metacontrast is that, under favourable stimulus conditions such as equal stimulus energy, the brightness or contour visibility of the target stimulus is optimally suppressed not at an SOA of 0 ms but rather when the mask follows the target by some 50–100 ms. Several mechanisms that have been proposed to account for this feature rely on the existence of a long latency, slow excitatory activity generated by the target and a short latency, fast inhibitory activity generated by the mask. These two activities are assumed to interact across space and it is also assumed that the interaction is optimal when they overlap maximally in time. Since there is a temporal delay between the fast inhibitory component of the mask and the slow excitatory component of the target, optimal interaction can occur only when, relative to the target, the mask is delayed by an interval equivalent to the latency difference between these two respective response components. If this interval is roughly 50–100 ms, the obtained, type B metacontrast function is characterized by little masking at target–mask onset synchrony (SOA = 0 ms), a progressive increase of masking as SOA increases until optimal masking is attained at an SOA of 50–100 ms, followed by a progressive decrease of masking as SOA increases further.

5.1.1. Alpern's retinal interaction mechanisms

Alpern (1953) proposed a modification of Crawford's (1947) *overtake hypothesis*. As mentioned, a version of this hypothesis originated with Stigler (1926) and was offered by Crawford to explain backward masking obtained in his luminance masking investigation. In this study the conditioning or mask flash generally was of a higher intensity than the threshold test–flash luminance which was used to index the response evoked by the mask flash. Since visual latency is inversely related to intensity the more intense mask flash can overtake a weaker test flash as both are being transmitted and processed along the visual pathway. Hence, the response to the delayed mask can still catch up with that of the target and suppress it. However, Chapter 4 showed that type B metacontrast can be obtained with masks whose energy is equal to or less than that of the target. Consequently, Crawford's version of the overtake hypothesis, although it may apply to masking by intense flashes of light, cannot apply to metacontrast.

Alpern's (1953) modification of the overtake hypothesis was to retain its logic but to restrict the mechanism to the retinal rather than to later stages in the visual pathway. Alpern proposed that a mask's fast inhibitory component, thought to be photochemical in nature, interacts with a slower excitatory component, believed to be of neural origin. Alpern's proposal is based on the prior finding (Fry and Alpern 1946) that the receptor response to a flash of light could be divided into a fast photochemical reaction which is almost instantaneous and a secondary (thermal) reaction with a measurable latent period. If one assumes that the effects of the secondary reaction activated by the target are suppressed at the point at which the neural response begins to be generated by the primary, photochemical reaction in adjacent receptors activated by the mask, type B metacontrast could result. Although this may be a workable mechanism, it cannot be the sole or most significant source of metacontrast, since strong metacontrast suppression, as noted in Chapter 4, can be obtained dichoptically and thus post-retinally. Moreover, to explain type B paracontrast, an extension of Alpern's proposed retinal mechanisms would require that a *slower neural inhibitory* component of the preceding mask interact laterally with a *faster excitatory photochemical* component. Since the distinction between target and mask is methodological but not physiological, it is hard to imagine why such a differential response valence and latency should occur as a function of arbitrarily designating one of the stimuli as target, the other as mask. On the other hand, as will be seen below, in several reported cases, activation of rod mechanisms by a preceding mask interacting with cone activity of a following target can contribute to the temporal characteristics of type B paracontrast.

Subsequently, Alpern (1965) proposed the following explanation of metacontrast. His earlier investigation of metacontrast (Alpern 1953) showed that optimal suppression of target brightness occurred when the spatially adjacent mask followed the target by approximately 50 ms. In one of the experiments the optimal target suppression occurred when the target flash excited only rods and the mask flash excited only cones. Alpern (1965) reasoned that the rod response latency might be sufficiently longer than the cone response latency so that the two receptor activities arise and interact optimally in time when the mask onset follows that of the target by about 50 ms. The interaction was hypothesized to be a suppression of the slower rod response to the target by the faster cone response to the mask. Alpern (1965) failed to find evidence of rod–cone interaction—a result which seemingly eliminated the role of such an interaction in metacontrast. Moreover, this particular explanation cannot account for type B paracontrast, unless one additionally assumes that here the mask activates the slower rod processes and the target the faster cone processes. As mentioned, under special situations this may be the case, but one would prefer a general model which can account for both type B para- and metacontrast when the target and mask activate the same as well as different chromatic mechanisms.

Besides the existence of rod–cone response latency differences, there is evidence that short-wavelength (blue) cone mechanisms are characterized by a longer response latency than long-wavelength (red) cones (Mollon and Krauskopf 1973). In their metacontrast study, Alpern and Rushton (1965) also failed to find evidence for interactions between these cone responses. This result apparently also eliminated the role of such cone interactions in metacontrast. However, we saw in Section 4.3.5, that more recent investigations (Foster 1976,1979), sampling a more extensive range of SOA values, found evidence for reciprocal interactions between both rod and cone mechanisms as well as blue- and red-sensitive cone mechanisms.

Although the latency differences and retinal interactions between rods and cones or between short- and long-wavelength cones can play a role in determining the optimal SOA at which suppression of target visibility occurs (Foster 19776,1979; see also Fig. 4.5 and Section 4.3.5), they none the less are not the sole source of metacontrast interactions. The reasons for this are as follows. For one, dichoptic type B metacontract can be obtained when the target isolates rod responses in one retina and the mask isolates cone responses in the other (Foster and Mason 1977). This implies that post-retinal, neural processes differentially sensitive to wavelength must also be used, in agreement with conclusions reached by Yellott and Wandell (1976). Moreover, if it is assumed that under monocular stimulation the latency differences between rod and cone responses are the sole source of type B metacontrast (the slower rod response to the target is suppressed by the faster cone response to the mask), then in the reverse situation in which the target activated cones and the mask activates rods, one would expect that optimal suppression of the target's cone response by the mask's rod response occurs only if the mask now leads the target; that is, when paracontrast prevails. However, although type B paracontrast effects are obtained under this condition, type B metacontrast is not eliminated; the SOA at which optimal metacontrast occurs is merely shifted to smaller (positive) values, whereas that at which optimal paracontrast occurs seems to be shifted to larger (negative) values relative to the condition in which the target and mask activate rods and cones, respectively (compare panels b and g of Fig. 4.5).

By a converse argument, if receptor processes alone were used, the type B paracontrast effect obtained when the target and mask isolate cone and rod mechanisms, respectively, ought to be eliminated when the two stimuli instead isolate rod and cone mechanisms, respectively. However, again, this is not the obtained result. Type B paracontrast is found under both the former and latter stimulus arrangement (see again, panels b and g of Fig. 4.5). If one were to rely on some form of latency difference between target and mask activity to explain metacontrast or paracontrast, the above arguments imply that in addition to rod–cone latency differences, both target and mask stimuli must also be able to activate a long latency *and* short latency post-receptor, neural response. Otherwise, type B para- and metacontrast, contrary to what in fact was found by Foster (1976), would be eliminated. The fact that, in Foster's (1976) experiment, type B para- and metacontrast were not eliminated when the target and mask activate rods and cones or else cones and rods, respectively, also implies that the latency differences between the post-receptor, neural responses usually is larger than that between rod and cone responses. This explanation, of course, assumes that the receptor and neural latency effects combine additively to determine the SOA at which optimal para- and metacontrast occurs.

Foster's (1979) investigation of metacontrast between blue- and red-sensitive mechanisms supports this additivity assumption. When the target and mask both consisted of either a long-wavelength flash (664 nm) or else a short-wavelength (421 nm) flash to which only the red-sensitive or else blue-sensitive cone mechanisms responded, optimal type B metacontrast was obtained at an SOA of 50 ms. Since in these experiments the target and mask activated the same cone mechanisms, receptor latency differences were eliminated. Hence, the type B metacontrast entailed interactions at post-receptor levels between slow (target) and fast (mask) neural responses. Presumably the latency differences between the fast neural response to the mask and the slow one to the target was 50 ms. In another condition a combination of a short-wavelength (421 nm) target flash and long-wavelength (644 nm) mask flash was employed. Here the optimal metacontrast SOA was 100 ms, 50 ms longer than that obtained in the prior condition. If we assume that the post-receptor, neural latency difference is again 50 ms, the relative SOA shift of 50 ms is due to the latency difference between the faster long-wavelength cone response and the slower short-wavelength cone response. If this is in fact the case, then using a long-wavelength target and a short-wavelength mask ought to shift the optimal SOA, relative to the 50 ms optimal SOA obtained in the first condition, to 0 ms. In other words, a type A metacontrast function should be produced.

This expected result was obtained by Foster (1979) in a third experimental condition in which a 664-nm target and a 421-nm mask were used. A related shift, shown in Fig. 4.5, was found by Foster (1976) when selectively activating rod and cone mechanisms in para- and metacontrast. When the test and mask stimuli activated the rod and cone mechanisms, respectively (panel g), the optimal paracontrast and metacontrast effects were obtained at SOA's of -100 and $+300$ ms. However, when test and mask stimuli activated cone and rod mechanisms, respectively (panel b), the optimal paracontrast and metacontrast effects were obtained at SOAs of -300 and $+100$ ms. That is to say, relative to the former condition, in the latter one, both para- and metacontrast peaks shifted to lower values by a constant 200 ms. We can infer from this finding, that the slower rod and faster, long-wavelength cone mechanisms were characterized effectively by a constant latency difference of 200 ms, a value four times as large as the value assumed by Alpern (1965). As mentioned in Section 4.3.5, this may be one of the possible reasons explaining Alpern's (1965) failure to obtain evidence for rod–cone interactions.

In summary, latency differences between receptor (rod–cone or cone–cone) as well as post-receptor, neural responses may be concluded to be used in determining the SOA value yielding the peak or optimum of the paracontrast and metacontrast function. Despite the importance of these wavelength effects, their investigation in metacontrast comprises a fairly restricted subset of all of the combined studies of lateral masking. In a majority of lateral masking studies, the wavelength composition of the target and the mask stimuli generally are the same. Therefore, in these studies one would expect only post-receptor, neural interactions to affect the masking function.

5.1.2. Weisstein's Rashevsky–Landahl two-factor neural net

Weisstein's (1968, 1972) model of metacontrast applies strictly to post-receptor, neural interactions in the visual pathway. In this neural-network model, each neurone's response was assumed to be a function of the difference between excitatory and inhibitory post-synaptic potentials which rise and decay exponentially in time. In the currently modified model (Weisstein *et al.* 1975) shown in Fig. 5.1 the neural network schematically consists of eight neurones, six of which are excitatory neurones and two of which are inhibitory neurones. In the figure the target or S1 stimulus and the mask or S2 stimulus are presented simultaneously. The S1 stimulus activates an excitatory pathway in which the neurones n_{11}, n_{12}, and n_{13} synapse in sequence. Similarly, the S2 stimulus also activates an excitatory pathway consisting of a primary (n_{21}), secondary, and tertiary neurone. The primary excitatory neurones in each pathway are assumed to be located in the periphery of the visual pathway; the secondary and tertiary neurones of each pathway are assumed to be central neurones. At the first synapse along each pathway, inhibitory interneurones are also activated by the primary excitatory neurones (n_{11} and n_{21}). These inhibitory neurones cross spatially and form inhibitory synapses on the tertiary neurone in each

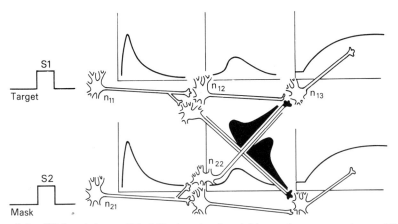

Fig. 5.1. Weisstein's modified Rashevsky–Landahl two-factor neural net. The behaviour of the net illustrates the hypothetical activity of target and mask neurones to simultaneous presentation of the target and mask stimuli. Activation functions are shown for the primary, secondary, and tertiary excitatory neurones in each pathway. Shaded regions correspond to the activation functions of the inhibitory interneurones. (From Weisstein, Ozog, and Szoc (1975).)

pathway. This makes explicit the spatial symmetry of the reciprocal inhibition that the target and mask stimuli can exert on each other. Earlier versions of the model (Weisstein 1968, 1972) only implicitly incorporated this symmetry. As can be seen, the inhibition is of a forward, non-recurrent type.

Each of the eight neurones is characterized by a temporally rising and decaying activity function. The crucial difference between activity functions is seen when comparing the output of the secondary excitatory neurones with that of the inhibitory interneurones. The rise and fall times of the former are longer than those of the latter. In effect the inhibitory interneurones respond faster than the secondary excitatory neurones. Since both types of neurones synapse on the tertiary neurones, the response of the tertiary neurones, in the case of target–mask simultaneity, is determined by the difference of a leading inhibitory post-synaptic potential and a lagging excitatory post-synaptic potential. Since, at target–mask simultaneity (SOA=0 ms), the two potentials are temporarily out of phase, the inhibitory potential will have dissipated appreciably by the time the lagging excitatory potential is generated. Hence little if any suppression of the target's tertiary neurone ought to occur. The target's visibility is not masked appreciably. However, at some positive SOA of the mask relative to the target, the generation of two antagonistic potentials ought to coincide in time. Here one would expect optimal suppression of the target's tertiary neurone's output and consequently optimal masking of the target's visibility. At still greater SOAs, the generation of the inhibitory potential will again be out of phase, occurring somewhat after the generation of the excitatory potential, and result in less than optimal target masking. Thus, by progressively increasing the SOA, a typical type B metacontrast function can be generated.

The merit of the model lies in its conceptual simplicity and in the fact that recent neurophysiological and psychophysical evidence, to be discussed in Chapters 6 and 7, indicates that the activity of relatively slow-responding visual neurones (sustained neurones) and that of relatively fast-responding ones (transient neurones) can reciprocally inhibit each other. A modification of the model incorporating also a slow inhibitory and a fast excitatory process could therefore also account for type B paracontrast (Weisstein *et al.* 1975). Moreover, since the model assumes that the excitatory–inhibitory interactions occur at central levels of processing, it can account for both monocular and dichoptic para- and metacontrast. However, the model, although conceptually simple, requires an extensive set of parameters. Each of the neurones in the set is a two-factor neurone and is characterized by four parameters describing the proportionality and rate constants of the excitatory and inhibitory factors.

In the example of metacontrast outlined above, the activity of five neurones (n_{11}, n_{12}, n_{13}, n_{21}, and n_{22}) must be parametrically specified in order to simulate a metacontrast function (thus, a total of 20 free parameters must be specified). Even with this number of parameters the model is only moderately successful. It can simulate some of the empirical type B metacontrast functions obtained by Weisstein and Haber (1965), Schiller and Smith (1966), and Alpern (1953) with a fair degree of success (see Weisstein 1972; Figs. 3, 5, and 7). Perhaps other prior findings could also be simulated fairly successfully. However, in simulations of metacontrast in which the target:mask (T/M) energy ratio was varied systematically it fared less well. Fig. 5.2 shows empirical and simulated results for three subjects and for five T/M energy ratios (1.0, 0.5, 0.2, 0.125, and 0.0625). The model correctly predicts a shift from type B to type A functions as the T/M energy ratio decreases. The empirical results also show this trend. None the less, in several cases there appears to be large deviations of the simulated from the empirical results, both in terms of the location of the SOA at which peak masking occurs and in terms of the overall magnitude of masking. Moreover, since only the n_{11} and n_{21} neurones shown in Fig. 5.1 respond to the target and mask, and since they presumably code only brightness information, the model of Weisstein *et al.* cannot really predict

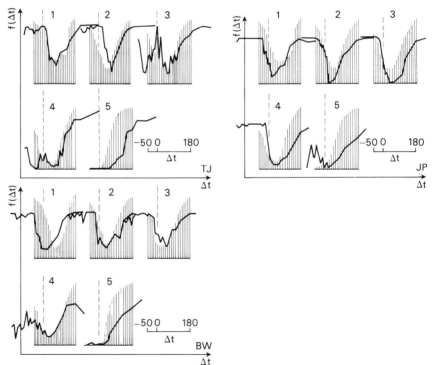

Fig. 5.2. Weisstein's simulated metacontrast functions (vertical lines) and empirical metacontrast functions (solid lines) at five target/mask energy ratios (1.0, 0.5, 0.2, 0.125, and 0.0625) in three subjects. (From Weisstein (1972).)

the *absence* of metacontrast effects when simple reaction time to, or forced-choice detection of, the target stimulus are required (Bernstein *et al.* 1973; Fehrer and Biederman 1962, Fehrer and Raab 1962; Harrison and Fox 1966; Schiller and Smith 1966). We shall see below that a similar critique applies to several other models of metacontrast. For now, however, one can state that further modification of the metacontrast model of Weisstein *et al.* seems to be indicated, and Chapters 7 and 9 show what some of these additional modification may comprise.

5.2. Spatiotemporal sequence models

Besides metacontrast, another related consequence of presenting two spatially adjacent stimuli in rapid succession is the appearance of stroboscopic motion. The activation of spatiotemporal sequence detectors irrespective of whether they give rise to a percept of motion also may be entailed in metacontrast. Two relevant models are reviewed below.

5.2.1. *Kahneman's impossible stroboscopic motion model*

In Chapter 4 it was noted that Kahneman (1967*b*) proposed a model of metacontrast in which stroboscopic motion plays a causal role in metacontrast. Specifically, metacontrast is a special case of impossible stroboscopic motion. For example, with the frequently employed metacontrast display in which a rectangle serves as a target and two spatially flanking rectangles serve as masks, the target–mask sequence activates two oppositely directed stroboscopic motion events of the same object—a physically impossible event. Kahneman (1967*b*) argues that the cognitive–perceptual system is unable to resolve these apparently opposite and contradictory motions of the same stimulus and thus suppresses its perception.

However, it also was noted that this formulation was flawed on several counts. For one, the empirical, spatiotemporal laws which characterize apparent motion diverge from those characterizing metacontrast (Breitmeyer and Horman 1981; Weisstein and Growney 1969). Moreover, as demonstrated by Stoper and Banffy (1977), split stroboscopic motion from the target to the flanking masks can be, but need not be, observed in metacontrast (Stoper and Banffy 1977). Consequently, the perceptual system can simultaneously accommodate such apparently contradictory motions as noted earlier by Wertheimer (1912). This perceptual accommodation to the metacontrast display may use processes responsible for the perception of disruptive events such as the rapid expansion and radial dispersion (explosion) of an object's surface (Gibson 1979). Finally, metacontrast can be obtained in a two-stimulus display (e.g. two adjacent rectangles or discs) in which the perception of stroboscopic motion is readily possible and present (Breitmeyer and Horman 1981; Breitmeyer *et al*. 1974, 1976). These results, as well as those of Stoper and Banffy (1977), rule out the causal necessity of any form of impossible or contradictory stroboscopic motion in producing metacontrast. As noted by Stoper and Banffy (1977) the mechanisms underlying stroboscopic motion and metacontrast are largely independent of each other although under some favourable stimulus conditions they may interact. In order to show that such interaction can yield impossible stroboscopic motion and thus also metacontrast, as required by Kahneman's (1967*b*) cognitive model, one may want to adopt the following strategy. What needs to be demonstrated is: (i) that each one of two flanking mask stimuli separately can induce a type B stroboscopic motion function when flashed at varying SOAs after the target stimulus (preferably, but not necessarily, without producing a type B metacontrast effect); but (ii) that in combination the two mask stimuli yield a type B metacontrast function *without* split stroboscopic motion. Although, in view of the findings reviewed above, such a demonstration would most likely hold only under highly specific, non-generalizable conditions, in hindsight it would none the less be more convincing than the equally non-generalizable and fortuitouus correlation between type B stroboscopic motion and metacontrast functions used by Kahneman (1967*b*) to support his cognitive model.

It also should be mentioned that Kahneman's (1967*b*) model cannot adequately account for the absence of type B (and type A) metacontrast when a forced-choice detection or simple reaction time measure is used by an observer. In fact, we noted that the split apparent motion (devoid of target contrast or figural information) generally observed in metacontrast is most likely a powerful source of information for detecting the mere presence or location of the target. Without such split apparent motion, i.e. with its suppression by the cognitive system, target detection or localization might be greatly impaired. Moreover, even if such motion information were suppressed by the cognitive system, it is hard to imagine how it could be recovered to produce apparent stroboscopic motion between a masked target and a third stimulus flashed at a locus laterally displaced from the target–mask area (Kolers 1963; von Grünau 1978*b*, 1979, 1981; see Section 4.4.1, Fig. 4.7 and 4.8).

5.2.2. *Matin's three-neurone model*

Matin (1975) proposed a model of metacontrast which is based on the existence of three classes of neurones. In a typical metacontrast experiment, a target is presented first, followed at a short interval by a laterally displaced mask. Matin assumed that the target activates one class of neurones which she calls T-neurones. Similarly the mask activates a second class of neurones which are termed M-neurones. Finally, at appropriate temporal intervals the target–mask sequence activates a third class of neurones which are called T-M neurones.

This last class of neurones, according to Matin (1975), could consist either of succession (motion) detectors used in analysing object motion or of neurones activated during relative image displacement produced by high-velocity saccades. In the former case, activation of succession detectors could, but need not, produce a sensation of (stroboscopic) motion. Via this assumption Matin avoids the shortcomings of Kahneman's (1967b) model of metacontrast which ascribes a causal role to stroboscopic motion. Saccade neurones comprise part of the latter class of T-M neurones and are tentatively identified with Y or transient neurones which are assumed to have a shorter response latency and higher conduction velocity than X or sustained neurones (see Chapter 6 for a review of response properties of transient and sustained neurones). The sustained neurones are further assumed to be T-neurones which are related to the perception of the target. Although not explicitly stated by Matin (1975) the M-neurones could also be identified with sustained neurones since the mask is typically perceived in a lateral masking experiment.

With this set of three neural classes, Matin (1975) can account for several findings obtained in lateral masking studies. For one, the existence of T-M neurones could explain the sensation of stroboscopic motion which generally, though not always (Stoper and Banffy 1976), accompanies lateral masking. Moreover, several kinds of empirically derived lateral masking findings can be adequately explained. These are type A and type B metacontrast as well as type A and type B paracontrast.

According to Matin (1975), pronounced type B metacontrast could be produced in one of two possible ways. For one, Matin notes that the suppressive effect that T-M neurones exert on T-neurones must be retroactive in order to produce type B metacontrast. She states that:

Although the T-M neurons do not fire until the presentation of the mask, the magnitude of response in these neurons would be expected to be greatest at some temporal interval between target and mask other than zero. It could therefore be argued that in those classes of metacontrast experiment for which the presentation of the target–mask sequence is an adequate stimulus for the T-M neurons, the psychophysical metacontrast function peaks at some interval other than zero *not per se because the target precedes the mask, but in spite of that fact and because the responses of the T-M neurons, which are a part of the total mask response, are greater at that interval than at some longer or shorter one* (Matin 1975; p. 457).

In this case, production of type B metacontrast uses that class of T-M neurones which are responsible for the detection of succession (with or without an accompanying sensation of motion).

In addition to this explanation, Matin (1975) offers a second one. The second explanation relies on the existence of high-velocity saccade neurones which are also activated by the T-M sequence. These saccade neurones are assumed to be fast-conducting transient cells which can suppress the activity of slower conducting sustained cells. Since the former are a class of T-M neurones and the latter comprises the set of T-neurones, the T-M (saccade) neurones would exert their optimal suppression of the T-neurones when the mask is delayed relative to the target, thus producing a type B metacontrast effect. Implicit in this additional specification of T-M neurones is the conclusion that here the metacontrast function peaks at some SOA other than 0 ms, not in spite of but because the target, activating slower T-neurones, precedes the mask and in conjunction with the mask activates faster T-M (transient) neurones. Consequently, Matin (1975) assumes that two mutually reinforcing mechanisms are used in producing type B metacontrast.

The existence of type A metacontrast, typically obtained when the mask energy is substantially higher than the target energy, is explained as follows. In type A metacontrast, peak masking occurs at an SOA of 0 ms. Here, due to the high energy of the mask, M-neurones could be activated more vigorously than T-M neurones. Since the T- and M-neurones are assumed to have similar conduction velocities, one would expect greatest suppression of T-neurones by M-neurones at target–mask synchrony, i.e. at an SOA of 0 ms. At progressively greater SOAs one would also expect that the suppression of T-neurone activity by M-neurones declines, thus producing a type A metacontrast function. An additional assumption must be made explicit here; namely, that at an SOA of 0 ms the suppression of T neurones by M neurones, when the T/M energy ratio is less than unity, is stronger than the maximal suppression of T neurones by T-M neurones at some optimal positive SOA value; otherwise a type B function would still persist.

Type B paracontrast or forward lateral masking effects, which typically are weaker than metacontrast masking effects (Alpern 1953; Kolers and Rosner 1960), can be explained by invoking an essential asymmetry between the interactions of T- and T-M neurones used in paracontrast and metacontrast. Recall that in metacontrast the activity of T-M neurones not only is optimal when the mask follows the target but also is conducted faster than the activity of T-neurones. In paracontrast the laterally displaced mask precedes the target in time. Here, T-M neurones are activated by the mask–target sequence. If these neurones (transient neurones), as is assumed, conduct more rapidly than T-neurones (sustained neurones), the suppressive effects of T-M neurones very likely dissipate to some extent by the time the activity of T-neurones arrives at the site where interaction between these two classes of neurones occurs. In fact, if the difference in conduction velocity were sufficiently great, the later activity of T-neurones could entirely escape the earlier suppressive effects of the T-M neurones; and consequently the paracontrast effect would then be produced only by the M-neurones. In either case of partial or complete escape of T-neurone activity from T-M neurone suppression, one would expect paracontrast to be weaker than metacontrast.

Although Matin (1975) does not make the following explicit claims regarding paracontrast, they are nevertheless implicit in the description of her model. In the case of partial escape of T-neurone activity from T-M neurone suppression, one would expect, relative to a powerful type B metacontrast effect, an attenuated type B paracontrast effect (see Fig. 4.2). Furthermore, if the above escape from T-M neurone suppression is total and only M neurones inhibit the T-neurones, paracontrast ought to be of type A, being optimal at an SOA of 0 ms and decreasing as the mask-to-target asynchrony increases. Finally, the activation of T-M neural activity preserves information specifying the presence and locus of the target even when target contrast and figural information processed by T-neurones is suppressed during metacontrast.

In so far as Matin's (1975) model adequately explains some of the major findings in a particular type of lateral masking which she calls the two-transient paradigm, it is successful. However, in Section 4.4.5 I discussed a form of lateral masking which may be more aptly termed the single-transient paradigm (Breitmeyer and Rudd 1981). Recall that in this paradigm the target was presented for 10 s and a 50-ms, transient mask was presented at an SOA of 2 s. Such a large temporal interval between the onsets of the target and the mask would certainly not activate the class of T-M succession detectors, since these are assumed to be optimally activated at SOAs as short as 50–150 ms, where peak metacontrast occurs. Despite this, one can obtain strong and prolonged suppression of the prolonged target, particularly at extrafoveal sites of stimulation (see Fig. 4.10).

Consequently, Matin's (1975) model cannot adequately deal with the single-transient paradigm. The existence of single-transient masking suggests the following possible modifications of Matin's model. Since the visibility of the prolonged or sustained target most likely depends solely on the activity of sustained neurones (see Chapter 6), the single-transient produced by the mask must suppress the sustained activity of the target. One way the mask can have this effect is to activate transient neurones which suppress the target's sustained activity. (In fact, this seems to be the most likely explanation since in an ancillary experiment performed by Breitmeyer and Rudd (1981), a sustained mask, presented throughout the 10-s target exposure, did not suppress the target's visibility.) If that is in fact the case, the brief, 50-ms mask itself must have activated transient neurones. Consequently, a T-M sequence at optimal SOAs is not necessary to activate transient neurones.

Although the finding of single-transient masking does not rule out the existence of T-M neurones, it none the less suggests that the brief mask alone can activate two types of neurones. One type may be sustained neurones which could be identical with Matin's M-neurones; the other type would be transient neurones, activated by the brief mask, which in themselves are sufficient to produce masking of the target. Since, in a typical two-transient masking paradigm, the target is also presented briefly, it also should activate these types of neurones besides the T-neurones. Hence, in addition to Matin's T-, M-, and T-M neurones, one would require two additional classes of neurones to account for the transient activity that is elicited by either the target or the mask alone. Section 5.1.2 showed that a related change is required in the model of Weisstein *et al.* (1975) to account for the absence of metacontrast when simple detection, rather than brightness or contrast rating, of the target is the task of the observer.

5.3. Two-process models

5.3.1. Ganz's model: interactive trace decay and random encoding time

It is well known from Heinemann's (1955) work on simultaneous brightness contrast that a more intense, spatially surrounding stimulus can appreciably reduce the apparent contrast of a central stimulus. Simultaneous brightness contrast is believed to be induced by the lateral inhibitory effect that the surrounding stimulus exerts on the central one. Ganz (1975) proposed a model of metacontrast based on these findings and this assumption, in which the temporally decaying traces (icons) of the target and mask stimuli interact laterally. As such, it is conceptually an adaptation of Stigler's (1910; see Section 1.3) persistence–lateral inhibition explanation of metacontrast. Ganz's analysis relies on results reported in a study of metacontrast conducted by Sukale-Wolf (1971). In that study the target and the mask were of equal energy, a condition which, as noted in Chapter 4, is conducive to the production of type B metacontrast.

Furthermore Ganz's (1975) model assumes that the following properties and events characterize metacontrast. Both the target and mask produce temporally decaying neural traces. Since the target and mask are of equal energy, the proportionality constants and the decay constants are assumed to be identical for the target and the mask. Moreover, since the mask is presented after the target, the neural trace of the mask ought to be stronger at any positive SOA than that of the target. Because the weaker trailing end of the target trace overlaps temporally with the stronger leading part of the mask trace, one in effect has a case of simultaneous brightness contrast induced via lateral inhibition by the stronger mask trace on the partially decayed target trace. Hence reduction of the target's visibility ought to progressively increase as SOA increases. (At an SOA of 0 ms, the two equally strong traces are activated simultaneously and decay at the same rate and little if any brightness contrast or masking ought to occur). This lateral, interactive trace–decay process would therefore be responsible for the rising part of the U-shaped metacontrast function as

SOA increases from 0 ms to that intermediate value at which optimal masking occurs.

In order to explain the descending portion of the type B metacontrast function for progressively still greater SOA values, Ganz (1975) incorporates a further assumption stating that the decaying trace of the target needs some minimal time, E, for its brightness or contrast to be fully encoded. Additionally, the duration of E is assumed to be distributed randomly in a Gaussian manner. Accordingly, since the encoding probability is very small at low SOAs and increases as SOA increases, one would expect the metacontrast function to rise initially from an SOA of 0 ms to some positive SOA at a rate which is determined by the mean and standard deviation of the Gaussian distribution of encoding time. At longer SOAs the likelihood of encoding becomes increasingly greater, and masking ought to decrease, in a statistical sense, after its maximum is attained.

Although this model adequately explains metacontrast when the target and mask stimuli are of equal energy, the question remains as to how well it would fare when the energies of the two stimuli are unequal. In the case of mask energies greater than target energy, it is not clear that it would correctly predict type A metacontrast (Breitmeyer 1978*b*; Weisstein 1972). Let us assume that the proportionality constant of the mask trace is twice that of the target trace. Then one ought to indeed obtain stronger masking at an SOA of 0 ms. However, since the target trace decays with time, one additionally obtains increasing masking as SOA increases, since the initial portion of the much stronger mask trace would inhibit a progressively weaker portion of the decaying target trace. In other works, the initial portion of the metacontrast function ought again to rise—indeed, more steeply—before the encoding process takes over to yield the later descending part of the metacontrast function. Consequently, contrary to the obtained type A metacontrast function, Ganz's (1975) model predicts a type B function. One possible way of circumventing this difficulty would be to incorporate a 'saturation' parameter so that whenever the mask energy becomes sufficiently large relative to the target energy, its trace inhibits the target trace equally well at the lower range of SOAs, thus yielding an approximation to a type A metacontrast function.

A greater difficulty may, however, arise with the model when the target energy is greater than that of the mask. Chapter 4 (see Fig. 4.3) showed that type B metacontrast functions can be generated when the mask energy is only one-half or one-fourth that of the target. Hence, according to Ganz's model, one would expect, on the basis of Heinemann's (1955) results, that no masking occur for a range of low SOAs which would first have to be exceeded before the weaker mask trace becomes effective in suppressing the trailing end of a decaying but stronger target trace. In fact it may be entirely possible according to Ganz's (1975) model that the target trace is encoded prior to any inhibitory effects that the mask trace may exert on the trailing end of the target trace. Thus, type B masking ought to be very weak (especially at SOAs ranging from 0 ms to some intermediate value) or entirely absent. The fact that empirical findings do not show this trend when mask energy is reduced by, say, one-half or one-fourth relative to target energy, but rather reveal a type B function characterized by a rising portion, a peak, and then a declining portion as SOA increases, militates against Ganz's model.

Moreover, Ganz's model predicts paracontrast effects which very likely do not match empirical results. Paracontrast brightness suppression, given equal energy target and mask stimuli, is a type B forward masking effect, however, generally weaker than type B metacontrast. If the target and mask stimuli are of equal energy, on would expect weak simultaneous brightness induction at an SOA of 0 ms, an expectation consistent with the typically weak metacontrast obtained at that SOA. What, according to Ganz's model, would be expected to happen as the mask leads the target at progressively greater SOAs? Since the neural trace of the mask and the target are characterized by the same proportionality and decay constants, one would expect that the trailing end of the leading mask's neural trace which overlaps in time with the leading end of the target's trace progressively weakens as the SOA increases. Hence, with target and mask stimuli of equal energy, little, if any, paracontrast should be obtained. If it is obtained, on would expect it to be at best a type A rather than a type B function, contrary to what is found (Alpern 1953; Kolers and Rosner 1960).

5.3.2. Reeves' model: temporal integration and segregation

Despite the failure of Ganz's (1975) model to account adequately for some major metacontrast and paracontrast findings, a recent model and results of metacontrast reported by Reeves (1982) generally seems to concur with and lend some credence to Ganz's (1975) account of metacontrast. In particular, Reeves (1982), adopting Weisstein's (1972) magnitude rating method, required his observers to (i) rate on each trial the brightness of the target as a function of target–mask onset asynchrony; and (ii) indicate whether the two stimuli were perceived as being simultaneous or successive. Although not specified by Reeves, perceived simultaneity would reflect an integrative process which, in turn, based on the discussion of integration in Section 3.6, most likely rests on the presence of visual persistence; perceived successions, however, would reflect a segregative process leading to temporally separate encodings and, thus, percepts of the two stimuli.

It should be evident that these two processes are analogous to Ganz's (1975) trace-decay and encoding stages. For, as one might expect, Reeves found that the proportion of simultaneity judgements decreased monotonically from a constant value of 1.0, obtained at SOAs of 0–40 ms, to 0.0 as SOA increased from 40 to 120 ms. Conversely, the proportion of successiveness judgements increased monotonically from 0 to 1.0 over the 40–120 ms range of SOAs. The former result indicates that temporal integration of target and mask stimuli (based on decaying visual persistence or visual traces of the target) extends with decreasing strength up to an SOA of 120 ms; whereas the latter result indicates that separate target and mask encoding processes occur with a probability of zero below an SOA of 40 ms, but thereafter increase monotonically up to a value of 1.0 at an SOA of 120 ms.

In line with these results, when Reeves separated the overall averaged metacontrast brightness ratings, characterized by the typical type B function of SOA on the basis of the type of concurrent temporal judgement, he found that the type B function could be approximated by two separable monotonic functions. One component showed progressively *increasing* target-brightness *suppression* over the corresponding SOA range of 40–120 ms (which yielded a progressive decrease of the proportion of simultaneity reports); the other, conversely, yielded progressively *decreasing* brightness suppression over the same SOA range (yielding a progressive increase of the proportion of successiveness reports). At face value this binary decomposition, and evident correlation between changes of target–mask temporal judgement and target brightness ratings agree with Reeves' two-process, integration–segregation account, and in particular with Ganz's more specific two-process model based on an initial stage of lateral, interactive, trace decay and a later perceptual encoding and segregation stage. However, we shall see below that these two-process models, in addition to a third version recently proposed by Navon and Purcell (1981), are characterized, as already evident from discussion of Ganz's (1975) model, by a lack of generalization or applicability to extant data outside the immediate scope of their investigations.

5.3.3. Navon and Purcell's model: integration and interruption

Employing a variety of chromatic and achromatic patterns consisting of target letters, e.g. F and a non-letter mask, ⊠, such that each letter target formed a subset of contours of the mask, Navon and Purcell (1981) obtained an *inverted* type A monotonic backward masking function over an SOA range of 0–50 ms. That is, masking of target letters was *minimal* at an SOA range of 0 ms and increased monotonically over the 50-ms SOA range. Such a function, by the way, is reminiscent of the ascending part of the type B metacontrast function found over approximately the same SOA range. In fact, Navon and Purcell (1981) maintain that this increase of masking cannot be due to an integration mechanism but rather must be due to what is commonly termed an interruption mechanism (Scheerer 1973).

Now the target–mask (T-M) spatial composites could be subdivided into two mutually exclusive components (MUT) and MUT̄), that is, that area of stimulation common to mask and target contours and that area stimulated only by mask contours. What Navon and Purcell (1981) proposed on the basis of their study is that since chromatic or achromatic integration of contrast makes MUT appear different from MUT̄, the target information is in fact *preserved* in the T-M composite rather than being masked (for a related study, see Schultz and Eriksen 1977). Because such integration generally decreases in type A fashion with target-to-mask SOA, whereas their obtained masking effect increased in type A fashion as SOA increased from 0 to 50 ms, Navon and Purcell (1981) further proposed that their obtained type B backward masking function is a composite of two separable and additive processes: (i) what they called 'fortunate' integration, which preserves target information, decreased monotonically at a rapid rate as SOA increased from 0 to, say, 50 ms; whereas (ii) masking by interruption, which destroys target information, decreased monotonically at a somewhat slower rate as SOA increased from 0 to, say, 100 ms. By adding these two countervailing type A processes, one obtains a U-shaped, type B backward pattern masking function, with optimal masking occurring at 50 ms.

Although this two-process model may account for the *particular* results reported by Navon and Purcell (1981), it is not consistent with the *general findings* (see Section 4.5) of type A *masking by integration* found (i) when noise masks are employed (Greenspoon and Erikson 1968; Kinsbourne and Warrington 1962*a*, *b*; Schiller 1966; Schiller and Smith 1965; Turvey 1973); and (ii) when structure masks having a higher energy than the target are employed (Hellige *et al.* 1979; Michaels and Turvey 1973; Purcell and Stewart 1970; Spencer and Shuntich 1970; Turvey 1973).

5.3.4. *General criticism of two-process models*

However, over and above these discrepancies, Navon and Purcell's (1981), as well as Reeves' (1982) and Ganz's (1975) two-process models, would make the wrong prediction regarding the expected shifts of the SOA at which optimal metacontrast occurs as background or stimulus intensity is altered.

The review of metacontrast in Section 4.3.2 showed that relative to lower background or stimulus luminances, higher ones effect a shift of the peak of the type B metacontrast function to larger SOA values (Alpern 1953, Purcell *et al.* 1974). We also know, on the basis of the inverse-background- and inverse-stimulus-intensity effect discussed in Section 3.1.2, that visual persistence, integration, and successive-flash resolution thresholds occupy a shorter duration at higher, relative to lower, intensities. The implication for Ganz's (1975) two-process models is that since the visual traces and, therefore, lateral interactions of target and mask stimuli are both curtailed in duration at higher background or stimulus intensities, the rising portion of the type B metacontrast function should terminate at a shorter SOA and thus shift the peak masking effect to a correspondingly shorter SOA value. Similarly, for Reeves' model, the expected shift of the upper limit of temporal integration and the lower limit of temporal resolution to lower SOA values would also predict that the two respective and correlated type A monotonic processes, comprising the overall type B metacontrast function, also shift to lower SOA values. Hence, again, peak masking ought to occur at a lower SOA when background intensity is increased. Finally, since the duration of temporal integration also is shortened at higher background intensities, according to Navon and Purcell's (1981) model we would again expect the additive combination of the target-preserving integrating function of the mask and its countervailing interrupting or masking function to produce a type B effect which peaks at a shorter SOA relative to lower backgrounds.

Thus, all of the above two-process models would predict a result which is contradicted by the findings of Alpern (1953) and Purcell *et al.* (1974) showing that the peak of the type B metacontrast function shifts to higher SOA values as stimulus or background intensity increases. The main problem with the above two-process models is that they apply in a restrictive manner only to the immediate data collected in support of them. Their explanatory extension to other extant data is, as shown, inadequate and refuted. Moreover, like the prior models, none of the two-process models make any specific attempt to explain the absence of masking obtained if, rather than using brightness ratings or figural identification as a response criterion, other ones such as simple reaction time or detection were employed. Ganz's (1975) model, based on lateral contrast induction between brightness response traces, could not account for the absence of metacontrast masking effects when the latter response criteria are employed, but rather, would predict a type B metacontrast function. It is not clear how Reeves' model would fare with the use of the latter, simple detection criteria, since no specific hypotheses were proposed regarding how the two components of his model, integration and segregation, affect simple detection. A similar uncertainty about the use of detection as opposed to identification criteria applies to Navon and Purcell's (1981) model. Here, without specifying additional hypothetical processes, it again is not clear how the two temporally overlapping processes of target-preserving integration and target-destroying interruption would affect simple detection of the target.

5.4. Distributed network models: Bridgeman's Hartline–Ratliff inhibitory network

The final model to be discussed in the present chapter is Bridgeman's (1971) model of metacontrast which is based on the existence of recurrent lateral inhibition among neurones in a distributed neural network. The equations specifying the inhibitory activity within the network are similar to those derived by Ratliff (1965). In Bridgeman's (1971) model the activity of the network, after presentation of a brief stimulus, is assumed to be a damped excitatory–inhibitory oscillation in space and in time. This is due, for one, to the fact that the process of recurrent lateral inhibition is assumed to iterate or repeat at a rate of once every 30 ms, a rate which approximates the latency of lateral inhibition of visual neurones found in the lateral geniculate of cat (Singer and Creutzfeldt 1970). Moreover, since the network is spatially distributed, that is, interconnected via lateral inhibitory processes, the activity of the network distributes spatially after a few iterations of the inhibitory activity.

This activity is assumed to store information about the brief stimulus temporarily; that is, it comprises a form of visual persistence or icon. The duration of the storage lasts for several iterations of the inhibitory process beyond which the spatiotemporal oscillations of the network fade away. In addition, the pattern of spatiotemporal oscillation in the network specifies the stimulus used in metacontrast. For instance, the target alone and the mask alone each activate a characteristic pattern of spatiotemporal oscillations in the network. Each pattern of activity thus corresponds to one of the stimuli.

In order to generate lateral masking or metacontrast functions, Bridgeman (1971) assumes that the activity of the target alone is stored in the neural network and is then compared with the activity produced by the target–mask combination. An example of the network activity for the

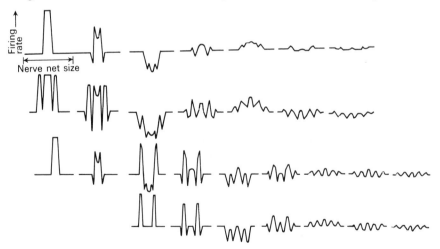

Fig. 5.3. The activity in Bridgeman's lateral inhibitory network in the presence of the target alone (uppermost panel), the simultaneous presentation of the target and mask (second panel), the presentation of the mask after the target (third panel), and of the mask alone (lowest panel). (From Bridgeman (1971).)

target alone, and for the target and mask at SOAs of 0 and 60 ms is shown in Fig. 5.3. Also shown for comparison is the neural network activity for the mask alone. The comparison process, although its exact mechanism is not specified, is realized via a simple Pearson r coefficient between the stored activity set up by the target, or else mask, alone and the activity set up by the target–mask sequence. As such it corresponds to a cross-correlation or template matching process of pattern recognition. High correlations correspond to good target recognition; lower correlations correspond to poor target recognition. Simulated masking functions, using a disk–ring paradigm, for a case of paracontrast, simultaneous masking, and metacontrast are shown in Fig. 5.4. The solid lines correspond to the magnitudes of the r coefficient obtained when comparing the disk activity with the disk–ring activity; the dashed lines correspond to the comparison between the ring activity and the disk–ring activity. Note that the disk's r coefficients are fairly high when preceded by the ring. This would correspond to a relatively weak paracontrast effect. At disk–ring synchrony the disk's r coefficients are somewhat lower, and when the ring follows the disk they are reduced substantially. The latter reduction corresponds to pronounced metacontrast at an SOA of 60ms. Roughly symmetrical effects are obtained when the ring activity is compared with the disk–ring activity.

One problem with this model, as noted by Weisstein *et al.* (1975), is that if one simulates metacontrast, using the r coefficients, over a wider range of SOAs, a pronounced temporal oscillation of the metacontrast function is obtained. Since empirical metacontrast functions generally do not yield such oscillations, a modification of the model seems to be required. Bridgeman (1977, 1978) incorporates a modification in which r^2 rather than r is used as a measure for comparing target with target–mask activity. This modification successfully eliminates temporal oscillations in the metacontrast function. Moreover, with varying degrees of adequacy, it also predicts the existence of iconic storage of pattern information, a shift from type B to type A functions as the target–mask energy ratio increases, the decrease of metacontrast magnitude as the spatial separation between the target and mask increases, the existence of type B paracontrast effects, as well as several other variations of the metacontrast paradigm (Bridgeman 1978). However, it also predicts a type A metacontrast function when the target and mask do not have the same sign of contrast, in particular when the target is dark and the mask is light. Section 4.3.2 showed that Breitmeyer (1978c) obtained type B metacontrast suppression of a target's contour detail under these stimulus conditions. Since in Bridgeman's (1977, 1978) modified model the network activity presumably represents the entire target stimulus and not only its contour information, the type B contour suppression may fail to be predicted by his model. A further modification of Bridgeman's model may be to incorporate separate processes for generating brightness or contrast suppression and suppression of contour detail.

Finally, Bridgeman's (1971) original and modified (Bridgeman 1977, 1978) models suffer from the same predictive flaw as the two-process models discussed in the prior section. In his simulations, Bridgeman, on the basis of Singer and Creutzfeldt's (1970) study of cells in cat lateral geniculate nucleus, assumed a fixed time constant of 30 ms specifying the latency of recurrent lateral inhibitions in his network. Thus, one iteration of the network's inhibitory process required 30 ms. Since the outputs of these successive, 30-ms iterations provide the input to the cross-correlating process, which determines the type B metacontrast function, the temporal characteristics of this function are in turn determined by the 30 ms latency of recurrent lateral inhibition. In Bridgeman's particular application, the peak metacontrast effect (the lowest cross-correlation) occurs at an SOA of 60 ms, i.e. after two network iterations of the target-inhibition activity. However, if we let the latency of recurrent lateral inhibition vary, we could expect to obtain correlated variations of the SOA at which peak masking occurs. For instance, if the inhibitory time constant is 15 ms, the peak of the metacontrast function should shift to an SOA of 30 ms; similarly if the time constant is 60 ms, the peak should shift to an SOA of 120 ms.

One way of varying the inhibitory time constant of the visual system is to change its light-adaptation level. Electrophysiological studies (Barlow *et al.* 1957*b*; Maffei, Cervetto, and Fiorentini, 1970; Poggio *et al.* 1969; Sasaki, Saito, Bear, and Ervin 1971; Virsu, Lee, and Creutzfeldt 1977) showed that the spatiotemporal response properties of single cells along

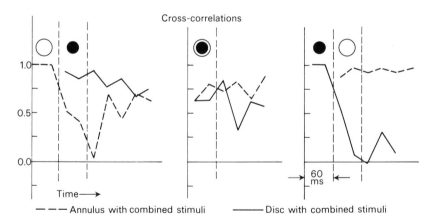

Fig. 5.4. Pearson r coefficients (ordinate) as a function of the temporal interval between the disk and ring presentations. Left panel: ring presented before disk. Middle panel: ring and disk presented simultaneously. Right panel: ring presented after disk. Solid lines correspond to r coefficients of the disk; dashed lines, to r coefficients of the ring. (From Bridgeman (1971).)

the entire retinogeniculostriate pathway change with the adaptation level of the visual system. Singer and Creutzfeldt's (1970) study of response latencies of lateral geniculate cells, on which Bridgeman based his lateral inhibitory time constant of 30 ms, was conducted at a fixed background luminance. However, Virsu *et al.* (1977) replicated Singer and Creutzfeldt's (1970) investigation of cat lateral geniculate neurones and found that, with progressive dark adaptation, response latencies increased. Moreover, response latencies increase as stimulus intensity decreases. Consequently, as we change from a high to a low background or stimulus luminance level, Bridgeman's model, like the two-process models discussed in the prior section, would predict corresponding peak metacontrast shifts from low to high SOA values. However, as I have already noted, the findings of Alpern (1953) and Purcell *et al.* (1974) show exactly the opposite trend.

Moreover, Bridgeman's model, although specifying a hypothesis linking the outcome of a cross-correlation of target and target–mask information reverberating in a distributed neural net to the magnitude of brightness or contrast suppression in metacontrast, also fails to specify the outcome of metacontrast tasks in which different criterion contents are employed. For instance, contour detail suppression may be very strong without accompanying brightness or contrast suppression (Breitmeyer *et al.* 1974, 1976; see Section 4.4.1). Furthermore, when simple target detection is used as a response criterion, no metacontrast occurs, indicating that some target information is immune to the suppressive effects of the following mask. However, these results pose no intractable problem to Bridgeman's model. On the one hand, to account for contour detail suppression, one need only assume that the *later* target–mask reverberations are cross-correlated with the network's stored representation of the later target reverberations. On the other hand, to account for failure of masking using a simple detection criterion, one need assume that the *initial* reverberation of the target–mask response is cross-correlated with the corresponding stored representation of the initial target reverberation. Here, we are also assuming that the *intermediate* reverberations of the target or the target–mask response correspond to brightness or contrast information. Of course, one would additionally need to posit the existence and properties of a central decision process according to which the particular information or criterion content satisfying an observer's task is selected.

The review and critique of the above models of lateral masking, although not completed, hopefully is sufficiently detailed to allow the reader a glimpse at some of their positive and negative aspects. Further evaluation of these models is deferred to Chapter 7 (Section 7.4.2 and 7.5), at which point another approach to masking is introduced and also evaluated in the context of the empirical reviews found in Chapters 2–4.

5.5. Summary

I have reviewed four basic types of mechanisms and models which have been proposed to account for many of the characteristics of lateral masking. One species of model assumes that type B metacontrast results when a slow target excitatory response is laterally suppressed by a faster mask inhibitory response. Among preferred mechanisms relying on slow excitation and fast inhibition are the lateral interaction: (i) between slow, secondary (thermal) reactions and fast photochemical reactions at the receptor level (Alpern 1953; Fry and Alpern 1946); (ii) between slow rod responses and fast cone responses (Alpern 1965; McDougall 1904a); (iii) between slow (e.g., short-wavelength) cone responses and fast (e.g., long-wavelength) cone responses (Foster 1976, 1979); and (iv) between fast and slow components of post-receptor, neural responses (Foster and Mason 1977; Weisstein *et al.* 1975; Yellott and Wandell 1976). In order, however, to explain type B paracontrast, the reverse response polarity (slow mask inhibition and fast target excitation) must be invoked. Although peripheral processes (i–iii) may contribute to the overall shape of the type B masking effects and the SOA at which peak masking occurs, it was argued that mechanism (iv), relying on post-receptor interactions between slow and fast neural responses, plays the major role in type B lateral masking. This conclusion was based on the fact that interactions among peripheral receptor processes cannot account for type B meta- or paracontrast when target and mask activate the *same* chromatic (π) mechanisms and, moreover, when target and mask are flashed *interocularly*. Although these mechanisms and models may account for type B contrast and contour suppression effects they do not apply, without making additional assumptions, to the lack of masking obtained when simple target detection is employed as a response criterion.

A second class of models relies on the activation of spatiotemporal sequence detectors. Kahneman's (1967b) model relies on a cognitive decision process which, because of its supposed inability to accommodate 'impossible', i.e. split, stroboscopic motions produced by a target–mask sequence, suppresses the contrast visibility of the target. It was noted, however, that:

1. Split apparent motion is a very possible percept (Wertheimer 1912; Stoper and Banffy 1977).

2. Metacontrast can be obtained under possible stroboscopic motion (Breitmeyer and Horman 1981; Breitmeyer *et al.* 1974, 1976):

3. Metacontrast can be obtained with or without the presence of split stroboscopic motion (Stoper and Banffy 1977).

4. The spatiotemporal relations governing stroboscopic motion diverge appreciably from those governing metacontrast (Breitmeyer and Horman 1981; Weisstein and Growney 1969).

Matin's (1975) neural model relies on two properties of spatiotemporal sequence (T-M) detectors: (i) their optimal activation at some intermediate SOA (50–150 ms); and (ii) in so far as fast, transient neurones comprise a major portion of these detectors, the existence of a shorter response latency of the T-M relative to T- (or M-) neurones. Although this model can account adequately for type A and type B meta- and paracontrast, it fails to account for target contrast suppression when a single- rather than two-transient masking paradigm is employed. In the former paradigm, no spatiotemporal *sequence* detectors can be activated.

Among the third, two-process, class of metacontrast models, I discussed Ganz's (1975) model based on an interactive (target–mask) trace decay and random encoding time, Reeves' model based on temporal integration and segregation, and Navon and Purcell's (1981) model based on 'fortunate' integration and ('unfortunate') interruption. Specific difficulties with Ganz's model are its failure to correctly predict a transition from type B to type A metacontrast as T/M energy ratio decreases below 1.0; the presence of type B metacontrast as the T/M energy ratio increases above 1.0, and the inability to account for type B paracontrast when the T/M energy ratio is 1.0. More generally, all the two-process models do not account for the results obtained when a simple detection criterion is employed. Moreover, they rely on response characteristics—in particular, temporal integration and segregation, which, unlike the prior models, predict shifts of optimal metacontrast SOA towards lower as opposed to higher values as background or stimulus (target *and* mask) energies are increased.

The last critique applies also to the fourth class of models; in particular, Bridgeman's (1971, 1977, 1978) distributed network model. Moreover, as with the other models, Bridgeman's model also does not specify the type of masking effects that would be obtained when in addition to brightness rating, either simple detection or contour-detail discrimination are used as response criteria. However, as noted, such specification is possible by incorporating a central decision process which selects information from the cross-correlation of the initial, intermediate, and later components of the stored target and the briefed target–mask responses when simple detection, brightness or contrast discrimination, and contour-detail discrimination, respectively, are employed as response criteria.

6 Sustained and transient neural channels

In Chapters 2–5 I reviewed and discussed empirical findings, methods, and theories of masking by light, visual persistence, and pattern masking, with the emphasis in the last topic on para- and metacontrast. In this cumulative review we, in particular, listed a variety of phenomena and stimulus and procedural variables systematically associated with the observed results. These variables and phenomena can be summarized by the three following classes:

1. *Spatiotemporal* variables, including (i) stimulus size; (ii) spatial frequency; (iii) orientation; (iv) contour or interstimulus separation; (v) onset rise time and offset fall time; (vi) retinal location and (vii) the phenomenon of stroboscopic motion.

2. *Intensity* variables, including (i) background intensity or adaptation level; (ii) stimulus intensity; and (iii) polarity of stimulus contrast relative to the background (brighter or darker than background).

3. *Chromatic* variables, including (i) wavelength composition of the background and (ii) wavelengh composition of the stimuli.

4. Stimulus *viewing condition*, including (i) monoptic; (ii) binocular; and (iii) dichoptic, with the dichoptic condition tapping only higher level processes occurring at or beyond the level of interocular response combination—in particular, the visual cortex.

The present chapter discusses these and additional variables, in one form or another, in the context of a review of physiological and anatomical properties which distinguish between two types of visual cells that presumably are differentially used in stimulus processing. The review may be of some difficulty to the psychologically or psychophysically oriented readers not familiar with these particular anatomical or physiological properties of the visual system. If that is the case, careful reading, perhaps even re-reading, is recommended, since at least minimal comprehension of the included topics is essential as a preliminary step to further understanding a theory of or an approach to visual information processing, outlined in the following chapter and successively elaborated and unfurled in Chapters 8–10.

The discovery by Enroth-Cugell and Robson (1966) of separate classes of X- and Y-cells in the cat retina and the subsequent studies and elaborations of response differences between these two types of cells has had a marked impact on recent theories of visual information processing. Enroth-Cugell and Robson found that X- and Y-cells differed, for one, on the basis of their responses to a sine-wave grating whose spatial phase was shifted relative to the centres of their respective receptive fields. X-cells were characterized by optimal positions in which either a strong excitatory or else inhibitory response was elicited and by null positions in which the introduction or withdrawal of the grating produced no response. Y-cells, on the other hand, responded to all positions or phases of the grating. From these results Enroth-Cugell and Robson concluded that whereas X-cells respond linearly to the sum of signals from different parts of the receptive field, Y-cells respond non-linearly. This response distinction between X- and Y-cells in cat retina has been analysed and elaborated further by Shapley and Hochstein (1975), Hochstein and Shapley (1976*a*, *b*), and Gielen, van Gisbergen, and Vendrik (1981).

Another property investigated by Enroth-Cugell and Robson (1966) which differentiated between X- and Y-cells was their response to the prolonged presentation of a grating. X-cells responded in a sustained manner throughout the presentation of the grating, wheras Y-cells responded primarily in a transient fashion to the abrupt onset or offset of the grating. For that and other experimentally based reasons X- and Y-cells have subsequently also been termed 'sustained' and 'transient' cells, respectively—a terminology I shall adopt for the remainder of this and subsequent chapters.* Both of the above response properties are illustrated in Fig. 6.1, which shows the response of (A) an off-centre X-cell and (B) an off-centre Y-cell to the prolonged presentation of a sine-wave grating at each of four spatial phases relative to the receptive field centres. Note, for instance, that at a phase angle of 90° and 270°, the response of the centre and surround receptive-field mechanisms of the sustained cell cancel each other, resulting in no total response change; whereas the transient unit is activated at all phase angles of the grating.

In recent years and on the basis of additional response properties (to be discussed below) separate classes of sustained and transient cells have been identified in a number of other mammals, including rats (Fukuda 1973;

* Rowe and Stone (1977) prefer to use the X/Y rather than the sustained/transient terminology. Their reasons for this preference are based on philosophical and empirical considerations. They claim that the sustained/transient terminology borders on a form of essentialism (see Popper 1962) in which one looks for essences rather than observable properties and in which one equates identification with classification. Another of their objections to the sustained/transient terminology is that the properties of sustainedness and transiency of visual cells are subject to change with dark adaptation. It will be noted below that transient cells become more sluggish; in fact, they show sustained responses under scotopic levels of illumination (Jakiela and Enroth-Cugell 1976; Zacks 1975). Moreover, as De Monasterio (1978*a*) has shown, the sustainedness and transientness of retinal neurones can depend on a number of other stimulus parameters. Being aware of these and other objections to the sustained/transient terminology (see also Shapley and Hochstein 1975; Hochstein 1979), I none the less prefer to use it. The reasons for this are simply that many other investigators use this terminology and that recent theories of human visual information processing have also adopted it rather than the X/Y terminology. Whether this is justifiable is a matter of debate; however, the terminology adopted in this chapter is merely conventional and meant to refer to response properties and not essences of visual cells.

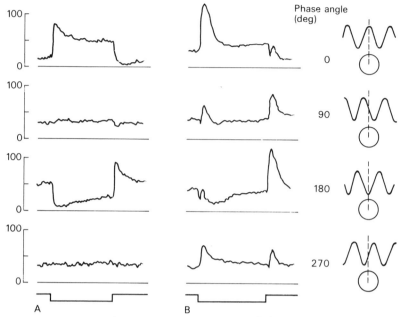

Fig. 6.1. Responses of an off-centre sustained cell (A) and an off-centre transient cell (B) to the onset and offset of a stationary sinusoidal grating. In the lowest panel, downward deflections in both A and B indicate offset of the grating; upward deflection indicate its onset. The phase angle of the grating relative to the midpoint of the receptive field centre is given at the right of the figure. (From Enroth-Cugell and Robson (1966).)

Fukuda and Sugitani 1974; Fukuda, Sugitani, and Iwama 1973; Fukuda, Sumitomo, Sugitani, and Iwama 1979; Hale, Sefton, and Dreher 1979; Lennie and Perry 1981), tree shrews (Norton, Casagrande, and Sherman 1977; Sherman, Norton, and Casagrande 1975) as well as primates (De Monasterio 1978a, b; De Monasterio and Gouras 1975; Dreher, Fukuda, and Rodieck 1976; Webb and Kaas 1976). This inter-species existence of sustained–transient cell dichotomy may also generalize to humans. In Chapter 7 I shall review the psychophysical evidence which points to the existence of these two distinct neural channels in human vision. The present chapter, however, reviews additional neurophysiological and anatomical properties of sustained and transient cells.*

* An additional review of recent work on parallel pathways in the mammalian visual system has been published by Lennie (1980a). Although the topics to be discussed below overlap substantially with those covered by Lennie (1980a), it would nevertheless benefit the reader—particularly if he or she is interested in a more extensive discussion of the anatomical, physiological, and interspecies comparison of parallel visual pathways—also to study Lennie's (1980a) review. Moreover, since some response properties of transient and sustained cells found in higher mammals, e.g. cat or monkey, are not found in lower ones, e.g. rat (see Lennie and Perry 1981), the following reviews of response properties are based on investigations conducted on cat and monkey where visual systems bear greater similarity to that of the human.

6.1. Spatiotemporal response properties

One of the readily determinable ways that sustained and transient cells differ is in terms of their respective receptive field dimensions. At a given location on the retina receptive field diameters of sustained cells are generally smaller than those of transient cells although overlap of the receptive field diameters of a few of the two types of cells may exist (Cleland, Dubin, and Levick 1971, Cleland, Harding, and Tulunay-Keesey 1979; Cleland and Levick 1974; Cleland *et al*, 1973; De Monasterio and Gouras 1975; Fukuda 1971; Famiglietti and Kolb 1976; Hammond 1974, 1975; Peichl and Wässle 1979; Sherman *et al*. 1975; Sherman, Wilson, Kaas, and Webb 1976; Stone and Fukuda 1974). Moreover, for both sustained and transient cells there is an increase of receptive field diameters as retinal eccentricity increases (Cleland and Levick 1974; Famiglietti and Kolb 1976; Hammond 1974, 1975; Peichl and Wässle 1979; Sherman *et al*. 1975; Stone and Fukuda 1974; Wilson and Sherman 1976). Both of the above properties of receptive field diameters are shown in Fig. 6.2.

The difference between receptive field diameters of sustained and transient neurones also is correlated with differences between their respective abilities to interact laterally with other receptive fields, their optimal stimulus sizes, their spatial frequency response profiles, and, in particular, their spatial resolution limits. Overall, the lateral influence which transient cells can exert on neighbouring cells extends over a larger distance than that of sustained cells (Ikeda and Wright 1972*a*). This difference between receptive field size and extent of lateral action has its correlate in corresponding, respective differences between dendritic field sizes of transient and sustained neurones (Wässle, Peichl, and Boycott

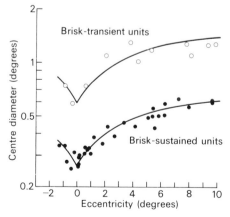

Fig. 6.2. Receptive field centre diameters of brisk sustained and transient cat retinal ganglion cells as a function of retinal eccentricity. (From Cleland, Harding, and Tulunay-Keesey (1979).)

1981*a*; Wässle, Boycott, and Illing 1981*b*). Sustained cells generally can be characterized as band-pass spatial filters with peak sensitivity at an intermediate spatial frequency and significant sensitivity attenuation at both higher and lower spatial frequencies; transient cells, on the other hand, lacking the attenuation of sensitivity at low spatial frequencies, can be characterized as low-pass spatial filters (Derrington and Fuchs 1979; Ikeda and Wright 1975*a*; Lee, Elepfandt, and Virsu 1981; Lehmkuhle, Kratz, Mangel, and Sherman 1980*a*). Moreover, the cut-off spatial frequency or the limit of spatial resolution is generally higher for sustained than for transient neurones. This relative advantage of sustained neurones holds along the entire visual pathway from the retina to the visual cortex (Cleland *et al.* 1979; Derrington and Fuchs 1979; Ikeda and Wright 1975*a*; Lehmkuhle *et al.* 1980*a*; Peichl and Wässle 1979). Moreover, Lehmkuhle *et al.* (1980*a*) as well as Cleland *et al.* (1979) report a significant correlation between receptive field diameter and spatial resolution. Whereas the former investigators reported a moderate correlation, the latter investigators found a strong correlation (r = 0.96) between the centre diameters of receptive field and the spatial period ((spatial frequency)$^{-1}$) of the cut-off frequency.

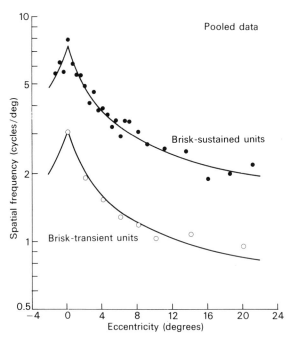

Fig. 6.3. Spatial cut-off frequency of brisk sustained and transient cat retinal ganglion cells as a function of retinal eccentricity. (From Cleland, Harding, and Tulunay-Keesey (1979).)

Since the receptive field diameters of both sustained and transient cells increase with retinal eccentricity, one would expect that their respective spatial resolutions, although differing predictably at any given retinal location, would jointly decrease as retinal eccentricity increases (Cleland *et al*. 1979; Lehmkuhle *et al*. 1980*a*). Both of these relationships are shown in Fig. 6.3. At any given retinal eccentricity sustained neurones are characterized by a higher spatial resolution limit than are transient neurones, and for both types of neurones there is a drop of about two octaves in their spatial resolution limits as retinal eccentricity increases from the fovea to the 20° periphery.

Although subcortical neurones, due to the approximate circular symmetry of their receptive fields, do not show orientation selectivity, cortical cells with their oriented receptive fields (Hubel and Weisel 1962, 1968) are characterized by orientation selectivity. Ikeda and Wright (1975*b*) found that both sustained and transient neurones in cat visual cortex showed orientation tuning. For both types of neurones, the orientation tuning curve was characterized by a mean half-height width of approximately 30°, indicating that transient neurones were as sharply tuned to orientation as were sustained ones. However, Stone and Dreher (1973), although also finding orientation selectivity for both types of cortical cells, in contrast report that sustained cells are more finely tuned than transient ones.

Besides the above spatial response properties, temporal response properties can also be employed to differentiate between sustained and transient cells. Recall that sustained cells respond in a maintained manner to the presence of a prolonged, adequate stimulus, whereas transient cells respond only to its abrupt onset and offset. In addition to this difference, the two types of cells can be distinguished on the basis of several other temporal response parameters.

For one, the responses of sustained and transient neurones differ to temporal modulation or flicker. Although some investigators (Lehmkuhle *et al*. 1980*a*; Lennie 1980*b*) report that the response sensitivity at low and intermediate temporal frequencies is similar for the two cell types, others (Ikeda and Wright 1975*a*) report that transient cells can be characterized as temporal band-pass filters showing a low- and high-frequency attenuation of sensitivity whereas sustained cells can be characterized as low-pass filters which do not show a dramatic low-frequency attenuation of sensitivity. Be that as it may, it is generally agreed that transient neurones are characterized by a higher critical flicker frequency (cff) or temporal resolution limit than are sustained neurones (Derrington and Fuchs 1978, 1979; Ikeda and Wright 1975*a*, 1976; Fukuda and Saito 1971; Lehmkulhe *et al*. 1980*a*). As shown in Fig. 6.4, this relative advantage of transient neurones over sustained ones is maintained at all but the largest retinal eccentricities (Lehmkuhle *et al*. 1980*a*).

The higher temporal resolution limit of transient neurones relative to

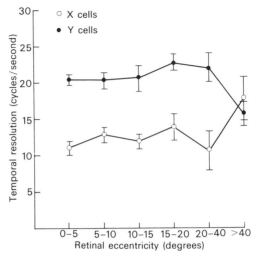

Fig. 6.4. Temporal resolution as measured by critical flicker frequency of sustained (X) and transient (Y) cells in the dorsal lateral geniculate of cat as a function of retinal eccentricity. (From Lehmkuhle, Kratz, Mangel, and Sherman (1980*a*).)

sustained neurones may correspond to marked differences between their respective temporal impulse responses (Cleland *et al.* 1973). Fig. 6.5 shows the responses of a sustained and a transient retinal neurone to a 2-ms flash whose intensity was adjusted to correspond to about 4–8 times each cell's threshold value. Note that relative to the transient cell the impulse response of the sustained cell is characterized by; (i) a longer duration; (ii) a slower rise time; and (iii) a slower decay time. Related results have been reported by Büttner *et al.* (1975) and De Monasterio (1978*b*). In fact, Büttner *et al.* (1975) have shown that the impulse response of transient cells is characterized by a damped oscillatory function composed of alternating excitatory and inhibitory phases whereas that of sustained cells is a monophasic excitatory response. The briefer, crisper, and oscillating impulse response of the transient cells may very well reflect their higher temporal resolution limit.

Sustained and transient neurones also are distinguishable on the basis of their selectivity to the velocity of movement. Sustained cells respond optimally to low-velocity movement whereas transient cells respond best at higher velocities (Bullier and Norton 1977; Cleland *et al.* 1971; Cleland and Levick 1974; Cohen, Winters, and Hamasaki 1980; Derrington and Fuchs 1979; Eckhorn and Pöpel 1981; Leventhal and Hirsch 1978; Sherman *et al.* 1975). Whereas the responses of sustained neurones attenuate dramatically for velocities in excess of about 50°/s (Cohen *et al.* 1980; Derrington and Fuchs 1979), transient cells show only minor response attenuation at velocities in excess of 200–300°/s (Cleland *et al.* 1971; Cohen *et al.* 1980).

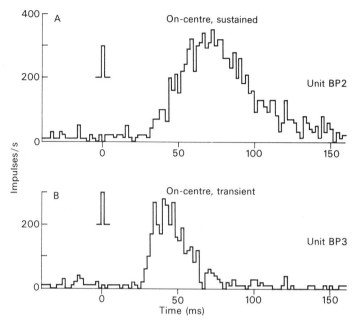

Fig. 6.5. Impulse responses of on-centre and off-centre sustained and transient cat retinal ganglion cell to a 2-ms flash of a spot falling on their receptive field centres. (From Cleland, Levick, and Sanderson (1973).)

Such high retinal motion velocities typically are not produced by objects moving across the visual field. However, since saccadic image displacement across the retina can attain velocities as high as 500°/s (Fuchs 1976), transient cells may be activated by and used in the detection of such retinal image displacement.

Additional properties which differentiate between sustained and transient neurones are their respective response latencies and fibre conduction velocities. Along the entire visual pathway, from retina to visual cortex, transient fibres have a higher conduction velocity than sustained fibres (Bullier and Henry 1979; Cleland *et al.* 1971; De Monasterio, Gouras, and Tolhurst 1976; Dreher *et al.* 1976; Dreher, Leventhal and Hale, 1980; Fukuda 1971, 1973; Fukuda and Sugitani 1974; Henry, Harvey, and Lund 1979; Hoffman *et al.* 1972; Sherman *et al.* 1975, 1976; Stone and Dreher 1973; Stone, Leventhal, Watson, Keens, and Clarke 1980). This advantage of transient fibres relative to sustained ones holds at all retinal eccentricities. Furthermore, as retinal eccentricity increases there seems to be a slight but significant trend towards shorter latencies and higher conduction velocities for both types of fibres (Cleland and Levick 1974; Kirk, Cleland, Wässle, and Levick 1975; Schiller and Malpeli 1978; Stone and Fukuda 1974).

Based on these findings of differential fibre conduction velocity, one may want to infer that the response latencies of transient neurones, particularly at later levels in the visual pathway, are shorter than those of sustained neurones. Although this inference may be generally true, a few experimental results have cast some doubt on its assertion (Ikeda and Wright 1972a; Lennie 1980b). For instance, Ikeda and Wright (1972a) found that in the retina of cat sustained ganglion cells yielded a shorter response latency to a flashed stimulus than did transient ganglion cells. However, this response latency difference was obtained under stimulus conditions which favoured the sustained cells. Whereas a near optimal stimulus diameter of 25' was used for activating sustained cells, a similar 27' diameter used to activate transient cells was far from optimal (due to the generally larger receptive field diameter of transient cells). Consequently, when optimal stimulus size was used for the transient cells, the response latency difference between the two types of cells vanished. In fact, when stimuli of superoptimal size were employed, the response latency of sustained cells increased substantially whereas that of transient cells continued to decrease, a result also reported by Bolz, Rosner, and Wässle (1982) and Singer and Bedworth (1973).

Lennie (1980b) claimed that even when sine-wave gratings of optimal spatial phase and frequency are used to activate either sustained or transient retinal ganglion cells, one can create stimulus conditions under which sustained cells have a shorter response latency than transient cells. In particular, Lennie (1980b) adjusted the contrast of the gratings so that small responses of peak amplitude at about 50 impulses/s above the spontaneous discharge rate were elicited. The contrast values used for both types of cells were thus near threshold. However, problems concerned with Lennie's (1980b) far-reaching interpretation of his particular results exist and need to be addressed. Lennie's interpretation that sustained responses are faster than transient ones rests on the assumption that results obtained with near-threshold contrasts can be extrapolated to suprathreshold results. But as Lennie himself notes, the findings reported by Shapley and Victor (1978) indicate that, as stimulus contrast is increased, the gain of transient cells increases at a faster rate as a function of increasing contrast than that of sustained cells. Hence, response latencies of transient cells decrease at a faster rate than those of sustained cells, as Ikeda and Wright (1972a) initially reported. Consequently, it is entirely likely that, at and above a given supra-threshold contrast value, transient ganglion cells have a shorter response latency than do sustained ganglion cells (Bolz *et al* 1982). Moreover, Lennie (1980b) did not specify the retinal locations of his samples of sustained and transient cells. Since the retinal distribution of sustained and transient cell receptive fields is characterized by foveal–peripheral gradients of activation latency and fibre conduction velocity (Cleland and Levick 1974; Kirk *et al.* 1975; Schiller and Malpeli

1978; Stone and Fukuda 1974), possible sampling biases may have contributed to Lennie's (1980*b*) findings. Consequently, without specifying and equating the retinal locations of the samples of sustained and transient cells, these findings are not unequivocally interpretable.

Additionally, Lennie (1980*b*) argued that the relative advantage in conduction velocity that transient fibres enjoy over sustained fibres (see above) is not sufficient to overcome the initial, retinal response latency difference between sustained and transient cells at later, post-retinal levels of visual processing. However, several studies indicate that this conclusion is wrong. Transient-type cells in the lateral geniculate nucleus (Ikeda and Wright 1976; Singer and Bedworth 1973) and in the visual cortex (Dow 1974; Ikeda and Wright 1975*b*) are characterized by a shorter response latency than are sustained-type cells. In particular, Dow (1974) found that transient-type neurones (his Class V neurones) in the visual cortex generally responded 50 ms or more faster to an optimal photic stimulus than did sustained-type neurones (his Class II neurones). Based on these results and pending further careful experimental analyses, one can tentatively conclude that at least for suprathreshold stimuli of optimal size transient cells generally respond faster to a flashed stimulus than do sustained cells.

6.2. Background and stimulus intensity

Both the spatial and temporal response properties of sustained and transient neurones change as a function of the prevailing background level of illumination. In regard to the spatial response, it has been established that the sensitivity of the antagonistic surround mechanism of retinal receptive fields falls off relative to that of the centre mechanism as a transition from photopic to scotopic levels of illumination is made (Barlow *et al*. 1957*b*). This trend has been identified in both sustained and transient neurones (Cleland *et al*. 1973; Hammond 1975) and is correlated with an increase of optimal stimulus size and a corresponding decrease of the cut-off spatial frequency or spatial resolution limit of both types of neurones (Cleland *et al*. 1973; Enroth-Cugell and Robson 1966).

The temporal and movement response properties of sustained and transient cells also changes with transitions from photopic to scotopic levels of illumination. Whereas the response of both types of cells is brisk and sharp under light-adapted conditions, there is a general blurring of the temporal and spatiotemporal or movement response of both types of cells under dark-adapted conditions. In particular, the responses to a stationary flashed stimulus become more sluggish for both types of cells when scotopic illumination levels are employed (Cleland *et al* 1973; Hammond 1975; Jakiela and Enroth-Cugell 1976; Zacks 1975), and their respective onset latencies and offset latencies (response persistences) also increases

(Hammond 1975, Ikeda and Wright 1972*a*; see also Fig. 3.9 for an illustration of sustained on and offset latencies at two differing background intensities). In addition, Cleland *et al.* (1973) found that the upper velocity limit of a moving stimulus at which a detectable response could be elicited by either type of cell decreased as a transition from light- to dark-adapted states was made.

Decrease of the sensitivity of the receptive field centre mechanism produced by increasing background illumination also differs between sustained- and transient-type units. For a given range of low background luminance, sensitivity does not change in either type of unit; however, the threshold value of the background luminance at which a decrease in the sensitivity of the centre mechanism is initially obtained is lower for transient than for sustained cells (Jakiela and Enroth-Cugell 1976). This differential light-adaptation effect may be due to the facts (i) that transient, relative to sustained, units generally are characterized by large receptive-field centres; and (ii) total luminous flux (area × intensity) rather than the level of retinal illumination determines the level of a unit's light-adaption (Enroth-Cugell and Shapley 1973*b*). The fact that the physiological summation area of transient cells is larger than that of sustained ones may also be related to the finding that at a fixed background luminance level, transient cells are activated at a lower stimulus intensity (Hoffman *et al.* 1972; Singer and Bedworth 1973) and, correspondingly, are characterized by a higher contrast sensitivity (lower contrast threshold) (Derrington and Fuchs 1979) than are sustained ones. Correspondingly, the suprathreshold response of transient neurones also saturates at lower luminance or contrast levels that that of sustained neurones (Ikeda, personal communication). This latter result, in turn, is consistent with Shapley and Victor's (1978) finding that the gain of transient cells increases at a faster rate with contrast than the gain of sustained cells.

Stimulus intensity also affects the temporal response characteristics of transient and sustained cells. As noted in Section 6.1, at a fixed background luminance the response latencies of both transient and sustained cells increase as stimulus intensity or contrast decreases. However, this rate of latency increase is greater for transient than sustained units; and along with this general increase of response latency as stimulus intensity decreases a decrease of response magnitude also occurs (Ikeda and Wright 1972*a*; Shapley and Victor 1978). However, one can maintain a fairly constant response magnitude by varying the stimulus intensity, Δ I, in direct proportion to background luminance, I, so that the Weber ratio Δ I/I remains constant (Büttner *et al.* 1975). Even here, response latency increases as overall (background and stimulus) luminance decrease (Ikeda and Wright 1972*a*). Since the rate of such response-latency increments with decreasing intensity is faster in transient as compared with sustained units, one would expect that at and below a given overall luminance level, the transient response latency converges on or even surpasses (Lennie 1980*b*) the sustained response latency.

6.3. Chromatic response properties

In monkeys which are known to have good colour vision and discriminability, sustained and transient cells are differentiable on the basis of their chromatic response properties. Gouras (1968) found that phasic or transient retinal ganglion cells receive input from both red and green cone mechanisms which provide input to both the excitatory centre and the inhibitory surround of their receptive fields. Consequently, these cells displayed spatial opponency between the antagonistically organized centre and surround response mechanisms of their receptive fields but no colour opponency within or between centre and surround regions. Tonic or sustained cells, on the other hand, were characterized by both spatial and colour opponency; they received centre excitatory signals from only one cone mechanism (red or blue) and surround inhibitory input from another cone mechanism. The non-opponent transient cells were found most frequently in the periphery of the retina whereas the opponent-colour sustained cells tended to predominate at and near the fovea. A similar classification of transient and sustained cells based on non-opponency and opponency of chromatic responses subsequently was confirmed not only for retinal ganglion cells (De Monasterio and Gouras 1975; Gouras 1969; Gouras and Zrenner 1981) but also for cells in the lateral geniculate (Creutzfeldt, Lee, and Elepfandt 1979; Dreher *et al.* 1976; Krüger 1977*a*; Marrocco and Brown 1975; Schiller and Malpeli 1978) and in the visual cortex (Bullier and Henry 1980).

However, although this classification of transient and sustained cells serves well as a first and approximate generalization, finer distinction between transient and sustained cells on the basis of chromatic response can be made. For instance, Schiller and Malpeli (1978) report findings which indicate that some sustained cells in the lateral geniculate nucleus may show broad-band or non-opponent colour responses—a result consistent with earlier findings reported separately by Wiesel and Hubel (1966) and Dreher *et al.* (1976). Some transient cells also are known to show colour-opponency. In particular, De Monasterio and co-workers (De Monasterio 1978*a*, *b*; De Monasterio *et al.* 1976; De Monasterio and Schein 1980) have demonstrated the existence of colour-opponent transient cells in the retina of macaque or rhesus monkeys. Whereas slow-conducting sustained cells found near the fovea showed typical colour-opponent responses in which the receptive field centre and surround were driven by separate cone mechanisms, some fast-conducting transient cells, also found at and near the fovea, also had colour-opponent properties. However, their opponent responses were in part mediated by the same types of cone mechanisms; for example, the centre response was driven by either red or green cone mechanisms whereas the surround response was driven exclusively by green or else red cone mechanisms. In addition, De Monasterio and Schein (1980) found a gradient of broad-band input to non-opponent transient cells as a function of retinal location. Although the centre and surround responses were driven equally by the same red-green, broad-band cone mechanisms in peripheral transient neurones, near the fovea transient non-opponent (as well as opponent) cells showed lower green sensitive cone input to the surround. As a consequence, Y-cells at or near the fovea show a reduced centre-sensitivity to red or long-wavelength stimuli as compared with green ones. This stronger red activation of the surround mechanism may be related to the finding of Dreher *et al.* (1976) that a steady red background suppresses the activity of these foveal or near-foveal transient cells.

One can summarize the findings as follows. Foveal and near foveal sustained cells by and large have narrow-band colour-opponent properties, characterized by distinct cone inputs to the centre and surround mechanisms. Peripheral transient cells are characterized by broad-band non-opponency, and transient cells near the fovea can be characterized as broad-band opponent cells, driven by two cone inputs in the centre of their receptive field and by either only one or by predominantly one cone mechanism in the surround of their receptive field.

6.4. Retinal distribution

The results of some of the above investigations (De Monasterio 1978*a*; De Monasterio *et al.* 1976; Gouras 1968) indicate that sustained and transient cells are not distributed uniformly over the retina. As will be seen below, the findings of several studies regarding the retinal distribution of sustained and transient cells are somewhat contradictory; moreover, significant inter-species differences in the retinal distributions of these two types of cells may exist.

Using the frequency of cells encountered by microelectrode recording techniques, several studies indicate that whereas the relative frequency of sustained neurones in cat is highest in the area centralis and decreases with retinal eccentricity, the opposite trend holds for transient cells. These two opposing trends are evident for retinal ganglion cells (Cleland and Levick 1974; Cleland *et al.* 1973), lateral geniculate cells (Hoffman *et al.* 1972), as well as cortical cells (Ikeda and Wright 1975*b*; Wilson and Sherman 1976). However, in these investigations systematic recording biases may have existed (Cleland, Levick, and Wässle 1975), and consequently their results can be regarded only as rough approximations.

Moreover, transient and sustained cells in cat retina, although clearly differentiable on the basis of several physiological response criteria, are also morphologically distinguishable on the basis of soma size, with transient cells being larger than sustained ones (Boycott and Wässle 1974; Famiglietti and Kolb 1976). Boycott and Wässle (1974) identified several morphological types of ganglion cells, among them, two types called alpha and beta cells, which on the basis of physiological criteria can be identified with transient and sustained cells, respectively (Cleland and Levick 1974; Cleland *et al.* 1975; Stone *et al.* 1980). In their count of cat retinal ganglion cells, Wässle, Levick, and Cleland (1975) and Wässle *et al.* (1981a) report that alpha or transient cells comprise only about 3–4 per cent of the total number, whereas beta or sustained cells comprise about 55 per cent of the total (Wässle, Boycott, and Illing 1981b). Results from individual cats differed. For some cats a slight nasotemporal gradient was evident. In at least one cat the distribution was characterized by a local minimum at the area centralis and maxima just outside it (Wässle *et al.* 1975). None the less, the above investigators concluded that the relative distribution of alpha or transient cells in cat is uniform over the entire retina (Wässle *et al.* 1975, 1981a). However, Fukuda and Stone (1974) applying both morpho-logical and electrophysiological criteria report that the relative distribu-tions of transient as well as sustained cells are not uniform across the retina. The percentage of transient cells was found to be minimal in the area centralis and to increase monotonically to the 20–30° periphery. On the other hand, the percentage of sustained cells was highest in the area centralis and decreased toward the periphery. In addition, Stone (1978) performed a count of cat retinal ganglion cells and found that although the overall percentage of transient cells comprised 4.0–6.3 per cent of the total number, the relative frequency of transient cells was minimal (1.6 per cent) at the areas centralis and increased progressively up to 5.5–6.9 per cent in the peripheral retina. Approximately similar respective values (2 per cent and 4 per cent) were recently reported on the basis of electrophysiological criteria by Peichl and Wässle (1979).

It thus seems that, despite several morphological studies revealing retinal uniformities, several other electrophysiological and morphological counts of cat ganglion cells, on the contrary, point to some non-uniformities of the retinal distribution of sustained and transient cells; the former, percentage-wise, more heavily concentrated in the area centralis, the latter more heavily concentrated in the periphery.

In a number or primates one also finds these non-uniformities of the retinal distribution of small- and large-soma cells. In owl monkey (Webb and Kaas 1976) and in galago (De Bruyn, Wise, and Casagrande 1980) small cells seem to concentrate percentage-wise in the fovea, whereas large cells concentrate in the periphery. However, as noted by Webb and Kaas (1976), the size distinction between functionally differentiable cell types, found in the periphery, is reduced or lost in the fovea. Consequently, one cannot infer unequivocally from these results that all of the small foveal cells are sustained ones. Moreover, both of these primates are nocturnal and may share more in common with cat than with diurnal primates. However, De Monasterio and Gouras (1975) report that non-uniformities in the distribution of sustained and transient cells in the retina of the cat also are found in the retina of the rhesus monkey, a diurnal primate.

Finally, sustained and transient cells also differ along other dimensions as a function of retinal eccentricity. Peichl and Wässle (1979) report that the ratio of the coverage factor (the product of cell density and the area of the receptive field centre (Cleland *et al.* 1975)) of sustained to transient neruones is highest in the fovea (about 6:1) and decreases in the periphery (about 3:1). In addition, Cleland and Levick (1974) note that, on the one hand, the *activity* of sustained cells was enhanced at smaller eccentricities relative to that of transient cells which, on the other hand, were relatively more easily activated at greater eccentricities. In so far as these clear, functional, or the above-mentioned, somewhat equivocal, distributional, non-uniformities between sustained and transient cells show a high interspecific similarity, they may also characterize the visual system of humans.

6.5. Excitatory and inhibitory interactions

The response of most visual cells can be characterized by the interaction of two types of response, one excitatory and the other inhibitory. In addressing the topic of antagonistic interactions, it is necessary to make the distinction between those which occur within a neurone and those which occur between neurones. The former type is realized in the functional architecture of receptive fields characterized by the centre and surround regions of a neurone's receptive field; the latter type, by inhibitory influences that one neurone, with its separate centre–surround organization, can exert on another.

Sustained and transient cells differ in the organization of their receptive fields and consequently yield quantitatively and qualitatively different antagonistic interactions. Pollack and Winters (1978) report that in retinal ganglion cells the amount of suppression which the surround region of the receptive field can exert on the centre region is about two times larger for sustained cells than for transient cells. Evidence for stronger inhibition within sustained cells also has been reported by several other investigators (Bullier and Norton 1977; De Monasterio 1978*b*; Fukuda and Stone 1976; Ikeda and Wright 1972*a*).

This quantitative difference between the inhibitory activities within sustained and transient cells may be correlated with related biochemical differences characterizing their surround mechanisms (Kirby 1979; Kirby and Enroth-Cugell 1976; Saito 1981) as well as differences between their receptive field structure. In regard to the latter difference, Ikeda and

Wright (1972*a*), Hammond (1975), and Bullier and Norton (1977) found that, when simultaneously stimulating centre and surround of receptive fields, sustained cells are characterized by a spatial sensitivity gradient which is much steeper than that of transient cells. Consequently, sustained cells are characterized by a significantly sharper boundary between the centre and surround. These results may relate to the finding that sustained cell activity is more sensitive (by being suppressed) to the optical blurring of stimuli than are transient cells (Ikeda and Wright 1972*b*). Hammond (1975) proposed the following schemes, illustrated in Fig. 6.6, for the organization of sustained and transient receptive field profiles of retinal ganglion cells. The diameter of the sustained receptive field is generally smaller than that of the transient one. For both types of cells the centre excitatory mechanism can be characterized by a unimodal response profile. However, the inhibitory surround of sustained cells is characterized by a bimodally distributed profile with little, if any, overlap at the very centre of the receptive field. On the other hand, the surround mechanism of the transient cell is a unimodally distributed profile which overlaps the entire centre region.

These differences in receptive-field organization may be related to the fact that the maintained discharge of transient cells, whose centre response is inhibited throughout by the spatially overlapping surround response, is lower than that of sustained cells (Cleland and Levick 1974; Cleland *et al.* 1971, 1973; Dow 1974; Hammond 1972; Ikeda and Wright 1972*a*). Related and similar schemes for the receptive field organization of sustained and transient cells have been proposed by Hickey *et al.* (1972, 1973), and De Monasterio (1978*b*). As a transition from photopic to scotopic levels of background illumination is made the strength of the surround mechanisms

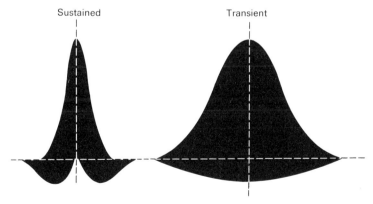

Fig. 6.6. Schemata of receptive field response profiles of sustained and transient cat retinal ganglion cells. Region above the horizontal dotted line corresponds to the centre response mechanism; regions below the line correspond to the surround response mechanism. (From Hammond (1975).)

in both types of cells declines (Cleland *et al*. 1973; Hammond 1975), thus resulting in a functionally larger centre. However, for sustained cells, this shift is less dramatic than for transient cells (Hammond 1975) suggesting that sustained cells retain a relatively sharp centre–surround boundary.

The temporal parameters influencing surround inhibition on centre excitation also seem to distinguish sustained from transient cells. Winters and Hamasaki (1975, 1976) found that for both types of cells, the maximal inhibition of the surround on the centre was obtained when the surround was activated several tens of milliseconds before the centre, a result consistent with findings reported by Poggio *et al*. (1969) in the lateral geniculate nucleus of cat. However, Winters and Hamasaki (1976) report that the lead-time of the surround activation required to obtain optimal inhibition of the centre response of sustained cells was shorter than that of transient cells. Moreover, for both types of cells, the lead-time decreased as the intensity of the surround stimulus decreased, a result to be expected since the surround activity is directly proportional to the intensity of the stimulus with which it is activated.

The existence of inhibitory interactions *between* sustained and transient neurones also has been established in recent years. Hoffman *et al*. (1972), Singer and Bedworth (1973), Singer (1976), and Tsumoto and Suzuki (1976) report evidence of reciprocal inhibition between sustained and transient cells in the lateral geniculate. Moreover, the results of the former investigators as well as those of Stone and Dreher (1973), Singer, Tretter, and Cynader (1975), and Tsumoto (1978) indicate that such interneural inhibition also exists in the visual cortex (area 17) of the cat. Furthermore, in the lateral geniculate nucleus of cat either on- or off-centre transient units can inhibit *both* on- and off-sustained units (Singer, personal communication). Consequently, at post-retinal levels transient-on-sustained neural inhibition is indifferent to the contrast polarity of the stimulus activating the unit. Since the transient response is faster than the sustained response, one would expect that the convergence of transient-activated inhibition on sustained-activated excitation is a relatively early event in the course of sustained neural transmission. Fig. 6.7 shows the response of a sustained cell in the lateral geniculate of cat to a light flash that activated it as well as transient cells. Note that the initial short-latency response is a *transient* inhibition followed by superimposed excitatory post-synaptic potentials generated at a latency of about 80 ms that reach their teminal firing level only 110 ms after the light is flashed. By contrast, the reciprocal inhibition which slower sustained cells exert on transient ones should be a rather long-latency effect since the excitatory post-synaptic potentials of sustained cells which activate inhibitory inter-neurones have a longer latency than those of transient cells. Moreover, since transient cells are activated by high-velocity image movement, Singer and Bedworth (1973) and Matin (1974*b*) suggest that the transient-on-

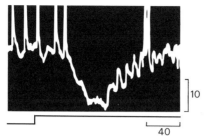

Fig. 6.7. The response of a cat lateral geniculate sustained cell to the onset of a light. Note the early transient inhibition followed at about 80 ms by the first subthreshold excitatory post-synaptic potentials. The firing level of the cell is reached only 110 ms after light on. (From Singer and Bedworth (1973).)

sustained inhibition may be a neural correlate of saccadic suppression. The reverse, sustained-on-transient inhibition may also have a functional role in vision and will be further discussed in Chapters 9 and 10.

6.6. Periphery and shift effects

McIlwain (1964, 1966) discovered that the response of cells in the cat retina and lateral geniculate nucleus could be affected by moving black-and-white patterns well outside their receptive field boundaries as defined by standard spot-mapping techniques (Kuffler 1953). An explanation of the periphery effect on the basis of stray light has been ruled out by Levick, Oyster, and Davis (1965) and by Barlow, Derrington, Harris, and Lennie (1977), indicating that its origin is neural. In particular McIlwain (1964, 1966) found that this so-called 'periphery effect' could produce the following types of response changes in cells: (i) the maintained discharge in the absence of a centre stimulus could be increased, and (ii) the response to a subthreshold centre stimulus could be enhanced. More recently, Jakiela (1975, 1978) and Fisher *et al.* (1978) found that in a few cases, in which suprathreshold centre stimuli were employed, the response was reduced (see Chapter 10, pp. 331–2, for a possible explanation).

In the cat retina the periphery effect is generally excitatory. Although Ikeda and Wright's (1972c) investigation indicated that the periphery effect can be found only among transient retinal cells, other investigators (Barlow *et al.* 1977; Cleland *et al.* 1971; Cleland and Levick 1974) also have found periphery effects among some sustained cells. However, Cleland *et al.* (1971), Barlow *et al.* (1977), and Derrington *et al.* (1979) report that although the periphery effect can be found for sustained retinal units, it generates a weaker and more sluggish response compared with the strong, brisk effects obtained in transient cells. Moreover, when significant periphery-effects occur, they are (i) found predominantly in the brisk transient and sustained cells but not the sluggish ones; and (ii) are stronger extrafoveally than at or near the fovea (Cleland and Levick 1974).

Krüger and co-workers (Fischer and Krüger 1974; Fischer, Krüger, and Droll 1975; Krüger 1977*b*; Krüger and Fischer 1973; Krüger, Fischer, and Barth 1975) have investigated the periphery effect, which they called the 'shift effect', not only in cat retinal ganglion cells but also in the cells of the retina and lateral geniculate nucleus of the rhesus monkey. Moreover, Wanatabe and Tasaki (1980) also have investigated the effect in retinal ganglion cells of the rabbit. The generally excitatory nature of the shift effect to adequate stimuli can be seen in Fig. 6.8, which shows the responses of on-centre and off-centre cells in the lateral geniculate of cat (Fischer and Krüger 1974), to either a prolonged light or dark spot in the centre of their receptive fields. Note that in both cells, an adequate centre stimulus (light spot for the on-centre cell, dark spot for the off-centre cell) evokes a transient discharge at the moment the peripheral grating is suddenly shifted through one-half cycle. An inadequate centre stimulus (dark spot for the on-centre cell, light spot for the off-centre cell) in contrast produces a transient inhibition, a result to be expected on the basis of classical spot-mapping techniques employed on visual receptive fields (Kuffler 1953).

The mechanisms and properties of the periphery effect have been studied extensively by several investigators (Barlow *et al.* 1977; Derrington *et al.* 1979; Fischer *et al.* 1975; Ikeda and Wright 1972*c, d*). Derrington *et al.* (1979) found that the mechanism which generates the periphery effect in transient cells is insensitive to the spatial phase of the peripheral, shifting

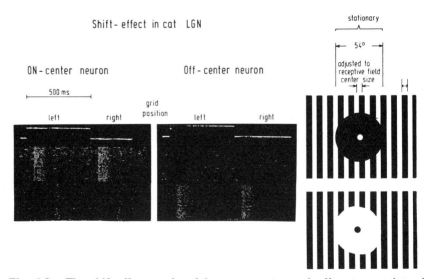

Fig. 6.8. The shift-effect produced in an on-centre and off-centre cat lateral geniculate cell by a suddenly shifting peripheral grating. Note the transient excitation and inhibition in either cell when adequate and inadequate stimuli, respectively, fall on their receptive field centres. (From Fisher and Krüger (1974).)

grating, and that it can resolve gratings of higher spatial frequency than can be resolved by the classically defined receptive field mechanisms of the affected transient cell. Moreover, they found that the mechanism producing the periphery effect extends to and accumulates over approximately 30–40° from the centre of the transient cells' receptive fields, with greater effects produced by peripheral patterns that are nearer to the receptive field centre.

Fischer *et al.* (1975) also investigated several parameters affecting the periphery or shift effect. When increasing the distance separating the receptive field centre from the peripheral, surrounding grating, they, like Derrington *et al.* (1979), found that the latency of the effect produced by a sudden grating shift increased. From this they deduced that the intraretinal conduction velocity of the periphery effect was about 0.35 m/s, which corresponds to 1600 deg/s, a value also approximated in the investigation of Derrington *et al.* (1979). The latency and strength of the periphery effect additionally were affected by the velocity with which the peripheral grating was shifted. Shift velocities varied from 6 to 600 deg/s. The latency decreased monotonically as shift velocity increased. Moreover, at shift velocities below 10 deg/s the strength of the effect is substantially lower than at higher shift velocities. The shift amplitude and the contrast of the peripheral grating also affected the strength and latency of the periphery effect. A weak but measurable effect could be obtained with shift amplitudes as small as 6' of arc. As a function of shift amplitudes varying from 6' to 5°, the strength of the effect initially increased when amplitude was increased from 6' to 1 or 2° and then remained at its asymptotic optimal value as amplitude increased further up to 5°. When grating contrast was increased, the strength of the effect also increased (see also Barlow *et al.* 1977) and its latency decreased, with asymptotic strength and minimal latency obtained at contrasts of 10–20 per cent luminance modulation.

Although the periphery effect is found to be generally excitatory at the retinal level, in the lateral geniculate nucleus it may display distinct excitatory and inhibitory effects. In particular, Krüger's (1977*b*) investigation of lateral geniculate neurones in rhesus monkeys revealed that although transient cells do display a strong excitatory periphery effect, sustained neurones displayed a weaker inhibitory effect. Fukuda and Stone (1976) also have reported this differential periphery effect on transient and sustained cells in neurones of cat lateral geniculate nucleus. One possible way in which this differential effect arises is through the convergence of inhibitory signals, transmitted via inhibitory interneurones activated by transient cells, onto sustained cells (Hoffmann *et al.* 1972; Singer and Bedworth 1973; Tsumoto and Suzuki 1976). Because such convergence is also found at the cortical level (Singer *et al.* 1975; Stone and Dreher 1973; Tsumoto 1978), the differential, excitatory and inhibitory effects, although not yet fully and carefully investigated at that level, may also exist for cortical transient and sustained cells (see Fischer, Boch, and Bach (1981) for preliminary and tentative negative results).

6.7. Cortical and subcortical projections

In the cat and monkey the existence of sustained and transient cells has been established along the entire visual pathway. They are found in the retina (De Monasterio 1978*b*; Enroth-Cugell and Robson 1966; Gouras 1968; Kaas, Huerta, Weber, and Harting 1978), in the lateral geniculate nucleus (Citron, Emerson, and Ide 1981; Dreher *et al.* 1976; Dreher, Leventhal and Hale 1980; Friedlander, Lin, and Sherman 1979; Friedlander, Lin, Stanford and Sherman 1981; Kaas *et al.* 1978; Kratz, Webb, and Sherman 1978; Schiller and Malpeli 1978; Wilson, Rowe, and Stone 1976), and the visual cortex (Citron *et al.* 1981; Dreher *et al.* 1980; Dow 1974; Ikeda and Wright 1974; Kulikowski, Bishop, and Kato 1979; Singer *et al.* 1975; Tretter, Cynader, and Singer 1975). Moreover, at progressively higher levels of the retinogeniculocortical pathway, and particularly at the cortical level (Dow 1974; Dreher *et al.* 1980; Schiller, Finley, and Volman, 1976), transient cells can be activated by stimuli of opposite contrast polarity (brighter or darker than the background). Hence, here, as in the lateral geniculate, transient-on-sustained inhibition (see Section 6.5) should be indifferent to the contrast polarity of the stimulus activating transient neurones. Although there is some electrophysiological and anatomical evidence supporting the presence of sustained-type cells in the superior colliculus (Cleland and Levick 1974; Fukuda and Stone 1974; Peck, Schlag-Ray, and Schlag 1980; Wässle and Illing 1980), of the two classes of cells, transient cells predominate there (Berson and McIlwain 1982; Fukuda and Stone 1974; Hoffman 1973; Marrocco and Li 1977; McIlwain and Lufkin 1976; Schiller and Malpeli 1978).[*]

In cat, sustained and transient cells project retinotopically to layers A, A_1, and C of the lateral geniculate nucleus (Fukuda and Stone 1974; LeVay and Ferster 1977; Leventhal 1979; Wilson *et al.* 1976). Moreover, transient retinal cells also project directly to the superior colliculus (Fukuda and Stone 1974; Hoffman 1973; Hoffman and Sherman 1975; McIlwain 1975). The direct, simultaneous projections of transient cells to the lateral geniculate nucleus and superior colliculus seem to be due to the bifurcation of transient optic tracts. From the lateral geniculate nucleus of the cat sustained cells project exclusively to area 17 of visual cortex

---

[*] It is plausible that the 'sustained' cells found in cat's superior colliculus may belong to a particular subclass known as W-cells (Fukuda and Stone 1974; Hoffmann 1973), in particular the tonic W-cells identified by Fukuda and Stone (1974). At the retinal level these tonic W-cells and the other subclass of phasic W-cells may correspond, respectively, to Cleland and Levick's (1974) classes of 'sluggish sustained' and 'sluggish transient' ganglion cells.

whereas transient cells project via two separate bifurcating pathways to areas 17 and 18 (Citron *et al.* 1981; Dreher *et al.* 1980; Hoffmann and Stone 1971; Singer *et al.* 1975; Stone and Dreher 1973; Tretter *et al.* 1975). Recently Citron *et al.* (1981) reported results indicating, in line with prior studies (Hoffmann and Stone 1971; Hoffman *et al.* 1972; Wilson and Sherman 1976), that cortical simple and complex cells primarily receive input from geniculate sustained and transient cells, respectively; although Movshon, Thompson, and Tolhurst (1978*a*, *b*, *c*), on the basis of their studies, suggest that whereas simple cells are generally sustained ones, complex cells can be either transient or sustained.

In area 17 of cat visual cortex, sustained and transient projections from the lateral geniculate nucleus also terminate in different cell layers. Sustained projection fibres terminate predominantly in layer 4c, whereas transient fibres terminate at the boundary of layers 3 and 4, and in layers 4ab, 5, and 6 (Bullier and Henry 1979; Ferster and LeVay 1978; Henry, Harvey, and Lund 1979; LeVay and Gilbert 1976; Leventhal 1979; Leventhal and Hirsch 1978). The fibres of transient cells in layer 5 may provide corticofugal input to the superior colliculus (Finlay, Schiller, and Volman 1976; Harvey 1980; Palmer and Rosenquist 1974; Rosenquist and Palmer 1971) and thus constitute an indirect transient projection to that structure (Hoffmann 1973).

In the monkey, due to anatomical differences as compared with the cat, the central projection of sustained and transient neurones take a different route. Sustained retinal cells, particularly those at and near the fovea, seem to project predominantly to the parvocellular layers of the lateral geniculate nucleus; transient cells, on the other hand, besides projecting to the superior colliculus (Leventhal, Rodiek, and Dreher 1981), project to the magnocellular layers (Bunt, Hendrickson, Lund, Lund, and Fuchs 1975; Dreher *et al.* 1976; Leventhal *et al* 1981; Marrocco 1976; Schiller and Malpeli 1978; Sherman, Wilson, Kaas, and Webb 1976), a finding which may also hold for the human lateral geniculate nucleus (Hickey and Guillery 1981). In fact, Dreher *et al.* (1976), Leventhal *et al.* (1981), and Sherman *et al.* (1976) claim that the segregation of sustained and transient projections to the parvo- and magnocellular layers, respectively, is complete, although more recent findings indicate only partial segregation (Kaplan and Shapley 1982; Marrocco, McClurkin, and Young 1982). Unlike the projections to the cortex of cat, the sustained and transient fibres originating in the lateral geniculate nucleus of monkey seem to terminate only in area 17 of visual cortex (but see Benevento and Yoshida (1981) and Yuki and Iwai (1981)). Sustained cells from the parvocellular layer of the lateral geniculate nucleus project predominantly to layer 4c and less so to a narrow band in layer 4a; transient cells, on the other hand, project predominantly to layer 4b, lying between layers 4a and 4c, with some extension into the upper part of layer 4c (Bullier and Henry 1980; Hendrickson, Wilson, and

Ogren 1978; Hubel and Wiesel 1972; Lund and Boothe 1975).

One can speculate about the functional role of this segregation of sustained and transient signals to area 17 of cat and monkey, Singer (personal communication) has suggested that one of the roles of the transient neurones in area 17 of visual cortex is to inhibit sustained neurones during saccades. Since layer 4c, containing sustained cells, is known to project to upper layers of area 17 (Lund, Lund, Hendrickson, Bunt, and Fuchs 1975) which are very likely involved in the coding of pattern information (Mountcastle 1978), the transient cells, lying above in layer 4b and activated by saccadic image displacements, could provide the source of inhibition which interrupts the flow of sustained signals from layer 4c to upper layers. Of course, another source of such interneural inhibition could exist in lateral geniculate nucleus of monkey similar to that found in cat (Singer and Bedworth 1973; Tsumoto and Suzuki 1976), provided that the segregation of parvo- and magnocellular layers in the lateral geniculate nucleus is not complete or that interactions occur at the boundaries of layers.

6.8. Developmental aspects

The ontogenetic development of the visual system has been investigated intensively in recent years. It is generally agreed that several response properties of neurones along the visual pathway are present at birth and that their subsequent development and refinement depend on maturational and experiential factors. Several methods exist for studying this postnatal development of single cells. One is to record their responses in a normally reared organism at various stages of development to ascertain their changes and refinements with age. However, using this method one cannot unequivocally differentiate between maturational and experiential sources of change. To overcome this problem additional methods have been developed to eliminate or distort early visual experience in order to study the effects of selective deprivation on neural development. Among the methods employed are dark-rearing and monocular or binocular eyelid suture, which effectively eliminate experience of pattern vision in one or both eyes. Other methods include paralysis and severing of extraocular muscles in order to induce abnormal binocular experience. Investigating the neural effects of such treatments allows one to evaluate the role of abnormal visual experience during development.

According to several investigators (Garey and Blakemore 1977; Lehmkuhle, Kratz, Mangel, and Sherman 1980b; LeVay and Ferster 1977; Norton *et al.* 1977; Sherman *et al.* 1972; Sireteanu and Hoffman 1979; Zetlan, Spear, and Geisert 1981) monocular deprivation exerts differential effects on the development of sustained and transient neurones found in the lateral geniculate nucleus. For instance, using electrophysiological recording techniques, Sherman *et al.* (1972), Norton *et al.* (1977), Sireteanu and Hoffmann (1979), and Lehmkuhle *et al.* (1980*b*) found an abnormally low sample of transient cells in the binocular segments of the lateral geniculate nucleus, whereas the number of sustained cells there appeared to be normal or only slightly reduced. In the monocular segment, both sustained and transient cells seemed to be unaffected. This finding has been corroborated anatomically by Garey and Blakemore (1977) and LeVay and Ferster (1977). These investigators found a selective effect of monocular deprivation on the development and growth of transient cells in the deprived laminae of the lateral geniculate nucleus. A severe arrest of cell growth and loss of cells in the deprived laminae appeared to affect mainly transient cells. The growth of sustained cells was affected modestly. This selective loss and developmental arrest may explain the above-mentioned reduction of recording frequency of transient cells in the binocular segments of the lateral geniculate nucleus.

Binocular deprivation also produces differential effects on sustained and transient cells in the lateral geniculate nucleus. Here transient cells are lost in both the binocular and monocular segments (Kalil and Worden 1978; Kratz, Sherman, and Kalil 1979; Sherman *et al.* 1972), whereas sustained cells are encountered at normal frequencies.

Besides these distinguishable anatomical effects, visual deprivation also seems to affect sustained and transient cells differentially at the physiological level. Lehmkuhle, Kratz, Mangel, and Sherman (1978) showed that monocularly deprived sustained cells in the lateral geniculate nucleus are characterized by a loss of spatial resolution, a finding also reported by Maffei and Fiorentini (1976) and Sireteanu and Hoffmann (1979). In subsequent investigation, Lehmkhule *et al.* (1980*b*) studied both sustained and transient cells and again found that sustained cells suffered this loss, whereas transient cells did not. Moreover, both of the two types of cells retained their normal temporal resolution.

On the basis of these studies, one would expect the number of transient fibres projecting from the lateral geniculate nucleus to the visual cortex to be abnormally low. Since the transient cells in visual cortex provide corticofugal input to the superior colliculus (Bunt *et al.* 1975; Finlay *et al.* 1976; Harvey 1980; Leventhal and Hirsch 1978; Palmer and Rosenquist 1974) one would also expect abnormal development of transient cells in the superior colliculus. Hoffmann and Sherman (1975) have investigated the effects of binocular deprivation on the superior colliculus. They found that the indirect pathway of transient cells to the superior colliculus, involving the geniculate–cortical loop, was entirely missing and that the direct, retinotectal input was unusually small. The former finding corresponds to the severe loss of transient cells in the binocular segments of the lateral geniculate nucleus produced by binocular deprivation (Kratz *et al.* 1979; Sherman *et al.* 1972). Moreover, Hoffmann and Sherman (1975) found that several response properties of those transient cells still found in the superior colliculus were affected by binocular deprivation. In normally reared kittens collicular cells are driven nearly equally well by either eye (Rosenquist and Palmer 1971; Sterling and Wickelgren 1969). However, in the binocularly deprived kittens, the input from the contralateral eye dominates, thus eliminating normal binocular integration. Additionally, sensitivity to direction and high velocity of motion were found to be reduced in collicular transient cells.

The above results revealing a selective loss of transient cells and of physiological function of sustained cells in the lateral geniculate nucleus must be accepted cautiously and tentatively in view of the contradictory findings recently reported by Shapley and So (1980) and Derrington and Hawken (1979, 1981). They showed that the proportion of transient cells encountered in monocularly deprived kittens was equal to that found in normally reared ones. The discrepancy between this finding and the prior ones mentioned above may, according to Shapley and So, be due to biases in electrode sampling of transient cells whose soma size is abnormally small (Garey and Blakemore 1977; LeVay and Ferster 1977). Moreover, Shapley and So (1980) as well as Derrington and Hawken (1981) found no reduction in the spatial resolution of either transient or sustained cells. These results contradict those reported by Lehmkuhle *et al.* (1978, 1980*b*) and Maffei and Fiorentini (1976).

With these caveats aside, let us turn to another form of visual deprivation in which the binocular correspondence of the inputs to the two eyes is eliminated. This, as stated, can be performed by either paralysing the extraocular muscles of one eye or surgically severing them to produce squint. Brown and Salinger (1975) and Salinger, Schwartz, and Wilkerson (1977) have shown that monocular paralysis results in a drastic reduction of sustained cells in the lateral geniculate nucleus of adult cats, which is innervated by the paralysed eye. The relative frequency of transient cells, as one would expect if they were not affected by the treatment, increased. However, here also, as in the types of visual deprivation discussed above, Winterkorn, Shapley, and Kaplan (1981) recently reported contradictory results indicating neither a selective loss of sustained cells nor any overall loss of spatial resolution in lateral geniculate cells of monocularly paralysed, adult cat. Winterkorn *et al.* (1981) suggest that any changes induced by monocular paralysis in the lateral geniculate cells of adult cat is more likely due to change in the strength, number, or pattern of intrinsic inhibitory connections of the lateral geniculate nucleus.

Although this may be the case in adult cats, in neonatal kittens, whose immature visual system, compared with that of the mature adult one, is much more modifiable by selective experimental intervention, a different constellation of results is produced by similar procedures. For example, under conditions of surgically induced squint in one eye, Ikeda and co-workers (Ikeda 1979; Ikeda *et al.* 1976; Ikeda and Tremain 1979; Ikeda and Wright 1975*c*, 1976) found a loss of spatial resolution and contrast sensitivity for both sustained and transient cells of the retina and the lateral geniculate nucleus of kitten innervated by the area centralis of the squint eye but not at retinal eccentricities larger than about 5°. The loss of spatial resolution of sustained cells was, moreover, much more severe than that of transient cells; and in both types of units of the squint eye it was correlated with an expansion of their receptive field centres and a weakening of the antagonistic surrounds. This effect again was more dramatic in sustained than in transient cells (Ikeda and Tremain 1979). The dramatic loss of contrast sensitivity and spatial resolution of sustained cells in the central retina (Ikeda and Tremain 1979) and the lateral geniculate nucleus innervated by the squint eye may provide a neurophysiological correlate of developmental amblyopia in which spatial resolution is lost only in the central 10° of the amblyopic eye but not at larger retinal eccentricities (Ikeda 1979).

Unlike spatial resolution, temporal resolution, as measured by critical flicker frequency, was not affected in sustained cells driven by the squint eye relative to those driven by the normal eye; however, it was reduced somewhat in transient cells (Ikeda and Wright 1976). Moreover, the response latency of central sustained cells in the lateral geniculate driven by the squint eye was found to be on the average about 40 ms longer than in cells driven by the normal eye; however, that of transient cells, although overall shorter than the latency of sustained cells, did not differ between the eyes (Ikeda and Wright 1976).

6.9. Summary

A review of the response properties of sustained and transient neurones reveals that they differ systematically along several dimensions. In the spatiotemporal domain, sustained relative to transient cells are characterized by smaller receptive fields and spatial range of lateral activity, higher spatial resolution, lower temporal resolution, a longer impulse response, selectivity for slower velocities, and a slower fibre conduction speed. At all but the lowest suprathreshold contrasts, sustained cells are characterized by a longer response latency. At cortical levels, they also may show finer orientation tuning than transient cells.

With changes of the light-adaption level the response properties of both types of cells also change systematically. At photopic luminance levels, both types of cells respond briskly; however, at scotopic luminance levels the response of both is characterized by a loss of spatial and temporal resolution. Responses are more sluggish, response latencies increase, and the upper limiting velocity to which either type of cell can respond is decreased.

Foveal and near-foveal sustained cells are generally characterized by narrow-band chromatic opponency with the centre of their receptive fields innervated by one cone process and the surround by another. Transient cells, especially in the peripheral retina, do not show any colour-opponency but rather show a broad-band response. However, at and near the fovea they are characterized by a form of broad-band colour opponency with the centre response driven by cone mechanisms and the surround response by only one of the two found to innervate the centre.

Sustained cells and their activities seem to be most heavily concentrated at and near the fovea, whereas transient cells and their activity are relatively more dominant in extrafoveal as compared with foveal areas. Moreover, the ratio of the coverage factor (the product of cell density and the area of the receptive field centre) of sustained to transient cells is significantly higher in the fovea than in the retinal periphery.

Both sustained and transient cells' receptive fields are characterized by centre–surround spatial antagonism. Typically the surround response of sustained cells is stronger than that of transient ones, making for a steeper spatial sensitivity gradient and sharper centre–surround boundary among sustained cells. This property may be correlated with the higher spatial resolution of sustained cells and a greater sensitivity to optical blur. For both types of cells, the optimal inhibition that the surround mechanism exerts on the centre mechanism occurs when the surround stimulation leads the central one by several tens of milliseconds. Moreover, besides this *intra*channel inhibition realized in the centre–surround antagonism of sustained or transient receptive fields, there also exists, at post-retinal levels, a reciprocal *inter*channel inhibition via which sustained cells inhibit transient cells and vice versa.

The periphery and shift effects, in which remote contour displacements can affect the response of a more localized receptive field as mapped by conventional spot-mapping techniques, are generally excitatory in transient cells but, particularly at post-retinal levels, inhibitory in sustained cells. Moreover, they are stronger in peripheral as compared with foveal areas. The magnitude of the periphery effect varies directly, and its latency varies inversely (up to limiting values) with the velocity, amplitude, and contrast of the displaced remote contours. The closer the remote contours are to the affected receptive field the greater its magnitude and the shorter its latency.

Sustained and transient cells project along separate pathways from the retina, via the lateral geniculate nucleus, to visual cortex. Moreover, transient cells also project directly from the retina and indirectly, via corticofugal fibres, to the superior colliculus. In the lateral geniculate nucleus of primate, sustained and transient cells are found in the parvo- and magnocellular layers, respectively. In area 17 of visual cortex, the two types of fibres terminate in separate layers: sustained fibres project mostly to layer 4c and less so to a narrow band in layer 4a; transient fibres project mostly to layer 4b, with some extension into the upper portion of layer 4c.

Although the reported differential effects of early monocular and binocular deprivation on sustained and transient pathways must be accepted with caution due to possible methodological and measurement confounds, it is generally agreed that squint induced in neonatal periods of development results in a selective loss of contrast sensitivity and spatial resolution, particularly among sustained pathways found within 5° of the fovea or area centralis. Transient cells show a somewhat reduced temporal resolution. Overall, the response latency of squint-affected sustained cells increases significantly whereas that of transient cells does not.

7 Implications of sustained and transient channels for human visual information processing

It was pointed out in the previous chapter that sustained and transient neurones have been identified in the visual systems of a number of subhuman mammals, including primates. Consequently, assuming phylogenetic continuity at the primate level, their existence in the human visual system may not seem surprising at all. In particular, I noted that Hickey and Guillery (1981) distinguished between parvocellular and magnocellular cell layers in the lateral geniculate nucleus of humans, similar to the cell-differentiated layers found in old-world primates and identified with sustained and transient neurones, respectively (Dreher *et al.* 1976; Kaas *et al.* 1978). Moreover, in the past decade evidence accrued from an increasing number of psychophysical and electrophysiological investigations also points to the existence of sustained and transient channels in human vision. This evidence, deriving from several paradigms, including spatiotemporal contrast sensitivity, visual reaction time, uniform-field flicker masking, and visual pattern masking, becomes relevant in explaining some of the major empirical findings reported in Chapters 2–4.

Before proceeding to review the theoretical and empirical aspects of human sustained and transient visual processing as studied by these paradigms, the following caveat is in order. Any claim for a clear functional dichotomization of sustained and transient channels is explicitly denied. As will become evident below, sustained and transient channels in the past, and often misleadingly, have been identified as 'pattern' and 'motion', 'high spatial frequency' and 'low spatial frequency', or 'pattern' and 'flicker' detectors, respectively. Although such distinctions serve as first and useful approximations, these distinctions are presently believed to be not absolute but relative, in the sense that in regard to a particular function or response property one type of channel merely outperforms the other one. For instance, Lehmkuhle *et al.* (1980*a*) argue that transient neurones, in addition to sustained ones, can perform some pattern analysis. This is not a radical statement; after all, Section 6.1 noted that transient as well as sustained cortical neurones are characterized by response selectivity for certain figural properties such as pattern orientation (Dow 1974; Ikeda and Wright 1975*b*; Stone and Dreher 1973). Recent psychophysical evidence (Derrington and Henning 1981; Green 1981*a*; Ronderos *et al.* in press) also indicates that transient channels can perform some rudimentary pattern analysis or conversely that sustained channels can perform some flicker detection (Breitmeyer *et al.* 1981*a*; Derrington and Henning 1981; Green 1981*a*) or motion analysis (Bonnett 1977; Green 1981*a*). Moreover, whereas transient and sustained channels generally are selectively sensitive to low and high spatial frequencies, respectively, both are capable of responding, albeit at a reduced sensitivity, to spatial frequencies outside their preferential range (Breitmeyer *et al.* 1981*a*; Green 1981*a*; Legge 1978; Mitov *et al.* 1981). Hereafter, when transient and sustained channels are identified as 'motion' and 'pattern' analysers, respectively, and so on, the reader should bear in mind that such functional distinctions are not strictly dichotomous or mutually exclusive but rather that they represent first approximation of heretofore conventional conceptual categories with some overlap and 'fuzzy' boundaries.

7.1. Spatiotemporal contrast sensitivity

I noted in the previous chapter that sustained and transient neurones differ in their spatiotemporal response characteristics. Sustained cells have smaller receptive fields and are more sensitive to spatial properties such as fine gratings and sharp contours; whereas transient cells have larger receptive fields and are more sensitive to motion and flicker. Human

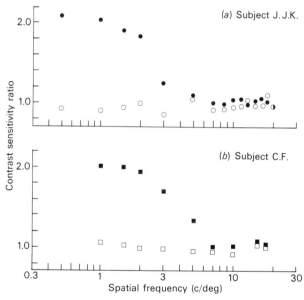

Fig. 7.1. The sensitivity ratio for threshold pattern recognition and flicker detection as a function of spatial frequency. The filled symbols represent the sensitivity ratio of flicker detection of counterphase gratings relative to pattern detection of stationary gratings; the open symbols represent the sensitivity ratio of pattern detection of counterphase gratings relative to pattern detection of stationary gratings. (From Kulikowski and Tolhurst (1973).)

spatiotemporal contrast sensitivity functions may reflect this difference between sustained and transient channels. For instance, Tolhurst (1973) investigated the contrast sensitivity to stationary vertical gratings and gratings either drifting or modulated sinusoidally in counterphase at a rate of 5 c/s. The spatial frequencies employed ranged from about 0.3 to 10.0 c/deg. On the assumption that transient channels are more selective for movement and for lower spatial frequencies than sustained channels, one would expect drifting or counterphase gratings to yield a higher contrast sensitivity at lower spatial frequencies than at higher ones. This expectation was confirmed in that relative to stationary gratings, presumably detected by sustained channels, drifting and counterphase grating yielded an increase of contrast sensitivity only at spatial frequencies of 4 c/deg and lower. Above 4 c/deg the contrast sensitivities obtained with stationary and non-stationary gratings did not differ.

Similar results have been reported by Kulikowski and Tolhurst (1973) and Kulikowski (1975). In these studies, subjects were asked to set two contrast thresholds of gratings modulated in counterphase at rates ranging from 3.5 to 8 c/s and ranging in spatial frequency from about 0.5 to 20.0 c/deg. One threshold setting required the subjects to detect any temporal change in the stimulus display, e.g. flicker or motion; the other required the subjects to detect the spatial structure or pattern of the counterphase grating. A third condition also was used in which subjects were asked to set the contrast threshold for stationary gratings. From Tolhurst's (1973) study it can be inferred that the latter threshold settings measure the contrast sensitivity of sustained channels. Fig. 7.1 shows the sensitivity ratios obtained for the movement and pattern thresholds of the counterphase gratings relative to the contrast threshold for the stationary gratings. Note that the pattern sensitivity ratio is nearly 1.0 at all spatial frequencies of the counterphase grating. However, the movement sensitivity ratio is roughly 2.0 at the lowest spatial frequencies and does not attain a value of 1.0 until spatial frequencies of about 5 c/deg and higher are reached. This indicates that movement or transient channels respond preferably to the lower range of spatial frequencies; whereas, sustained channels respond preferably to intermediate and higher spatial frequencies, although they also can respond to the lower range of spatial frequencies (see Section 7.3 below). Similar conclusions have recently been reached by Nagano (1980) who measured the duration threshold of gratings as a function of their spatial frequency and contrast. He also used the two threshold criteria employed by Kulikowski and Tolhurst (1973) and found that the sensitivity ratio of the transient to the pattern criterion was well above unity for spatial frequencies ranging from 0.25 to 4.0 c/deg and at unity for higher spatial frequencies. The above experiments employed the method of adjustment in which the observer adjusts the contrast of a test grating until it appears to give rise to a flicker sensation or else to a pattern sensation, depending on which criterion content is adopted. Recently Burbeck (1981) reported that, when using a criterion-free, forced-choice procedure, pattern thresholds were actually *lower* than flicker thresholds, contrary to what Tolhurst (1973) and Kulikowski and Tolhurst (1973) had reported. Burbeck's (1981) criterion-free procedure employed the following rationale. In order to establish pattern detection thresholds of a counter-phase flickering grating, observers were asked to discriminate between the counterphase grating and a uniform field flickering at the same temporal frequency and set slightly above threshold contrast. Conversely, to establish flicker thresholds of a counterphase flickering grating, observers were asked to discriminate between the counterphase grating and a stationary grating of the same spatial frequency also set slightly above threshold.

Let us examine the pattern threshold procedure first. In my own observations, a counterphase flickering (or moving) grating *does not* mimic uniform field flicker unless the grating is of sufficiently low spatial frequency; in fact, at higher frequencies one usually sees a 'ripple' effect at threshold spread throughout the viewing field rather than uniform-field flicker. Since Burbeck (1981) used an 8° diameter field and a lowest spatial frequency of only 0.5 c/deg, most of the test gratings would have produced a ripple rather than a uniform-field flicker effect. In fact, since the two thresholds seem to converge at the lowest spatial frequencies used by Burbeck (1981), particularly when higher flicker rates are employed, it seems likely even lower ones, e.g. 0.25–0.125 c/deg, may actually mimic uniform-field flicker and thus reverse the advantage of pattern relative to flicker thresholds, since here ripple effects are no longer capable of being employed to discriminate a counterphase grating from a uniform-field flicker. Hence, with the range of spatial frequencies (0.5–8.0 c/deg) used by Burbeck (1981), the ripple effect, rather than either uniform-field flicker or spatial structure *per se*, could have been used to discriminate the counterphase grating from a comparison stimulus comprised of uniform-field flicker. To overcome this difficulty, it may have been a better procedure to present as a control stimulus not a uniformly flickering field but rather one containing a two-dimensional counterphase grating of the same spatial frequency as the test counterphase grating but composed of random orientations or of two orthogonal gratings each at 45° inclination relative to the fixed orientation of the test grating. Setting this control grating at or slightly above flicker/ripple threshold, one could then have observers discriminate the *spatial orientation* (i.e. pattern component) of the test pattern from this flicker/ripple effect.

With regard to the flicker threshold procedure, an observer was given, as a control stimulus, a barely visible stationary grating of the same spatial frequency as the counterphase flickering test stimulus. The task was to discriminate the counterphase test grating from the control grating.

However, one might ask here whether the threshold discrimination of the test grating was done only on the basis of a perceived flicker/ripple effect. After all, observers were required to discriminate the test stimulus from a stationary control stimulus whose pattern or spatial structure was already, albeit slightly, above threshold contrast. It is not clear whether such a procedure is indeed criterion-free as claimed or biases the observer to a more conservative criterion based on pattern *plus* flicker/ripple detection as opposed to only on flicker/ripple detection. Given this bias, one would expect that (i) the 'flicker' threshold may in fact have been based on a flicker/ripple *plus* pattern criterion, and (ii) as noted in the prior paragraph, the 'pattern' threshold might in fact have been using a ripple rather than pattern criterion relative to a uniform-field flicker criterion. Future research, using a more extensive signal detection analysis, may better establish whether the effects reported by Burbeck are due to threshold sensitivity changes free of criteria or to a biasing of criteria used in her procedure. Until such research is done, the claim of criterion-free pattern and flicker thresholds is questionable.

A psychophysical result related to those of Nagano (1980), showing differential contrast sensitivity of sustained and transient channels in human vision, was reported by Breitmeyer and Julesz (1975; see also Kelly 1973, and Tulunay-Keesey and Bennis 1979). These investigators measured contrast sensitivity to vertical, sinusoidal gratings under two presentation conditions. In one condition the gratings were presented for 480 ms with an abrupt onset and offset; in the other condition the gratings were presented with a slow, 200-ms long ramped onset and offset, thus eliminating temporal transients. Relative to the latter condition, the former one, which retained abrupt transients at on- and offset, produced about a two-fold increase of contrast sensitivity at spatial frequencies ranging from 0.5 to 4.0 c/deg. At higher spatial frequencies no difference of contrast sensitivity between the two conditions was observed. Here presumably only sustained channels were used in the detection of the gratings. This result corroborates those of the above investigators. In summary, the experiments cited above are in agreement regarding the range of spatial frequencies to which transient and sustained channels *preferably* respond. Moreover, the results indicate that at low spatial frequencies transient channels are characterized by a lower contrast threshold than sustained channels, results consistent with the single-cell studies reported by Derrington and Fuchs (1979), Hoffman *et al.* (1972), and Singer and Bedworth (1973) (see Section 6.1).

In another study, King-Smith *et al.* (1976) measured the contrast threshold of a narrow, $0.05 \degree \times 2.0\degree$, vertical line as a function of the oscillation frequency with which it moved to-and-fro in a left-to-right direction over an amplitude of $0.05\degree$. Relative to the contrast threshold of a stabilized line, contrast threshold was lowered at oscillation frequencies of about 0.1–8 c/s, but increased at higher ones. This range of oscillation frequencies corresponds to velocities ranging roughly from 0.01 to 0.8 deg/s. Since the target line was only 3′ wide, for the most part its fundamental and higher spatial frequency components were all above 10.0 c/deg. Consequently, on the basis of the above studies, it should activate mainly sustained channels. Given this reasonable assumption, it can be seen that sustained channels activated by the line are selectively sensitive to a fairly low range of velocities, whereas transient channels responding, say, to a 0.3 c/deg grating drifting at a rate of 8 c/s, would be selectively sensitive to a velocity of at least 26.6°/s; a conclusion consistent with the psychophysical findings reported by M. G. Harris (1980; see also Butler *et al.* 1976). M. G. Harris (1980) found that pattern sensitivity exceeded flicker sensitivity at drift velocities of a grating below 1°/s, whereas at higher velocities the reverse was true. This differential sensitivity of human sustained and transient channels to velocities of retinal image motion is consistent with the single-cell results reported in Section 6.1, particularly that of Eckhorn and Pöpel (1981) who showed that sustained ganglion cells in the area centralis of cat are specifically tuned to low velocities characterizing the destabilizing retinal image drifts required to maintain pattern vision.

Moreover, several electrophysiological studies on humans point to the existence of sustained and transient channels. Kulikowski (1974) measured the cortical visually evoked potential to a counterphase grating before and after prolonged adaptation to the same stationary grating. Since the latter grating contained no temporal transients, it selectively adapted sustained channels but left transient channels unaffected. Consequently, Kulikowski (1974) found little, if any, difference between the shape and form of the counterphase evoked potential produced prior to and after selective adaptation of sustained channels. Related and similar psychophysical results have been reported by Bodis-Wollner and Hendley (1977, 1979).

Based on the above results showing that transient channels prefer rapid movement and abrupt onsets and offsets, it can be inferred that other temporal transients such as flicker also affect sustained and transient channels differentially. Tulunay-Keesey (1972) measured flicker and pattern thresholds for a 0.067°× 1.0°, vertical line at flicker frequencies ranging from 0.3 to 30.0 c/s. Here subjects were required to detect flicker (regardless of pattern detail) and pattern detail (e.g. line orientation), respectively. She found that flicker sensitivity was higher than pattern sensitivity for practically the entire range of flicker frequencies. Flicker sensitivity was greatest at frequencies ranging from about 2.0 to 15.0 c/s. At lower and higher frequencies the two sensitivity functions tended to converge. This result again shows that flicker or transient channels have a lower threshold than pattern or sustained channels, particularly for intermediate to high flicker rates. Sustained channels, in contrast, seem to prefer low flicker frequencies or else higher ones nearing or exceeding the critical flicker frequency of transient channels, above which the flickering line at threshold would eventually appear as being sustained and non-flickering.

The spatial response profiles of flicker and pattern channels also can be assessed psychophysically by using subthreshold summation techniques. King-Smith and Kulikowski (1973, 1975) measure the effect that two parallel, narrow, 1.2′ wide subthreshold lines had on the threshold detectability of a central line of the same width. All three lines were flickered in phase at a rate of 12 c/s. Separate flicker and pattern thresholds were obtained as a function of the spatial separation of the two flanking lines from the central test line. Fig. 7.2 shows the obtained results. Note again in Fig. 7.2(a) that the sensitivity of the flickering line detector is greater than that of the pattern line detector. Furthermore, the spatial response profiles of both detectors are characterized by a central summative or excitatory region flanked by two symmetrical subtractive or inhibitory regions, similar to what is found in single-cell studies of visual receptive fields. However, note also that the spatial response profile of flicker or transient detectors has an overall greater spatial extent and a relatively weaker surround inhibitory region than the response profile of pattern or sustained detectors. This difference corresponds to the difference between the sizes and response gradients of transient- and sustained-cell receptive fields discussed in Section 6.1. Figure 7.2(b) shows

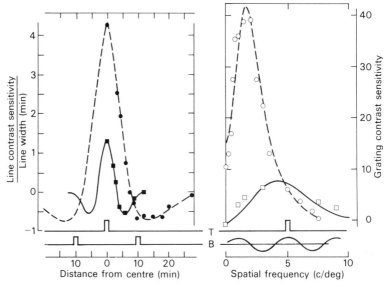

Fig. 7.2. Spatial response profiles (*a*) and spatial frequency sensitivity (*b*) of flicker (dashed lines) and pattern (continuous lines) detectors. (From King-Smith and Kulikowski (1975).)

the respective spatial contrast sensitivity functions derived from the response profiles in Fig. 7.7(a) by applying the Fourier transform. Note again that the flicker or transient channel is more sensitive at lower spatial frequencies than at higher ones, with a peak at a frequency of about 1.0 c/deg. On the other hand, the pattern or sustained channel's sensitivity peaks at a spatial frequency of about 5.0 c/deg, also consistent with the corresponding electrophysiologically measured differences between sustained and transient neurones discussed in Section 6.1.

Moreover, employing the same subthreshold technique, King-Smith and Kulikowski (1975) also psychophysically measured the spatial response profiles of pattern and flicker detectors as a function of flicker frequencies ranging from 1.0 to 24.0 c/s. Although no variation of the pattern detectors' response profiles was observed as flicker frequency varied, the response profiles of flicker detectors decreased in magnitude and increased in width as flicker frequency increased. This suggests that in human vision flicker or transient channels sensitive to higher flicker frequencies also have larger receptive fields. Such a trend has not yet been identified at the single-cell level.

However, a psychophysical trend which is consistent with known electrophysiological results (Section 6.1) is that human transient mechanisms increases in size as a function of retinal eccentricity (Wilson 1980). As a consequence of King-Smith and Kulikowski's (1975) findings, it can be inferred that flicker sensitivity also becomes greater at larger eccentricities. Recently Hartmann *et al.* (1979) found that the critical flicker frequency, especially for larger stimuli, increased as the locus of stimulation is moved from the fovea to about the 10–20° periphery, and thereafter declines. Addditional psychophysical evidence indicates that sustained pattern detectors also increase in size with retinal eccentricity. Enoch *et al.* (1970*a*, *b*) and Ransom-Hogg and Spillmann (1980) found that the sustained Westheimer function (Section 2.1.5) increases in spatial extent with eccentricity. Moreover, Rijskijk *et al.* (1980) report that the contrast sensitivity to stationary gratings decreases with eccentricity, and correspondingly Lie (1980), among others (Kerr 1971; Wertheim 1894), replicated the by-now well-known finding that visual pattern resolution decreases as eccentricity increases.

Another psychophysical means of demonstrating the existence and differential properties of human sustained and transient channels is to measure spatial frequency contrast sensitivity as a function of grating duration. Such contrast sensitivity functions have been reported by Schober and Hilz (1965), Nachmias (1967), and Legge (1978). Generally speaking, contrast sensitivity at all spatial frequencies increases up to a limiting value as duration increases. However, whereas a low-frequency attenuation of sensitivity is present at longer durations it is absent at shorter ones, where only a high-frequency attenuation prevails. If one were to measure changes in contrast threshold as a function of duration for a range of spatial frequencies, one could obtain the duration–contrast reciprocity function which specifies temporal integration in spatial frequency channels according to Bloch's Law. This empirical law states that at threshold, contrast and duration can be traded off reciprocally up to a critical duration. Breitmeyer and Ganz (1977) measured duration–contrast reciprocity functions for vertical, sinusoidal gratings at spatial frequencies of 0.5, 2.8, and 16.0 c/deg. The results, shown in Fig. 3.5, indicate that the critical duration for which duration–contrast reciprocity holds is determined by spatial frequency. At frequencies of 0.5, 2.8, and 16.0 c/deg the respective critical durations are 60, 150, and 200 ms; above these respective values threshold contrast seems to level off or decrease only at a low rate (Legge 1978). What these results indicate is that low spatial-frequency, transient channels are characterized by a shorter integration time than are higher spatial frequency, sustained channels; a result also reported by Legge (1978) and consistent with Hood's (1973) finding that optically blurred stimuli yield a shorter integration time than sharply focused ones. This may correspond to a broader temporal impulse response of sustained relative to transient neurones (Cleland *et al.* 1973; see Section 6.1). Similar psychophysical results derived from monkey have been reported by Harwerth, Boltz, and Smith (1980). Since, as discussed in Chapter 6, monkeys have anatomically and physiologically distinct populations of sustained and transient cells in their visual system (Bullier and Henry 1980; Bunt *et al.* 1975; Dreher *et al.* 1976; Hendrickson *et al.* 1978; Hubel and Wiessel 1972; Lund and Boothe 1975; Schiller and Malpeli 1978; Sherman *et al.* 1976), these and other analogies (see below) between human and monkey psychophysical results lend the interpretation of the human psychophysical results in terms of the sustained–transient channel approach greater credence.

Psychophysical evidence also indicates that human transient channels are characterized by different impulse responses than those of sustained channels. Kelly (1971*a*) measured uniform-field flicker sensitivity functions and mathematically, via the Fourier transform, derived the impulse response corresponding to the flicker detector. He found that at high background luminance the impulse response could be characterized by initial excitatory phase followed by an inhibitory one, which in turn was followed by a smaller excitatory phase. The temporal interval, i, between the primary and secondary excitatory phases of the damped oscillating impulse response defined the temporal frequency, i^{-1}, at which peak flicker sensitivity was attained. However, as background luminance was decreased progressively, the impulse response and i increased in duration. Moreover, the inhibiting phases eventually attenuated until, at scotopic levels, only a monophasic excitatory component, like that characterizing the impulse response of pattern detectors, was evident. In other words, the transient flicker detectors become more sluggish or 'sustained' as background luminance decreases, a result consistent with the single-cell recordings of transient cells made by Cleland *et al.* (1973), Hammond (1975), Jakiela and Enroth-Cugell (1976), and Zacks (1975).

Rashbass (1970), using a subthreshold summation technique similar to that employed by Ikeda (1965) (see Section 3.1.2), measured the interaction of two impulse responses produced by two consecutively flashed subthreshold stimuli. He also found that the impulse response to transient changes of luminance can be characterized by a triphasic, exitatory–inhibitory oscillation. Similar oscillating functions have been empirically derived by Grossberg (1970) and Ueno (1977) using visual reaction time to two sequentially and briefly flashed light stimuli. On the assumption that, in human, transient channels are the primary flicker and transient detectors, their impulse response ought to oscillate from excitation to inhibition as described above. Moreover, by varying the spatial frequency of grating from low to high values one can, by employing the subthreshold technique, obtain measures of transient and sustained channels' impulse responses.

Breitmeyer and Ganz (1977) and Watson and Nachmias (1977) used this technique and found distinct differences among impulse responses at low and high spatial frequencies. The results of Watson and Nachmias' study are shown in Fig. 7.3. Note that at the lowest spatial frequency the interaction of the two subthreshold impulse responses can be characterized as an initial summative or excitatory phase followed by a weaker subtractive or inhibitory phase. As spatial frequency increases, the inhibitory phase attenuates and the impulse response assumes a monophasic, excitatory shape. A similar result was obtained in Kelly's (1971*b*) investigation of impulse response functions mathematically derived from psychophysical data on contrast sensitivity to temporal counterphase modulation of gratings. With increases of spatial frequency, the low temporal-frequency attenuation characterizing uniform-field flicker and counterphase modulation at low spatial frequencies (Kelly 1972) drops out. The corresponding effect on the impulse response is to eliminate the inhibitory phases and thus the oscillation of the temporal impulse response as spatial frequency increases. From these results, it can be inferred, in line with single-cell findings reported by Büttner *et al.* (1975; see Section 6.1), that human transient channels show a multiphasic oscillation of excitation alternating with inhibition, whereas sustained channels show only a single excitatory phase.

7.2. Visual reaction time

In Section 6.1, I noted that, at least above a certain minimal contrast level, transient neurones should respond at a shorter latency than sustained

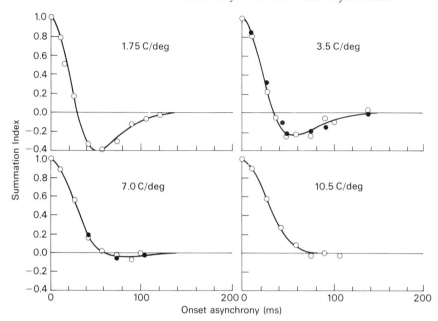

Fig. 7.3. Subthreshold impulse–response summations curves for four grating spatial frequencies. Note the biphasic facilitatory–inhibitory response at the lowest spatial frequency. The inhibitory phase attenuates at higher frequencies and is not evident at the highest one. (From Watson and Nachmias (1977).)

ones, whereas below that critical contrast sustained cells respond faster.

Psychophysical evidence supporting this interaction between contrast level and reaction time derives from two recent studies. In particular, Burbeck and Kelly (1981) recently demonstrated that human transient channels (activated by test gratings of 2 c/deg or lower), as compared with sustained ones (activated by test gratings ranging from about 4 to 12 c/deg), are characterized by a gain that increases at a faster rate as a function of contrast. However at contrasts below 5 per cent, the gain characteristics did not differ significantly between transient and sustained channels. On the whole, these results, as indicated by Burbeck and Kelly (1981), are consistent with the electrophysiological study of the gain characteristics of cat retinal ganglion cells reported by Shapley and Victor (1978). As argued in Section 6.1 the implication of these findings is that at low contrasts sustained channels may actually respond at a shorter latency than transient ones (Lennie 1980*b*), but that this relation reverses at higher contrasts.

Harwerth and Levi (1978) investigated reaction time to vertical sine-wave gratings as a function of, among other variables, spatial frequency, and contrast. For 500-ms flashes of gratings, the reaction time at low spatial frequencies (e.g. 0.5 c/deg) and high spatial frequencies (e.g.

12 c/deg) generally decreased continuously and exponentially with increases of contrast ranging from respective threshold values to 45 per cent. However, for intermediate spatial frequencies (1–8 c/deg) the decrease of reaction time with similar increases of contrast was characterized by a discontinuity revealing that one exponentially decaying function dominated up to a contrast value of about 5–10 per cent, followed by another function which dominated at higher contrast values. Analogous psychophysical findings have been reported by Harwerth *et al.* (1980) in their study of monkey vision. These similar psychophysical findings in human and monkey vision indicate that the low and high spatial frequencies presumably isolated transient and sustained channels, respectively; whereas the intermediate spatial frequencies isolated sustained channels at contrasts below 5–10 per cent and transient channels at larger contrast values (Harwerth and Levi 1978; Levi, personal communication). Hence, the response latency advantage enjoyed by transient units holds only above some critical contrast value of say 5–10 per cent, a value which corresponds closely to the value of 5 per cent, estimated by Burbeck and Kelly (1981), above which the gain or contrast response of transient channels increases at a faster rate than that of sustained channels.

Besides a faster response latency at higher contrasts or intensities, the fibres of the transient neurones have higher conduction velocities along the retinogeniculocortical pathway than do sustained ones. Consequently, one would expect that reaction time to sufficiently intense visual stimuli ought to be faster when selectively activating transient as compared with sustained channels. Breitmeyer (1975) measured the visual reaction time to the onset of a 50-ms presentation of sinusoidal gratings at a contrast of about 60 per cent and at spatial frequencies ranging from 0.5 to 11.0 c/deg. He found that reaction time over this range increased monotonically from about 200 to 350 ms. Since contrast sensitivity declines at higher spatial frequencies, Breitmeyer (1975) also measured reaction time to the same gratings but with their individual physical contrasts adjusted so that their apparent contrasts were equal. Despite this cancelling out of differences in contrast sensitivity, Breitmeyer (1975) none the less found that the reaction time increased by about 40 ms over the same range of spatial frequencies employed. Similar and related psychophysical results have subsequently been reported by several other investigators (Breitmeyer *et al.* 1981*a*; Levi, Harwerth, and Manny 1979; Lupp, Hauske, and Wolf 1976; Tartaglione, Goff, and Benton 1975; Vassilev and Mitov 1976). These findings are indicative of a shorter response latency of low spatial frequency, transient as compared with high spatial frequency, sustained channels and support one of the major assumptions of the approach to visual masking outlined below in Section 7.4. Moreover, similar increases in reaction time as a function of spatial frequency can be obtained when a subject is asked to respond to the sudden offset or contrast reversal of a grating (Breitmeyer *et al.* 1981*a*; Long and Gildea, 1981; Parker 1980).

A more telling set of experiments which indicated that sustained and transient channels differ in response latency characteristics as a function of spatial frequency has been reported by Lupp (1977) and Lupp *et al.* (1978). The latter investigators used the following technique to measure visual reaction time to the onset of a sinusoidal grating. A grating at a spatial frequency of 1.0–16.0 c/deg and with a contrast 1.6 times the threshold value was presented abruptly for 500 ms. The abrupt onset of a 1000-ms grating of the same respective spatial frequency and at a contrast of 0.6 its threshold value occurred at temporal intervals ranging from 500 ms prior to the onset of the suprathreshold grating to 250 ms after its onset. It was reasoned that the addition of the subthreshold grating would facilitate or increase the rate of response of the respective spatial frequency channels and thus facilitate or decrease reaction time. The *changes* of reaction time as a function of onset asynchrony of the subthreshold gratings are shown for the 1.0, 2.0, 5.3, and 16.0 c/deg gratings in the upper to lower panels, respectively, of Fig. 7.4.

Note that at a spatial frequency of 1.0 c/deg, the subthreshold grating facilitates reaction time only transiently at asynchronies ranging from −200 ms (subthreshold grating onset precedes suprathreshold grating onset) to about 50 ms (subthreshold onset follows suprathreshold onset). The largest facilitation was obtained at an asynchrony of about −30 to −40 ms. On the other hand, with higher spatial frequencies, the facilitation became more sustained; in particular, with the 5.3 and 16.0 c/deg gratings, reaction time facilitation was sustained over the entire 500 ms interval preceding the onset of the suprathreshold grating and was eliminated at an asynchrony of about 100 ms. These psychophysical results clearly demonstrate (i) that the low spatial frequency, subthreshold grating activated transient channels, whereas the high spatial frequency, subthreshold gratings activated sustained channels; and (ii) that the subthreshold summation or integration time of transient channels, in line with the results reported by Breitmeyer and Ganz (1977) and Legge (1978) is shorter than that of sustained channels (see Section 7.1).

The above psychophysical reaction-time results also have been corroborated in several electrophysiological studies. Sphelmann (1965) demonstrated that the late-component waves of the human cortical visually evoked response (VER) shifted to longer latencies as the size of the elements in a flashed checkerboard pattern became smaller. Subsequently, several investigators (Jones and Keck 1978; Kulikowski 1977; Parker and Salzen 1977*a*, *b*, 1982; Vassilev and Strashimirov 1979) have investigated the latency of the VER to onset and offset of sinusoidal gratings varying in spatial frequency. Typical results showed that major positive wave components of the evoked potential (e.g. P_1, P_2) increased by about 100

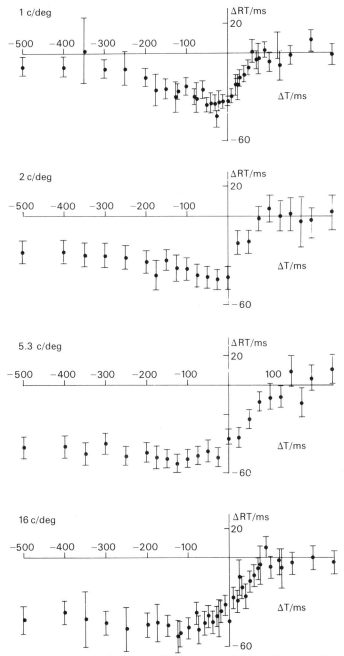

Fig. 7.4. Reaction time facilitation (ΔRT) produced by a subthreshold grating on a suprathreshold grating at four spatial frequencies and as a function of the interval of onset of the subthreshold grating relative to the suprathreshold grating. (From Lupp, Hauske, and Wolf (1978).)

ms in latency as spatial frequency increased from 0.5 to 10 or more c/deg. The relatively weaker $N_0 - P_0$ components, particularly interesting since they are the earliest waves in the potential, can be evoked strongly only by low spatial frequency gratings (Jones and Keck 1978; Kulikowski 1977), and may therefore reflect the earlier activity of more sparsely distributed transient channels (see Section 6.4).*

The fact that physiological measures of visual latency correspond well to psychophysical measures has been demonstrated by Williamson, Kaufman, and Brenner (1977, 1978). Instead of measuring the cortically evoked neuroelectric potential, these investigators measured the latency of the human cortical neuromagnetic response to 66%-contrast gratings as a function of spatial frequency ranging from about 0.2 to 11.0 c/deg, contrast and spatial frequency ranges approximating those employed by Breitmeyer (1975). Figure 7.5 shows the latency of the neuromagnetic response (left ordinate) and the reaction times (right ordinate) obtained in Breitmeyer's (1975) study. Note the close correspondence between the neuromagnetic and psychophysical results. Both functions increase monotonically by about 150 ms over spatial frequencies ranging from 0.2 to 11.0 c/deg. Moreover, the neuromagnetic responses overall were about 120 ms shorter than the reaction times. This indicates that the motor component in Breitmeyer's (1975) study comprised a constant 120 ms of the total reaction time at all spatial frequencies. Consequently, the motor component cannot be a source of the variation of reaction time with spatial frequency.

All of the above results, indicating that human transient channels respond and conduct faster than do sustained ones, buttress one of the major assumptions incorporated into the approach to visual masking outlined below in Section 7.4.

7.3. Effects of transient and flicker adaptation

Section 7.1 showed that adapting to a stationary grating does not affect transient mechanisms (Kulikowski 1974). However, since transient channels respond selectively to abrupt, brief stimuli and flicker, and since the human visual system contains flicker-selective pathways (Nillson *et al.* 1975; Pantle 1971; Regan 1970; Sternheim and Cavonius 1972), one would expect that masking or adapting the visual system with such stimuli would selectively effect transient mechanisms. Legge (1978) used a procedure in which a brief, 20-ms mask flash of a grating immediately preceded and followed the same grating flashed at durations ranging from 20 to 3000 ms.

---

* The fact that the latency of later components of the cortical VER increase with spatial frequency indicates that they reflect the longer-latency processing, sustained-channel information. In this regard, cortical VER studies of metacontrast (Schiller and Chorover 1966; Vaughn and Silverstein 1968) show that, relative to the VER elicited by a target disk, the VER elicited by a disk–ring metacontrast sequence is characterized by attenuation of the later components, especially at intermediate disk–ring SOAs of 30–60 ms.

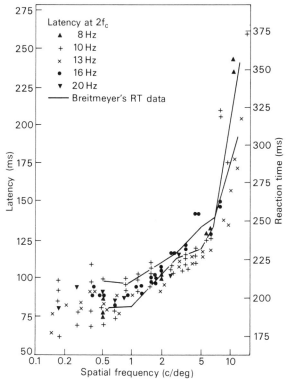

Fig. 7.5. The latency of the human cortical neuromagnetic response (+, left ordinate) and psychophysical reaction time functions from Breitmeyer (1975) (solid lines, right ordinate) as a function of spatial frequency. (From Williamson, Kaufman, and Brenner (1978).)

In this way, Legge (1978) measured the effects of the preceding and following transient mask flash on the duration–contrast reciprocity function, at threshold, of the intervening test grating presentation. He found that compared with the no-mask condition, the critical duration, for which duration–contrast reciprocity held, increased particularly at low spatial frequencies. This result indicates that the preceding and following, 20-ms flashes of the grating effectively attenuated or masked the transient responses at test on- and offset, thus leaving the sustained mechanisms, characterized by a longer integration time, to determine the critical duration of the duration–contrast reciprocity function.

Related results have been reported by Breitmeyer *et al.* (1981*a*). These investigators studied the masking effects of 6 c/s uniform-field flicker on a variety of psychophysical responses including onset and offset reaction time, visual persistence, contrast sensitivity, and reaction time to near-threshold gratings. The spatial frequencies tested ranged from 0.25 to 16.0 c/deg. Some of the main results can be summarized as follows. Compared with the no-mask condition in which a non-flickering, steady background field was used, the 6 c/s uniform-field flicker mask selectively and dramatically increased both onset and offset reaction times for spatial frequencies ranging from 0.25 to about 4.0 c/deg. The effect was obtainable under both monoptic and dichoptic viewing conditions, indicating that the flicker mask affected central, cortical mechanisms (Green 1981a; Lipkin 1962; Thomas 1954). Visual persistence, as measured by the phenomenal continuity procedure used by Meyer and Maguire (1977; see Section 3.2.1), also increased dramatically and selectively for the same range of low spatial frequencies. Moreover, the very existence of an increase of visual persistence with spatial frequency in unmasked conditions (Breitmeyer *et al.* 1981a; Corfield *et al.* 1978; Meyer and Maguire 1977) also suggests that sustained channels have a longer response persistence. In addition, movement or counterphase contrast sensitivity also decreased significantly and selectively for that range of spatial frequencies, consistent with related flicker-adaptation effects reported by Green (1981a). These results indicate that when transient channels are masked by the background flicker, sustained channels characterized by a longer reaction time, longer response persistence, and higher contrast threshold at lower spatial frequencies are left unmasked and responsive to the range of frequencies otherwise preferably activating transient channels.

In the final experiment of this series, Breitmeyer *et al.* (1981a) measured reaction times to a 500-ms presentation of a 0.5 c/deg grating at a contrast 0.15 log unit above threshold. This technique was employed previously by Tolhurst (1975), who showed that reaction times to a 0.2 c/deg grating distributed themselves probabilistically and bimodally at intervals corresponding to the abrupt onset and offset of the grating. From these results, Tolhurst (1975) inferred that the reaction times were determined by transient channels selectively activated by the abrupt grating on- and offsets. However, using the flicker masking technique, Breitmeyer *et al.* (1981a) found that sustained channels are also activated by low spatial frequencies. The results of their experiment are shown in Fig. 7.6. The upper two panels show reaction time distributions to the presentation of the near-threshold grating when no flicker mask is used. Note that, here, as in Tolhurst's (1975) study, the reaction times appear to be distributed probabilistically and bimodally, with the modal reaction times being about 500 ms apart, thus corresponding to the transient onset and offset of the 0.5 c/deg grating. The lower panels, on the other hand, show the probabilistic distribution of reaction times when a 6 c/s uniform-field flicker mask is used. Note that, here, the reaction time distribution is unimodal; moreover, this mode is displaced to longer latencies relative to the initial mode in the unmasked condition. This latter finding indicates that when transient channels are selectively masked, the unmasked sustained channels characterized by a longer response latency, determine reaction

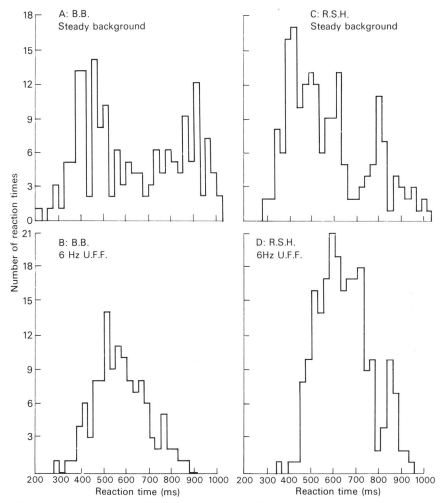

Fig. 7.6. Reaction time distribution to a 0.5 c/deg grating at 0.15 log unit above threshold. Upper panels, distributions obtained without uniform-field flicker masking; lower panels, distributions obtained with 6 c/s uniform-field flicker masking. (From Brietmeyer, Levi, and Harwerth (1981*a*).)

time. It also suggests that the reaction times in the unmasked condition, although bimodally distributed, with the first and last mode corresponding to transient activity at onset and offset, are additionally affected by sustained activity which yields intermediate reaction times bracketed by the two extreme modes.

Related and similar findings have been reported by Harwerth *et al.* (1980) in monkey. As in Tolhurst's (1975) study, bimodally distributed reaction times, corresponding to on- and offsets of a 500-ms, near-threshold gratings, were obtained at spatial frequencies below 2 c/deg;

however, above 2 c/deg, clear unimodal reaction times were obtained. Moreover, whereas the modal reaction times of the first of the two bimodal distributions at spatial frequencies below 2.0 c/deg were about 400 ms, those of the unimodal distributions obtained at 4 and 8 c/deg were about 530 and 600 ms respectively. These corresponding results indicate that in monkey, as in human, (i) sustained channels have a greater response latency than do transient channels; and (ii) among sustained channels response latency increases as spatial frequency increases.

7.4. Visual masking

7.4.1. Outline of masking model

To set the stage for the following discussion I shall present a theory of visual masking based on the existence of sustained and transient channels in human vision. Two other theories of masking, that of Matin (1975) and Weisstein *et al.* (1975), reviewed in Chapter 5, have already incorporated the existence of these channels in one way or another. The theoretical explanation of visual masking adopted in this chapter is a modification of that outlined by Breitmeyer and Ganz (1976). The main assumptions of this theory are based on the single-cell data reviewed in the prior chapter and the human psychophysical and electrophysiological results discussed preliminarily in the prior three sections. The following assumptions or conditions are explicitly stipulated.

1. Both the brief target and mask in the target–mask stimulus sequence activate long-latency sustained as well as short-latency transient channels.

2. Within a class of channels, inhibition is realized via the centre–surround antagonism of receptive fields. I shall call this type of antagonism intrachannel inhibition.

3. Between the classes of neurones there exists mutual and reciprocal inhibition. I shall call this type of antagonism interchannel inhibition.

4. Masking can occur in one of three ways: (i) via intrachannel inhibition (particularly realized in sustained channels); (ii) interchannel inhibition (particularly the transient-on-sustained channel inhibition); (iii) the sharing of common sustained or else transient pathways by the neural activity generated by the target and mask when they are spatially overlapping. Implicit in this last assumption also is the sharing or prior common peripheral receptor activity by both stimuli.

5. Transient channels primarily signal the location and presence of stimuli or their rapid changes of location (displacement, motion) over time; sustained channels primarily signal pattern aspects such as brightness, contrast, and contour.

Other, for now implicit, empirical boundary conditions affecting these masking mechanisms will be made explicit in the following discussion. The major assumptions, however, must be sufficient to explain the main features of visual masking as shown in Fig. 4.2. In Fig. 4.2(a), type A forward and backward masking effects are schematized. Fig. 4.2(b) shows the typically weaker type B paracontrast effect and the stronger type B metacontrast effect. These masking effects and prior theories were reviewed in Chapters 4 and 5.

The basic properties of the model are illustrated in Fig. 7.7. Fig. 7.7(a) indicates the types of interactions that can occur when the mask (M) precedes the targer (T) in forward masking. We can distinguish between two general types of forward masking: (i) masking with spatially overlapping patterns such as masking by structure or noise; and (ii) masking with spatially adjacent patterns as in paracontrast (see Fig. 4.1). In both types of forward masking, the mask's transient activity, indicated by the short-latency, spike-like response, cannot interact in any way with the target activity; however, reciprocal inhibition between the target's

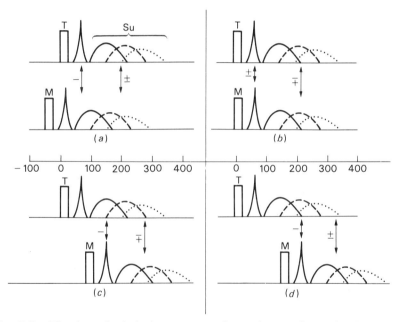

Fig. 7.7. The hypothetical time course of transient and sustained channels activated by the target (T) and the mask (M) at various target–mask asynchronies. In (*a*) the mask onset precedes that of the target; in (*b*) the target and mask onsets are synchronous; in (*c*) and (*d*) the mask onset follows that of the target at increasing temporal asynchronies. The transient response is represented by the short-latency spike-like function; the sustained responses at increasing spatial frequencies is indicated by the symbol Su in (*a*). The inhibitory and excitatory interactions within and between the two types of channels is indicated by two-way arrows signed negatively and positively, respectively. (From Breitmeyer and Ganz (1976).)

transient activity and the mask's sustained activity can occur as indicated by the left, negatively signed arrow. Moreover, for the former type of spatially overlapping stimuli, two mechanisms of forward masking are possible. For one, mask sustained channels and target sustained channels may mutually inhibit each other (intrachannel inhibition), as indicated by the negative sign of the right arrow; and secondly, mask sustained activity can integrate with target sustained activity via the sharing of common retinotopically organized pathways (intrachannel integration), as indicated by the positive sign of the right arrow. In the other, paracontrast type of forward masking, in which the mask and target patterns do not overlap spatially, intrachannel inhibition of target sustained channels by mask sustained channels constitutes the major masking mechanism.

Fig. 7.7(b) demonstrates the neural interactions at target–mask synchrony. With spatially overlapping masks, one would again expect both intrachannel inhibition and intrachannel integration of activity in common target and mask sustained (or else transient) pathways to contribute to the overall masking effect. In fact, at temporal synchrony masking by intrachannel integration ought to be optimal. However, when using spatially adjacent target and mask stimuli, as in meta- or paracontrast, one would expect only intrachannel inhibition to contribute to the overall masking effect.

Fig. 7.7(c) schematizes the neural interactions which occur when the mask onset follows that of the target by 100 ms. Again here we can distinguish two general types of backward masking: (i) masking with spatially overlapping patterns (masking by noise or structure); and (ii) masking with spatially adjacent patterns (metacontrast). In both types of backward masking the transient activity of the mask and the earliest sustained activity of the target can interact via mutual interchannel inhibition. However, only in the former type of spatially overlapping masking paradigm would integration or sharing of sustained activity in common sustained pathways provide an additional masking mechanism. Basically the same argument holds for the neural interactions illustrated in Fig. 7.7(d), except for the fact that the earlier transient or sustained activity of the mask interacts with progressively later sustained activity of the target.

The three masking mechanisms outlined above—intrachannel inhibition among spatially adjacent sustained (or transient) pathways, interchannel inhibition between transient and sustained pathways, and integration or sharing of common activity within sustained (or transient) pathways—are intended to account, respectively, for the main features of the masking phenomena shown in Fig. 4.2: type B forward masking or paracontrast, type B backward masking or metacontrast, and type A forward and backward masking. Before discussing more specific empirical properties of these three masking effects in the context of the theoretical approach introduced above, it should be mentioned that although the designations 'mask' and 'target' serve a valid methodological distinction, physiologically, in terms of the possible interactions outlined above, that distinction is lost. Consequently, as will become apparent in several cases below, one must consider mutual interactions between the physiological activities generated separately by the target and the mask stimuli irrespective of the temporal order of their presentation.

7.4.2. Explanatory scope of masking model

The extensive explanatory scope of the present model is developed in the context of (i) empirical data reviewed in Chapters 1–4 and (ii) some additional results introduced in this chapter. I ask for the indulgence of readers thoroughly familiar with the previously discussed empirical findings. For the novice, however, such abbreviated recapitulation seems a helpful feature. The intent is not only to impress on the reader the scope of the model but also to compare this model's explanatory power with that of prior ones reviewed in Chapter 5. Furthermore, several shortcomings of the model will also be discussed, and suggestions will be offered for its future developments and improvements.

(1) In Chapter 1, Section 1.3.1, I noted that Sherrington (1897) and subsequently Piéron (1935) reported a facilitatory effect on critical flicker frequency (cff) of the first of two spatially adjacent stimuli flashed in recycled sequence *relative* to an inhibitory effect on cff of the second of the two stimuli. As noted, the latter effect is consistent with a form of paracontrast suppression; however, the former result is not consistent with a form of metacontrast suppression, but rather with its opposite, metacontrast facilitation. However, unlike Sherrington (1897), Piéron (1935) additionally investigated the effect of such repetitively-cycled sequential and spatially adjacent stimuli on brightness perception. With this indicator response, a metacontrast suppression was obtained. The current model can account for these differential, response criterion-dependent findings in the following manner:

1. Metacontrast suppression of the brightness of a preceding flash of light by a following, spatially adjacent one is a direct consequence of the faster transient activity generated by the lagging flash inhibiting the slower sustained response of the leading flash.

2. Conversely, the paracontrast suppression of cff of the second of two spatially adjacent repetitively flashed stimuli is a direct consequence of the slower sustained activity generated by the first stimulus reciprocally inhibiting the faster transient (flicker) activity generated by the second stimulus. Hence, relative to the inhibited, i.e. lower, cff of the second stimulus, a higher cff of the first one.

These results may be related to psychophysical studies revealing effects on the cff of an intermittently flashed stimulus when a second, steady or sustained neighbouring stimulus is present. Several investigators (Berger 1954; Fry and Bartley 1936; Geldard 1932, 1934; Graham and Granit 1931) have reported enhancement of cff of an intermittently flashed test stimulus when the brightness of the *steady*, neighbouring mask stimulus increased to slightly above the Talbot level of the flickering one, followed by a suppression of the cff as the brightness of the steady mask stimulus increased further.

What processes contribute to this non-monotonic effect of steady surround or adjacent luminance on cff of a flickering stimulus? In my opinion, there may be at least two processes which contribute to the initial rise in cff as the surround luminance increases up to slightly above the flickering stimulus's Talbot brightness. For one, as the intensity of the steady stimulus increases, one increases the local light adaptation level of neural summation pools (Rushton 1965; Rushton and Westheimer 1962; Westheimer 1968) which intrude into the areas spatially adjacent to the steady stimulus. Since, according to the Ferry–Porter law, cff is directly proportioned to the light-adaptation level, one would expect the cff to increase as the luminance of the steady stimulus increases. For another, despite an increase in cff, the apparent brightness of the flickering stimulus simultaneously decreases as the luminance of the steady stimulus increases. Since, according to the present model, longer persistent sustained channels are primarily involved in determining brightness or contrast of a stimulus, the more sluggish sustained activity of the flickering stimulus, persisting from one flash to another, would be increasingly inhibited, via simultaneous brightness induction (Heinemann 1955), i.e. via lateral intrachannel inhibition by the adjacent steady stimulus. This in turn should result in a reduced interchannel, sustained-on-transient inhibition produced between successive flashes of the flickering test stimulus. Such a disinhibition of transient channels also would be expected to result in an increase of their activity as reflected in the initial increase of cff.

However, as a countervailing process, as the luminance of the steady surround increases, it also exerts progressively more lateral, interchannel sustained-on-transient inhibition on the flickering stimulus. Thus, as the brightness of the steady stimulus exceeds the Talbot brightness of the flickering stimulus, the transient activity generated by the latter stimulus also ought to be laterally inhibited. On the assumption that this latter, countervailing lateral *inter*channel inhibition of transient channels eventually outweighs (i) the facilitatory effects on flicker of light-adaptation; and (ii) the simultaneous lateral disinhibitory effect produced by lateral *intra*channel sustained inhibition, one would expect an initial rise or facilitation of cff followed by its suppression as the luminance of the steady stimulus increases.

(2) In Section 2.2 I reviewed masking of pattern by uniform light flashes. In particular, sudden luminance changes produced by the abrupt

onset and offset of a prolonged masking flash are known to produce transient masking of pattern such as wide-stroke letters (Boynton and Miller 1963). Green (1981*b*) showed that these transient masking effects near the on- and offset of a prolonged luminance flash are specific to the spatial frequency of a sinusoidal test grating. In particular, he found that a brief, 30-ms test grating of 1.0 c/deg yielded transient masking overshoots at abrupt on- and offset of a 700 ms luminance flash at 58.4 cd/m². However, for the same masking flash, no transient mask overshoots were obtained when a 7.8 c/deg test grating was employed; here sustained and equal masking was obtained for the duration of the flash. If we assume that (i) the on- and offsets of the luminance flash activate peripheral, e.g. retinal, transient channels and (ii) that the 1.0 c/deg and the 7.8 c/deg grating predominantly activate low spatial-frequency transient and high spatial-frequency sustained channels, respectively, then the peripheral transient activity generated by the mask flash adds 'noise' to the 'signal' of transient channels activated by the 1.0 c/deg test grating, but not to the sustained channels activated by the 7.8 c/deg grating. In effect, in transient channels the internal signal-to-noise or Weber ratio is reduced, yielding transient masking overshoots at on- and offset. Conversely, since the sustained response component of the mask flash (recall Assumption 1 of the current model, that a stimulus can activate both transient and sustained channels) adds 'noise' only to the 'signal' in the sustained channels activated by the 7.8 c/deg test grating, only a sustained masking effect without transient overshoots at on- or offset of the mask flash ought to occur.

One may wonder, here, why the transient activity generated at on- and offsets of the luminance mask do not inhibit, via interchannel inhibition, the sustained activity generated by the 7.8 c/deg test grating, and thus also produce transients near the abrupt luminance transitions of the mask. Recall first of all from Sections 2.1.5 and 2.1.6 that masking by light, devoid of contour interactions, is of peripheral, prechiasmic origin (Battersby and Wagman 1962; Battersby *et al.* 1964). From our review of inhibitory interactions between transient and sustained neural pathways in Sections 6.5 and 6.7 of Chapter 6, we have evidence that, at best, such inhibitory interactions occur no earlier than the lateral geniculate nucleus. In fact, according to Lennie's (1980*a*) additional review, in primates such interchannel inhibition is most likely found only at cortical levels. Hence, if contour interactions do play a role in masking by light as shown in Sections 2.1.5 and 2.1.6 and since these contour effects can be obtained dichoptically as well as monoptically one would expect transient-on-sustained interchannel inhibition to play a role in masking by light only if the spatial separation between contours of the test pattern and the border-edge of the uniform mask flash are sufficiently small (Battersby and Wagman 1962; Markoff and Sturr 1971; Weisstein 1971). Hence, since the stimulus field used by Green (1981*b*) was 4.5° in diameter, visibility of the 7.8 c/deg test grating may have been appreciably suppressed by transient channels activated at the border the field; however, in the interior of the test field, such suppression would be absent or, at best, highly attenuated.

(3) In order to produce contour proximity and overlap of mask and target stimuli Mitov *et al.* (1981; see Section 4.5) altered Green's (1981*b*) technique by employing the masking-by-pattern paradigm in which a 500-ms flash of one grating and a 20-ms flash of another grating served as mask and test stimuli, respectively. Their results can be summarized as follows: when the spatial frequency of the mask was 6 c/deg or lower, the magnitude of the transient overshoots at mask on- and offsets were inversely proportional to the spatial frequency of the test grating; specifically, the magnitude was largest for a 2 c/deg test grating, smaller, yet pronounced, for a 6 c/deg test grating, and smallest for an 18 c/deg test grating. One would expect such a result (i) if transient channels are preferentially activated at spatial frequencies at and below 6 c/deg and (ii) if the magnitude of transient activation decreases with spatial frequency whereas that of sustained channels correspondingly increases. Since these experiments were conducted under binocular viewing which uses peripheral luminance, as well as central, contour masking mechanisms, the role of more central transient-on-sustained interchannel inhibition cannot be assessed. A similar study, but one using dichoptic viewing of target and mask stimuli, may reveal such target–mask inhibitory interactions.

Moreover, the study of Mitov *et al.* (1981) dovetails neatly with results reported by Teller *et al.* (1971) and Matthews (1971). Teller *et al.* (1971), employing scotopic luminance, and Matthews (1971) employing photopic luminance, reported that conditioning or mask flashes of relatively small diameter (e.g.< 30′) produced no transient masking overshoots at their on- or offset; however, with relatively large-diameter mask flashes (e.g.> 60′), such transient overshoots were obtained. Since larger diameter masks, like lower spatial frequency gratings, are optimal for activating transient channels, one would expect pronounced transient activity at on- and offset of a large conditioning flash to produce such transient masking overshoots. However, smaller or high spatial frequency stimuli are either suboptimal for or incapable of activating transient channels (Enroth-Cugell and Shapley 1973*b*; Ikeda and Wright 1972*a*; see Section 6.1); hence attenuation or absence of transient mask overshoots.

Moreover, based on Breitmeyer and Julesz's (1975) and Tulunay-Keesey and Bennis's (1979) related results (see Section 7.1), one can infer that the transient overshoots at on- and offset of the mask flashes employed by Green (1981*b*) and Mitov *et al.* (1981) depend on the rise and fall times of the mask at its on- and offsets. In particular, since slowly ramped, relative to abrupt, on- and offsets attenuate the transient response, one would expect a corresponding attenuation of the transient masking overshoots reported by Green (1981*b*) and Mitov *et al.* (1981), similar to the attenuation of transient masking overshoots reported by Matsumura (1967*b*; see Section 2.1.1, Fig. 2.4) in his investigation of masking of light by luminance increments and decrements.

(4) In the review of pattern masking in Chapter 4, it was noted that type B paracontrast masking is optimal when the spatially flanking mask onset precedes that of the central target by several tens of milliseconds (Alpern 1953; Kolers and Rosner 1960; Pulos *et al.* 1980; Weisstein 1972). The mechanism responsible for this effect is intrachannel inhibition within sustained pathways. Recall from Section 6.4 (Winters and Hamasaki 1975, 1976) that the surround activity of sustained neurones lags the centre activity by several tens of milliseconds. Consequently, to have an optimal inhibitory effect on a sustained neurone, the stimulus activating the surround must precede the centre stimulus by a corresponding asynchrony. At longer or shorter asynchronies, less than optimal inhibition of the centre response is produced. Consequently, in paracontrast an optimal type B masking effect should also be produced when the mask precedes the target by several tens of milliseconds.

(5) The transition from type B to type A metacontrast as mask energy increases relative to target energy also can be explained by a correspondingly increasing effectiveness of sustained intrachannel inhibition. Breitmeyer (1978*b*) (see Section 4.3.2, Fig. 4.3) showed that at a target duration of 16 ms, metacontrast shifted from type B to type A as the mask duration increased from 2 to 32 ms. Beyond mask duration of 8 ms, the mask became progressively more effective at lower onset asynchronies (SOAs) whereas the peak metacontrast effect obtained at an SOA of 56 ms did not change. Breitmeyer (1978*b*) was able to determine the masking threshold at the lowest SOA of 16 ms and at the optimal type B metacontrast SOA of 56 ms by plotting the strength of masking at each SOA as a function of mask energy or duration. The results are shown in Fig. 7.8. Note that the masking effects for the optimal type B SOA has a low threshold which increases linearly from the 1-ms, up to the 8-ms mask duration and asymptotes thereafter. On the other hand, at the shortest SOA of 16 ms, the masking effect has a higher mask threshold. No masking is obtained at mask durations of 1, 2, 4, and 8 ms, beyond which increasing amounts of masking are obtained. This difference in masking contrast thresholds corresponds to the lower thresholds of transient as compared with sustained channels as shown electrophysiologically in single neurones as well as psychophysically in humans (see Section 7.1). On that basis, we can infer that the transition from type B metacontrast, produced by low-threshold, interchannel transient-on-sustained inhibition, to type A metacontrast when mask energy increases relative to that of the target, is in turn produced by a high-threshold intrachannel, sustained-on-sustained inhibition superimposed on the former interchannel inhibition.

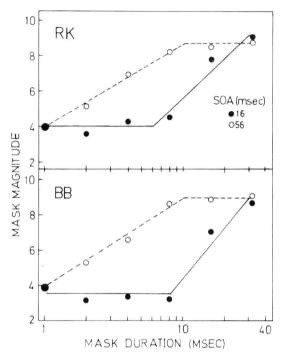

Fig. 7.8. Metacontrast magnitude functions at SOAs of 16 ms (filled circles, solid line) and 56 ms (open circles, dashed line) as a function mask duration for a target flashed for 16 ms. Note the lower masking threshold at a stimulus onset asynchrony (SOA) of the 56 ms relative to the 16-ms SOA. (From Breitmeyer (1978*b*).)

(6) In Section 4.5 (Greenspoon and Eriksen 1968; Schiller and Smith 1965; Turvey 1973) it was also noted that dichoptic type A forward masking by noise or structure is typically weaker than type A backward masking. This can be explained by invoking an essential asymmetry between forward and backward masking, as shown in Fig. 7.9, dependent on the effectiveness of the *target's* transient activity in inhibiting the mask's sustained activity. In forward masking by structure or noise (see Fig. 7.7(a)) post-retinal transient activity generated by the target can locally inhibit post-retinal sustained activity of the mask. Therefore, the intrusion at cortical levels of mask sustained activity into the sustained channels shared with the target ought to be less than maximal and hence less masking ought to result (Fig. 7.9, open circles). On the other hand, in backward masking by noise or structure (see Fig. 7.7(c) and (d)), not only does the sustained activity of the mask intrude unobstructed into the sustained channels of the target, but also, at post-retinal levels, the transient activity of the mask can inhibit the sustained activity of the target and facilitate this intrusion; thus, more masking ought to result (see Fig.

7.9, filled circles). Moreover, since these interactions are obtained dichoptically, they very likely exist at central, cortical levels.

(7) However, under monoptic viewing conditions type A forward masking is typically stronger than type A backward masking (Kinsbourne and Warrington 1962*a, b*; Scharf and Lefton 1970; Schiller and Smith 1965; Turvey 1973). The explanation for this difference between dichoptic and monoptic masking by structure or noise is attributable to the fact that under monoptic conditions integration of target and mask activity can occur as early as the photoreceptor level and at post-receptor neural levels prior to the centrally located, sustained–transient inhibitory interactions. In Section 2.1.2, we noted that in masking by light the monoptic forward masking effects are more persistent than backward masking effects (Sperling 1965; see Fig. 2.5). The inference that these masking effects entail integration of target and mask information in common peripheral pathways prior to the later inhibitory interactions among sustained or transient channels is consistent with the finding that these powerful and long-lasting peripheral on-effects produced by mask onset are not obtained dichoptically (Battersby *et al.* 1964; see Section 2.1.6; Turvey 1973) whereas interactions among and between sustained and transient channels can be demonstrated dichoptically (see Explanation 9).

(8) The fact that transient-on-sustained inhibition is used in backward masking by noise and structure has been demonstrated by Michaels and

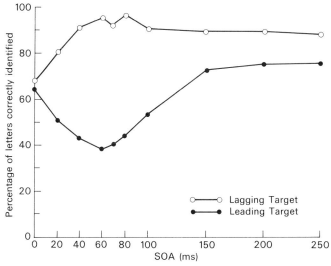

Fig. 7.9. The effect, as a function of stimulus onset asynchrony (SOA), on visibility of a consonant-trigram target, measured in percent correct letter identification, produced by a spatially overlapping pattern mask when the target temporally lags the mask (open circles) or else leads the mask (filled circles). (From Michaels and Turvey (1979).)

Turvey (1973, 1979) and Turvey (1973). In these studies the energy or contrast of the masks was appreciably lower than that of the target. Here one would expect not only weak peripheral (photoreceptor) on-effects produced by the mask but also a weak activation of the sustained channels of the mask relative to the target, and consequently little masking by integration in common post-receptor, sustained pathways ought to occur. On the other hand, based on Breitmeyer's (1978*b*) study (see Fig. 4.3), we know that even with monoptic viewing lower energy masks can yield type B metacontrast functions which are produced by transient on sustained inhibition. Michaels and Turvey (1973) and Turvey (1973) also reported type B backward masking functions when a low-energy, structure or noise mask was used, indicating that this form of interchannel inhibition also was involved in producing their results. This interpretation was given additional clarification by Michaels and Turvey (1979). They showed that with a target:mask energy ratio of 1/2, a type B function was generated under dichoptic viewing whereas a type A function was generated under monoptic viewing. Presumably here the monoptic type A effect, especially at low SOA values, was due to peripheral processes such as (i) photoreceptor on-effects produced by the mask and (ii) common integration of weak-target and strong-mask sustained activity, whereas the dichoptic type B effect was produced by a more central mask transient-on-target sustained inhibition isolated from the peripheral, energy-dependent (Battersby *et al.* 1964; Turvey 1973) integrative effects obtained monoptically.

In this connection, Purcell *et al.* (1975) showed that in backward masking by pattern, moving the retinal location of the target and mask stimuli from the fovea to extrafoveal areas can facilitate production of type B masking functions. Section 6.4 indicated that transient and sustained activities increase and decrease, respectively, with retinal eccentricity (Cleland and Levick 1974). Hence, in comparison with foveal stimuli, extrafoveal ones should yield weak type A masking by common integration of target and mask sustained pattern information but relatively stronger type B masking produced by transient-on-sustained inhibition.

(9) Type A forward and backward pattern masking as well as type B para- and metacontrast are obtained dichoptically and monoptically (Alpern 1953; Kinsbourne and Warrington 1962*a*; Kolers and Rosner 1960; Michaels and Turvey 1973; Schiller and Smith 1965; Schiller and Smith 1968; Turvey 1973; Weisstein 1972; Werner 1940). We know that sustained and transient neurones exist in the visual cortex where visual cells receive primarily binocular innervation (Hubel and Wiesel 1962, 1968). Consequently, both dichoptic and monoptic masking effects due either to integration in common sustained pathways or to interchannel inhibition should be obtainable.

(10) Type B metacontrast effects decrease in magnitude as the spatial

separation between the target and mask stimuli increases (Alpern 1953; Breitmeyer and Horman 1981; Breitmeyer *et al.* 1981*b*; Weisstein and Growney 1969). This is explainable on the basis of the spatially restricted receptive fields of sustained and transient neurones and the topographic mapping from the retina to the visual cortex (Brooks and Jung 1973). Although transient receptive fields are larger and exert their influence over greater distances than sustained channels (Ikeda and Wright 1972*a*) at increasingly larger distances between stimuli, their inhibitory effects on sustained neurones ought to attenuate. In particular, at the cortical level interchannel inhibition should be maximal within a cortical column of cells and should decrease as the physical separation of retinotopically organized columns, subserving different regions of visual space (Hubel and Wiesel 1974*a*, 1977), increases.

(11) Metacontrast is relatively weak in the fovea (Alpern 1953; Bridgeman and Leff 1979; Kolers and Rosner 1960; Lyon *et al.* 1981; Saunders 1977). At progressively larger eccentricities, the metacontrast effect becomes more robust (Bridgeman and Leff 1979; Kolers and Rosner 1960; Lyon *et al.* 1981; Saunders 1977; Stewart and Purcell 1970) and can be obtained at target–mask separations exceeding 1° visual angle (Alpern 1953; Breitmeyer *et al.* 1981*b*; Growney *et al.* 1977; Weisstein and Growney 1969).

The relative weakness of the effect in the fovea is consistent with and explainable by the following facts.

1. The fovea is characterized by a stronger activity, higher coverage factor, and higher concentration of sustained channels than transient channels (Cleland and Levick 1974; Peichl and Wässle 1979; see Section 6.4).

2. The precipitous decrease in the response strength and relative frequency of sustained channels with retinal eccentricity accompanied by an increase in the relative response strength and relative frequency of transient channels is consistent with more robust metacontrast as retinal eccentricity of stimulation increases. Moreover, since transient as well as sustained receptive fields increase with retinal eccentricity one should (i) expect stronger metacontrast with smaller target and mask stimuli in the fovea than in the parafovea or periphery (Bridgeman and Leff 1977; Lyon *et al.* 1981); and (ii) the inhibitory effect of transient channels on sustained channels ought to extend over larger target–mask spatial separation in the parafovea as compared with fovea (Alpern 1953; Breitmeyer *et al.* 1981*b*; Kolers and Rosner 1960; Weisstein and Growney 1969).

(12) Type B contour masking and brightness suppression is obtained during stroboscopic motion (Breitmeyer and Horman 1981; Breitmeyer *et al.* 1974, 1976). Specifically, as in the metacontrast paradigm, only the first of two stroboscopic stimuli is masked in accordance with a type B function (provided that the spatial separation between the two stimuli does not

exceed about 1.0–1.5°). The second stimulus is not at all masked (Breitmeyer *et al.* 1976; Breitmeyer and Horman 1981). Since low-spatial frequency, transient channels are most likely used in detecting rapid motion, a type B suppression of high spatial frequency contour information or contrast information carried in sustained channels activated by the first stimulus would be expected during stroboscopic motion, just as in metacontrast.

(13) In Section 4.3.1, I noted that simple reaction time to the target (Bernstein *et al.* 1973; Fehrer and Biederman 1962; Fehrer and Raab 1962; Harrison and Fox 1966; May and Grannis, unpublished observations) or forced-choice detection of target location (Schiller and Smith 1966) are unaffected in metacontrast masking. However, when a pattern discrimination task is used, type B reaction-time functions can be generated (Eriksen and Eriksen 1972; May and Grannis, unpublished observations). The current model states that, since optimal intrachannel inhibition is obtained when the mask *precedes* the target, transient-channels activated by the target are immune to the intrachannel inhibition by the following transient activity generated by the mask (see Fig. 7.7 (b–d)). Moreover, due to the longer response latency of sustained as compared to transient channels, the sustained activity of the later flashed mask cannot inhibit the transient activity of the preceding target. Since both intra- and interchannel inhibition of the target's transient response are effectively eliminated, the activity generated by the abrupt onset of the target could therefore be easily detected by transient channels in the visual cortex or by transient channels in the superior colliculus, activated either via direct retinocollicular projections (Hoffmann 1973; McIlwain and Lufkin 1976) or indirectly via corticofugal projections from transient cells in the ipsilateral cortex (Bunt *et al.* 1975; Finlay *et al.* 1976; Leventhal and Hirsch 1978; Palmer and Rosenquist 1974; Rosenquist and Palmer 1971). On the other hand, the type B reaction time functions obtained when pattern discrimination is employed as found by Eriksen and Eriksen (1972) and May and Grannis (unpublished observations) is consistent with the existence of interchannel, transient-on-sustained inhibition in metacontrast. These results again indicate that whereas figural information is carried in sustained channels, location information is carried in transient channels. This conclusion is further corroborated by the fact that in backward masking by noise, location information is masked for a shorter interval than is pattern or figural information (Breitmeyer, unpublished findings; Smythe and Finkel 1974; see Section 4.5), indicating that location information is carried in the faster and briefer persisting transient channels whereas pattern information is transmitted in the slower and longer persisting sustained channels.

(14) Kolers (1963) and von Grünau (1978*b*, 1981) report that when the shape or contour of either of two spatially separated and sequential stimuli is masked by metacontrast, it nevertheless can contribute to the perception

of stroboscopic motion when the second stimulus follows the first at optimal SOAs. Explanation 13 stated that transient activity generated by a given stimulus can escape the masking effects produced by the following metacontrast mask. Furthermore, since transient channels are presumably used in spatiotemporal sequence detection (Matin 1975) or the detection of stroboscopic motion (see Explanation 12), Koler's (1963) and von Grünau's (1978b, 1981) results are readily explainable in terms of the transient activity generated equally by the two sequential stimuli.

(15) Both on- and off-centre sustained and transient cells can be found along the entire visual pathway. Moreover, as one progresses up the visual pathway, transient neural responses become progressively more indifferent to the sign of the stimulus contrast (Citron *et al.* 1981; Schiller *et al.* 1976). Additionally, since transient channels characteristically respond to stimulus onset as well as offset (Enroth-Cugell and Robson 1966; see Fig. 6.1), one ought to be able to obtain metacontrast contour masking (i) irrespective of the sign of mask contrast as shown by Breitmeyer (1978c); and (ii) with mask offset as demonstrated by Breitmeyer and Kersey (1981) and suggested earlier by Turvey *et al.* (1974).

(16) Werner's (1935) investigation of metacontrast indicated that the magnitude of metacontrast suppression of a target pattern is inversely related to the orientation-difference between target and mask stimuli. Alternatively, his results indicated that metacontrast is orientation-specific. This finding is readily explained by the facts (i) that cortical transient as well as sustained neurones are orientation selective (Ikeda and Wright 1975a; Stone and Dreher 1973); and (ii) that mutual inhibition among cortical orientation-selective cells is itself orientation selective (Benevento, Creutzfeldt, and Kuhnt 1972; Blakemore, Carpenter, and Georgeson 1970; Blakemore and Tobin 1972; Creutzfeldt, Kuhnt, and Benevento 1974; Nelson and Frost 1978), extending up to orientation differences of about 30–40°.

(17) To obtain type B metacontrast functions, the present model specifies that transient channels activated by the mask must inhibit sustained activity generated by the target. Transient channels, relative to sustained ones, are insensitive to high spatial frequencies. Consequently they are also insensitive to image blur, whereas sustained channels are sensitive to it by yielding a weaker response with greater blur (Ikeda and Wright 1972b). Therefore, the prediction follows that blurring of the mask should not substantially reduce metacontrast of a non-blurred target. Growney (1976) has shown that mask blurring does not reduce the magnitude of metacontrast appreciably.

(18) Since high spatial frequency, sustained channels have a longer response latency than intermediate or low spatial frequency ones, type B metacontrast ought to peak at longer SOAs for high spatial frequency targets than for lower spatial frequency targets (see Fig. 7.7(b–d)).

Rogowitz (1983) investigated metacontrast using flanking gratings as a mask and a central grating as a target. She found, as expected, that the SOA at which metacontrast was optimal increased as the spatial frequency of the target stimulus increased.

Moreover, as spatial frequency increases, the transient response ought to decrease in magnitude and increase in latency. Consequently, as the spatial frequency of the flanking mask increases, type B metacontrast ought to decrease in magnitude with its optimal value shifting to lower SOAs. Rogowitz (1983) also reported the presence of both of these trends.

(19) As noted in Section 4.4.4, target suppression can also be obtained using a single-transient paradigm (Breitmeyer and Rudd 1981) in which a brief mask suppresses the visibility of a prolonged, sustained peripheral target for several seconds. Since the prolonged target stimulus activated only sustained channels, the reduction of its visiblity was a result of their activity being inhibited by the transient channels activated by the mask. Moreover, as also noted in Chapter 4, the single-transient paradigm indicates that any single transient stimulus can activate transient-on-sustained inhibition. Therefore, despite the methodologically necessitated use of a two-transient paradigm in metacontrast, activation of target–mask (T-M) neurones, contrary to Matin's (1975) claim, is not required; sufficient is the activation of transient neurones by the mask alone.

(20) Metacontrast also can be obtained with transient hue substitution rather than achromatic contrast or brightness transients (Reeves 1981). Reeves (1981) obtained type B metacontrast when parafoveal target and mask stimuli—either white, yellow, blue, or red—were briefly substituted against a green background of equivalent brightness. Optimal masking was obtained when red target and mask stimuli were substituted against the green background. This may correlate with the fact that at and near the fovea, most broad-band, colour-opponent transient cells as well as narrow-band, colour-opponent sustained cells are maximally sensitive to red and green light (De Monasterio 1978a, b; De Monasterio and Gouras 1975; De Monasterio, Gouras, and Tolhurst 1976; De Monasterio and Schein 1980; Gouras 1968, 1969; Gouras and Zrenner 1981). Sustained cells are for one, organized as follows: the centre–surround antagonism is specified either as R+/G− or G+/R−. On the other hand, that of parafoveal, colour-opponent, transient cells is specified either as R+G+/G− or R+G+/R− with a weaker green surround and red centre response for foveal cells (De Monasterio and Schein 1980). Let us see how Reeves' (1981) results might be explained. A steady, sustained, green background would first of all inhibit, by surround action, the R+/G− organized sustained cells. Thus, when a red light is substituted in the target area, the R+, centre response of these sustained cells would be attenuated. Moreover the centre response of R+G+/G− transient cells, due to the fact

that a sustained rather than transient green background is employed, may not be optimally inhibited by the green surround when the red mask is transiently flashed into the centre region. Also, the centre response of R+G+/R− transient cells should not be inhibited at all by the green background; consequently, they should generate a large centre response when the red mask is substituted against the green background. As a result, due to the relatively weak sustained channel output of the target and the relatively strong transient channel output of the mask, optimal type B metacontrast ought to prevail for these colour combinations of target–mask stimuli and background. This would also hold in the case where parafoveal green target and mask stimuli are presented against a red background (Reeves, personal communication). What may happen when the stimuli used by Reeves (1981) are centred foveally is yet to be determined in light of De Monasterio and Schein's (1980) finding of protan-like, reduced, red spectral sensitivities of foveal transient cells.

(21) Section 4.3.2 indicated that background and stimulus luminances affect the locus of the peak metacontrast on the SOA axis. In particular, at lower relative to higher background luminances (Purcell *et al.* 1974) or stimulus luminances (Alpern 1953), the peak effect occurs at shorter SOAs. Section 6.2 showed that as overall (background and stimulus) luminance level decreases, one would expect the faster increasing response latency of transient channels to converge on the slower increasing one of sustained channels. Hence, according to the present model of masking, the SOA at which peak metacontrast masking occurs should correspondingly shift toward a value of 0 ms. It may be worth while reiterating (see Sections 5.3.4 and 5.4) that the models proposed by Bridgeman (1971, 1977, 1978), Ganz (1975), Navon and Purcell (1981), and Reeves (1982) predict the opposite shift of the SOA at which optimal metacontrast occurs.

Moreover, Matin's (1975) model, discussed in Explanation 20, faces difficulties here. With a reduction of background luminance, optimal velocity sensitivity to real and apparent motion shifts towards lower values (Breitmeyer 1972, 1973; Crook 1937; Oyama 1970). This may relate to the facts (i) that the fusion velocity, i.e. the velocity at or above which a moving grating is no longer distinguishable from a uniform field, decreases linearly with decreases of log background luminance according to the Ferry-Porter law (Crook 1937; Oyama 1970), originally formulated to account for similar background-luminance dependence changes of the critical flicker or fusion frequency (Ferry 1892; Kelly 1961); and (ii) that the respective upper-limit velocities at which sustained and transient neural responses are barely activated decreases as background luminance decreases (Cleland *et al.* 1973; see Section 6.1). Therefore, since the optimal 'velocity' characterizing stroboscopic motion as well as the optimal temporal resolution decreases as background luminance decreases, the SOA at which one obtains either optimal stroboscopic motion or optimal

T-M neurone activation correspondingly increases. Accordingly, of Matin's two proposed T-M neurone properties contributing to metacontrast, the first specifying their optimal response to a given SOA, would predict incorrectly that the corresponding optimal metacontrast effect also shifts to a higher SOA as background luminance decreases, whereas their second property—response latency—would predict correctly a shift to a lower SOA. Here, and possibly also in other situations, Matin's (1975) model faces the problem of stating which property dominates under high as compared with low background luminance and how such shifts from presumably bilateral, co-operative, to unilateral, countervailing interactions of the two properties arise. It would seem from a standpoint of parsimony that the latter, response-latency property by itself readily accounts for these background-dependent results without the complications arising when the former, optimal-SOA property is also invoked. Parenthetically, it also may be worth mentioning that Kahneman's (1967*b*) impossible-motion formulation of type B metacontrast suffers on the same grounds as Matin's (1975) formulation based on optimal, SOA-dependent, response magnitude of T-M neurones. Of course, deciding to eliminate this latter process in explanations of metacontrast does not, as yet, necessitate its elimination in Matin's (1975) explanation of type B paracontrast. A systematic, heretofore unperformed, investigation of type B paracontrast as a function of background luminance should yield the relevant answers.

(22) Pantle (1971) and Nilsson *et al.* (1975) demonstrated the existence of selective flicker adaptation (see Section 7.3 above). The latter investigators found in particular that flicker frequencies ranging from about 8.0 to 24.0 c/s produced the largest adaptation effect. Since transient cells are selectively sensitive to that higher range of flicker frequencies, such flicker adaptation ought to produce a subsequent weaker response in transient channels (Breitmeyer *et al.* 1981*a*; Green 1981*a*). Therefore, if flicker adaptation is used prior to a metacontrast presentation, the magnitude of transient-on-sustained inhibition and, hence, metacontrast ought to decrease. Recently, Petry *et al.* (1979) confirmed this prediction in their study of metacontrast.

(23) Averbach and Coriell (1961) studied metacontrast using a single-element target display or a multi-element target display. They found weaker metacontrast in the former than in the latter condition. On the assumption that transient channels, for instance, in the superior colliculus (Wurtz and Albano 1980), carry information about stimulus location to which attention can be directed (Breitmeyer and Ganz 1976; Ikeda and Wright 1972*a*; Wässle *et al.* 1981*a*), these results are readily explainable. The location of a single target is entirely predictable if it falls at the same location as in Averbach and Coriell's (1961) study. Hence, attention can be directed to the location of the target even prior to a metacontrast presentation. However, in a multi-element display the location of the

relevant target is not known until the surrounding masking stimulus appears. Since the mask is delayed relative to the target, attention to the mask-designated one of several target locations is delayed. Moreover, since visual pattern sensitivity to stimuli falling at a given spatial location can be enhanced by the direction of attention to that location (Bashinski and Bacharach 1980; Parasuraman 1979), the earlier allocation of attention to the single target display should enhance sensitivity in sustained pattern channels relative to the later allocation of attention to the target in the multi-element display. Consequently, metacontrast will be weaker in the former than in the latter situation.

(24) Since sequential blanking (Mayzner and Tresselt 1970; Newark and Mayzner 1973; see Section 4.4.2) can be identified with type B metacontrast (Piéron 1935; Hearty and Mewhort 1975; Mewhort *et al.* 1978), it also falls easily into the explanatory scope of the present model of visual masking.

(25) A further implication of the sustained-on-transient approach to human visual information processing for iconic or visible persistence concerns the previously noted decrease of visible persistence both as stimulus duration increases (Haber and Standing 1970; Bowen *et al.* 1974; see Section 3.2.1 and Fig. 3.4) and as the spatial frequency of a prolonged stimulus decreases (Bowling *et al.* 1979; Breitmeyer *et al.* 1981*a*; see Sections 3.3.2 and 3.5.2; Corfield *et al.* 1978; Meyer and Maguire 1977). Section 2.1 noted that for luminance flashes over and above about 40-ms duration, transient on- and as well as off-effects should be observed (Ikeda and Boynton 1965). Thus, as the duration of a stimulus increases one should obtain a clear off-transient effect. In section 4.4.3, and in Explanation 15, it was furthermore shown that the pattern stimulus offset can produce type B backward masking (Breitmeyer and Kersey 1981) which, according to the model presented in Section 7.4, is due to transient-on-sustained inhibition. Consequently, as stimulus duration increases, transient channels activated at offset of the stimulus (Enroth-Cugell and Robson 1966; see Fig. 6.1) should retroactively inhibit sustained channels and thus curtail their response persistence beyond the offset of the stimulus. A related line of reasoning applies to the decrement of visual persistence of prolonged, low spatial frequency stimuli. Since transient cells are selectively sensitive to low spatial frequencies, their activation at stimulus offset again would retroactively suppress the activity of low spatial frequency sustained channels and thus curtail their persistence beyond stimulus offset. At high spatial frequencies this inhibitory mechanism at grating offset would be absent or greatly attenuated since transient channels are relatively insensitive to these spatial frequencies.

Moreover, Meyer *et al.* (1975, 1979) found orientation- and spatial frequency-adaptation to decrease visible persistence selectively (see

Section 3.5.2). Since selective adaptation for prolonged periods of time reduces the sensitivity of sustained channels but leaves that of transient channels unaffected (Bodis-Wollner and Hendley 1979; Kulikowski 1974; see Section 7.1), visible persistence in sustained channels ought to decrease since, due to their lower response after adaptation, they can be more effectively inhibited by the non-adapted transient channels activated at grating offset.

(26) The explanation of visual masking outlined above also has implications for other theories of pattern masking. According to several investigators (Haber 1969; Kolers 1968; Scheerer 1973; Turvey 1973; Uttal 1971) backward masking functions, in particular type B functions, are due to an interruption of transfer from the cortical iconic level of processing to some post-iconic categorical state, or, as Uttal (1971) claims, they are due to some fairly late interruption of cognitive processes. The theory of masking outlined in the present section claims otherwise. Type B metacontrast is the result of transient-on-sustained inhibition occurring perhaps as early as the lateral geniculate nucleus, but certainly no later than the earliest stages of visual processing in area 17 (see Section 6.6). Since the elaboration of contrast and pattern information is most likely carried out in the upper layers of visual cortex and beyond (Mountcastle 1978), the cortical transient-on-sustained inhibition most likely found in layer 4 of area 17 would prevent the sustained pattern and contrast information from being transferred to the upper layers for further pattern processing and formation. Hence, type B backward masking prevents the sustained channel information from ever transferring to the stage of visible iconic pattern synthesis. The inhibitory interactions between transient and sustained channels, presumably occuring no later than layer 4 of area 17, would occur at a previsible or pre-perceptual level of processing. This is suggested on theoretical grounds (Hebb 1949) and by empirical evidence (Blake and Fox 1974) indicating that sensory registration does, but perception of pattern does not, occur as early in the visual processing chain as area 17. This point will again become relevant in Chapters 8 (Section 8.3) and 10 (Section 10.3).

(27) The present theoretical explanation, moreover, can make some claims about the nature of the icon. Although the icon is presumably stabilized in visual space, temporally it is not a static representation but rather an ongoing, dynamic process. Inspection of Fig. 7.7 and empirical evidence discussed in Sections 7.1 and 7.3 reveal that sustained channels are characterized by an increasing response latency, integration time, and response persistence as spatial frequency increases. Consider the brief flash of a sharply contoured pattern which contains spatial frequency components ranging from low frequencies to high ones. According to the remarks above, the faster conducted low spatial frequency information arriving at the visual cortex should enter the iconic level earlier than the progressively

later and more persistent intermediate and high spatial frequency information. Transfer of sustained channel information would be expected to mirror this temporal sequence. Hence, the forming of contrast and pattern information at the iconic level should be a dynamic, temporally extended process of spatial frequency information sequentially transferred from the preiconic level to the iconic level. As will be shown in the following chapter, evidence exists indicating that the iconic pattern-forming process is indeed characterized by such a temporal extension.

7.5. Further assessments, comparisons and critiques

The exposition of the current model of pattern masking and related spatiotemporal phenomena in human vision and its explanatory coverage of the many extant as well as some anticipated empirical findings is now fairly complete. Discussion of further extensions and elaborations of the scope of the present model are appropriately relegated to specific topics covered in the following three chapters. The extensive explanatory scope of the model illustrated in the current chapter already exceeds that of the other major models reviewed in Chapter 5. The current model, based on five fundamental initial conditions, besides adequately explaining the main aspects or motifs of pattern masking illustrated in Fig. 4.2, can also give adequate accounts for many variations on those motifs which reflect the effects of correlated variations of boundary conditions such as wavelength, intensity, spatial frequency, retinal eccentricity, and so on, on the responses of transient and sustained cells. The effects of each of these variables can be specified on the basis of either electrophysiological or related psychophysical results.

Of the models reviewed in Chapter 5, the closest and most easily modifiable approximation to the current model is the revision by Weisstein *et al.* (1975) of her (Weisstein 1968, 1972) original Rashevsky–Landahl neural-net simulation of metacontrast. The older model was a fast-inhibition, slow-excitation model which in the newer version is realized, for metacontrast, by transient-on-sustained inhibition of the non-recurrent forward type (negative feedforward as opposed to recurrent inhibition or negative feedback) and for paracontrast by the reverse sustained-on-transient inhibition, implicitly also of the non-recurrent, forward type. These assumed reciprocal inhibitory mechanisms correspond to Assumption 3 specifying the current model (see Section 7.4.1). Moreover, the recent version by Weisstein *et al.*, as noted in Chapter 5, explicitly incorporates the symmetrical physiological masking effects that the stimuli arbitrarily designated as 'target' and 'mask' can exert on each other. This assumption of the model of Weisstein *et al.* corresponds to Assumption 1 of the current model. Where the models differ is in combined Assumptions 2 and 4, which in the current model states that paracontrast is realized via intrachannel inhibition, effected particularly in sustained channels, rather than the corresponding assumption of Weisstein *et al.* of interchannel, sustained-on-transient inhibition. Only future research can decide on which of the two alternatives is more viable in explaining paracontrast brightness suppression. Even if the current model is correct, this is not to imply that sustained-on-transient inhibition fails to manifest itself in pattern masking; in fact, as shown in Section 9.1, this particular interchannel inhibition can manifest itself in the recovery or disinhibition of the target (heretofore not discussed) rather than in target-masking effects. Moreover, as argued in Explanation 1 above, it may also be involved in a suppression of cff of the second of two spatially adjacent, repetitively flashed stimuli.

However, as it stands now, the model of Weisstein *et al.* cannot adequately account for the absence of type B metacontrast when simple reaction time (RT) or detection rather than brightness perception are used as criterion responses. That is so, because her initially activated, primary, target and mask neurones, n_{11} and n_{21} (see Fig. 5.1), produce a single response to each stimulus and therefore cannot differentiate between fast transient activity and separate slow sustained activity. Moreover, at the level of the secondary neurones where the first differentiation of fast (transient) and slow (sustained) activity is specified, the fast activity acts only to laterally inhibit the central, tertiary, slow activity without itself being separately channelled to a central, tertiary, detector level. Hence, the fast activity serves only to produce type B metacontrast brightness or contrast suppression, but does not serve to detect the mere presence or location of the target, as would be required to account for the fact that simple RT and target detection tasks yield no metacontrast. Of course, minor modifications of the model of Weisstein *et al.* could incorporate these required features.

The similarity between the current and Matin's (1975) models is more remote, particularly in regard to her model's required activation of T-M neurones. In so far as Matin identifies T-M neurones with transient ones and T neurones with sustained ones, her assumption of a shorter response latency of T-M as compared with T neurones does bear a similarity to the current model. In fact, this latter hypothesized response latency differ-ence combined with interchannel inhibition is essentially equivalent to the Assumptions 1 and 3 of the present model and the fast-inhibition hypothesis of the model of Weisstein *et al.* (1975). Consequently, based on these similarities, the three models would, at least qualitatively, make the same predictions of type B metacontrast variations with experimental variations of the stimulus conditions listed in the relevant explanations reviewed in Section 7.4.2.

Despite Uttal's (1981) recent claim that all the 'neuro-reductionist' models are conceptually similar, the current model—adapted after the model of Breitmeyer and Ganz (1976)—and also the models of Matin (1975) and Weisstein *et al.* (1975) differ most significantly from Bridgeman's (1971, 1977, 1978) neural-network model, Ganz's (1975) trace decay–lateral inhibition model as well as the non-neural models of Reeves (1982) or Navon and Purcell (1981) discussed in Chapter 5. None of the latter neural or non-neural models incorporate the distinction between fast, transient response components and slow, sustained ones which can reciprocally inhibit each other.

Specifically, Bridgeman's model is based on a single-channel rather than dual-channel approach. His Hartline–Ratliff neural net model assumes a spatially and temporally isotropic, excitatory–inhibitory interactive network, e.g. like that characterizing the eye of *Limulus*. A time constant, specifying the latency of lateral inhibition, and a space constant, specifying the variations of lateral interaction with variations of spatial separation between the target and mask, characterize the lateral inhibitory network. The role of lateral inhibition is not to suppress the target's pattern information, and thus decrease its signal-to-noise ratio, as in the dual channel models, but rather to distribute and store target–mask pattern information in the network's briefly persisting spatiotemporal activity or icon. As a consequence, the role of the mask activity, rather than to suppress the target activity, is to bury or alter it in the persisting spatiotemporal distribution of combined target- *and* -mask activities, thus reducing the target's signal-to-noise ratio by increasing the mask-generated noise. It is the subsequent process of template matching or cross correlation of this combined iconic representation (Bridgeman 1978) with a more permanently stored target representation, rather than any latency differences between antagonistic target and mask response components, which yields the type B metacontrast (and paracontrast) functions. Hence, it should be apparent that the dual-channel neural models and Bridgeman's single-channel model of lateral masking are conceptually quite distinct.

Despite conceptual differences, Uttal's (1981) other claim, that the formulation of these models, particularly in a quantitative sense such as those of Bridgeman and Weisstein *et al.*, are characterized by sufficient degrees of freedom so that they cannot be differentiated on quantitative or goodness-of-fit criterial grounds, seems more plausible, although again debatable. However, the goodness-of-fit between empirical and predicted results, no matter how typically and frequently it is employed, is nevertheless a conventional and therefore arbitrary criterion. Its application, in the past, led to a variety of quantitative models of psychological processes and has very often effected a host of questionable, low-yield exercises in comparative 'chi-squaremanship'. No matter how precise a model is from a quantitative viewpoint, such precision does not *per se* imply the model's validity; nor, without some appeal to intuitively or conceptually buttressed notions of *which* and *how, realistically*, representative psychophysiological processes can be quantitatively modelled, does such precision allow a choice of models.

Although one could quantify the current model by incorporating relevant response latency differences, reciprocal intrachannel centre–surround as well as interchannel antagonisms, response persistence, and so on, the value of such an exercise, due to the number of free parameters that would have to be specified, presently seems, in agreement with Uttal's (1981) claim, questionable. Hence, the at-best ordinal quantitative predictions, as well as the qualitative ones, of the current approach are taken as sufficient criteria for making comparisons among models, not from a precise quantitative viewpoint but rather from a conceptual, realistically representative one.

Moreover, since all the above neural models are (i) implicitly premised on prior receptor activity and (ii) none of them explicitly take into account the relevant receptor processes (see Sections 4.3.5 and 5.1.1), they must also incorporate these latter processes in order to account adequately for wavelength-dependent effects on masking. It seems, however, that the lack of such incorporation does not pose a crucial problem, provided that we can simply append known explanations of receptor activity to explanations of subsequent neural activity. An example of this, discussed in Section 5.1.1, are the additive effects of rod–cone and transient–sustained latency differences in determining the SOA at which optimal metacontrast occurs. However, whenever receptor-specific properties, e.g. wavelength specificity, *interact* with neural properties as evidenced by the differential chromatic sensitivity of the receptive field centre and surround of transient broad-band opponent cells, the presence of these wavelength-dependent properties can pose problems for models of masking based solely on post-receptor, neural interactions. One then has to introduce ancillary boundary conditions which specify how receptor properties interact with properties of neural network to yield variations of magnitude or shape of the lateral masking functions. A similar argument holds for stimulus and background-intensity variables, since both manifest themselves as early as receptor levels (see Sections 3.4 and 3.5.1). As an example, recall the critique of the two-process models (Ganz 1975; Navon and Purcell 1981; Reeves, 1982) and of Bridgeman's (1971, 1977, 1978) neural network models of metacontrast discussed in Sections 5.3.4 and 5.4.

Moreover, if we allow other criteria such as the explanatory range or scope of a model or its goodness-of-fit, not according to precise quantitative or less precise qualitative criteria, but rather, according to its definable role and consistency in a broader, more inclusive and integrative theoretical approach to visual behaviour, I can and will defend in the following chapters the claim that the current model is by far the most inclusive and the most accurate in the sense of corresponding to the relevant facts we know about visual behaviour. Accordingly, and here in agreement with Uttal's (1981) claim, the transient–sustained approach to masking and other visual phenomena introduced in the present chapter serves perhaps less as a formal model than as one of two heuristic devices guiding specific research as well as specific theoretical speculation. On a more global level, another heuristic device—relying on an ecological approach (Gibson 1979; Neisser 1976) to visual perception—employed and discussed in Chapters 10 and 11, serves to guide not only specific research but also more global theoretical formulations of visual function. Chapters 8 and 9 introduce new findings, not only to supplement the already extensive explanatory scope of the current transient—sustained approach to visual masking but also to serve in a preparatory manner as the empirical context for the topics introduced and discussed in Chapter 10.

7.6. Summary

A review of psychophysical studies of spatiotemporal properties of human vision, as measured by separate pattern and movement or flicker thresholds, temporal integration and persistence, reaction time, and the effects of flicker adaptation as a function of spatial frequency, points to the existence of sustained and transient channels in the visual system of humans.

This review, as well as that of the prior chapter, formed the empirical basis for outlining a heuristic model of visual information processing—in particular, of visual masking—based on the different spatiotemporal response properties of sustained and transient channels.

The masking model is based on five major assumptions or initial conditions:

1. Both target and mask stimuli activate sustained and transient channels.

2. Within channels, inhibition is realized via the centre–surround antagonism of neural receptive fields.

3. Between transient and sustained channels exists mutual and reciprocal inhibition.

4. Masking occurs in one of three ways: (i) via intrachannel inhibition; (ii) via interchannel inhibition; and (iii) via the sharing in common sustained or else transient pathways of the respective neural activities generated by spatially overlapping target and mask stimuli.

5. Whereas transient channels primarily signal the sudden appearance at a given location or the sudden change of location (motion) of a stimulus, sustained cells process its figural attributes such as brightness, contrast, and contour.

These major assumptions, when combined with specific knowledge of experimental conditions, can give an adequate qualitative account of the wide range of visual masking phenomena discussed in Chapters 1, 2, and 4, and bear on our understanding of certain spatiotemporal properties of visual persistence discussed in Chapter 3. Moreover, additional critiques of the masking models discussed in Chapter 5 were made in the context of comparing them with the heuristic approach outlined in the present chapter. When possible, suggestions for additions to or changes of any of the models were made to either give them a wider range of empirical applicability or to render them more parsimonious or rigorous.

8 Extension and applications of the sustained–transient approach: I. Higher order processes and abnormal vision

Uttal (1981) has proposed the following taxonomy of levels of visual processes. Level 1 refers to processes attributable to photoreceptor activity; Level 2 processes are based on post-receptor, neural interactions; Level 3 processes are concerned with spatial and figural organization; and Level 4, with perceptual relativism and interdimensional interactions or, for short, contextual effects. Up to now, our discussions in Chapters 5 and 7 of mechanisms and models of masking have emphasized primarily Level 2, and secondarily Level 1, processes. As such, these models can claim justifiably to be models of photoreceptor and neural activity used in masking, but they cannot claim to be photoreceptor or neural models *of* masking. As will become evident below, masking phenomena—for example, the magnitude or shape of a masking function—depend also on Level 3 and 4 processes.

Some of the contextual, Level 4 processes that are known to affect, for instance, metacontrast and backward masking by structure (Bernstein 1978; Jacobson and Rhinelander 1978; Merikle 1977) are discussed by Uttal (1981, pp. 902–13). In this chapter I shall supplement Uttal's (1981) discussion by reviewing several other investigations which recently have applied and extended the transient-sustained channel approach to visual masking to the study of higher order perceptual and cognitive processes. These include organizational and contextual, Level 3 and 4, processes used in the object-superiority, connectedness and grouping effects investigated by Weisstein and collaborators (see below), as well as processes used in the formation of and information transfer from the visual icon. Additionally, we shall see how the transient–sustained channel approach may aid our understanding of eye-dominance effects in masking and of spatiotemporal processes, including metacontrast, in abnormal, amblyopic vision.

8.1. Object- and connectedness-superiority effects (Level 4)

In several psychophysical investigations Weisstein and co-workers (Berbaum, Weisstein, and Harris 1975; Weisstein and Harris 1974; Weisstein, Harris, and Ruddy 1973; Weisstein and Maguire 1978; Weisstein, A. Williams, and Harris, unpublished observations; A. Williams and Weisstein 1978; see also McClelland 1978) demonstrated that a briefly flashed line segment was more accurately identified when it was part of a drawing that appeared unitary and three dimensional than when the line was embedded in one of several less coherent, flatter-appearing drawings. This object-superiority effect is analogous to the word-superiority effect (Baron and Thurstone 1973; Egeth and Gilmore 1973; Johnston and McClelland 1974; Reicher 1969; Smith and Haviland 1972; Wheeler 1970) in which a single briefly flashed letter is typically recognized better when it is part of a word than when it is part of a meaningless and unpronounceable string of letters. Typical examples of context patterns varying in perceived three-dimensionality are shown in Fig. 8.1. Figure 8.1(a) shows context patterns that appear strongly three dimensional. Fig. 8.1(b) and (d) shows context patterns that appear progressively less three dimensional; in fact, the patterns in Fig. 8.1(d) look flat. In Fig. 8.1(c) the diagonal target lines which comprise part of the context patterns shown in Fig. 8.1(a), (b), and (d) are represented without any context pattern. Using these stimuli, A. Williams and Weisstein (1978) investigated the effects of the various context patterns (including no context) on the perceptual identifiability of the diagonal test lines when the display was flashed for 20 ms. The differences of performance in the last three stimulus conditions displayed in Fig. 8.1(b–d), relative to the condition displayed in Fig. 8.1(a), are shown for four separate experiments in Fig. 8.2. Note the progressive decline in performance as the context pattern shifts from a strong apparent three-dimensionality to a flat pattern. In fact, in Experiments 1 and 2, the flat-context condition yielded poorer performance than the no-context

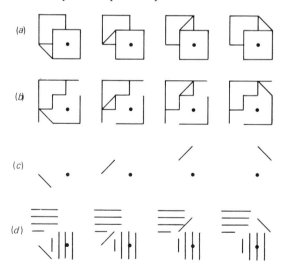

Fig. 8.1. Typical stimuli employed to study the object-superiority effect. Observers were required to identify one of the four line segments either alone as shown in (c) or in a context of decreasing apparent three-dimensionality or depth as shown in (a), (b), (d). (From A. Williams and Weisstein (1978).)

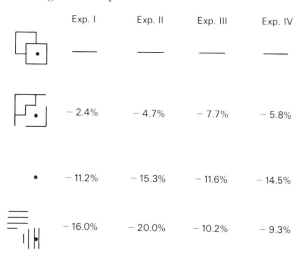

Fig. 8.2. The mean differences in accuracy of identifying the line segments between the object or highest apparent depth context (Fig. 8.1.(a)) and the other three contexts (Fig. 8.1.(b–d)). Separate results are shown for four experiments. (From A. Williams and Weisstein (1978).)

condition, indicating that instead of aiding identification of the target lines, in contrast, the context acted as a noise source and impaired identification.

The performance results of the first two display conditions in which the contexts in combination with the target lines appear three dimensional was significantly better than in the no-context condition. The interpretation of these results suggests some interesting possibilities for theories of pattern recognition and perception. Many recent and current theories of pattern recognition (e.g. see Neisser 1967) view the process of pattern recognition as constructional and, in part, sequential. They assume that the first step is the extraction and identification of component features of a pattern, followed by interpretative and constructional operations which take the output of the feature coding stage, construct the pattern, and determine what relations these features have to each other in the constructed pattern. Such an approach might appropriately be called the bottom–top approach. However, the fact that three-dimensional appearing contexts or objects facilitate identification of a target line relative to the no-context conditions indicates, in turn, that the output of the target line from the feature-encoding level is facilitated by the perception of the three-dimensional object context. This suggests that the global perceptual construction of the object occurs either in parallel with or prior to the encoding or transfer of local, feature information to subsequent levels. That is to say, a global pattern processing at a different or higher stage may facilitate the encoding or transfer of local features. Consequently, the perceptual process may also incorporate, besides the bottom–top sequence, a feedback, top-down sequence or else a temporally parallel operation between perceptually organized objects and local features. These possibilities seem particularly plausible since, as Berbaum (unpublished observations) recently demonstrated, a variation of the object-superiority effect, called the vertex-superiority effect (Berbaum *et al.* 1975), can be obtained retroactively when the context follows the test line.

The time-course of these and related effects has also been investigated by using the metacontrast paradigm (Weisstein, unpublished observations; A. Williams and Weisstein, unpublished observations; M. Williams 1980; M. Williams and Weisstein 1980*a*, *b*, 1981). Here the brief, diagonal target lines are followed at varying SOAs by the context pattern. In one series of studies, M. Williams (1980) investigated the effects of temporal blurring on metacontrast obtained with depth-context and connectedness-context patterns serving as masks. Figure 8.3(a) and (b) shows the six depth and three connectedness patterns, respectively, with their associated ratings of apparent depth and connectedness. A sharp temporal window was

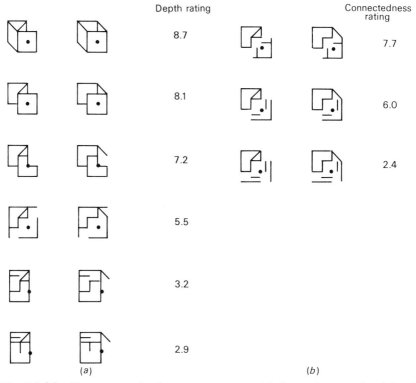

Fig. 8.3.(*a*)   Six apparent depth context patterns with decreasing associated depth ratings from top to bottom as shown. (*b*) Three apparent connectedness context patterns with decreasing connectedness ratings from top to bottom as shown. (From M. Williams and Weisstein (1980*a*).)

produced by flashing the test and context patterns for a 20–ms duration characterized by abrupt on- and offsets. A blurred temporal window of equivalent time-integrated contrast, on the other hand, consisted of a similar 20–ms duration characterized, however, by a 10–ms ramped onset and offset, thus eliminating or attenuating the temporal transients contained in the high temporal frequency components. Since transient channels are selectively affected by abrupt on- and offsets (Breitmeyer and Julesz 1975; Breitmeyer *et al.* 1981*a*; Tolhurst 1975; Tulunay-Keesey and Bennis 1979) and high temporal frequency components (see also Sections 6.1 and 7.1), the elimination or blurring of these components should result in a reduction in transient response magnitude and an increase in transient response latency. As a consequence of temporal blurring, overall masking magnitude and the optimal-masking SOA values should decrease for both the depth and connected-context patterns.

Typical depth- and connectedness-pattern results are shown for one of three observers in Figure 8.4(a) and (b) respectively. Both of the predicted effects were confirmed for the depth and connected contexts. Moreover, an especially striking effect of temporal blurring is obtained when a connectedness context is employed. Here the highly significant increase in masking magnitude (at a fixed SOA) as connectedness decreases, obtained with use of the sharp temporal window is, in contrast, entirely eliminated with the blurred or ramped temporal window. This indicates that with the sharp temporal window, the transient response and hence the transient-on-sustained inhibition generated by the high temporal frequency components of the connectedness-context pattern effectively becomes stronger as connectedness decreases. The fact that the differential effects produced by connected versus fragmented patterns are eliminated when the transient response component is diminished suggests that this component is necessary for the perceptual distinction between connected and fragmented patterns when briefly flashed. Moreover, by the same token, since higher temporal frequency selectivity of transient channels is related to a lower spatial frequency selectivity (see Section 7.1), spatial blurring, with its consequent shift to lower spatial and correlated higher temporal frequency among transient channels, should not eliminate the increase in masking magnitude for less connected context patterns.

In another, related series of studies (M. Williams 1980; M. Williams and Weisstein 1980*a*), the effects of apparent depth, connectedness, and spatial blur on metacontrast were investigated. The patterns shown in Fig. 8.3 could be flashed in one of three blur conditions: sharp image, moderate blur, and extreme blur. The last two conditions eliminated spatial frequency components above 15 and 5 c/deg, respectively.

Typical results obtained with the depth and connectedness-context patterns are shown for a second of three observers in Fig. 8.5(a) and (b), respectively. Several aspects of the portrayed results are noteworthy.

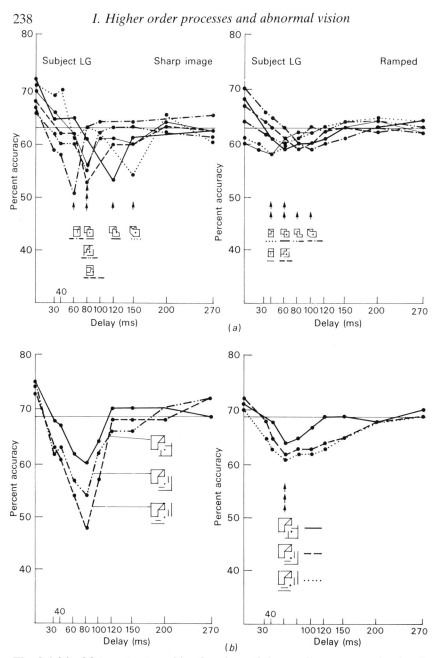

Fig. 8.4.(*a*)   Metacontrast masking functions of the test line segments by the six depth context patterns shown in Fig. 8.3(*a*). Left and right panels show results for a sharp and a ramped or blurred temporal window, respectively. (*b*) Metacontrast functions of the test line segments by the three connectedness context patterns shown in Fig. 8.3(*b*). Left and right panels show results for a sharp and a ramped or blurred temporal window, respectively. (From M. Williams (1980).)

Inspecting first the sharp-image condition, the effect of increasing the apparent depth of the context pattern is twofold:

1. The magnitude of the facilitation effect at SOAs less than about 50 ms successively increases as the perceived depth of the context patterns increases.

2. The SOA or mask delay at which optimal masking occurs shifts to successively larger values as the perceived depth of the context patterns increases.

These effects imply, respectively, a stronger sustained component of response to the context pattern–target combination and either (i) a faster rise time and/or latency of the mask's transient responses; or (ii) greater dependence on slower, high spatial frequency, sustained-channel responses, as apparent depth increases. The effect of increasing the apparent connectedness of the context pattern is to decrease the magnitude of masking at a fixed optimal SOA, implying either (i) a successive decrease in the strength of the inhibition exerted by transient channels on sustained ones; or (ii) an increasing sustained response as the perceived connectedness of the patterns increases. Next, comparing the sharp- to the blurred-image conditions reveals, for one, that for both types of context patterns the facilitation effect found at low SOA values in the sharp-image conditions is greatly attenuated or eliminated in the blurred-image conditions. Moreover, as blur increases for the connectedness contexts, the overall masking magnitude increases and the optimal masking SOA decreases uniformly for all three degrees of connectedness, thus preserving the relative ordering of masking magnitude across the levels of apparent connectedness. Similarly, for the depth contexts, greater blurring not only increases the overall masking magnitude but also compresses the range of optimal SOAs to progressively lower values, thus eliminating the differential effects of three-dimensional versus flat patterns. The fact that the differential effects produced by three-dimensional versus flat patterns are eliminated when the high spatial frequency, sustained response is diminished suggests that this component of response, as indicated by options (ii) above, is necessary for the perceptual distinction between three-dimensional and flat patterns.

Under both types of contexts increasing blur shifts the spatial frequency content of the test lines and context patterns to lower values. Thus, based on Rogowitz's (1983) results showing that lower target and mask spatial frequencies shift the optimal masking effect to lower and higher SOAs respectively (see Section 7.4.2, Explanation 18), one would expect a shift of optimal masking SOA, on the one hand, toward lower values, due to the lower spatial frequency content of the masked test lines, and, on the other, toward higher values, due to the lower spatial frequency content of the context or masking pattern. The fact that the overall shift is towards lower SOA values shows that the former influence on the optimal SOA

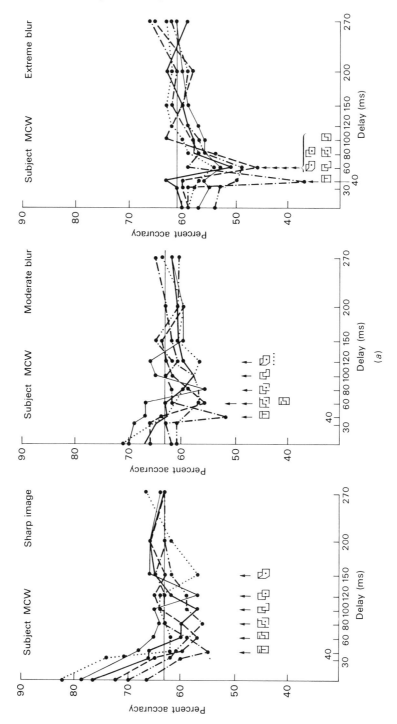

(a)

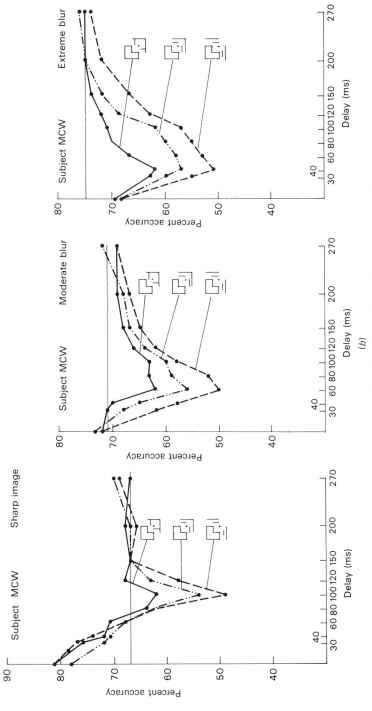

Fig. 8.5.(*a*) Metacontrast masking functions of test line segments by the six depth context patterns shown in Fig. 8.3(*a*). Left, middle, and right panels show results obtained under sharply focused, moderate blur, and extreme blur viewing conditions, respectively. (*b*) Metacontrast masking functions of test line segments by the three connectedness context patterns shown in Fig. 8.3(*b*). Left, middle, and right panels show results obtained under sharply focused, low blur, and high blur viewing conditions, respectively. (From M. Williams and Weisstein (1980*a*).)

shift outweighs the latter influence, indicating that a decrease of spatial frequency decreases sustained channel latency to the test lines more than it decreases transient channel latency to the context pattern.

Confirmation of this interpretation derives from results reported by M. Williams and Weisstein (1981). In their study of depth context or object-superiority, the effect of blur again was measured using the metacontrast technique. As in the previous study, increasing the blur of both the test lines and the context patterns (i) decreased the facilitation effect at low SOAs; (ii) increased overall masking magnitude; and (iii) compressed the range of optimal masking SOAs to low values. In fact, for the highest blur condition, the optimal SOA was 0 ms. Now let us compare the results produced when (i) the target alone was progressively blurred and the mask was sharp and (ii) when the mask alone was blurred and the target was sharp.

1. In the first, but not the second condition the facilitation effect at low SOAs dropped out, indicating that the higher spatial frequency components of the target lines but not those of the context patterns are required to maintain facilitation effects at low SOA values. As a corollary, the faster responding and, hence, earlier global, low-spatial frequency processing of the depth-context pattern is sufficient to facilitate, perhaps via positive feedback, as suggested above, the later local, high spatial frequency processing of detailed features.

2. In the former, blurred-target condition overall masking magnitude again increased; however, in the latter, blurred-mask condition masking magnitude did not change appreciably, indicating that a shift towards lower spatial frequencies of either the target or the mask greatly decreases the overall sustained channel response magnitude but not that of transient channels (see also Growney 1976; and Section 7.4.2, Explanation 17).

3. Furthermore, as would be expected from Rogowitz's (1983) results, in the blurred-target condition the SOAs at which optimal masking occurred were compressed into a narrower range and *shifted from high values* (approximately 100 ms) to 0 ms; however, in the blurred-mask condition, only a small (about 20 ms) shift, if any, *towards overall higher* optimal SOA values was observed for all context patterns. These differential results again indicate that a shift to a lower spatial frequency composition significantly lowers the response latency of sustained channels activated by the test line but only slightly lowers the response latency of transient channels activated by the context pattern.

4. Finally, since response latency and spatial-frequency selectivity of sustained channels are directly related (see Figs. 7.5 and 7.7), the very fact that, with a sharp target, the optimal masking SOA increases as the apparent depth of the context pattern increases (see Fig. 8.5(a), sharp image) implies that the output of target-activated sustained channels responding to progressively higher spatial frequencies is utilized in-

creasingly as apparent depth of the mask increases. Such depth-context dependent utilization of the test line's spatial frequency components also is consistent with the aforementioned increasing facilitation of test-line recognition at low SOAs as the depth-rating of the mask increases.

How might this facilitation, at low SOAs, of the high spatial frequency components of the target arise through the presence of global, low spatial frequency object-like masks? Based on M.Williams' and Weisstein's (1981) study, the following scheme seems plausible and in one way or another shares commonalities with other recent and current studies and theories of pattern perception (Alwitt 1981; Beck and Ambler 1973; Breitmeyer and Ganz 1976; Broadbent 1977; Hoffman 1980; Kinchla and Wolfe 1979; Marr 1978, 1982; Marr and Poggio 1979; Martin 1979; Miller 1981*a*, *b*; Navon 1977, 1981; Neisser 1967; Todd and van Gelder 1979). The main notion in all these theories is that, in response to presentation of the visual pattern, a low spatial frequency, global system has temporal priority to a high spatial frequency, local system. Spatially the latter system is nested within the former one. Moreover, the low spatial frequency, global system signals the occurrence and location of a relatively large area to which attention is directed in order to select subsequently the high spatial frequency, detailed information contained within the globally defined area. In Neisser's (1967) terminology, these two systems are preattentive and attentive processes, respectively.

How are we to envisage these systems in the context of the sustained–transient channel dichotomy? Since the local and focal system is concerned with the processing of pattern detail, one could, without raising too much controversy, identify it with the sustained channels, particularly those tuned to intermediate and high spatial frequencies. The global, low-spatial frequency system, however, may or may not correspond uniquely to one or the other neural channels. To which type of channel it corresponds could depend on at least two pattern-presentation conditions discussed below.

When a pattern display is presented to view for a prolonged period of time (e.g. a page of print to be read; a photograph or a projected slide to be inspected), one would primarily activate sustained channels, with the active role of transient channels restricted to brief saccades separating the longer fixation intervals. Under these circumstances (or even under more restricted ones requiring steady fixation of a prolonged pattern display) diffusely attended, global processing (Carpenter and Ganz 1972; Ginsburg 1972, 1976, 1978; Levinson and Frome 1979), particularly, of extrafoveal areas would be tied to low-spatial frequency sustained channels. By subsequently directing the fovea, via a saccade, to the locus of the previous extrafoveal stimulus a more focally attentive analysis of the detailed spatial aspects of the stimulus can be made. Although here we have temporal priority of global pattern processing, the priority is not due to latency differences between low- and high-spatial frequency sustained channels but rather due to the sequential nature of inspecting an extended visual display when an attentionally local, foveal pattern analysis is required to supplement a prior, global and attentionally diffuse, extrafoveal pattern processing. Now let us suppose that a pattern is flashed briefly as in a typical tachistoscopic display employed in studies of visual masking, persistence, and so on. Here, the low spatial frequency, global system could serve several purposes. For one, as suggested by Breitmeyer and Ganz (1976), low spatial frequency, transient channels could preattentively signal the location of a sudden event to which foveal attention subsequently is directed.

These transient channels would not perform any preliminary pattern analysis but simply alert the visual system to a potentially pattern-informative area of the visual field. Such transient, attention-capturing and directing channels may, for instance, project (directly or indirectly; see Section 10.2.2) from the retina to the superior colliculus (Wurtz and Albano 1980) and aid in determining the direction and amplitude of saccades to extrafoveal targets (Frost and Pöppel 1976, Pöppel, Held, and Frost 1973). Concurrently with signalling the location of a potentially informative area of the visual field, a preliminary global processing of form at that location could be performed in separate, low spatial frequency transient pathways projecting from the retina to the visual cortex. The incorporation of transient channels in a relatively global, elementary processing of form has been indicated by electrophysiological (Dow 1974; Ikeda and Wright 1975a; Lehmkuhle *et al.* 1980a, Singer *et al.* 1975; Stone and Dreher 1973; Tretter *et al.* 1975), as well as psychophysical studies (Derrington and Henning 1981; Green 1981a; Ronderos *et al.*, in press).

Although low spatial frequency, sustained channels most likely provide the only global processing of form information under prolonged stimulus viewing, under tachistoscopic presentation the possible relative contributions of these same channels, as compared with low spatial frequency, transient ones, to the early, global analysis of pattern is yet to be determined. I noted above that with *brief* presentations and at SOAs not exceeding approximately 50 ms, a concurrent or aftercoming depth-context pattern, which optically can be either sharp or blurred, facilitates detection of only an optically sharp target-line. In other words, whereas the early processing of low spatial frequency, global form information of the depth-context pattern is a sufficient condition for producing the object-superiority effect, the later analysis of high spatial frequency, local detail of the target-line is a necessary condition.

However, in our laboratory Mary Williams and I also have investigated the effect of a *prolonged* presentation of a depth-context pattern on visibility of a brief target-line. Since only sustained channels respond during the presence of a prolonged stimulus, any effect that the depth-context pattern has on the target-line visibility must be due to pattern information processed by sustained channels. Our results, as yet unpublished, showed that for five observers progressive increases of the apparent depth of the context pattern produced progressive *decreases* of the target-line's visibility. These results suggest that the global components of the context patterns processed by sustained channels do not facilitate target-line visibility. As a corollary and, by extension, the results tentatively indicate that with *brief* flashes the early global processing of form information leading to facilitation of the target-line's visibility is performed by low spatial frequency transient, as opposed to sustained, channels. Of course, additional research is required to determine more precisely if and how the, as yet, suggestive findings obtained under prolonged context–pattern presentations can be translated to related findings obtained with brief context–pattern presentations.

8.2. Perceptual grouping effects (Level 3)

In another series of experiments, M. Williams and Weisstein (1980*b*) investigated the effects of Gestalt or perceptual grouping (Wertheimer 1958) on metacontrast. In one particular experiment the observers were required to detect the position of one odd, mirror-reversed pattern element (the target) among three normally oriented ones, each of which was a rectilinear capital letter C, as shown in the four examples of Fig. 8.6(c). Here we already have a grouping effect, produced by the Gestalt organizational factor of similarity, which distinguishes, although rather weakly (Beck 1966; Julesz, Gilbert, Shepp, and Frisch 1973), the target element from the other three elements. The metacontrast mask patterns also consisted of two possible arrays of rectilinear C-like elements as shown in the left-most panels of Fig. 8.6(a) and (b). The respective combinations of these two mask arrays with each of the four possible target arrays yield the pattern arrays numbered 1 through 4 in Fig. 8.6(a) and (b). Note first that the combined arrays in Fig. 8.6(a) form more distinguishable groupings than those in Fig. 8.6(b). This readily apparent difference was experimentally corroborated by M. Williams and Weisstein (1980) through independently obtained ratings of perceptual grouping according to the technique developed by Pomerantz and Garner (1973). The enhanced grouping effects illustrated in Fig. 8.6(a) as compared with Fig. 8.6(b) and (c) are due to the pitting of the organizational factor of similarity (produced by the combinations of the odd, mirror-reversed, target element with the nearest similar mask element) against the factors of symmetry and closure (produced by the remaining three combinations of normal target and mirror-reversed mask elements).

Besides varying the perceptual grouping level, Williams and Weisstein (1980*b*) also introduced spatial blurring as another variable. Both target

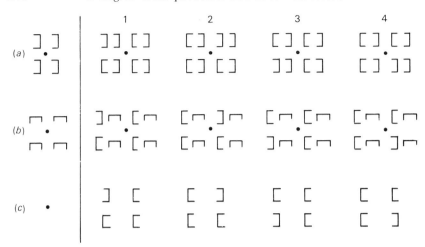

Fig. 8.6. Two mask patterns (leftmost patterns of panels a and b) and four target patterns (panel c, 1–4) used to study perceptual grouping effects in metacontrast. In panels a(1–4) and b(1–4) are shown the respective composites of the four target patterns with each of the two mask patterns. (From M. Williams and Weisstein (1980*b*).)

and mask stimuli could be presented under either a sharp image condition (no blur), a low degree of blur (elimination of spatial frequency components above 15 c/deg), or high-degree blur (elimination of spatial frequency components above 5 c/deg).

Typical results are shown for one observer at all levels of blur in Fig. 8.7. The solid horizontal line corresponds to the proportion of correct responses made when the target arrays were displayed alone. Note that in the presence of each of the two mask arrays a type B metacontrast function is obtained. However, the masking magnitude, particularly at lower SOAs, varies directly with the degree of perceptual grouping. Moreover, although masking magnitude increases overall as the level of blur increases, as would be expected on the basis of Result 2 of M. Williams and Weisstein's (1981) investigation discussed in Section 8.1 (see also Fig. 8.5), the direct relation between masking magnitude and level of perceptual grouping is maintained.

Although M Williams and Weisstein (1981; see Section 8.1) found that whereas blur of the target, but not the mask, in this depth-context study compressed and shifted the optimal metacontrast range from about 100 ms towards 0 ms, no such shift occurs in the present perceptual grouping study. Both levels of perceptual grouping yield a peak masking effect at a low SOA of 40 ms *irrespective of the level of blur*. This indicates that, although higher spatial frequency components of a target are required to yield depth-context dependent facilitation of its visibility, perceptual grouping is not affected by the high spatial frequency components of either

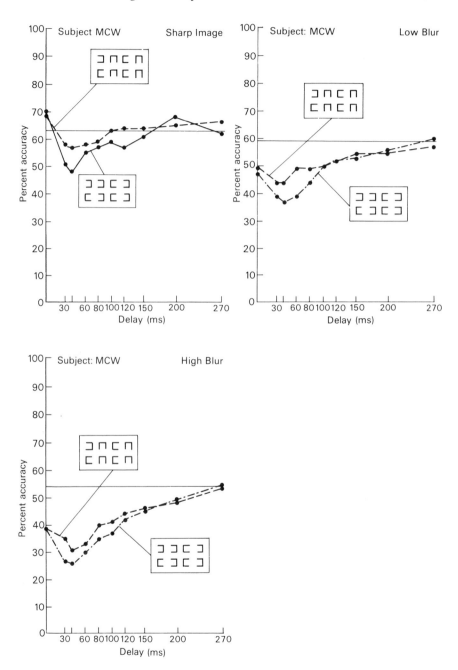

Fig. 8.7. The metacontrast functions produced by the two masks shown in Fig. 8.6 under two corresponding grouping conditions and three degrees of spatial blur as indicated at the top of each panel. (From M. Williams and Weisstein (1980*b*).)

the target or the mask stimuli. Otherwise, here also one would expect to obtain spatial frequency-specific increases of target visibility at low SOAs and shifts of the optimal-metacontrast SOA toward higher values, as the level of blur decreases or, equivalently, as more high spatial frequency components are added to the target. The obtained results dovetail neatly with Ginsburg's (1976, 1978; Ginsburg *et al.* 1972) findings of spatial frequency-specific effects of perceptual grouping. By selectively filtering out bands of low, intermediate, or high spatial frequencies, Ginsburg demonstrated, among other phenomena of form perception such as illusions and subjective contours, that Gestalt factors of figural organization depend only on the presence of low to intermediate, but not high, spatial frequency components of a pattern.

8.3. Icon formation and read-out

In the previous chapter (Section 7.4.2, Explanation 26 and 27) it was argued that the visual construction of the cortical icon, when it is not prevented by visual masking, occurs at a stage later than the earliest levels of afferent input into area 17 of the visual cortex. Moreover, it was noted that the icon can be characterized as a dynamic, temporally extended process of the integration of output from area 17 sustained channels varying in response latency and persistence as a function of increasing spatial frequency. Consequently, the iconic process must be a higher-order one than the more peripheral masking processes.

8.3.1. Icon formation

In an ingenious series of experiments, Michaels and Turvey (1979) were able to obtain a measure of the central or cortical iconic integrative process. Their rationale is based on the following line of reasoning. They asked whether iconic synthesis of two briefly, interocularly flashed visual displays, temporally separated by SOAs ranging from 0 to around 100 ms, represents pattern information equally from both fields, i.e. whether iconic synthesis is symmetrical about an SOA of 0 ms. One way of answering this question, as indicated by Michaels and Turvey (1979) would be to identify some mask that integrated with the target, but had neither a transient response, which would inhibit transfer of sustained information into the icon process, nor an independent iconic representation which would divert attention from the read-out of the target–icon information. A spatially uniform and extensive, non-patterned flash might serve as such a mask. However, since masking of pattern by light can be obtained only monoptically (Battersby *et al.* 1964), temporal contrast or luminance summation in peripheral channels would confound with the central iconic integration process. Unless one knew the time course of temporal integration of luminance or contrast at peripheral levels of visual processing for a flash of a given intensity, the time of iconic synthesis could not be factored out.

Michaels and Turvey (1979) used another, albeit more complicated, rationale. The target consisted of black letter-trigrams and the mask consisted of a spatially overlapping string of eight backward printed letters. The target and mask were each presented for 10 ms. Both forward and backward masking functions were measured over SOAs ranging from −80 to +150 ms. Since dichoptic viewing was employed, any target–mask interaction could occur only at central levels at or beyond the site of binocular combination. Suppose, as did Michaels and Turvey (1979) that one can isolate the central iconic integration process by comparing two masking functions which differ *only* in the magnitude of masking produced at *this* iconic level. In particular, assume that one of two dichoptically flashed visual displays (i) dominates in integration through common iconic synthesis; but (ii) does not dominate in the preiconic transient-on-sustained inhibition which prevents (interrupts) transfer of preiconic sustained pattern information to the iconic integration process. It follows that the difference between two masking curves—one obtained when the mask display dominates the iconic integration stage, the other when the mask display instead is dominated at the same stage by the target display—will factor out the equal masking effects produced by transient-on-sustained inhibition and consequently will yield a positive-valued iconic masking function symmetrical about an SOA of 0 ms. The absolute values of the positive or negative SOA at which this difference function just attains a value of zero would be an estimate of the duration of the central iconic synthesis (Maffei and Fiorentini 1972) of target and mask pattern components in this particular experiment. One way of actualizing this experimental rationale is to take advantage of eye-dominance effects in visual masking (Breitmeyer and Kersey 1981; Michaels and Turvey 1979; see Section 4.4.3). Implicit in exploiting these specific effects is the additional assumption that they manifest themselves in terms of a corresponding dominance of sustained pattern information integrated into the central iconic process rather than in transient-on-sustained inhibition (see Section 8.3.2 below).

Michaels and Turvey (1979) investigated forward and backward dichoptic visual pattern masking in several subjects showing pronounced eye-dominance effects. With dichoptic viewing, the obtained masking effects were most likely occurring at the visual cortex, in particular, at the level of cortical iconic integration or else the prior level of cortical transient-on-sustained inhibition. Two conditions were applied. In one condition the target and mask were presented to the dominant and non-dominant eye, respectively; the second condition was the reverse of the first. The results of the experiment are shown in Fig. 8.8(a). Note that when the target–letter-trigram is presented to the dominant eye, virtually no forward masking is obtained, whereas a pronounced monotonic forward masking effect is obtained when the target is flashed to the non-dominant eye. Since the former result indicates that masking by common iconic integration did not occur when the target, presented to the dominant eye,

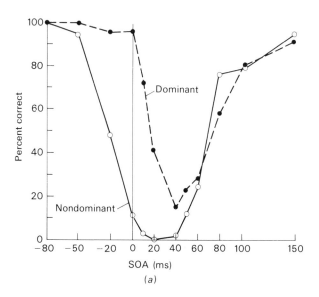

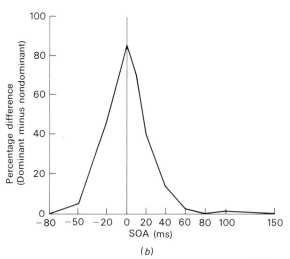

Fig. 8.8.(*a*)   Target masking curves showing target identifiability as a function of mask SOA for dichoptic forward and backward masking with the target flashed to either the dominant or else the non-dominant eye. (*b*) The computed difference between the dominant and non-dominant masking functions shown in (*a*). (From Michaels and Turvey (1979).)

was preceded by a mask-flash (forward masking) to the non-dominant eye, we also would expect that such masking would not occur when the non-dominant mask follows the dominant target (backward masking). Note further, in line with this expectation, that in the case of backward masking, a relatively weaker U-shaped, type B effect and a stronger J-shaped type B effect is obtained, at least up to an SOA of 60 ms, when the target is flashed to the dominant and non-dominant eye, respectively. Hence, according to the rationale, the relatively weaker U-shaped type B effect isolates the masking mechanism produced by transient-on-sustained inhibition; whereas the latter J-shaped type B function obtained when mask and target were flashed to dominant and non-dominant eyes, respectively, additionally manifests the masking produced by common central, iconic integration of dominant-eye and non-dominant-eye pattern information.

The difference between the stronger and weaker forward masking functions thus would indicate that common central iconic integration of the dominated target- and the dominant mask-pattern information prevailed when the target was presented to the non-dominant eye. The question Michaels and Turvey (1979) posed is whether or not, as assumed, the same difference between the two functions held in the case of backward masking. To obtain the answer, Michaels and Turvey (1979) plotted the difference function obtained by subtracting the masking function obtained when the mask was flashed to the non-dominant eye (weaker forward and backward masking) from that of the masking function obtained when the mask was flashed to the dominant eye (stronger forward and backward masking). The difference function is shown in Fig. 8.8(b). Note that the function is symmetrical about an SOA of 0 ms, indicating that masking by common cortical iconic synthesis of the sustained pattern information of the target and mask occurred equally in the backward masking and forward masking. Moreover, in agreement with Michaels and Turvey's (1979) interpretation, these findings also add credence to the similar interpretation offered above (Section 7.4.2, Explanation 26) that the cortical iconic process occurs at a later level than the cortical transient–sustained inhibitory interactions. In this regard, the other noteworthy point evident from inspection of Fig. 8.8(b) is that the cortical iconic integrative process, as indexed by this particular experiment, lasts approximately 60–80 ms, indicating additionally that it is a dynamic, temporally extended synthesis of sustained channel information (again, see Explanation 26 and 27 in Section 7.4.7).

In an independent study, DiLollo and Woods (1981) employed the following technique to investigate these dynamic and other additional properties of visible or iconic persistence. Two separate 2.5-ms displays, each composed of twelve 3×3 dot elements of a 5×5 square matrix as shown in Fig. 8.9 were flashed consecutively at interflash intervals (IFIs) ranging from 15 to 155 ms. The composite two-flash pattern consisted of 24 elements with one randomly chosen element position left blank. The task of the observer was to determine which matrix location was blank. Since iconic persistence decays as a function IFI, the number of errors made in determining the blank matrix location was expected to increase monotonically as IFI increases.

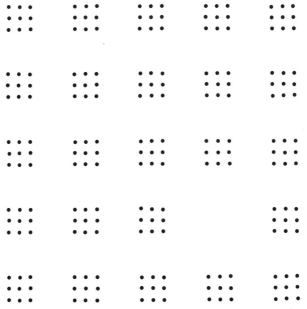

Fig. 8.9. The 5 × 5 matrix display showing one missing element. (From DiLollo and Woods (1981).)

Besides varying IFI, DiLollo and Woods (1981) also varied the spatial frequency content of the matrix display. This was accomplished by placing optometrist spectacles in front of the observers' eyes using four negative diopter (−D) lens values of −OD, −2D, −4D, and −6D. The consequent blurring or defocusing of the matrix display, in terms of highest transmittable spatial frequencies, corresponded to 60.00, 3.49, 1.71, and 1.41 c/deg, respectively.

Fig. 8.10 shows the error curves of the four blur conditions as a function of IFI for four observers. Note the following trends. First, for all blur conditions the proportions of error, as expected, increased monotonically as IFI increased. Furthermore, the error functions separate for the four blur conditions; as a function of IFI, they increase progressively later as the spatial frequency composition of the matrix display is characterized by progressive additions of higher spatial frequency components. These results indicate, for one, as illustrated in the masking model outlined in the previous chapter (see Section 7.4.1) and confirmed by previous psycho-

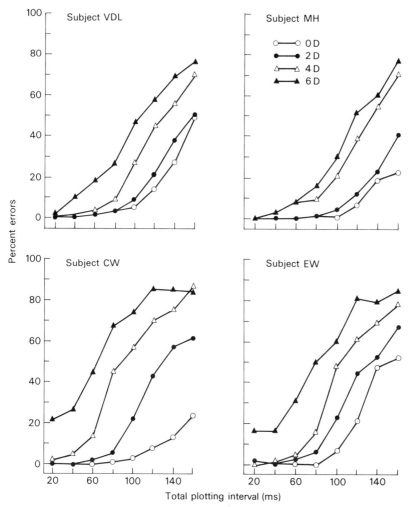

Fig. 8.10. Percent errors in correctly locating the blank matrix element as a function of interflash interval separating the two complementary flashes composing the 5 × 5 matrix pattern. Curve parameters shown in upper left panel correspond to degree of defocusing using negative diopter lenses. (From DiLollo and Woods (1981).)

physical results (Breitmeyer *et al.* 1981*a*; Corfield *et al.* 1978; Meyer and Maguire 1977; see also Sections 3.3.2, 3.5.2, and 7.3) that visible persistence increases with spatial frequency.

Moreover, the results show that the continuous temporal decay of the iconic information began progressively later as more high spatial frequency information was added to the matrix display, indicating progressively higher spatial frequency components arrive later and reside progressively longer at the iconic level than lower ones. This order of decay also probably reflects the temporal order and continuity characterizing the transfer of sustained-channel spatial-frequency information from the preiconic to the iconic level as schematized in Fig. 7.7. Finally, since the decay curves are aproximately parallel to each other, the rate of decay from the iconic process must be the same for all spatial frequency components. These properties additionally reinforce the claim that the iconic integrative process is temporally dynamic rather than static.

8.3.2. Further comments on eye dominance effects in masking

Recall that Michaels and Turvey's (1979) rationale for estimating the symmetrical forward and backward masking mechanism attributable to common, central iconic synthesis of sustained pattern information of the target and mask was premised on the assumption that eye dominance affected only this mechanism but not the temporally prior, cortical transient-on-sustained inhibition. However, an alternative assumption which nevertheless has equivalent consequences is that the latter, central, interchannel inhibition *as well as* the strength of sustained activity is affected by eye-dominance. For instance, when, in forward masking, the preceding mask and aftercoming target are flashed to the non-dominant and dominant eye, respectively, the dominant transient activity of the target stimulus, as already indicated in Section 7.4.2 (Explanation 6, and Fig. 7.9), would exert an extra-strong, local inhibition of the mask's non-dominant (relatively weaker) sustained activity. This central, preiconic inhibition would severely attenuate the 'noise' strength of the mask's sustained activity subsequently transferred into the composite iconic representation of target and mask pattern information. Here the target's 'signal-to-noise' ratio is greatly enhanced, and as a consequence its pattern representation dominates in the composite icon. Thus, no type A forward masking should result. By the same token, in the reverse situation in which the mask and target are presented to the dominant and non-dominant eyes, respectively, the 'noise' produced by the mask's sustained activity entering into the composite central icon is greatly enhanced whereas the 'signal' of the target's preiconic sustained activity is now severely attenuated. Hence, in regard to the target's pattern information residing at the iconic level, the 'signal-to-noise' ratio is also severely attenuated, and strong type A forward masking should result.

This line of reasoning also applies to backward masking by structure. Here the mask, when flashed to the dominant eye, due to the extra-strong transient-on-sustained inhibition it exerts on the non-dominant sustained activity of the target, would additionally facilitate the transfer of its relatively stronger preiconic, sustained pattern information and thus dominate at the iconic level. In the reverse situation, when the mask is flashed to the non-dominant eye, the resulting weaker transient-on-sustained inhibition exerted by the mask in conjunction with the dominant and relatively stronger sustained activity of the target should eliminate and reverse the dominance of the mask-pattern information in the composite target–mask iconic representation. Hence, here, only (or primarily) the transient-on-sustained inhibition should contribute to backward masking.

The reasons one cannot rule out this possibility are, for one, that target and mask are methodological designations to which the neurophysiology of the visual system is indifferent. Hence, since both target and mask stimuli (by Assumption 1 of the model outlined in Section 7.4.1) activate transient and sustained channels, mutual intra- as well as interchannel interactions are used and therefore must be given account. Secondly, as Breitmeyer and Kersey (1981) showed (see Fig. 4.9 and Section 4.4.3) analogous eye-dominance effects are also evident in type B metacontrast in which, due to spatial separation of target and mask stimuli, common integration of pattern information at the central iconic level is not a source of masking (unless, as suggested for instance by Bernstein (1978), later cognitive, decisional processes based on comparing the apparent brightness of the target and mask icon also affect measures of metacontrast). However, in terms of the processes used up to and including the iconic integration stage, such sensory (pre-cognitive) integration *per se* cannot be a source of masking.

Consequently, by elimination what remains as possible processes contributing to eye-dominance dependent, type B, metacontrast functions are either a stronger, central, transient-channel or a stronger, central, sustained-channel response activated by the stimulus flashed to the dominant relative to the non-dominant eye. In either or both cases the effective transient-on-sustained inhibition and, hence, type B metacontrast ought to be relatively stronger when the mask and target are flashed to the dominant and non-dominant eyes, respectively, than when the mask's and target's ocular inputs are reversed.

The fact that Breitmeyer and Kersey's (1981) results yielded no statistically significant interaction between dichoptic viewing condition and SOA indicates that metacontrast was *uniformly* stronger at all SOAs when the mask and target were flashed to dominant and non-dominant eyes, respectively. Moreover, since this uniform difference was maintained at the smallest and largest SOAs, where one expects little transient-on-sustained inhibition, the more likely candidate for expressing eye-dominance effects in noise or structure masking, as implied in Michaels and Turvey's (1979) interpretation, seems to be sustained channel pattern activity. Informal observations made by the observers, including myself, in the Breitmeyer and Kersey (1981) study was that the 10-s mask annulus presented to the dominant as compared with non-dominant eye had a greater interocular suppression effect on a 50-ms test disk, even when the test flash occurred temporally at the midpoint of the 10-s mask presentation, when neither on- or offset transients of the mask could effectively mask the target. Since, at this point of target presentation, the mask was effectively activating only sustained channels, the interocular suppression of target contrast by the mask would have to be due to either binocular rivalry or the intrachannel suppressive effect (e.g. simultaneous brightness contrast) occurring among central sustained channels. However, more experimental results must be collected before one can establish the relative contributions of these two possible mechanisms, or of other sustained and transient activities, to interocular suppression under brief, tachistoscopic viewing of stimuli.

8.3.3. Iconic read-out

Let us return now to an additional important finding reported by Michaels and Turvey (1979). Michaels and Turvey reasoned that in dichoptic backward masking, besides common integration at the cortical iconic level and the transient-on-sustained inhibition at the cortical preiconic level, another source of masking is the diversion of a central attention mechanism from the scanning and read-out of target-icon information produced by the aftercoming mask. Such scanning of visual information at the iconic level can be extremely rapid as shown by Sperling, Budianski, Spivak, and Johnson (1971) and is affected by attention (Sperling and Melchner 1978). In one experiment, Michaels and Turvey (1979) conjectured that the recognition of target words and non-words depends on different cognitive recognition processes occurring later than the stage of transient-on-sustained inhibition or iconic integration. Consequently, because words and non-words may engage different post-iconic recognition algorithms, but are not, due to their spatial similarity, differently affected at the preiconic or iconic levels of masking, a differential effect of word and non-word, target and mask, stimuli might be found in the latter post-peak portion of the type B masking function where performance progressively improves as SOA increases. According to the rationale associated with discussion of the results shown in Fig. 8.8 above, this portion of the type B backward pattern masking function should be prone to differences in *post-iconic* scanning or read-out strategies. Four target–mask combinations were employed: word–consonant trigram, word–word, consonant trigram–consonant trigram, and consonant trigram–word. The obtained masking functions are shown in Fig. 8.11. Note that with words as target, the ascending, latter portion of the type B masking functions occurs earlier than that of the masking functions when non-words are used as target. Presumably, in the former case, the higher-order recognition algorithm can focus attention or selective processing to the iconic read-out of the word targets earlier and faster than of the non-word consonant trigrams. Related and similar results have been reported by Taylor and Chabot (1978) and earlier by Toch (1956).

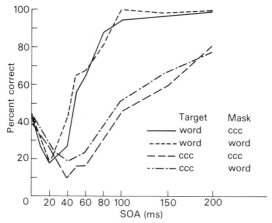

Fig. 8.11. Average masking curves (for four observers) showing letter identifiability as a function of SOA for three-letter word or consonant trigram targets followed dichoptically by similar pattern masks as indicated in the figure. (From Michaels and Turvey (1979).)

Additional evidence for this interpretation derives from studies conducted by Spencer (1969) and Spencer and Shuntich (1970), who, similar to Averbach and Coriell (1961) (Section 7.4.2, Explanation 23), investigated backward pattern masking for a single target letter and for a twelve target-letter array. Recall that attention to a given location increases perceptual sensitivity (Bashinski and Bacharach 1980; Parasuraman 1979; see Section 7.4.2, Explanation 23). In the former case the location of the lone target was known as soon as the target flash registered, and consequently attention could be focused on it relatively early. Hence, the attention allocated to the spatial location and subsequent iconic read-out of the target was taxed relatively little by the aftercoming mask. However, in the twelve-letter array displayed around an imaginary circle centred at the fovea (Spencer 1969; Spencer and Shuntich 1970), a single randomly selected target letter was signalled, via a delayed visual pointer, for report by the observers. Here, attention to spatial location or to the iconic read-out process could not be allocated as early as in the single-letter display but rather was delayed until the marker was flashed and, therefore, would be more severely taxed by the aftercoming mask. For example, Spencer and Shuntich (1970) investigated backward pattern masking (under binocular or monocular, as opposed to interocular, viewing) at three mask-energy levels. With the single-letter target, the magnitude of backward masking increased as mask energy increased. Moreover, the masking effect, composed of a variable sensory magnitude and a *weak* attentional magnitude, was eliminated at a target–mask asynchrony of about 100 ms. However with the twelve-letter array, all three mask energies yielded overall more pronounced and extensive backward masking effects, comprised of a stronger, variable sensory and a *stronger* attentional magnitude, extending up to a target–mask asynchrony of 300 ms. From this three tentative conlcusions may be drawn:

1. One masking component included early peripheral, energy-dependent, sensory integration (Turvey 1973) and extended to a target–mask asynchrony of around 100 ms.

2. A second masking component included the allocation of attention to target location and interacted with the early sensory component. Relative to the single-letter target, the allocation of attention to the cued letter of the twelve-letter target array occurred later and, therefore, less effectively enhanced sensitivity of sustained pattern channels at preiconic cortical or subcortical levels. As a consequence, relative to the single-letter target, in the twelve-letter target array the integration of the cued target and the mask pattern information at the cortical iconic level of processing would favour the mask and, hence, again effect an increase of masking at early SOAs ranging from 0 to 100 ms (Michaels and Turvey 1979; see Fig. 8.8). It follows that iconic read-out and post-iconic encoding of the more degraded pattern information should be more difficult in the twelve- as compared with single-letter target display.

3. Since (i) allocation of central attention to these central, cognitive processes is more crucial for the cued target in the twelve-letter array as compared with the single-letter target; and (ii) the aftercoming mask disrupts the allocation of this attention, here one would also expect to obtain more extensive masking due to interruption of the later iconic read-out and post-iconic encoding processes.

In another experiment, Michaels and Turvey (1979) used three-letter words or consonant trigrams as targets and three-letter trigrams as masks to study dichoptic masking effects as a function of the visual field to which the targets were flashed. The results, averaged across four observers, are displayed in Fig. 8.12. Note, first of all, that again word targets overall produce faster rising masking functions at later stages of masking than do trigram targets. Moreover, there is a visual-field effect: masking functions rise faster in the right as compared with left visual field. Since pattern information from the left and right visual fields is processed by the right and left cortical hemispheres, one would expect verbal material such as words or trigrams to be better recognized by the left or speech-dominant hemisphere (for recent reviews of asymmetric brain function see Kinsbourne (1978)).

According to Swanson, Ledlow, and Kinsbourne (1978) one way of realizing such a left-hemisphere verbal processing bias involves the roles of cognitive orientation and selective attention. Their model is based on Sherrington's (1906) notion of reciprocal innervation applied to the direction of lateral orientation. Presumably when the focus of attention is straight ahead, e.g. directed towards a fixation point of a target–mask

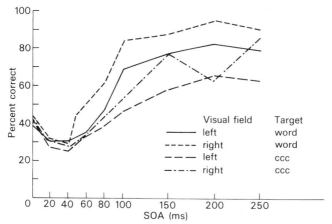

Fig. 8.12. Masking curves showing identifiability of letters as a function of SOA for three-letter word and consonant trigram targets under dichoptic masking with the target presented either to the left or right visual field as indicated in the figure. (From Michaels and Turvey (1979).)

display, the contralateral gaze centres are in a state of mutually inhibitory balance. A laterally displaced or eccentric stimulus elicits a corresponding, covert, or overt orientation response by the affected hemisphere which inhibits the contralateral orientation response. Consequently, the processes or responses compatible with the orientation response, such as verbal processes or responses in the left hemisphere should be attentionally favoured. On the other hand, non-verbal spatial responses or processes aroused along with the orientation response and controlled by the right hemisphere should not attentionally favour verbal material presented to it: the pattern information would first have to be transmitted cross-callosally to the left hemisphere. Consequently, there would be better and faster verbal processing in the left, speech-dominant hemisphere than in the right.

Given such an attentional model, the results shown in Fig. 8.12 are readily explainable in terms of a more efficient and rapid read-out of iconic information and its subsequent processing by the verbal recognition algorithm in the left hemisphere. This interpretation is given additional support by the findings reported by DiLollo (1981), Erwin and Nebes (1976), and Marzi, Stefano, Tassinari, and Crea (1979), all of which indicate the existence of hemispheric or visual hemifield *symmetry* in the duration of *pre-categorical*, visible persistence, although predictable hemispheric asymmetries do occur when the information must be read out and cognitively categorized by the observer (Erwin and Nebes 1976; Marzi *et al.* 1979).

On the basis of the results illustrated in Fig. 8.11 and 8.12, Michaels and Turvey (1979) arrived at the following model of *central*, *cortical* masking effects, as schematized in Fig. 8.13. In this model of central masking, three processes are involved. One is the integration of common sustained or pattern information of the target and the mask at the iconic level (see Fig. 8.8(b) and the discussion relating to it). This integration would therefore give rise to type A forward and backward masking effects as shown. A second process is the preiconic, transient-on-sustained inhibition which gives rise to the typical type B backward masking function. Finally, a third and latest process incorporates the icon read-out and post-iconic recognition algorithms, both of which require allocation of central, selective attention. As can be seen from inspection of Fig. 8.13, the icon read-out and algoristic recognition of the target information are affected most at intermediate and longer SOA values, at which the mask begins to exert progressively less transient inhibition of sustained input to the iconic level but also effects the interrupting or diverting of attention from these processes (Scheerer 1973; Spencer and Shuntich 1970). This interpretation is highly plausible in view of psychophysical results reported by LaBerge (1973) and Posner and Cohen (1980), indicating that such diversion or switching of attention from one stimulus to another requires at least about 50–100 ms, a range of values which coincides with the range of SOAs at which optimal type B metacontrast occurs. As SOA progressively increases, more and more of the target pattern information is attentively read out from the icon and processed by the algoristic recognizer before attention is diverted by the mask pattern; hence, progressively less

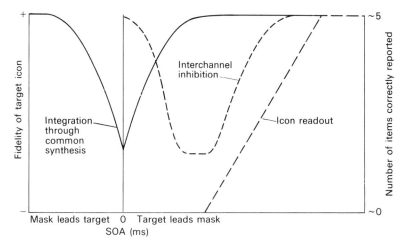

Fig. 8.13. A three-process model of central, cortical pattern masking, showing hypothetical target fidelity (left ordinate) and number of items correctly reported or read out from icon (right ordinate) as a function of mask SOA. Target pattern fidelity is affected by central integration through common synthesis or by the prior central transient-on-sustained (interchannel) inhibition. (From Michaels and Turvey (1979).)

masking occurs at these post-iconic levels. Moreover, whereas the first two masking processes entail integration and interaction at sensory levels of processing, the latter, central, attentional masking effects occur at a higher cognitive level, and hence, one can appropriately call them a form of cognitive masking (Walley and Weiden 1973).

8.4. Application of the sustained–transient channel dichotomy to amblyopic vision

I noted in Section 6.7 that abnormal visual experience early in the development of organisms can produce anatomical and physiological deficits in sustained and transient neurones. In particular, it was noted that non-corresponding binocular visual input produced by monocular paralysis or surgically induced squint produces a selective loss of sustained and transient neural function. This loss, as shown by Ikeda and co-workers (Ikeda 1979, 1980; Ikeda *et al.* 1976; Ikeda and Tremain 1979; Ikeda and Wright 1975*c*, 1976), was particularly pronounced for sustained cells of the retina and lateral geniculate nucleus representing the central 10° of the visual field.

In human visual development, amblyopia also is often characterized by non-corresponding binocular visual inputs. Consequently, one would expect losses of transient and of sustained function, particularly in central vision. In fact, Ikeda (1979, 1980) notes the close correspondence between functional loss of visual resolving power of central sustained cells in cat's visual system and psychophysically measured loss of visual acuity in central vision of humans. This correspondence is illustrated in Fig. 8.14. The top panel shows the typical relative visual acuity of a human amblyope in his or her normal and amblyopic eye. Note the selective loss of acuity from the fovea and up to the 5° periphery for the amblyopic eye. At progressively larger eccentricities the visual acuities of the two eyes are equal. In the bottom panel are shown visual acuities of cat sustained cells mapped from the normal eye and the eye in which squint was surgically produced. In the squint eye, the pronounced and selective loss of acuity of visual cells representing central (up to 5° eccentric) vision contrasts the high acuity of area centralis cells in the normal eye. Moreover, in correspondence with the human acuity curves, beyond a 5° eccentricity the acuities of the sustained cells from the normal and squinting eye are equal.

As noted by Ikeda (1979,1980), amblyopia is often associated with a loss of binocular vision. However, the bases of amblyopia and binocular vision deficits, although associated, most likely are different (Blakemore and Eggers 1978). The physiological basis of amblyopia seems to be due to cell dysfunction at lower than binocular levels of the visual pathway as indicated by the selective, marked loss of central visual acuity of retinal ganglion sustained cells (Ikeda and Tremain 1979). This selective loss is

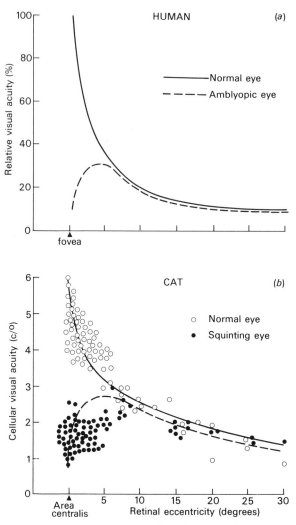

Fig. 8.14.(*a*)   Relative visual acuity as a function of retinal eccentricity for the normal and amblyopic human eye, as indicated. (*b*) The cellular visual acuity of cat sustained ganglion cells in the normal and amblyopic squinting (esotropic) eye as a function of retinal eccentricity. (From Ikeda (1979).)

also found in the lateral geniculate nucleus (Ikeda and Wright 1976).

The selective loss of central visual acuity in the amblyopic eye has its correlate in a selective loss of grating contrast sensitivity. Levi and co-workers (Levi and Harwerth 1977, 1980; Levi *et al.* 1979) measured the contrast sensitivities of the normal and amblyopic eyes of human amblyopes. In particular, Levi and Harwerth (1980) in one experiment investigated the contrast sensitivity of both eyes as a function of grating duration and spatial frequency. The results are shown in Fig. 8.15. At all durations, the contrast sensitivity is higher for the normal (open circles) as compared with the amblyopic (filled circles) eye. In particular, in the first condition, in which the gratings were presented continuously and thus lacked temporal transients, only sustained channels were used in detecting the gratings. Note that here the difference in the contrast sensitivity between the two eyes, as indicated by the contrast sensitivity ratio (x's), is particularly large at spatial frequencies greater than 2.0 c/deg. At progressively shorter grating durations contrast sensitivity for both eyes declines (Nachmias 1967; Schober and Hilz 1965) and the contrast sensitivity ratio also decreases. These psychophysical findings have been corroborated via electrophysiological measures of the cortically evoked visual response (Levi and Harwerth 1978*a*, *b*).

From the results of the above experiment, Levi and Harwerth (1980) determined the duration–contrast reciprocity threshold functions in the

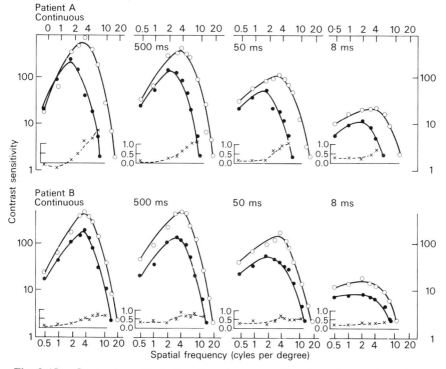

Fig. 8.15. Contrast sensitivity functions for two amblyopic patients. Open circles show data for the non-amblyopic eye; closed circles show data for the amblyopic eye. The stimulus duration, from left to right, is continuous, 500, 50, and 8 ms. The curves marked by x's show the log of the ratio of contrast sensitivity of the amblyopic to the non-amblyopic eye. (From Levi and Harwerth (1980).)

normal and amblyopic eye at spatial frequencies of 0.5 and 5.0 c/deg. The results are displayed in Fig. 8.16. For one, note that the critical duration,as shown previously by Breitmeyer and Ganz (1977) and Legge (1978) (see Section 3.3.2, Fig. 3.5) is longer for the higher than the lower spatial frequency irrespective of the eye being tested. Also note that whereas the normal and amblyopic eye are characterized by the same critical duration of about 60–70 ms at the 0.5 c/deg spatial frequency, the respective critical durations at a spatial frequency of 5 c/deg are approximately 150 and 300 ms for the normal and amblyopic eye. Moreover, whereas the slopes of the duration reciprocity functions are the same for both eyes at a spatial frequency of 0.5 c/deg, they differ, with a smaller slope for the amblyopic eye, at the higher, 5.0 c/deg, spatial frequency. The presence of critical duration and slope differences between the two eyes at higher but not at lower spatial frequencies may correlate with the finding that low spatial

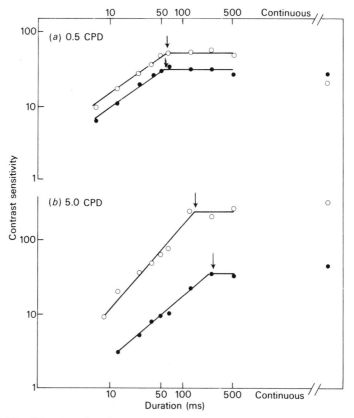

Fig. 8.16. Stimulus duration-versus-contrast sensitivity functions for the non-amblyopic eye (open circles) and amblyopic eye (closed circles) of one amblyopic patient. (*a*) 0.5 c/deg grating (*b*) 5.0 c/deg grating. (From Levi and Harwerth (1980).)

frequency, transient channels are not affected as much by amblyopia as are the high spatial frequency sustained channels. As a correlate, one can say that the higher spatial frequency sustained channels of the amblyopic are more sluggish as evidenced by a longer integration time over which duration–contrast reciprocity holds.

In a second experiment to be discussed here, Levi and Harwerth (1980) also investigated the contrast sensitivity function for flicker and pattern detection in the normal and amblyopic eye. The results, as shown in Fig. 8.17, show clear differences not only in the pattern threshold (in accord with the prior experiments) but also in the flicker threshold at all spatial frequencies tested; both amblyopes yielded a lower flicker contrast

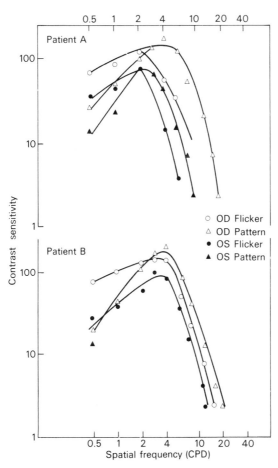

Fig. 8.17. Contrast sensitivity functions for two amblyopic patients for gratings flickering at 10 c/s. Flicker and pattern thresholds are shown for the non-amblyopic eye (open symbols) and amblyopic eye (closed symbols). (From Levi and Harwerth (1980).)

sensitivity in the amblyopic than the normal eye (see also Wesson and Loop (1982) for corroborative results). Since flicker detectors are most likely transient channels (King-Smith and Kulikowski 1973, 1975; see Section 7.1) this result suggests that spatial contrast sensitivity and resolution is lost not only in sustained but also in transient channels of the central region of the amblyopic eye. Ikeda and Wright (1976), as noted in Section 6.7, in fact reported such a loss of contrast sensitivity and spatial resolution of both sustained and transient neurones representing the central visual field.

Visual masking also reveals differences between the normal and amblyopic eye of human amblyopes. Tytla and McAdie (1981) measured metacontrast fuctions in both eyes of amblyopes. They found, as illustrated in Fig. 8.18, not only a greater type B metacontrast effect in the amblyopic eye relative to the normal one, but also a shift of the SOA at which peak masking occurred towards higher values. This result, according to the visual masking model outlined in Section 7.4.1 (see Fig. 7.7) indicates that

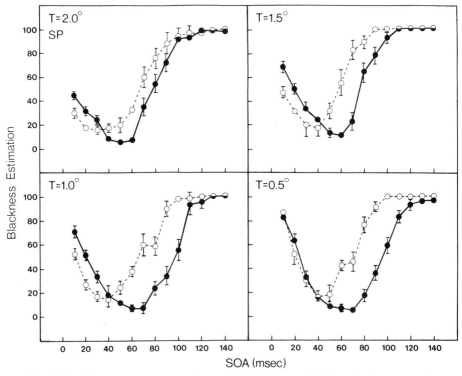

Fig. 8.18. Metacontrast suppression of a black target disk by a black surrounding ring for the non-amblyopic (open circles) and amblyopic (filled circles) eyes of a human amblyope. The results shown in the four panels are obtained at four target eccentricities, as indicated. (From Tytla and McAdie (1981).)

the response latency of sustained channels in the amblyopic eye is longer than that of sustained cells in the normal eye, as independently shown through the use of visual reaction time measures employed by Levi *et al.* (1979). Additional, independent corroboration for this conclusion derives from Ikeda and Wright's (1976) (see Section 6.7) investigation of sustained and transient cells in the lateral geniculate nucleus of cat in which surgical squint was produced in one eye. Relative to the non-squint eye, the sustained cells driven by the squint eye were characterized by a longer response latency; whereas no latency differences were found between the transient cells driven by either eye.

8.5. Final comment

The studies and their associated results discussed in the current chapter have shown that the approach to visual information processing based on the transient–sustained dichotomy can be fruitfully applied and extended beyond the specific empirical phenomena reviewed in Section 7.4.2 of the prior chapter. However, it should be mentioned that the topics and results covered in the present chapter by no means constitute confirmation of the approach or any of its specifically derived models. Again I have merely used the approach as an investigative tool or a heuristic device in our explanation and understanding of additional visual phenomena. In so far as the main investigators reviewed in this chapter also subscribed to the transient–sustained approach as a heuristic device, I have not done them injustice by either recapitulating or additionally elaborating on their explanations and interpretations of specific experimental results. In the next chapter, I shall show how the current approach can be further applied and extended to two other important visual masking phenomena, the discussions of which—along with those covered in the present and prior chapters—lay the preliminary groundwork for Chapter 10.

8.6. Summary

The application of the sustained–transient model of visual masking can be extended beyond its explanation of results listed in Chapter 7. In the present chapter, it was applied in the context of studying the time course of the object-superiority, connectedness-superiority, and perceptual grouping effects. Generally, the greater the object or three-dimensionality rating of a context pattern, the more it facilitates recognition of a target line at short target-context SOAs and the longer the SOA at which optimal type B target masking occurs. For context patterns of greater connectedness, target facilitation also is greater at short SOAs, and although the SOA at which optimal masking of the target occurs is fixed, the magnitude of masking varies inversely with connectedness.

Temporal blurring of target and object or connected-context stimuli reduces the magnitude but increases the latency of the transient response component. Overall masking magnitude decreases and the optimal masking SOAs shift to lower values. Moreover, the increase of masking magnitude as connectedness is decreased is eliminated, implying (i) that without temporal blurring, the magnitude of transient response component increases as connectedness decreases; and (ii) that, in the two-flash studies, the transient response component is necessary for a perceptual distinction between connected and fragmented patterns.

With spatial blurring the facilitation of target-line recognition at low target-context SOAs is reduced in direct proportion to the degree of blur. This holds for object as well as connected-context patterns. Furthermore, for both contexts, the SOAs at which optimal type B target masking occur shift toward lower values and the overall masking magnitude increases as spatial blur of the target and context stimuli increases. In regard to the object context, its differential effects on target-line recognition are not eliminated as it is increasingly blurred, indicating that the low spatial frequency components are sufficient to specify the three-dimensionality of an object. However, if the target line alone is blurred, the object contexts no longer facilitate target-line recogntion at low SOAs and the magnitude of target masking increases. Hence, the facilitatory effect an object context has on a target line depends on the presence of sustained responses to its high spatial frequency components.

On the other hand, perceptual grouping effects produced by Gestalt factors such as similarity, symmetry, and closure are not drastically changed by spatial blur of target and context. Whereas spatial blur produced a shift of optimal masking SOAs to lower values for the object- and connectedness-context patterns, it did not change the optimal SOAs in the perceptual grouping condition. This indicates that the effects of perceptual or Gestalt grouping on metacontrast depends only on the lower spatial frequency sustained components of the stimuli.

A further extension and application of the sustained–transient approach to visual masking allowed us to distinguish among three central or cortical processes of pattern masking. One consists of common integration of target and mask sustained pattern information at a central iconic level and results in symmetrical type A forward and backward masking. The icon formation or integration is a temporally dynamic process whose rate of pattern integration depends on the pattern's spatial frequency composition. Central iconic integration of sustained pattern information extends over approximately 80 ms. A second mechanism is the transient-on-sustained channel inhibition which (i) occurs centrally at preiconic levels of processing and thus (ii) interrupts or prevents the transfer of sustained pattern information *into* the later icon-synthetic level of pattern processing and (iii) causes type B backward pattern masking effects. The third mechanism entails a form of cognitive, specifically, attentional masking, in which the aftercoming mask, particularly at longer SOAs, diverts attention from, or interrupts, the read-out of information *from* the iconic level to later ones where it is encoded at a more permanent non-visual, e.g. verbal, level of analysis.

Finally, the sustained–transient model of visual processing was extended to account for several psychophysical studies of amblyopic vision in humans. The obtained psychophysical results are consistent with the neurophysiological and neuroanatomical studies of abnormal visual de-velopment induced by various types of visual deprivation and squint during an organism's neonatal periods of development. In particular, it was shown that in amblyopic eyes type B metacontrast peaks at longer SOAs than in normal eyes, consistent with the neurophysiological finding that squint-induced amblyopia produces an increased response latency among foveal and near-foveal sustained channels, but not among transient channels.

9 Extensions and applications of the sustained–transient approach: II. Target recovery and long-range masking effects

Section 7.4.1 outlined a theory of visual masking based on sustained–transient interactions and demonstrated its explanatory scope, at least qualitatively, for a variety of relevant empirical findings. In addition to these empirical findings, the following sections introduce applications and extensions to two further psychological phenomena: (i) target recovery and enhancement in masking; and (ii) long-range masking effects.

9.1. Target recovery and enhancement in visual masking

9.1.1. Target recovery

Target recovery or disinhibition phenomena in human vision can be demonstrated through the use of several paradigms. For instance, Mackavey *et al.* (1962) demonstrated that reduction of the brightness of a central stimulus produced by a spatially adjacent one in simultaneous brightness induction (Heinemann 1955), can be counteracted by the addition of a stimulus in turn surrounding the contrast inducing one. The explanation of this disinhibitory effect invoked processes similar to the recurrent lateral inhibion found among optic fibres of the *Limulus* (Hartline and Ratliff 1957, 1958). Other related spatial disinhibitory effects can be obtained in the detection of lines (Rentschler and Hilz 1976; Wilson *et al* 1979) and in orientation-specific tilt inductions (Magnussen and Kurtenbach 1980; O'Toole 1979). Of course, in these studies stimuli were presented simultaneously and for prolonged periods. Hence, these results, although related to those of visual masking studies, do not, strictly speaking, apply to them.

However, several investigations have focused on a variety of target recovery effects during visual masking. For instance, in orientation-specific masking (Gilinsky 1967, 1968; Gilinsky and Cohen 1972; Gilinsky and Mayo 1971; Houlihan and Sekuler 1968; Sekuler 1965), one can obtain target disinhibition which, in turn, is orientation-specific (Long and Scheirlinck 1981). In orientation-specific masking, the visibility of a line or grating of a given orientation can be suppressed optimally by either a preceding (Gilinsky 1967, 1968), concurrent (Campbell and Kulikowski

1966; Kulikowski 1973), or a following (Sekuler 1965) line or grating of similar orientation. As the orientation difference between test and mask patterns increases from 0°, the masking effect declines monotonically until a difference of 20–30° is attained, beyond which little or no masking is evident. Similarly, the disinhibitory effect, which a second mask applied to the first one exerts on target visibility, is maximal when the first and second mask have similar orientation and again declines monotonically to an asymptote as the orientation difference between them increases to about 20° (Long and Scheirlinck 1981).

These orientation-specific masking and disinhibition effects are believed to be due to lateral inhibition between cortical orientation detectors (Benevento *et al.* 1972; Blakemore, Carpenter, and Georgeson 1970; Blakemore and Tobin 1972; Creutzfeldt *et al.* 1974; Hubel and Wiesel 1962, 1968; Nelson and Frost 1978), the response of which can be affected by patterns aligned within ±15–30° of their optimal orientation (Bisti, Clement, Maffei, and Mecacci 1977; Campbell, Cleland, Cooper, and Enroth-Cugell 1968; Creutzfeldt *et al.* 1974; Henry, Bishop, Tupper, and Dreher 1973; Ikeda and Wright 1975*a*; Stone and Dreher 1973). The fact that progressively weaker interactions occur between patterns of progressively greater orientation difference is tied to the orderly mapping of cortical orientation colums in which, say, two orientations of progressively greater difference are mapped into two different orientation columns of progressively greater separation in cortical space (Hubel and Wiesel 1974*a*, 1977).

Besides these orientation-specific masking and recovery phenomena, target disinhibition also can be obtained in the several types of masking paradigms reviewed in Chapters 2 and 4. For example, several investigators reported disinhibition of a target (T) when a uniform-light mask (Long and Gribben 1971; Purcell and Dember 1968; Robinson 1966, 1968; Stewart, Purcell, and Dember 1968) or else a pattern mask (M_1) (Dember, Schwartz, and Kocak 1978; Purcell and Stewart 1975; Tenkink and Werner 1981) following the T was followed in turn by a second uniform-light mask (M_2). Optimal disinhibition occurs when (i) the interstimulus interval (ISI) between M_1 and M_2 varies from about 10 to 20 ms (Robinson 1968; Tenkirk and Werner 1981), and moreover, when (ii) T–M_1 ISIs range from 50 to 75 ms (Robinson 1968). As shown by Robinson (1968) optimal target recovery at these M_1–M_2 and T–M_1 ISIs can be obtained with binocular viewing but is not observed when T and M_1 are presented to one eye and M_2 is presented to the other eye. In contrast to binocular viewing, under such dichoptic viewing the optimal target recovery occured at M_1–M_2 ISIs of 75–200 ms, with, if anything, additional target suppression produced at M_1–M_2 ISIs of 10 and 20 ms. The most probable explanation of these effects depends on considerations of masking by light reviewed in Sections 2.1.5 and 2.1.6. Under binocular (or monocular) viewing, the first light-mask, M_1, reduces the visibility of targets by luminance summation–contrast reduction as hypothesized by Eriksen (1966). The second light-mask, M_2, flashed, say, 20 ms later in turn reduces the contrast of M_1 and consequently M_1's masking effect on the target is reduced; hence, target recovery. The fact that this particular recovery under binocular viewing cannot be obtained dichoptically (Robinson 1968) (i) is in accord with the findings of Battersby *et al.* (1964) that masking by light cannot be obtained interocularly and (ii) implies that it is due to peripheral mechanisms of brightness or contrast reduction of M_1 by M_2.

Therefore, under dichoptic viewing the optimal target recovery produced when the M_1–M_2 ISI ranges from 75–200 ms cannot be due to peripheral luminance summation but rather depends on central contour interactions between M_1 and M_2 (Battersby and Wagman 1962; Battersby *et al.* 1964, Weisstein 1971; see Section 2.1.6). In Robinson's (1968) study, T, M_1, and M_2, each flashed for 5 ms, had diameters of 23′, 46′, and 92′, yielding T–M_1, T–M_2, and M_1–M_2 contour separations of 12′, 35′, and 23′, respectively. Dichoptic disk–disk metacontrast suppression obtained under these circumstances (Weisstein 1971) seems to be a prime candidate for producing target recovery at larger M_1–M_2 ISIs and additional target suppression at small M_1–M_2 ISIs. Both effects should be especially marked at the optimal T–M_1 ISIs of 50–75 ms or SOAs of 55–80 ms. Here, an M_1–M_2 ISI of 10–20 ms (SOA of 15–25 ms) corresponds to T–M_2 SOAs of 70–105 ms. Relative to T, the SOAs of both M_1 and M_2 are within the range of optimal values (50–150 ms) for producing type B metacontrast suppression. On the other hand, with respect to M_1 the M_2 SOAs of 15 and 25 ms would be significantly less than optimal. Here, M_2 exerts a greater suppression of T than of M_1; hence, overall greater target masking. At the larger M_1–M_2 ISIs of 75–200 ms, the corresponding T–M_2 SOAs range from 135 to 285 ms, overlapping very little with the 50–150-ms range yielding optimal metacontrast. However, the correspondingly shorter M_1–M_2 SOAs ranging from 80 to 205 ms overlap to a greater extent the same optimal range of 50–150 ms. Here M_2 suppresses M_1's contrast significantly more than it suppresses T's contrast. Hence, an overall greater contrast recovery, rather than contrast suppression, of T. Since transient-on-sustained neural inhibition is the primary mechanism for producing type B metacontrast, the dichoptic target recovery obtained at larger M_1–M_2 SOAs (Robinson 1968) implicates the contributory role of transient–sustained channel interactions in this and other target disinhibition phenomena, a conclusion also reached by Dember *et al.* (1978) on the basis of their study of target recovery under binocular viewing. The consequences of this implication are pursued below, after several discrepant findings are dealt with.

Schurman and Eriksen (1969) failed to replicate Robinson's (1968) later finding, and Barry and Dick (1972) only partially replicated Robinson's

(1966) prior disinhibition effects. These discrepancies are more apparent than real and are related to clearly specifiable differences of procedure and criterion content (see Section 4.3.1). As noted, Robinson (1968) and more recently Tenkink and Werner (1981) have shown that the binocular disinhibition effect produced by a uniform-light M_2 on T's visibility is optimal when the M_1–M_2 ISIs range between 0 and 20 ms. Schurman and Eriksen (1969) in fact employed an optimal M_1–M_2 ISI of 20 ms. However, in Robinson's (1968) more extensive study, it was additionally shown that the effectiveness of M_1–M_2 ISI interacts with the T–M_1 ISI. In other words, optimal target disinhibition was obtained only when the M_1–M_2 ISI ranged between 0 and 20 ms and *additionally* when the T–M_1 ISI ranged between about 50 and 75 ms. At T–M_1 ISIs of 0–20 ms, little if any disinhibition was obtained at any M_1–M_2 ISI. However, in their Experiment II, Schurman and Eriksen (1969) employed a fixed and non-optimal T–M_1 ISI of 20 ms. Hence, both M_1's and M_2's masking effects may, in fact, have combined superadditively (Boynton 1961) rather than subadditively, to produce overall greater rather than smaller masking relative to that produced by M_1 alone. Moreover, whereas Robinson (1968) required that the observers use responses 1, 2, or 3 in order to indicate which of each equally intense 5–mL, *light* flash (T, M_1, or M_2) they saw, Schurman and Eriksen (1969), in their Experiment I, instead employed a forced-choice detection criterion in which target and mask intensities were 0.6 and 1.2 mL, respectively. Thus, different T–M_1 intensity relations and a more liberal response criterion which is insensitive to apparent brightness reversals of T produced by M_1 (Barry and Dick 1972; Purcell and Dember 1968; Stewart *et al.* 1968; see p. 104 footnote), may have eliminated the measure of T's disinhibition found under Robinson's (1968) more stringent response criterion.

In fact, Barry and Dick (1972) found that Robinson's (1966) disinhibition effects are dependent on response criteria which interact not only with brightness reversal phenomena but also with retinal locus of stimulation. In Barry and Dick's (1972) study, which employed the same, low-stimulus intensities as Robinson (1966), brightness reversals occurred only in the 5° periphery but not in the fovea. However, this result may itself be specific to their study, since Purcell and Dember (1968) obtained foveal brightness reversal with more intense stimuli. At any rate, Barry and Dick (1972) reported disinhibition effects under peripheral viewing when the response criterion used by the observers was based on detecting a *light* T-flash as in Robinson's (1966) study. However, when a more liberal detection criterion was used in which brightness reversed stimuli also qualified as being detected, no disinhibition effects were obtained. On that basis, we can conclude that disinhibition phenomena in masking of light or pattern by light are obtained only when brightness reversal (or suppression) occurs and when a stringent, contrast-dependent ('light' or else 'dark') response criterion is employed. This conclusion is no more than a recapitulation of the conclusion which Purcell and Dember (1968) arrived at in their study of target recovery. According to these investigators, masking by light when using a stringent, contrast-dependent response criterion is due to either suppression or reversal of T's contrast by M_1; whereas recovery results from a de-suppression or re-reversal of T's contrast produced, in turn, by M_2's contrast reversal of M_1. Since metacontrast also yields brightness reversal effects (Brussel *et al.* 1978; see p.104 footnote), its primary neural mechanism, transient–sustained channel interaction, again is implicated in target recovery phenomena.

As a less ambiguous illustrative example, let us analyse, from the standpoint of the transient–sustained channel approach to visual masking, the results of another study of target recovery reported by Robinson (1971). In this study T consisted of, say, a large capital letter N composed of small upper-case O's. M_1 consisted of an arrangement of lower-case O's, spatially embedded between but not on the diagonal and vertical lines of the N target. When these two stimuli were flashed in temporal proximity, a composite visual representation was formed in which the arrangement of lower-case O's served as noise, obscuring identifiability of the N target. Since pattern recognition was used as a criterion, the pattern processing at some stage entailed activity in sustained afferent pathways. The second mask, M_2, consisted of an array of upper-case O's, each of which surrounded in metacontrast fashion a lower-case O of M_1. When M_2 in turn followed the T–M_1 sequence, the N target was easily identified. The explanation of this target recovery effect can be readily made given the existence of transient-on-sustained inhibition. In particular, if transient channels activated by M_2 inhibit sustained channels activated by M_1, these latter channels can no longer provide pattern information which temporally integrates with and thus masks (Michaels and Turvey 1979) at the cortical iconic level the sustained pattern information of the target; neither, due to their late arrival, can be sustained channels activated by M_2; hence, the N target should be identified better in the T–M_1–M_2 sequence than in the T–M_1 sequence.

A similar explanation can be applied to target sensitization by annular surrounds reported by a number of other investigators (Alexander 1974; Dember and Purcell 1967; Kristofferson *et al.* 1979; Purcell, Stewart, and Hochberg 1982; Schiller and Greenfield 1969; Sturr and Teller, unpublished observations; Teller 1971), and to the target recovery effects in sequential blanking studies reported by Mayzner (1970) and Tresselt *et al.* (1970). For the most part, these studies differed in procedure from Robinson's (1971) study in that either a spatially uniform or patterned M_1 overlapped the target. Consequently, with this procedure sustained pathways activated by the leading target and the lagging M_1 would, due to temporal response integration and persistence form a composite, noisy representation carried in common, preiconic sustained channels. Since the transient activity produced by a spatially surrounding, temporally lagging M_2 inhibits the sustained activity produced by M_1, this integration of sustained channel activity in common channels no longer occurs, thus leading to an increased visibility of the target pattern.

For the same reasons, but because of different procedure, Uttal (1970) failed to obtain target recovery when using mask and target stimuli similar to those employed by Robinson (1971). Here, again, the target consisted of a large, capital letter, say N, composed of smaller, circular dots. Type A backward masking was obtained, as in Robinson's (1971) study, when the target flash (< 3 ms) was followed at ISIs varying from 0 to 100 ms by a single, 3-ms noise-mask (M_1) composed of a spatially random array of 100 dots. In order to study target recovery, Uttal (1970) set the T–M_1 ISI at 20 ms and varied the ISI separating M_1 and a second, 3–ms noise-mask (M_2) (again composed of 100 randomly arrayed dots), from 20 to 100 ms. No target recovery was obtained at any M_1–M_2 ISI; on the contrary, at all of these ISIs the suppression of the target's visibility by M_1 was enhanced superadditively by the presence of the M_2 flash. Note that, whereas M_2, in Robinson's (1971) study, consisted of non-random, systematically arranged capital O's surrounding in metacontrast fashion and thus suppressing each lower-case O composing the noise-mask M_1, the random array of dots composing M_2 in Uttal's (1970) study combined with, rather than suppressed, the noise introduced by M_1's common iconic synthesis with the signal-representation of T (Michaels and Turvey 1979). Under these circumstances, when one type A backward masking effect is superadded to another, one should not be surprised to obtain enhanced masking of target visibility rather than its recovery.

Besides the target disinhibition *produced by M_2's metacontrast suppression* of M_1 as in Robinson's (1971) or, say, Schiller and Greenfield's (1969) studies, another form of target recovery specifically incorporates target disinhibition on metacontrast where the target (T) and mask stimuli (M_1 and M_2) do not overlap spatially. The rationale used here is generally the same as those used in the above studies. By measuring the effect of M_2 on the T–M_1 sequence, one can make several inferences about the mechanisms responsible for target recovery in metacontrast. Breitmeyer (1978a) employed the following procedure (based on Westheimer and Hauske's (1975) study) and rationale for investigating target recovery during metacontrast. A vertical Vernier acuity target, as shown in Fig. 9.1, was flanked by larger vertical mask bars; these in turn could be flanked by two additional bars. The vertical offset of the upper part of the Vernier target could be to the right, as shown in Fig. 9.1, or else to the left. On any masking trial, the observers were required to say whether the offset was to the right or left. Proportion of error as a function of SOA was taken as a measure of metacontrast magnitude.

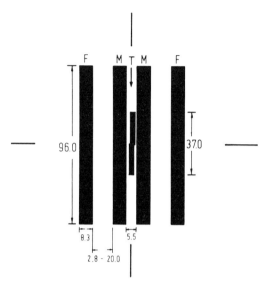

Fig. 9.1. Spatial lay-out of stimulus display used to study target recovery. Dimensions are in minutes of arc. T= Vernier target; M= mask; F= flank. Vertical and horizontal lines were used as fixation markers. Observers were required to fixate on the imaginary intersection point of the lines falling in the fovea. (From Breitmeyer (1978a).)

Three conditions were used. In condition 1 only the Vernier target and the two flanking bars were presented. This should yield a typical type B metacontrast function. In condition 2 the outer flanks were presented prior to, throughout, and after the target–mask sequence. Since the outer flanks were presented continuously at a spatial separation of 4' from the inner flanking masks, they should activate only sustained channels. Chapter 6 (Hoffman *et al.* 1972; Singer 1976; Singer and Bedworth 1973; Tsumoto and Suzuki 1976; see Section 6.5) and the masking model outlined in Section 7.4.1 showed that the inhibitory interactions between sustained and transient channels are reciprocal. Hence, in this second condition, the sustained channels activated by the outer flanks ought to inhibit the transient channels activated by the inner flanking mask and thereby produce disinhibition of the visibility of the Vernier target. In condition 3, the outer flanks were separated by 20' from the inner flanks. If the reciprocal, sustained-on-transient inhibition, due to the relatively small receptive and dendritic field size of sustained neurones, is a highly local phenomenon (Ikeda and Wright 1972a. Wässle *et al.* 1981b; see Section 7.1) which does not extend beyond a few minutes of arc in the fovea, no target recovery ought to occur in this condition. As shown in Fig. 9.2, the expectations were borne out by the experimental results. Condition 1 and 3 yielded typical type B metacontrast functions; however, condition 2, in

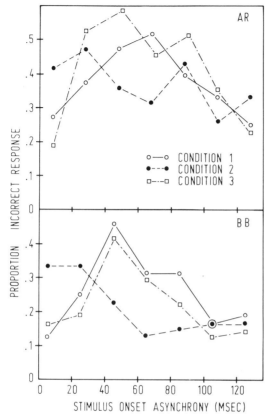

Fig. 9.2. The proportion of incorrect identification (masking magnitude) of the Vernier target as a function of stimulus onset asynchrony and experimental conditions in two observers. Higher error proportions correspond to greater masking. (From Breitmeyer (1978a).)

comparison, yielded a significantly smaller masking effect, i.e. it produced target-recovery. Consequently, target disinhibition produced by stationary flanks adjacent to the mask stimulus employed in metacontrast can be taken as an indicator of sustained-on-transient inhibition.*

---

* In an independent set of experiments von Grünau (1978a) also demonstrated the existence of sustained-on-transient inhibition. He investigated the perceived strength of stroboscopic motion as a function of sharply contoured stimuli and stimuli which, via a 4 diopter lens, had their higher spatial frequencies removed by defocusing or blur. It was found that weaker apparent stroboscopic motion, particularly in foveal vision, resulted in the former as compared with the latter condition. Presumably, the weaker perceived motion was due to the higher spatial frequency content of the sharply contoured stimuli, which, especially at the fovea (Cleland and Levick 1974), strongly activate the relatively higher proportion of sustained cells that, in turn, inhibit the transient neurones detecting the stroboscopic motion. With defocusing, which eliminates the higher spatial frequency components of the stimuli and also attenuates the response of sustained but not transient neurones (Ikeda and Wright 1972a), this inhibitory effect on transient detectors ought to be weaker, hence a stronger motion percept.

In a further series of experiments, Breitmeyer *et al.* (1981*b*) investigated the temporal and spatial parameters of sustained–transient inhibitory interactions more systematically by using a target recovery technique similar to that described in the previous studies. The stimulus arrangement, as illustrated in Fig. 9.3, consisted of a black disk serving as the target (T), and two surrounding annuli, the smaller of which served as the mask (M_1) of the target and the larger of which (M_2) was used to mask M_1. In a preliminary, baseline metacontrast experiment, using only the T followed by M_1, Breitmeyer *et al.* (1981*b*) found that optimal type B metacontrast suppression of T was obtained at SOAs of 45–65 ms. In one of the main experiments the time course of target recovery produced by M_2 was investigated. In this experiment the T–M_1 SOA was fixed at a value of 45 ms which, as determined in the baseline experiment, optimized metacontrast. Consequently, here the transient-on-sustained inhibition exerted by M_1 on T should be maximal. By measuring the difference between the amount of target contrast suppression obtained at the fixed, optimal T–M_1 sequence and the variable (T–M_1)–M_2 sequence, the degree of target recovery produced by M_2 can be factored out. M_2 either preceded or followed the fixed T–M_1 sequence at variable T–M_2 SOAs.. In addition, the masking effect of M_2 on M_1, as a function of M_1–M_2 SOA, was also measured to determine whether suppression of M_1's contrast visibility by

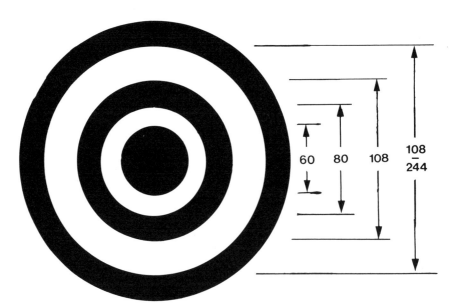

Fig. 9.3. Spatial lay-out of stimulus display used to study target recovery. Dimensions are given in minutes of arc. Target= inner disk; mask 1= smaller annulus; mask 2= larger annulus. The inside diameter of mask 2 could vary from 108 to 244' as shown. (From Breitmeyer, Rudd, and Dunn (1981*b*).)

M_2 is either a sufficient or necessary condition to obtain target recovery in this particular paradigm. Recall from our above analysis of, for instance, Schiller and Greenfield's (1969) and Robinson's (1971) studies, that such suppression of M_1's contrast by metacontrast-mask M_2 *was* sufficient to obtain target recovery. However, in those studies M_1 overlapped or added noise to T; whereas in the present experiments no such spatial overlap or noise addition existed. Hence, the present results and conclusions apply only when spatially non-overlapping stimuli are employed.

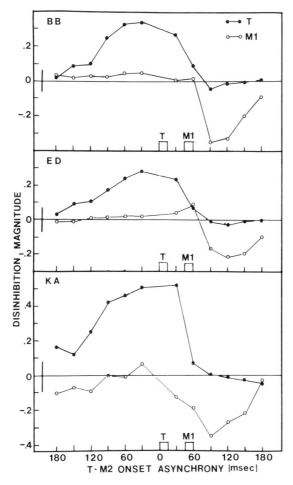

Fig. 9.4. The magnitude of target recovery (filled circles) and contrast masking of mask 1 by mask 2 (open circles) as a function of the onset asynchrony between the target–mask 1 sequence and mask 2. The onset delay between the target and mask 1 was fixed at 45 ms, as shown, which produced optimal metacontrast masking of the target. (From Breitmeyer, Rudd, and Dunn (1981*b*).)

Figure 9.4 shows results obtained from three observers as a function of the asynchrony of M_2 relative to the onset of T or equivalently the *start* of the T–M_1 sequence. Target disinhibition data are given by the filled circles and the magnitude of masking of M_1's contrast visibility by M_2 is given by the open circles. Note first of all that the target disinhibition is not obtained at positive T–M_2 SOAs when M_2 follows the T–M_1 sequence. Hence retroactive disinhibition effects as found in Robinson's (1971) or, for instance, Schiller and Greenfield's (1969) studies were not obtained here. However, pronounced *proactive* target recovery was obtained at negative T–M_2 SOAs when M_2 precedes the T–M_1 sequence. In particular, the target recovery is minimal at T–M_2 SOAs of -180 ms and increases to its maximal value at SOAs ranging from -60 to $+30$ ms after which it is no longer obtained.

This and related (Long and Scheirlink 1981) proactive target disinhibition effects can be explained as follows. According to the model presented in Section 7.4.1 (see Fig. 7.7), the latency and persistence of sustained channels is longer by several tens of milliseconds than that of transient channels. Since sustained channels activated by M_2 exert their greatest inhibitory effect on transient channels activated by M_1, when these neural activities are temporally synchronous, M_2 must precede M_1 by some optimal SOA so that M_1's transient activity, inhibited by M_2's sustained activity, in turn, is effectively prevented from inhibiting the target's sustained activity. Estimations of this optimal interval, obtained by subtracting the fixed T–M_1 SOA of 45 ms from the T–M_2 SOAs of -60 to $+30$ ms, at which target recovery is optimal, range from -105 to -15 ms, with an intermediate optimal value of -60 ms.

Moreover, over this negative range of M_1–M_2 SOAs, the additionally measured paracontrast suppression (negative disinhibition values indicate suppression) of M_1 by M_2 (open circles in Fig. 9.4) is barely, if at all, evident. Only when M_2 follows M_1 does one obtain a type B metacontrast suppression of M_1 (again indicated by negative disinhibition values). These combined results are of interest for several reasons. For one, they demonstrate conclusively that the suppression of M_1's contrast visibility is neither sufficient nor necessary to obtain targe disinhibition. Note again that at negative M_1–M_2 SOAs no significant paracontrast suppression of M_1 is produced; yet here the largest target recovery is obtained. This disconfirms the necessity criterion. At positive M_1–M_2 SOAs type B metacontrast suppression (negative disinhibition) of M_1 by M_2 is obtained, with optima at M_1–M_2 SOAs of 45–60 ms. However, despite this metacontrast suppression of M_1 by M_2, no concurrent target recovery is produced. That is to say M_1, despite the suppression of its contrast, is still able to suppress T. This result, in turn, disproves the sufficiency criterion. In addition, the fact that (i) M_2 at positive M_1–M_2 SOAs produces contrast suppression of M_1 when (ii) M_1 itself, at the optimal T–M_1 SOA of 45 ms, produces contrast suppression of the target indicates, as Piéron (1935) originally demonstrated, that one can obtain sequential metacontrast in a series of spatiotemporally displaced patterns.

Given the disconfirmation of the above necessity and sufficiency criteria how is it that (i) when M_2 suppresses the contrast of the preceding M_1, M_1 can nevertheless still suppress or mask the contrast of the preceding T; and (ii) inversely—in the opposite temporal order—when M_2 fails to suppress the contrast of the following M_1, M_1's ability to suppress the contrast of the following T is attenuated?

The answer to the first question can be stated as follows. Although M_2's faster transient activity can suppress M_1's sustained activity when M_2 follows M_1 at optimal metacontrast SOAs it cannot suppress M_1's transient activity (see Fig. 7.7(c) and (d)). Moreover, M_2's slower sustained activity cannot suppress the faster transient activity of M_1. Since pattern contrast information is transmitted in sustained channels, M_2 can suppress M_1's contrast without, however, also suppressing M_1's transient response, which, in turn, still can suppress the pattern-contrast of the preceding T. This explanation is similar to that applied in Explanation 14 of Section 7.4.2. Here it was noted that although an aftercoming metacontrast mask can suppress the slower form and contrast information of the first of two flashed stimuli used to produce stroboscopic motion, the mask cannot suppress its faster transient response, which in conjunction with the transient response of the second stimulus results in a percept of stroboscopic motion (Kolers 1963; von Grünau 1978a, b, 1979, 1981).

The second query can be answered by a reverse form of the first question's answer. Here the preceding M_2's slower sustained activity can inhibit M_1's faster transient activity, however, the paracontrast effect produced by intrachannel inhibition of M_1's sustained activity by M_2's surrounding sustained activity was minimally, if at all, evident. The reduction of M_1's transient activity by M_2, in turn, disinhibits the visibility of the prior T's contrast; but since M_1's sustained channel response to contrast is not appreciably suppressed by M_2, M_1's reduced masking effectiveness on T's contrast occurs despite its own contrast being essentially left unmasked. Since the magnitude of paracontrast increases as the preceding mask's energy increases (Alpern 1953; Weisstein 1972), perhaps with an M_2 of higher energy than M_1, M_1's contrast *can* be more appreciably suppressed. None the less, the results of Breitmeyer *et al.* (1981b) indicate that such suppression of M_1 *need not* occur in order to produce target recovery.

Besides the involvement of interchannel, sustained-on-transient inhibition in producing target recovery in metacontrast, Fotta (1979) argued on the basis of his investigations that within-channel transient-on-transient inhibition may also be used. At first glance this seems reasonable, for, as was shown in Section 6.5, the inhibitory influence of the surround mechanism of transient neurones on the centre response is optimal when the surround stimulus precedes the centre one by several tens of milliseconds (Winters and Hamasaki 1975, 1976). Thus in the study reviewed in the above paragraphs, M_2 would again have to precede the T–M_1 sequence in order to disinhibit the target maximally. An additional set of experiments reported by Breitmeyer *et al.* (1981*b*) makes Fotta's (1979) account seem plausible.

In one experiment an estimate of the spatial range of transient-on-sustained and sustained-on-transient inhibition was determined. To this end, target disinhibition produced by a continuous, sustained M_2 stimulus was measured as a function of M_1–M_2 spatial separation. M_2, presented prior to, throughout, and after the T–M_1 sequence, which was again set at optimal metacontrast SOAs of 45 and 65 ms, would only activate sustained channels and thus inhibit the transient activity of M_1. The spatial separations ranged from 0′ to 68′. Variations of target recovery with spatial separation between M_1 and M_2 would thus yield an estimate of the spatial range of sustained-on-transient inhibition. Moreover, by letting M_1 serve as the target and M_2 serve as the mask, again using optimal metacontrast SOAs of 45 and 65 ms, Breitmeyer *et al.* (1981*b*) similarly determined the spatial range of transient-on-sustained inhibition produced by M_2 on M_1.

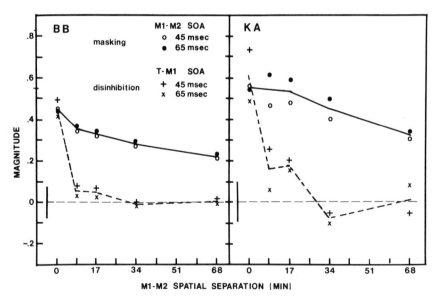

Fig. 9.5. Masking magnitude of mask 1 by mask 2 (circles) at two optimal mask 1–mask 2 asynchronies, and target disinhibition magnitude (+, ×) produced by a stationary mask 2, when the target and mask 1 were presented at two optimal masking delays, as a function of mask 1–mask 2 spatial separation. (From Breitmeyer, Rudd, and Dunn (1981*b*).)

The results for both procedures are shown in Fig. 9.5. Note that the magnitude of masking produced by M_2 on M_1 is optimal at an M_1–M_2 spatial separation of 0′ but is obtained at significant levels at all spatial separations, declining only slightly at progressively larger ones. On the other hand, the disinhibition of T produced by a stationary, sustained M_2 stimulus is also maximal at an M_1–M_2 spatial separation of 0′, but drops precipitously to 0 at spatial separations ranging approximately from 8.5′ to 25.5′. These results indicate (i) that sustained channels activated by the stationary M_2 stimulus inhibit the transient activity produced by M_1 and thus produce target recovery and, moreover (ii) that the spatial range of the sustained-on-transient inhibition is significantly shorter than that of transient-on-sustained inhibition. This conclusion seems reasonable in view of the aforementioned facts: (i) that transient-cell receptive and dendritic fields are larger than those of sustained cells (Wässle *et al.* 1981*a*, *b*); and (ii) that the spatial extent of transient-neurone activity, as reported by Ikeda and Wright (1972*a*), is larger than that of sustained ones.

The question remains as to whether, as Fotta (1979) hypothesized, intrachannel transient-on-transient inhibition also is a source of target recovery. To answer this question Breitmeyer *et al.* (1981*b*) employed the

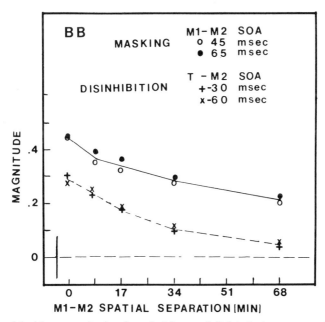

Fig. 9.6. Masking magnitude of mask 1 by mask 2 (circles) at two optimal mask 1–mask 2 asynchronies, and target disinhibition magnitude (+, ×) produced by a flashed mask 2, when it preceded the target–mask 1 sequence at optimal disinhibition asynchronies, as a function of mask 1–mask 2 spatial separation. (From Breitmeyer, Rudd, and Dunn (1981*b*).)

following rationale. In the prior experiment, M_2 was stationary and thus could not activate transient channels during or prior to the T–M_1 sequence. In the present experiment, the above investigators compared the optimal M_2–on–M_1 masking curves, measured a a function of M_1–M_2 spatial separation in the prior experiment, with disinhibition magnitudes produced by a transient M_2 presentation at optimal disinhibition SOAs, separating the onsets of M_2 and the T–M_1 sequence by -30 and -60 ms (see Fig. 9.4 and accompanying discussion). Since the M_2 was flashed transiently, it should activate transient as well as sustained channels. On the assumption that transient activity is just as effective over spatial separation in producing intrachannel inhibition as it is in producing interchannel inhibition, one would expect the masking and disinhibition functions produced by transient-on-sustained and transient-on-transient inhibition, respectively, to run parallel to each other as M_1–M_2 spatial separation increases. As shown in Fig. 9.6, this expectation was borne out by the obtained data. Note that for both the metacontrast masking and the disinhibition functions, optimal values were obtained at a spatial separation of $0'$, with parallel decreasing values at larger spatial separations. These results, according to the rationale outlined above, indicate that whenever M_2 activates transient channels it can produce target recovery via intrachannel inhibition of M_1's transient excitatory response by the surrounding M_2's transient inhibitory effect.

9.1.2. Target enhancement

The target recovery effects discussed in the preceding section required the presence of a second mask, which, in one way or other, decreased the first mask's ability to suppress target visibility. In the current section we shall review findings of target enhancement, rather than masking, produced during the conventional target-mask sequence employed in metacontrast. These target-enhancement effects are particularly evident in metacontrast paradigms in which the target stimulus, rather than being of uniform contrast, contains a variable number of internal contours (Dember, Mathews, and Stefl 1973; Dember and Stefl 1972; Dember, Stefl, and Kao 1974).

I have already noted that Werner (1935; see Section 7.4.2, Explanation 16) introduced internal target contours as a variable in his qualitative metacontrast investigation. According to Werner's observations, internal target contours reduced the overall masking effect, particularly when the orientation of the target contours was orthogonal to that of the flanking mask contours. In more recent experiments, the typical means of introducing internal contours was to use targets composed either of radial gratings—for example, a disk composed of alternate black and white wedge-shaped sectors—or of linear gratings—for example, a disk or rectangle composed of a vertical grating. For instance, with the former radial grating as target, it turns out that as the number of sectors (spatial frequency) increases—provided that they remain clearly resolvable—the target initially becomes less susceptible to the masking effects of an immediately aftercoming, annular black mask; and, more notably, beyond a given number of sectors its visibility often is actually *enhanced* by the mask relative to that obtained in a no-mask viewing condition (Arand and Dember 1978; Cox and Dember 1970; Dember and Stefl 1972; Dember *et al*. 1973; Ellis and Dember 1971).

In contrast to these results, Lefton (1974) and Lefton and Griffin (1976) also employing a radial grating and black annulus as target and mask, reported the opposite results; targets were more rather than less susceptible to the masking effect of the aftercoming annulus as their radial spatial frequency increases. These discrepant findings are related, in turn, to the use of discrepant background luminances (Arand and Dember 1978). Whereas Lefton (1974) and Lefton and Griffin (1976) employed scotopic (0.12–0.16 cd/m^2) background luminances, Dember and co-workers always employed photopic ones ranging from 34 to 85 cd/m^2. However, it is not entirely clear how or why variations of background luminance affect backward masking of sectored-disk targets. One likely possibility is that visual acuity and contrast sensitivity is severely limited at scotopic luminance levels, implying that target masking and enhancement effects interact with the resolvability of the radial (or linear) target grating. In fact, Dember and Stefl (1972) and Dember *et al*. (1974) found that, even at photopic background luminance, target enhancement was obtained only as the angular or linear spatial frequency of the target grating increased up to an intermediate value. In contrast to this enhancement, at still higher spatial frequencies, target *masking* was observed. Here the response to the contrast of the individual target sectors or bars is progressively attenuated as spatial frequency increases; hence, the overall response to the target may be determined by its lower, space-averaged contrast rather than the higher contrast of the individually unresolvable sectors or bars.

Let us assume, for the sake of the following discussion, that the internal target sectors or bars are clearly resolvable and that photopic background luminance prevails. How or by what process does target enhancement occur under these optimal conditions? It is clear from Werner's (1935) observations and those subsequently made by Dember and co-workers that a target becomes less susceptible to masking as the number of its internal contours increases. However, whereas an increase of the number of internal target contours may be a sufficient condition for reducing its susceptibility to masking, it *per se* is not sufficient to produce target enhancement.

The studies by Dember and co-workers (Cox and Dember 1970; Dember and Stefl 1977; Dember *et al*. 1973; Ellis and Dember 1971) reporting significant enhancement effects employed radial gratings as targets. However, in subsequent studies employing disk-like targets composed of linear, vertical gratings, Dember *et al.* (1974) and Lefton and Hernandez (1977) reported only slight and no target-enhancement, respectively, despite the fact that these latter investigations were conducted under photopic background luminances and included the same range of internal target contours (0–32) as the former investigations. The difference between these two sets of studies is very likely related to differences between target and mask contour relations. With the radial target gratings, no matter along which diameter we scan, local orthogonalities between internal target contours and the inside contour of the annular mask are obtained everywhere. However, for the target composed of a disk-like, vertical grating local orthogonalities occur only if we scan along diameters at or near the vertical one. As the diameter along which we scan approaches the horizontal orientation, progressively more of the internal target contours are aligned in *parallel* with the local contoured arcs of the annulus. Consequently the latter linear-grating targets are overall more maskable than the former radial ones which contain only loca orthogonalities. This suggests that, relative to a no-mask condition, the visibility of gratings, be they radial or linear, (i) is enhanced in direct proportion to the number of internal contours which are perpendicular to the mask's contour and (ii) is masked in direct proportion to the number of internal contours which are parallel to the mask contours. Indeed, Gilinsky (1967) reported such orientation-specific masking and enhancement effects using spatially *overlapping* target and mask gratings. However, further research, of course, is required with non-overlapping target and mask stimuli; firstly, to test the plausibility of the above explanation and, secondly, to reveal how orthogonal contours interact in a mutually facilitatory rather than inhibitory manner in the visual system.

9.2. Long-range masking: the jerk effect

In section 6.6 (see Fig. 6.8) the existence of the neural periphery or shift effect was discussed. The periphery effect is produced in visual neurones by sudden, abrupt movement of patterns, e.g. a grating, in areas which spatially are so remote that by conventional spot-mapping techniques (Kuffler 1953) they have no effect on the classically defined, spatially circumscribed receptive fields of visual neurones (McIlwain 1964). It also was noted that whereas the effect is predominantly excitatory in transient neurones it is, in contrast, inhibitory in sustained neurones (Fukuda and Stone 1976; Krüger 1977*b*).

In recent years several psychophysical and electrophysiological investigations have provided evidence for the existence of an analogous effect in human vision. For example, MacKay (1970*a*) showed that a sudden displacement of a 10°–diameter, uniform field (and, hence, its border), can suppress the visibility of a centrally flashed test spot. Moreover, MacKay (1979b) subsequently showed that this suppressive effect transfers inter-ocularly, only tentatively indicating a central level of interactions, since a more recent investigation by Dortmann and Spillmann (1981) failed to yield such interocular effects. However, similar monocular or binocular effects have been reported by Mitrani and co-workers (Mateeff, Yaki-moff, and Mitrani 1976; Mitrani, Mateeff, and Yakimoff 1971).

This suppressive effect has also been studied in several varieties of vertebrates, besides those (cat, monkey, and rabbit) discussed in Section 6.6. Brooks and Holden (1973) demonstrated that the proximal negative response (PNR) of the pigeon retina to a centred light flash is suppressed by the sudden displacement of a peripheral pattern. The same reduction in PNR amplitude has been reported in goldfish retina by Seim, Sørensen, and Valberg (unpublished observations) when a peripherally surrounding grating was oscillated back and forth in square-wave fashion. These investigators also showed that the suppression is specific to the oscillation frequency of the peripheral pattern, with optimal PNR suppression at a frequency of 4–6 c/s. Moreover, Jeannerod and Chouvet (1973) showed that the evoked potential to a centrally flashed stimulus recorded at suprageniculate levels in cats is suppressed by the sudden movement of a surrounding peripherally displaced square-wave grating pattern.

In view of these generally found, suppressive effects of a peripheral pattern displacement, it is somewhat surprising that Sharpe (1972) correlated the neural periphery-effect with psychophysically enhanced visibility. Sharpe studied the visibility of entoptic shadows of retinal blood vessels which normally are not visible because the are stabilized (Yarbus 1967). One way of moving these shadows is to scan the plane of the dilated pupil with a spot of light brought to focus with a convergent lens to produce Maxwellian viewing. Another way, used by Sharpe, is to shine the spot of light in the eye at low scanning rates (below 2 c/s) which do not produce visibility of the entoptic shadows, but which do become visible at the border of the shadow produced by an opaque card moving to-and-fro in fromt of the pupil at rates of about 10 c/s. Sharpe associated this restoration of visibility of semi-stabilized entoptic images with the generally excitatory periphery effect found in transient retinal and post-retinal neurones.

However, as noted by Ikeda and Wright (1972c), Sharpe's techniques most likely were not conducive to the psychophysical study of the periphery effect. Moreover, Sharpe made the implicit, but unproved assumption that excitation in visual neurones is a correlate of increased visibility. However, this need not be true. For example, interneurones in the lateral geniculate nucleus can be excited directly by retinal afferents (Dubin and Cleland 1977) and at the next synaptic level these inter-

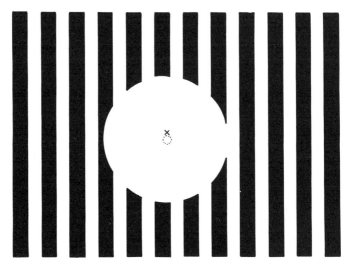

Fig. 9.7. The stimulus display used to investigate psychophysical correlates of the periphery effect. See text for details. (From Breitmeyer, Valberg, Kurtenbach, and Neumeyer (1980).)

neurones may very well produce inhibitory effects at the lateral geniculate. In addition, Neumeyer and Spillman (unpublished observations), exploiting the parafoveal Troxler effect, which results in the fading or retinally semi-stabilized images, found that a grating oscillating back and forth in the periphery accelerated the fading of the visibility of the Troxler stimulus, thus directly contraverting Sharpe's finding and interpretation of the psychophysical correlate of the periphery effect.

Recently, several investigators have attempted to study the psychophysical correlate of the periphery effect more systematically and quantitatively. Breitmeyer, Valberg, Kurtenbach, and Neumeyer (1980) employed the stimulus display shown in Fig. 9.7. The peripheral stimulus consisted of a square-wave grating $26.0° \times 19.5°$ surrounding a centrally illuminated white disk, $7.0°$ in diameter. A fixation cross appeared in the centre of the disk as shown, and the target spot, $0.38°$ in diameter, indicated by the dashed outline was flashed fo 100 ms either against an entirely uniform field (no peripheral grating) or one containing the grating, oscillating back and forth through one-half cycle ($0.94°$) at varying frequencies. By taking the logarithm of the test threshold without the grating in the surround relative to the threshold with the oscillating grating, one can obtain a measure of the effect of the grating oscillations. Negative log values denote target suppression.

In one experiment, Breitmeyer *et al.* (1980) studied the effects of oscillation frequency on test threshold changes. McIlwain (1964) noted that the magnitude of the periphery effect depended on the frequency with which a hand-held peripheral pattern was moved back and forth. He estimated the optimal frequency to be about 3.0 c/s. Under more tightly controlled conditions, Fischer (personal communication) found that the shift effect was greatest at sinusoidal oscillation frequencies ranging from 4.0 to 6.0 c/s. In view of these findings and those of Brooks and Holden (1973) and Seim *et al.* (unpublished observations) discussed above, one might also expect oscillation frequency-dependent threshold changes measured psychophysically. Typical results obtained in this experiment for sinusoidal and square-wave oscillation frequencies of the peripheral grating are shown in Fig. 9.8. Note that an optimal suppression of the test spot's visibility is obtained at about 3.0–4.0 c/s for both the square-wave and sinusoidal oscillations. No significant threshold change is obtained for either a stationary grating or a grating oscillating at frequencies higher than about 8.0 c/s. These psychophysical results dovetail neatly with the single-cell finding of McIlwain and Fischer.

In another experiment the effects of drift rate of the grating were compared with its oscillation rate. Here the peripheral grating drifted to and fro past the central uniform disk-background at rates determined by drift velocity and spatial frequency (0.53 c/deg) of the grating (drift rate = velocity × spatial frequency). The range of drift velocity was chosen to produce drift rates from 1.0 to 60.0 c/s. To generate smooth motion, the

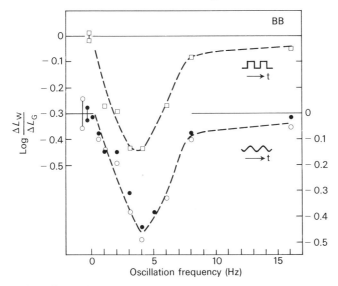

Fig. 9.8. The effect of oscillation frequency of a peripheral grating on the threshold luminance ΔL_G for a foveal test spot relative to the threshold luminance ΔL_W obtained in the presence of a uniform, white outer surround. Negative values of log ($\Delta L_W/\Delta L_G$) indicate suppression of the test flash visibility produced by the oscillating grating. The grating was oscillated either in a sinusoidal or square-wave manner. (From Breitmeyer, Valberg, Kurtenbach, and Neumeyer (1980).)

grating moved back and forth through roughly 15° in a triangle-wave manner; hence, it jerked suddenly at each direction reversal of movement. The triangular oscillation frequencies varied from 0.01 to 4.0 c/s. The results of the experiment are shown in Fig. 9.9. On the abscissae are shown drift velocity, drift rate (lower abscissa), and oscillation frequency (upper abscissa). Although in the prior experiment an oscillation frequency of 3.0–4.0 c/s produced maximal test suppression, at a corresponding drift rate no appreciable suppression was obtained. Test suppression increased with drift rate and notable suppression was obtained only at drift rates of 50–60 c/s. These correspond to optimal oscillation frequencies of 3.0–4.0 c/s as obtained in the prior experiment. In fact, the dashed curve in Fig. 9.9 reproduces the test suppression effects obtained as a function of oscillation frequency in the prior experiment. The close correspondence between this curve and the results of the present experiment demonstates that a jerk or sudden reversal of direction of motion of the grating pattern produced by its triangular oscillation is necessary to obtain test suppression. Consequently, this psychophysical effect can be appropriately called the *jerk effect*.

Inspection of the results of Fig. 9.9 in terms of the drift velocity shows that the suppressive effects of the drifting, peripheral grating is not obtainable until a drift velocity of at least 5–10°/s is attained. This finding

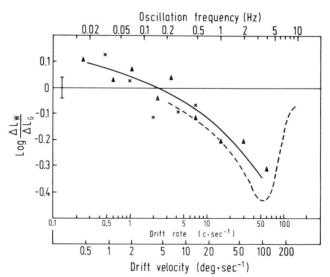

Fig. 9.9. The magnitude of the jerk effect produced by a drifting peripheral grating oscillating in triangle-wave fashion. Upper abscissa gives the triangular oscillation frequency. Lower abscissae give the corresponding drift rates and velocities of the grating. Dashed line represents average data from Fig. 9.8 for the square-wave oscillation. (From Breitmeyer, Valberg, Kurtenbach, and Neumeyer (1980).)

corresponds well to the results of the single-cell study of the shift effect reported by Fischer *et al*. (1975). They reported that the shift-effect reached threshold at drift velocities of approximately 10°/s and progressively increased up to velocities of several hundred degrees per second. A corresponding increase in the jerk effect with increasing velocity is also evident in the results displayed in Fig. 9.9.

In a third experiment Breitmeyer *et al*. (1980) measured the jerk effect as a function of the central disk radius or the spatial separation between the test spot and the inside border of the peripheral grating. For this experiment an optimal oscillation frequency of 4.0 c/s was employed. One subject (AV) was tested at a central disk luminance of 5.75 cd/m^2; the other subject (BB) was tested at a luminance of 25.66 cd/m^2; which, as determined in a separate series of experiments (Valberg and Breitmeyer 1980; see below), yielded optimal test suppression. The results, in line with those reported by Mateef *et al*. (1976) revealed that the strength of the jerk effect decreased as the radius of the central background disk increased. Optimal suppression was obtained at a radius of 1.75° with progressively less suppression at larger radii. Extrapolation of the two curves indicated that the jerk effect was not obtainable at radii in excess of 10.0°. These psychophysical results also correspond to the single-cell findings on the periphery effect reported by McIlwain (1964) and Fischer *et al*. (1975) Both studies showed that the periphery effect decreases in magnitude as the distance between a neurone's receptive field centre and the peripheral grating increased.

Given this suppressive jerk effect, what the underlying mechanisms might be remains to be determined. At least two likely alternative hypotheses can be formulated. For one, the excitatory response generated by the central test flash may be inhibited by the activity generated by the oscillating, peripheral grating. The second, noise hypothesis relies on the finding that the periphery effect is generally excitatory in transient neurones. The oscillating grating could increase the firing level of transient neurones representing the area of the test flash and thus introduce an effective increase in their background noise. This would increase their internal Weber ratio and consequently decrease their sensitivity to the test flash. Another version of the noise hypothesis states that as background disk luminance is increased, an increasing amount of stray light invades the area of the test spot and consequently also raises the Weber ratio. To determine which hypothesis is correct, Valberg and Breitmeyer (1980) adopted a mathematical model of the response characteristics of a hypothetical luminance threshold detecting mechanism developed by Valberg (1974), which assumes that the level of activation of the detector mechanism increases as the luminance of a uniformly illuminated surface increases (for quantitative development and details of the model see Valberg and Breitmeyer (1980). The oscillation frequency of the outer grating was set at the optimal value of 4.0 c/s. The luminance of the central background disk was varied from 0.8 to 1850.0 cd/m². The obtained test suppression curves as a function of background-disk luminance are shown in Fig. 9.10. Test suppression was a U-shaped function of background luminance with optimal suppression at about 6.0 cd/m² and 25.0 cd/m² for observers AV and BB, respectively.

From these data, hypothetical response curves, shown in Fig. 9.11, of the assumed luminance threshold detecting mechanism were generated. These derived functions revealed that the response of the mechanism detecting the test flash against a background without the grating and against a background with the grating both increased monotonically with central disk luminance. However, in the latter as compared with the former condition the rising response curve was displaced to higher values of disk luminance; moreover, its increase as a function of disk luminance was characterized by a shallower slope than in the former condition. The first, displacement, effect, according to the model of the hypothetical response mechanism, indicates that the peripheral activity generated by

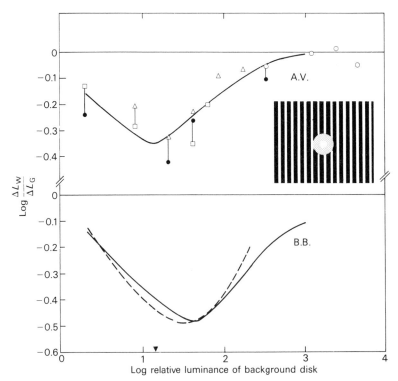

Fig. 9.10. The magnitude of the jerk effect, produced at an optimal oscillation frequency of 4 c/s, as a function of background-disk luminance (in relative log values). (From Valberg and Breitmeyer (1980).)

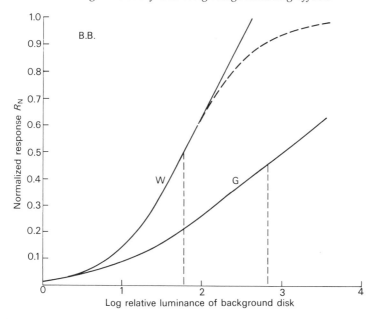

Fig. 9.11. Hypothetical normalized response of subject BB's luminance threshold detecting mechanism as a function of the log relative luminance of the background disk in the presence of a uniform white surround (W) or an oscillating grating (G). (From Valberg and Breitmeyer (1980).)

the oscillating grating had an inhibitory effect on the excitatory response of the threshold detecting mechanism, thus confirming the inhibition hypothesis. The latter effect, in turn, indicates that whereas the sensitivity of the threshold detecting mechanism was reduced by the oscillating grating, its dynamic range of brightness response varied over a wider luminance interval. Thus, an increase of response range is sacrificed for response sensitivity. (I shall note the implications of this result for visual search performance in Section 10.3.1).

To rule out the two versions of the noise hypothesis, Valberg and Breitmeyer (1980) assumed that the stray-light effects of increases in the luminance of the background disk and the possible excitatory effects of the oscillating grating were additive. In either case, in order to maintain a constant Weber ratio, $\Delta L_W/L = [\Delta L_G/(L + L_N)]$; where ΔL_W is the increment threshold obtained without the grating, ΔL_G is the increment threshold obtained with the oscillating grating, and L_N is the amount of equivalent light generated either by excitation produced by the periphery effect in transient channels responding to the test spot or by stray light. As a function of background disk luminance, Fig. 9.12 shows the expected results, at three values of L_N (solid lines), and the obtained results (dashed lines). Note that whereas the obtained results fit a non-monotonic

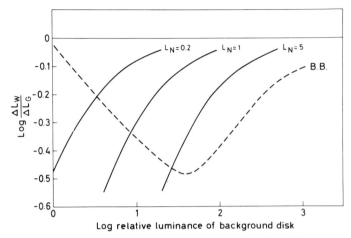

Fig. 9.12. The predictions for the jerk effect as a function of background-disk luminance of two forms of a noise hypothesis (solid lines; see text for details) as a function of three noise levels. Dashed line gives the averaged empirical results. (From Valberg and Breitmeyer (1980).)

function, the results expected on the basis of the noise hypothesis are given by highly deviating monotonic functions. These discrepancies disconfirm the two versions of the noise hypothesis.

Given that the inhibition hypothesis is correct, the task of offering a plausible neurophysiological correlate remains. Recall that transient channels inhibit sustained ones. In addition, the periphery effect is a long-range mechanism through which brief excitation is produced in transient channels by spatially remote contour shifts (see Fig. 6.8). Accordingly, as an initial approximation to what will become a more specific explanation, the following conjecture may be proposed. The remote, oscillating grating locally excites transient neurones, the activity of which—summed globally over the area stimulated by the grating—is relayed via the periphery effect to the transient cells in the retinal area representing the test spot and there, via local, transient-on-sustained inhibition, suppresses the activity of sustained cells otherwise freely responding to the test spot. This explanation, of course, additionally assumes that increment thresholds for a longer, say, 100–ms test-spot duration are determined by sustained channel activity as argued by Ikeda and Wright (1972b; see also below).

Recall that the periphery effect is stronger outside the fovea than at or near it (Cleland and Levick 1974; see Section 6.6). Consequently, given the above inital explanatory scheme, the psychophysical jerk-effect ought to be weak in the fovea and increase with retinal eccentricity. Breitmeyer and Valberg (1979) investigated the magnitude of the jerk effect as a function of retinal locus by displacing the fixation point, shown in Fig. 9.7, progressively farther to the right of the test spot. Again an optimal oscillation frequency of 4.0 c/s was employed at all fixation eccentricities. The results, shown in Fig. 9.13, surprisingly show, contrary to expectation, that the inhibitory jerk effect is *maximal* at the fovea and is not obtainable at eccentricities beyond approximately 2.0°. The initial reaction to this counter-intuitive result is to question the validity of the neurophysiological mechanisms offered above as an explanation of the jerk effect. However, I shall amend this explanation slightly but significantly.

The transient excitation produced by the neurophysiological periphery effect, either locally or remotely, anywhere on the non-foveal retina reflects merely one of many linkages in the activity of transient neurones which channel their combined excitation *en masse* centripetally toward the fovea where, at the concentred terminus of this extra-foveal excitation, foveal transient channel activity is enhanced to produce optimal inhibition of foveal sustained channels. This explanation of the restriction of the inhibitory jerk effect to the fovea is discussed further and given greater

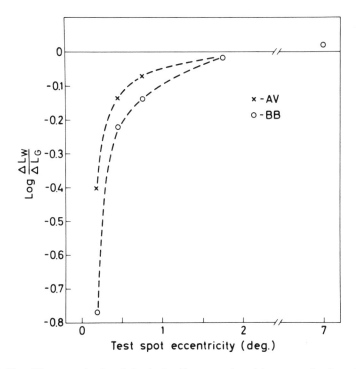

Fig. 9.13. The magnitude of the jerk effect, produced by an optimal oscillation frequency, as a function of test spot eccentricity. (From Breitmeyer and Valberg (1979).)

plausibility in Section 10.3.1. The foveal suppression of test flash visibility produced by an oscillating peripheral grating also has been reported in an electrophysiological study conducted by Valberg, Olsen, and Marthinsen (1981). These investigators showed that the cortical visually evoked response to a suprathreshold foveal flash was inhibited by the contrast reversal or a sudden one-half cycle shift of a peripheral, square-wave grating.

An unreported result obtained by Breitmeyer and Valberg (1979) was a tendency for the test spot's visibility to be increased by the oscillating grating relative to a uniform field at eccentricities beyond those shown in Fig. 9.13. Ikeda and Wright (1972c) hypothesized that a facilitatory correlate of the neural periphery effect should be obtainable since, as they and McIlwain (1974) demonstrated, the sensitivity to a near-threshold test stimulus falling on the centre of a transient cell receptive field is increased by the periphery effect. At larger eccentricities where the relative frequency and response strength of transient cells increases (Cleland and Levick 1974) one may expect such a facilitatory effect.

The possibility of transient channel response facilitation was systematically investigated by Dortmann and Spillmann (1979, 1981). These investigators studied the jerk effect as a function of: (i) the duration of the central, 8'-diameter, foveal test spot; (ii) its asynchrony relative to the onset of a jerk of the peripheral grating; (iii) the diameter of the central uniformly illuminated disk; and (iv) dichoptic and binocular viewing. At a test-spot duration of 100 ms (also used in the previously discussed studies) and a central-disk diameter of 9°33', they observed, as displayed in Fig. 9.14(a), an inhibitory effect which was maximal at test spot-grating jerk asynchrony of 0 ms. However, for a 3-ms test flash, as shown in Fig. 9.14(b), a more complicated set of functions was obtained. When the grating jerk followed the test spot (denoted by negative delay intervals), facilitation was observed peaking at a delay of about −35, −30, −25, and −15 ms at central-disk diameters of 1°50', 4°04', 9°33' and 15°12', respectively. At the same respective diameters, suppression of test spot visibility was obtained when the jerk preceded the test spot of intervals ranging from 20 to 30 ms.

Recall that the response of transient channels to a brief stimulus can be characterized by an oscillating excitatory–inhibitory function (Breitmeyer and Ganz 1977; Büttner *et al.* 1975; Watson and Nachmias 1977). This oscillation is also produced by the periphery or shift effect. Careful inspection of Fig. 6.8 shows that the initial phasic excitatory response of transient cells produced by the shift effect is followed by a brief period of suppression. This is particularly evident when an adequate stimulus is presented to the ON-centre neurone, shown in Fig. 6.8, which may be responsible for signalling brief luminance increments (Jung 1973). For several reasons, a brief, 3-ms test flash near threshold may be a more

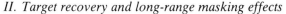

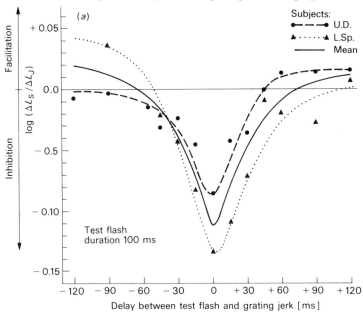

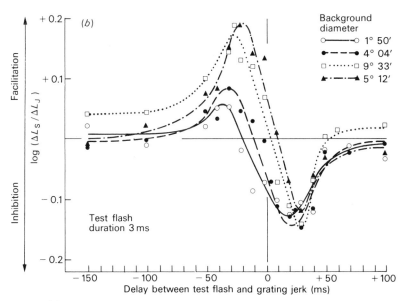

Fig. 9.14.(*a*) The magnitude of jerk effect as a function of the asynchrony between the onset of a 100-ms foveal test flash and the jerk onset of a peripheral grating. Central-disk diameter was 9°33′. (*b*) Same as in (*a*) except that a 3-ms test flash and four central-disk diameters, as indicated, were used. Negative delays indicate test onset was prior to jerk onset; positive delays indicate test onset followed jerk onset. (From Dortmann and Spillmann (1981).)

adequate stimulus for transient channels than for sustained ones. For one, sustained cells are characterized by a longer integration time and thus prefer stimuli of longer duration. Moreover, transient cells, preferring such brief stimuli, have a lower contrast threshold than sustained cells. Consequently, a given intensity of light may provide sufficient time-integrated energy at 3 ms to activate transient but not sustained channels at threshold. Finally, Kulikowski (1977), using a visually evoked potential measure, showed that, at very brief stimulus durations, transient mechanisms determine detection threshold. Hence, the results displayed in Fig. 9.14(b) may be thought of as subthreshold summation curves, which yield a measure of cross-correlation between the oscillatory transient impulse responses produced by the grating shift and the 3-ms central test flash.

For this early facilitation or positive cross-correlation to be optimal the initial subthreshold oscillatory activity produced at the central foveal transient channels by the test spot must be temporally coincident and, therefore, in phase with the subthreshold oscillatory activity produced, again at the same foveal transient channels, by the peripheral jerk. By the same token, in order to produce the later optimal suppression or negative cross-correlation of the foveal subthreshold oscillation to the test spot with that produced at the fovea by the peripheral grating, the grating onset must occur relatively earlier by an interval separating the onset of the primary excitatory and inhibitory phases of the subthreshold oscillatory responses.

From inspection of Fig. 9.14(b), we can see that this interval, obtained by taking the difference between optimal facilitation and optimal inhibition SOAs, is a *constant* value of about 55 ms at all central-disk diameters. Implied by this conclusion, and again apparent from inspection of Fig. 9.14(b), is the corollary that when the positive phase of the subthreshold summation curves shifts by a certain interval, as a function of central-disk diameter, so does the negative phase. In other words, the entire subthreshold summation index is shifted uniformly as a function of central-disk diameter.

Two additional aspects of the results portrayed in Fig. 9.14(b) require discussion: one is the fact that the initial facilitation or positive sub-threshold summation occurs only when the peripheral jerk *follows* the foveal test spot. A second one is the particular rightward or positive shift in the target-jerk delay as central-disk diameter increases.

Turning to the first aspect, we must recall (Sections 6.1, 6.2, and 7.2) that the response latency of transient channels to stimuli at or near contrast threshold is several tens of milliseconds longer than to high-contrast, suprathreshold stimuli (Harwerth and Levi 1978; Harwerth *et al.* 1980; Ikeda and Wright 1972*a*; Lennie 1980*b*). Therefore, taking into account the fact that the responses of extrafoveal transient channels activated by the jerk of the high-contrast, peripheral grating must be transmitted to the fovea, these fast suprathreshold responses nevertheless arrive earlier at the fovea than does the slow response to the near-threshold foveal test spot. Hence, in order to produce optimal summation of excitatory responses, the peripheral jerk onset can occur *after* the foveal test flash by a value which depends on the diameter of the central disk.

This brings us to the second aspect. The fact that the subthreshold summation indices shift uniformly to the right or to decreasing delays between the jerk onset relative to the onset of the test spot can be accounted for as follows. The closer the border of the peripheral grating to that of the foveal test spot, the shorter the delay between the fast suprathreshold response of peripheral transient channels and their facilitating effect at the foveal transient channels. Hence, the smaller the central-disk or alternatively the closer the grating and test contours, the greater the delay—of the peripheral jerk onset relative to the foveal test flash—required to obtain optimal positive subthreshold summation.

Given the validity of this argument, we can additionally estimate, to a first approximation, peripheral-to-foveal transmission velocity of the jerk effects in the following manner. From smallest to largest, the radii (r) of the four central, background disks are 0.917, 1.854, 4.817, and 7.600°, respectively. Their respective optimal subthreshold facilitation of the test spot's visibility occur at correspondingly decreasing jerk delays (d) of 0.035, 0.030, 0.025, and 0.015 s. A linear-regression estimate of the relation between these two variables yields the following function: $d(s) = 0.036(s) - [0.0028 \text{ (s/deg)} \cdot r(\text{deg})]$ or, alternatively, $r(\text{deg}) = 12.86(\text{deg}) - [358.1(\text{deg/s}) \cdot d(s)]$. The slope of the latter equation is an estimate of the propagation velocity of the jerk effect measured under Dortmann and Spillmann's (1981) experimental conditions. Electrophysiological measures of the propagation velocity of the shift-effect in cat retina, made by Fischer *et al.* (1975), yielded an estimate of approximately 1600 deg/s; moreover, the investigation by Derrington *et al.* (1979) of the shift effect in cat retina indicated that the propagation velocity must be greater than 1500 deg/s. These estimates exceed by a factor of four the roughly 360 deg/s estimate of the jerk effect's propagation velocity derived from Dortmann and Spillmann's (1981) data. None the less, the estimated, 360 deg/s, propagation velocity of the jerk effect is already orders of magnitude larger than the 0.65–6.0 deg/s estimates of *short-range, local* lateral propagation velocities obtained psychophysically by Motokawa, Iwama, and Ebe (1954), Katayama and Aizawa (1956), Smith and Richards (1969), and van der Wildt and Vrolijk (1981).

Besides these above results, Dortmann and Spillman (1979, 1981) also demonstrated that both the facilitatory effect as well as the inhibitory effect, as previously found by Breitmeyer and Valberg (1979), were confined to foveal test spots.

Recently Brooks and Impelman (1981) showed that the foveal specificity of the jerk effect may depend on the size of the target flash. These investigators employed either a 0.5°- or 5.0°-diameter test disk flashed concentrically against a 11.0°-diameter uniform white background, in turn surrounded by a 37° × 25° peripheral grating. The centre of the test disks fell either on the fovea or 5° in the periphery. For the 0.5° test disk, suppression was obtained only foveally; however, for the 5.0° test disk, both foveal and extrafoveal suppression was obtained. However, several aspects of Brooks and Impelman's (1981) procedures merit closer analysis. For one, the test-disk duration was very likely *less than 1 ms* since a Grass S88 two-channel stimulator was used to time its presentation. Moreover, the test disk was flashed 5 ms *after* the termination of a peripheral grating-jerk. These two procedural parameters maximize the likelihood that, as in Dortmann and Spillman's (1981) study, the peripheral jerk affected only the responses of foveal or extrafoveal transient channels but not those of sustained ones. In particular, at the 5-ms delay, the oscillating, subthreshold, transient responses produced by the peripheral jerk and the foveal test spot should be such that their opposing phases combine; hence the peripheral jerk ought to index the inhibitory phase of the response of foveal transient channels to the test disk and vice versa. In addition, that the visibility of the 5°-diameter test-disk was suppressed when flashed extrafoveally may partly be due to the fact that although the fixation point was removed 5° from the centre of the test disk, the nearest edge of the disk was within 2.5° of the fixation point as compared with a 4.75° value when a 0.5°-diameter test disk was employed. Finally, whereas the separation of the nearest test-disk contour to the remote grating was 5.25° for the 0.5°-diameter test disk, it was only 3.0° for the 5.0°-diameter test disk. Although Brooks and Impelman's (1981) results indicate that the degree of foveal specificity of the jerk-effect may be affected by the size of the test stimulus, it does so only tentatively until (i) the role of other, possibly confounding variables, noted above, are ruled out; and (ii) their experimental manipulations can be shown to affect sustained as well as transient channels responding to the test stimulus.

The result of Dortmann and Spillman (1979,1981) adds support to the physiological interpretation of the jerk effect offered earlier. It was assumed there (i) that the jerk effect concentred at and excited foveal transient neurones via the long-range periphery effect initially activating transient neurones in the periphery; and (ii) that these excited foveal transient neurones in turn inhibit foveal sustained neurones. The joint facts that extrafoveal transient neurones are more effectively excited by the periphery effect (Cleland and Levick 1974), yet that no suppression—of either the sustained activity or the oscillating transient activity generated by an extrafoveal test spot—occurs is readily reconcilable on the basis of these two assumed properties of the periphery effect.

Furthermore, Dortmann and Spillmann (1979, 1981) demonstrated that as the duration of the foveal test flash was increased from 3 to 150 ms the oscillatory subthreshold effect declined to zero by about 50 ms after which the purely inhibitory effect prevailed increasingly as duration increased. This finding (see also Breitmeyer and Ganz (1977) and Legge (1978)) corroborates the present interpretation that transient channels determine the foveal increment threshold at brief test durations whereas sustained channels determine it at longer ones.

Additionally, the finding may shed some light on how to relate the results illustrated in Fig. 9.14(a) and (b). Figure 9.14(a) shows that the inhibition of the sustained channels activated by a 100-ms foveal test spot is optimal when the onsets of the test spot and the peripheral jerk coincide. From Fig. 9.14(b) we can see that the response of foveal transient channels activated by the strong, suprathreshold, long-range peripheral jerk would be optimal at a time about 30 ms prior to the transient response elicited by a foveal, subthreshold test spot. At this negative asynchrony of −30 ms, however, since the sustained channels respond faster to near-threshold test flash of adequate duration than do the transient channels activated by the peripheral jerk, the initial generation of the foveal sustained response would precede that of the foveal transient response produced by the jerk and thus the former response would not be suppressed optimally by transient-on-sustained inhibition. In order to optimize this interchannel inhibition, the onset of the foveal test spot, relative to the peripheral jerk, must be delayed and therefore shifted from an asynchrony of −30 ms to a higher one. In the results depicted in Fig. 9.14(a), the optimal transient-on-sustained channel inhibition occurred at a test-jerk onset asynchrony of 0 ms, indicating that the required shift of onset asynchrony was about 30 s. (This value by the argument outlined above, also would correspond to the latency difference between the faster near-threshold sustained response and the slower near-threshold transient response to the onset of an adequately long foveal (> 50 ms) test flash (Harwerth and Levi 1978; Harwerth *et al.* 1980; Ikeda and Wright 1972*a*; Lennie 1980*b*; see Sections 6.1, 6.2, and 7.2)).

Regarding possible roles of the jerk effect, Valberg and Breitmeyer (1980) noted that it may have its functional application in amblyopic vision. It is known that patients with functional amblyopia have an impaired sensitivity called the crowding phenomenon (Burian 1969) analogous to that produced by the jerk effect. These patients are capable of discriminating much smaller patterns when they are presented in isolation rather than when embedded within multi-pattern display. Since latent nystagmus, unsteady fixation, and irregularities of saccades are frequently observed in these patients, it is reasonable that saccade or jerk-like displacements of patterns in the peripheral retina contribute to the inhibitory crowding phenomenon.

This example already illustrates one form of application of the extended theory of visual masking based on transient–sustained interactions to dynamic, although abnormal vision. In the following chapter, the additional application of this extended model to normal dynamic viewing—characterized either by visual pursuit of a moving object or by the serial inspection of a visual scene and effected by alternating and continual sequences of fixations followed by saccades—will be discussed at some length.

9.3. Summary

Partial recovery of target visibility when a second mask (M_2) is introduced into the typical target–mask $(T-M_1)$ display sequence can be obtained in masking-by-light and masking-by-pattern paradigms. In general the presence of M_2 reduces the suppression effect exerted by M_1 on T's visibility. In masking-by-light, under binocular or monocular viewing, optimal target disinhibition is obtained at short M_1-M_2 SOAs *and* at intermediate $T-M_1$ SOAs. However, when M_1 and M_2 are flashed to separate eyes, the M_1-M_2 SOAs at which optimal target recovery occurs are relatively long; moreover, contour interactions between M_1 and M_2 also must prevail. These findings are consistent with those, discussed in Chapter 2, indicating that masking-by-light occurs at peripheral levels of visual processing, whereas central masking-by-light requires the presence of contour proximity and, thus, contour interactions between M_1 and M_2.

In particular, when M_1 either adds pattern noise to the target display or overlaps the target and M_2 constitutes a mask which spatially surrounds and follows M_1, one can obtain target recovery by M_2's metacontrast suppression of M_1's contrast or pattern information, thus preventing it from integrating with that of T's contrast and pattern information. As a result, T's visibility is disinhibited by M_2.

If T and M_1 comprise the spatially adjacent stimuli in a typical metacontrast masking paradigm, the degree of type B contrast suppression of T by M_1 can be attenuated by a continuously present M_2 which spatially flanks M_1. Here, since M_2 activates only sustained channels, it exerts an interchannel inhibition on the transient activity of M_1. Since M_1's transient activity is now attenuated, its ability to suppress the contrast of T via the interchannel transient-on-sustained inhibition is, in turn, attenuated; hence, T's contrast visibility, masked by M_1, recovers when M_2 is also present.

By varying the onset interval separating a brief or transient M_2 presentation from a temporal $T-M_1$ metacontrast sequence which yields optimal suppression of T's contrast, one can measure the time course of M_2's disinhibitory effect on T's visibility. T's recovery of visibility is optimal when M_2 precedes the $T-M_1$ sequence; here, however, M_1's visibility need not be affected. This implies (i) that M_2's slower sustained activity inhibits M_1's faster transient activity; and (ii) that suppression of

M_1's contrast visibility is not necessary for obtaining recovery of T's visibility. Moreover, when M_2 follows the T–M_1 sequence, M_1's contrast visibility is reduced by M_2; however, T's visibility is not, in turn, disinhibited. This implies (i) that M_2's transient activity suppresses M_1's sustained activity; and (ii) that suppression of M_1's contrast is not sufficient for obtaining recovery of T's contrast, since T's contrast can still be suppressed by M_1's intact transient activity.

Target enhancement in metacontrast paradigms can be obtained when an optimal number of internal contours of the target are locally orthogonal to the contours of the surrounding mask. This is related to the finding that metacontrast masking magnitude diminishes as the difference between the contour orientation of the target and mask increases. However, the target *enhancement*, i.e. an increase of the target visibility in the presence of the mask relative to its absence, would require a direct facilitatory effect of the mask on the target in addition to the attenuated masking effect obtained as the number of internal target contours orthogonal to the mask contour increases. The nature of such a facilitatory effect is, as yet, unexplored.

Long-range effects on foveal increment test thresholds, known as jerk effects, can be produced by peripheral contour shifts. Based on several parametric similarities between psychophysics and neurophysiology, these effects may be correlates of the neural periphery and shift effects. At longer test durations (> 50 ms) the long-range effects on foveal increment thresholds are strictly suppressive; at shorter test-flash durations the same effects are facilitatory and inhibitory depending on the timing of the test flash relative to the onset of the peripheral contour shift. The former finding indicates that foveal sustained channels are inhibited by peripheral contour shifts whereas the latter finding indicates that the oscillatory response of foveal transient channels summates with oscillatory transient activity generated peripherally by contour shifts. Accordingly it was argued that peripheral transient activity generated by contour shifts concentres at the fovea, facilitates transient channel responses there, which in turn inhibits the activity of local, foveal sustained channels. It was suggested that the jerk effect may be related to problems of pattern vision in amblyopes also known to display oculomotor unsteadiness. In the following chapter, both the jerk effect and the target recovery effects discussed above are shown additionally to relate to our understanding of normal visual behaviour in extralaboratory settings in which eye movements are the rule rather than the exception.

10 Extensions and applications of the sustained–transient approach: III. Dynamic viewing

10.1. Introduction

The title of this book promised that an integrative approach to visual masking would be adopted. The previous chapters have only partly fulfilled this promise. Chapters 1–6, besides outlining historical and empirical reviews of visual masking and cognate psychophysical and neurophysiological phenomena, also noted several interrelations between these phenomena. Chapters 7–9 presented a heuristic model of visual information processing and masking based on the existence of separate transient and sustained channels and applied and extended it to several other relevant topics. The model is comprehensive and can adequately account for and conceptually integrate a large number of human psychophysical findings. Within the limits of this heuristic approach, the book has already fulfilled a large part of its promise.

According to some perceptual psychologists, for instance, Sekuler (1973), the complex phenomena of visual masking, in particular metacontrast, do not deserve serious attention. What has been demonstrated in the previous chapters is that such phenomena do deserve such attention. For instance, visual masking not only can be used to investigate the temporal stages and processes of human pattern recognition and visual information processing (Breitmeyer and Ganz 1976; Michaels and Turvey 1979; Turvey 1973) but also, given an adequate theoretical model, it can provide a powerful methodological tool for studying the role of sustained–transient interactions as well as high-order (e.g. attentional and Gestalt) and abnormal processes in human vision. Visual masking phenomena are not merely striking parlour tricks; on the contrary, in an integrative theoretical context they essentially play meaningful, interesting, and important roles.

In a similar vein, Turvey (1977) offers the following critical metaphor regarding the existence and study of iconic persistence in vision. Consider that in an attempt to understand how a telescope magnifies, attention is diverted to an investigation of chromatic aberration manifested by the optical elements which comprise the machinery of the telescope. The object of study is no longer intrinsic to the powers of magnification but rather to a property of the optical machinery which supports magnifying. Although the machinery may itself be necessary for the process of magnifying, the property of chromatic aberration is entirely incidental to that process. In a like manner, the implied role of iconic persistence is merely as a property of the mechanisms supporting seeing and basically incidental, rather than intrinsic, to the process of seeing. However, in contrast to Turvey's (1977) position, what shall become evident in a later part of this chapter (Section 10.3.1) is that iconic persistence is not merely incidental to the process of seeing, but rather comprises an essential, constitutive spatiotemporal element in that process.

In the previous chapter (see Section 9.2), it was noted that the long-range jerk effect, a visual masking phenomenon, may have applications to the interpretation of the defective visual inspection of multi-pattern scenes in functional amblyopes. In the present chapter, normal dynamic visual viewing, characterized either by visual pursuit of a moving, foveated object or by the serial inspection of large visual scenes via saccade-fixation sequences, also will be discussed in the context of the topics covered in the previous four chapters. Thus, the limited laboratory phenomena of visual masking, in which visual persistence and visual attention play leading roles, will be placed in the broader framework of the behaviour of an organism in its visual ecology. Below, it will become particularly clear how the model of visual masking and cognate phenomena outlined and reviewed above provide the basis for an evermore inclusive integrated and ecologically valid theoretical approach—as recently advocated by Neisser (1976) and Gibson (1979)—to an extensive range of visual phenomena. As such we shall be edging still closer to fulfilling the promise of this book.

Prior theories of visual masking, like other accounts of visual processing, for the most part, have been based either on an information processing approach (Turvey 1973) or on a neurophysiological approach (Breitmeyer and Ganz 1976; Bridgeman 1971, 1978; Matin 1975; Weisstein *et al.* 1975). These approaches have contributed substantially to our understanding of how visual masking phenomena arise, in particular, to our understanding of the possible underlying mechanisms and processes. As noted by Neisser (1976), although fairly successful, such theories and the empirical data collected in rather artificial and visually impoverished laboratory settings to support them bear little relevance or resemblance to events occurring in more naturalistic and enriched settings characterizing our everyday, non-laboratory visual behaviour. To complement the largely ecologically inapplicable (if not ecologically invalid) empirical findings and theoretical accounts of visual masking, the current approach attempts to specify the functional, ecological significance of visual masking.

Although the functional or behavioural significance of masking phenomena is at least tacitly or fleetingly acknowledged by many investigators, few have made such an acknowledgement explicit. Exceptions have been Matin *et al.* (1972), Matin (1974*b*), Breitmeyer and Ganz (1976), Volkmann *et al.* (1978*b*), and recently Breitmeyer (1980) and Brooks *et al.* (1980), all of whom have suggested that a particular type of visual masking, namely, metacontrast, may contribute to afferent neural mechanisms of saccadic suppression. This suggestion is taken up, extended, and expanded in the present chapter, the principal aim of which is to demonstraté a fruitful relation between explanations of visual masking based, on the one hand, on specific neural mechanisms and, on the other hand, on more molar, functional aspects of visual behaviour. Implicit in attaching functional significance to visual masking mechanisms and phenomena is the assumption that they serve some definable purpose in visual behaviour. The notion of purposive behaviour or function is not novel (Granit 1972*a*, *b*, 1977; Lorenz 1974, 1977; Tolman 1967; Tolman and Brunswik 1935); none the less, it has suffered second-rate status among scientists committed to radical mechanistic or so-called objective accounts of behaviour (e.g. Hull 1943). The justifications for either the mechanistic, the purposive or both approaches to the study of behaviour are discussed among other topics in the Epilogue following this chapter.

The approach to be outlined below advocates the application of both mechanistic and purposive explanations to vision. To this end it is basically a problem–solution oriented one. Discussion of the purpose or functional significance of masking mechanisms or processes is placed in the context of the following question: 'What problems existed in the organism–environment relation, i.e. the ecosystem of which the organism is a part, that a particular organ, physiological mechanism, or behaviour was designed to solve?' At this point it is necessary to make a distinction between two general sets of problems facing an organism in its behavioural ecology. The challenges posed to an organism by the external environment or the *milieu exterieur* comprise a set of biological survival problems that we can conveniently call the set of exogenous problems.

The solution to these problems generally consists of appropriate behavioural adjustments and adaptations (learning) made by the organism. Both the recognition of these problems and the eventual discovery of their solutions depend to an appreciable extent on the integral function of the organism's sensori-motor apparatus, that is, on its ability to pick up and adaptively act on environmental information. In this informational sense the problem–solution dyad is akin to Gibson's (1977, 1979) notion of the ability of an organism to recognize 'affordances'. Provided that the organism is endowed with sufficiently adaptable and flexible sensori-motor mechanisms, the behavioural adjustment (the exploring and exploiting of affordances) can be made within the functional limits of the existing senori-motor apparatus. Often, the required behavioural adjustments evolve prior to the biological evolution of new sensori-motor apparatus (Lorenz 1977); however, should the former exceed the latter's functional limits, changes in the functional architecture of the sensory and motor organs must evolve within the constraints of the extant design of these organs. These constraint characteristics of the organism's internal environment or *milieu interieur*, of its inherited structure and its ontogeny, comprise a set of biological survival problems which can be conveniently called endogenous problems. These constraints may be as important in directing evolutionary change as are the external constraints providing the basis of Darwinian (natural) selection (Gould 1982).

What evidence exists for the fact that such behavioural adaptability requirements can be functionally accommodated by an organism? Biological evolution of species through selective adaptation, of course, provides one long-term means of providing such functional accommodation. Moreover, other evidence derives from the study of more short-term postnatal development and ontogenesis of individuals within a species. In particular, on the one hand, it is known that the innate functional architecture of the mammalian visual system is intricately organized at birth (Wiesel and Hubel 1974); on the other, it is also a highly plastic structure subject to ontogenetic, environmental influences (Blakemore and Cooper 1970; Hirsch and Spinelli 1970; Hubel *et al.* 1977; Singer 1978; Spinelli and Jensen 1979). As a consequence, the innate 'hypotheses' carried within an organism's functional visual architecture, in particular in its visual cortex, about the general structure of the world may be behaviourally modifiable to appreciable extents during critical periods of ontogenetic development (Spinelli and Jensen 1979) in order to accommodate the behavioural significance or meaning of objects in the organism's ecology. To this end, as noted by Singer (1978), the selection processes, through which visual experience fosters the modification of the functional architecture of, for example, the developing visual cortex, are specified by the following two conditions: (i) the local pattern of retinal activity which (ii) must be adequate to the requirements of more integral sensori-motor processes found in the total behavioural context. What may be of significance here is that these specific conditions are suggestive of and consistent with Piaget's (1971) notions of accommodation and assimilation during the initial period, i.e. the sensori-motor period, of human cognitive development.

For the most part, the forthcoming account of visual masking focuses on the latter, endogenous, set of problems concerned with the functional design of organs and on their solutions that presumably evolved phylogenetically through the processes of selective adaptation and ontogenetically through postnatal visually based commerce with the events and objects in the organism's ecosystem. To the extent that the former, evolutionary, phylogenetic process, as we know it, is a singular process, some of the scenarios to be used in advancing the current thesis are admittedly speculative; however, others are firmly grounded in our knowledge of visual behaviour in a naturalistic setting. As stated above, the approach taken here attempts to combine explanations of visual masking based on neural mechanisms, i.e. on *how* it is produced, and based on its purpose or functional significance, i.e. on *why* it works as it does. Since the 'why?' question is posed in the context of dynamic visual behaviour characterized by eye movements, the following section focuses on the 'how?' question of visual masking; in particular, on underlying neurophysiological mechanisms in masking and the control of eye movements.

10.2. Neural processes in visual masking and the control of eye movements

10.2.1. Visual masking mechanisms

In the previous chapters I outlined several mechanisms of visual masking: The ones relevant to the development of the present chapter are again briefly outlined below. For one, it was noted in Section 7.4.1 that transient-on-sustained inhibition is a neural mechanism producing the type B metacontrast masking effects. Psychophysical investigations of the strength of perceived stroboscopic motion as a function of the spatial frequency composition of stimuli (von Grünau, 1978*a*, *b*) and of target recovery in metacontrast (Breitmeyer 1978*a*; Breitmeyer *et al.* 1981*b*), also demonstrated the existence of the reciprocal, sustained-on-transient inhibition (discussed in Section 9.1.1). Thirdly, in Section 9.2, I introduced the psychophysical jerk effect (Breitmeyer *et al.* 1980; Valberg and Breitmeyer 1980), which very likely is a correlate of the neural periphery effect. Although the periphery effect is excitatory among transient neurones, it was shown (Fukuda and Stone 1976; Krüger 1977*b*) to be inhibitory, particularly among post-retinal sustained neurones. Psychophysically, the jerk effect can be demonstrated to be a purely inhibitory (Dortmann and Spillman 1981; Valberg and Breitmeyer 1980) one in human sustained channels and to be restricted to the foveal region of retinal space (Breitmeyer and Valberg 1979).

A fourth masking effect, briefly noted in Section 3.2.1., but not yet discussed, incorporates the role of visual persistence in forward pattern masking. According to the visual masking model outlined in Section 7.4.1, the response latency and persistence of sustained channels increase with the spatial frequency of a stimulus. It is common knowledge that the foveal region of the visual field is characterized by a high degree of spatial resolution relative to the more eccentric, extrafoveal regions. In particular, the fovea can resolve gratings of high spatial frequency that non-foveal regions fail to resolve. This fact is consistent with the aforementioned neurophysiological findings (see Section 6.4) showing that sustained neurones, characterized by small receptive-field diameters, are most highly concentrated in the fovea but are relatively sparse in extrafoveal regions; moreover, sustained-cell activity and coverage is stronger in the fovea than in extrafoveal regions (Cleland and Levick 1974; Peichl and Wässle 1979).

Consequently, foveal sustained channels are characterized by a higher spatial resolution than non-foveal ones (Cleland *et al.* 1979; Hoffmann *et al.* 1972; see Fig. 6.4). Finally, sustained neurones respond with a longer latency and slower conduction velocity in the fovea than in extrafoveal regions (Cleland and Levick 1974; Stone and Fukuda 1974). On that basis, one might expect longer visual persistence in the foveal than in extrafoveal regions of visual space. In fact, employing the phenomenal continuity method of assessing visual persistence (see Section 3.2.1), Breitmeyer and Halpern (1978) showed that visual persistence for gratings of variable spatial frequencies was generally longer with foveal than with extrafoveal stimulation.

In addition and of special relevance to the present chapter, Breitmeyer and Halpern (1978) also employed a forward masking technique to investigate visual persistence as a function of retinal location. The mask stimulus, flashed for 100 ms, consisted of one of two vertical square-wave gratings at either 11.0 or 5.5 c/deg. The grating display was approximately 9.0° × 6.0°—sufficiently large to stimulate foveal as well as extrafoveal regions of the visual field. The targets, of variable duration, consisted of vertical, black-on-white bars having dimensions of 5.5′ × 90.0′ for the 5.5 c/deg mask grating or else 2.70′ × 90.0′ for the 11.0 c/deg mask. In each case, the target bar was equal in width to, and thus optically superimposable on, a given black bar of the respective mask gratings. The target was either centred in the fovea or centred along the horizontal meridian, 3° to the left of the fovea. The target was flashed from 0 to 210 ms after the offset of the mask. Using a method of descending limits, target duration thresholds were measured in the presence of the mask and in its absence. The logarithm of the mean target threshold durations with the mask relative to the no-mask condition was taken as an index of masking magnitude. Higher positive log relative threshold changes indicated greater masking magnitude.

The rationale of this forward masking procedure was as follows: if sustained channels in the foveal location are characterized by a longer response persistence than sustained channels in extrafoveal locations, then the sustained activity generated in the fovea ought to obscure the visibility of the corresponding target, via common integration in sustained neural channels, for a greater post-offset time interval than in extrafoveal regions. Moreover, since sustained channels and their activity are most heavily concentrated at the fovea, the forward masking effect ought to be overall stronger in the foveal than the extrafoveal areas. Figure 10.1 shows typical results. Both expectations were confirmed. At each spatial frequency of the mask, the obtained masking functions in the foveal condition generally decay at a slower rate than in the extrafoveal condition. In addition, the masking effect also is stronger in the fovea than in the extrafovea. These results are consistent with the above-mentioned high concentration of sustained activity in the fovea relative to the weaker sustained activity found outside the fovea. In summary, sustained response persistence (i) decays more slowly and (ii) is stronger and lasts longer in the foveal than in the extrafoveal regions of visual space. Moreover, such extensive persistence, as measured by forward masking, is primarily an early, retinal phenomenon, since as noted previously (see Section 7.4.2, Explanations 6 and 7) they are appreciably attenuated when target and mask stimuli are flashed to opposite eyes.

10.2.2. Neural mechanisms subserving the directive control of eye movements and visual attention

Eye movements are usually employed when one wishes to change visual, foveal attention from one part of an extended, visual display to another. The eye movements used in this case are saccades which can rapidly change the locus of visual fixation. In addition, eye movements are used in the maintaining of foveal attention on moving objects or on a stationary object while an observer is moving. This is actualized via smooth pursuit and compensatory movements of the eyes. The former, saccadic eye movements can be employed in one of several ways. For instance, an observer is attending to or foveating a particular object in a more extensive visual scene. A new object is suddenly introduced in the periphery of the visual

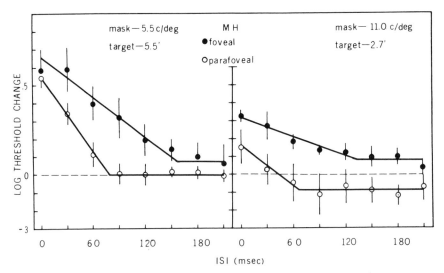

Fig. 10.1. The log threshold change as a function of the temporal interstimulus interval (ISI) separating the target bar from the preceding, 100-ms mask grating. The spatial frequency of the mask and the width of the target are as indicated. Positive values of log threshold change indicate a decrease in visibility of the target produced by the mask. (From Breitmeyer (1980).)

field or an up-to-now stationary object in the extrafoveal visual field suddenly moves. Both situations require a saccade to effect a rapid shift of foveation to the peripheral event. We shall refer to such saccades and shifts of attention as *event-triggered*. Another example of saccade utility entails visual search or inspection of a static scene, such as attempting to find a given location in a city road map or reading text. The search or inspection is guided by cognitive and extrafoveal, sensory, information (Hochberg 1978; Neisser 1976) and is performed by sequential static foveations lasting several hundred milliseconds, interrupted by directive saccades which last on the order of tens of milliseconds. We designate these saccades and associated shifts of attention as *information-guided*. On the other hand, smooth pursuit or compensatory movements are employed when one wishes to maintain foveal attention on a moving object, for example, a running ground animal or a bird in flight, by continually tracking it with foveal vision, or when one is in motion but wishes to maintain fixation of a stationary object. I shall refer to these and similar types of attentive foveations as *dynamic*. These three types of eye movements can, of course, combine in two- or three-way fashion to accommodate the attentive pick-up of information in more complex behavioural settings.

In each of these three basic hypothetical settings, some visual processing must occur prior to the shift of attention to, or the maintenance of attention on, a given object. As noted by Robinson and Goldberg (1978*a*) such processing requires analysis of the visual object in terms of at least the following three questions: where is it, what is it, and is it behaviourally significant? Along with Robinson and Goldberg (1978*a*), I shall concentrate on the following four visual areas of the brain: the superior colliculus, the visual cortex, the posterior parietal cortex, and the frontal eye fields. All of these visual areas, and several others to be mentioned more briefly, have been implicated in the directive control of eye movements and visual attention (Albano and Wurtz 1981; Butter, Weinstein, Bender, and Gross 1978; Crommelinck, Roucoux, and Meulders 1977; Goldberg and Bushnell 1981; Goldberg and Robinson 1978; Guitton 1981; Guitton, Crommelinck, and Roucoux 1980; L. R. Harris 1980; Latto 1978*a*, *b*; Luria 1959; Luria, Karpov, and Yarbuss 1966; Mohler and Wurtz 1976, 1977; Pentney and Cotter 1978; Robinson and Fuchs 1969; Roucoux and Crommelinck 1976; Roucoux, Crommelinck and Guitton 1981; Roucoux, Guitton and Crommelinck 1980; Schiller and Stryker 1972; Schiller, True, and Conway 1979; Sparks and Mays 1980, 1981; Stein, Goldberg and Clamann 1976; Straschill and Takahashi 1981; Stryker and Schiller 1975; Wurtz and Albano 1980; Wurtz and Goldberg 1971, 1972; Wurtz, Goldberg, and Robinson 1980*a*). In addition I shall discuss the brain-stem reticular formation as an area directly involved with eye movements and visual attention.

Superior colliculus and visual cortex In recent years both collicular and cortical influences on eye movements and directed attention have been established. Robinson and Goldberg (1978*a*) assign the following general functions to the superior colliculus and the visual cortex. The superior colliculus is used in detecting the presence and localizing a visual stimulus which potentially may be informative and behaviourally significant, but not in a detailed qualitative analysis of the stimulus. For the most part, the visual cortex, on the other hand, is used in the fine localization of a stimulus (e.g. Vernier acuity) and in analysis of its qualitative or figural aspects. (For the latter function one could also include the visual association areas, e.g. the circumstriate and inferotemporal areas which, like the visual cortex, also project efferents to the superior colliculus (Goldberg and Robinson 1978; Kuypers and Lawrence 1967; McIlwain 1977)). However, unlike the superior colliculus, the visual cortex, except for area 19 (Fischer and Boch 1981*a*, *b*, *c*; Fischer *et al.* 1981) seems not to respond selectively to behaviourally relevant as compared with irrelevant stimuli. This differential functional assignment to the superior colliculus and visual cortex concurs in part with the differential roles ascribed to these structures by Trevarthen (1968, 1978), Schneider (1969), and Humphrey (1974).

The superior colliculus is comprised of three general subdivisions: the superficial, the intermediate, and the deep layers. The cells in the superficial layer are visually responsive; their receptive fields are relatively large, and their response is *selectively enhanced* (i.e. if the stimulus is behaviourally relevant) from approximately 200 ms *prior to and at the onset* of a goal-directed saccade made to a stimulus falling in their receptive fields but not elsewhere (Goldberg and Wurtz 1972*b*; Robinson and Goldberg 1977; Schiller and Koerner 1971; Wurtz 1976; Wurtz and Mohler 1974, 1976*a*). This may be a neurophysiological event corresponding to the psychological phenomenon of shifts of selective attention or target selection (Goldberg and Wurtz 1972*b*) which, as found in several human psychophysical investigations (Del Pezzo and Hoffmann 1980; Kolers and von Grünau 1977; Posner 1980; Posner, Snyder, and Davidson 1980; Remington 1980; Shulman, Remington, and McLean 1979) can be dissociated from saccades.

Moreover, as noted by Posner (1980) and Posner, Nissen, and Snyder (1978), when shifts of visual attention are associated with eye movements such as saccades, the attentional shift, like the selective response enhancement of superficial collicular cells, occurs prior to the eye movements. Since the superior colliculus projects to the pulvinar nuclei (Benevento and Fallon 1975; Benevento and Rezak 1976; Lin and Kaas 1979, 1980), which in turn, project to the primary visual cortex (area 17) (Benevento and Rezak 1976; Cooper, Kennedy, Magnin, and Vital-Durand 1979) and other visual association areas (circumstriate, inferotemporal and frontal cortex (Diamond 1980; Diamond and Hall 1969; Jones 1974; Trojanowski and Jacobson 1974), an enhanced collicular response to a stimulus at a given location of the visual field could render that location more salient to the pulvinar (Keys and Robinson 1979; Perryman, Lindsley, and Lindsley 1980) and thus to its cortical recipient areas (Chalupa, Anchel, and Lindsley 1973; Chalupa, Coyle, and Lindsley 1976; Goldberg and Robinson 1978; Gross, Bender, and Roch-Miranda 1974; Rezak and Benevento 1979). Immediately *after the saccade* is executed, the same collicular cells frequently are characterized by a *response suppression and reduction of visual sensitivity* (Goldberg and Wurtz 1972a; Richmond and Wurtz 1980; Robinson and Wurtz 1976). Such suppression is adaptive since after a saccade one wants to *maintain* foveal attention in the newly inspected object rather than divert attention from it.

Cells in the deep layers of the superior colliculus are not visually responsive but, rather, respond prior to saccadic eye movements and via ascending fibres may provide the enhancing influence to the superficial layer cells (Goldberg and Robinson 1978; Mohler and Wurtz 1976; Wurtz and Mohler 1976a). In fact, on physiological and cytological grounds, Edwards, Ginsburgh, Henkel, and Stein (1979) and Edwards (in press) argue that the deeper cells ought to be classified as reticular rather than collicular. As will be seen, reticular activity generally seems to have a facilitatory effect on activity in the geniculostriate pathway. In the intermediate layers of the superior colliculus, most cells combine the properties of the superficial and deep layer cells; that is, they respond selectively prior to goal-oriented saccades and are visually responsive. Most likely they integrate information from superficial and deep collicular layers and may constitute an important source of sensori-motor efference from the superior colliculus (Wurtz and Mohler 1976a).

The superficial layers of the superior colliculus, as noted in Section 6.7, are innervated indirectly by transient corticofugal fibres from the visual cortex and directly by transient afferents from the retina. In both pathways visual receptive fields are topographically organized, thus serving as a sensory map of the retinal surface (Cynader and Berman 1972; Goldberg and Wurtz 1972a; McIlwain 1973a, b, 1975; McIlwain and Lufkin 1976; Schiller and Koerner 1971). The foveal representation of this map by and large depends on the indirect cortical afferents; extrafoveal and peripheral representations depend both on direct retinal and indirect cortical inputs (Bunt *et al.* 1975; Wilson and Toyne 1970). These separate inputs to the superior colliculus may be related to separate programming modes of saccades to stimuli within, as compared with outside, a 10–15° eccentricity (Frost and Pöppel 1976; Henson 1979). Since a large portion of these collicular inputs are comprised of short-latency, high-velocity transient fibres (Hoffman 1973; Leventhal and Hirsch 1978; McIlwain and Lufkin 1976; Palmer and Rosenquist 1974), cells in the superficial layers are particularly sensitive to onsets and offsets of stimuli and to rapid movement (Moors and Vendrick 1979a). However, in monkey, unlike cat, and presumably also in other primates including humans, relatively few of these cells are directionally selective (Cynader and Berman 1972; Goldberg and Wurtz 1972*a*; Marrocco and Li 1977; Moors and Vendrik 1979*b*; Schiller and Koerner 1971). One can argue that the transient collicular inputs comprise an 'early warning signal' which alerts the visual system to the presence and location of novel objects suddenly introduced or set into motion in the visual field. Thus, these inputs may be required to generate and control event-triggered saccades and associated shifts of foveal attention to the location of the novel object.

Does any evidence exist for this interpretation of alertness? Goldberg and Wurtz (1972*b*) and Oyster and Takahashi (1975) found that the responses of cells in the colliculus habituate to repetitive stimuli applied to their receptive fields. Similarly, Robinson, Baizer, and Dow (1980) reported such habituation in a small (20 per cent) proportion of prestriate neurones of rhesus monkey. That is to say, an initially novel stimulus is recognized as a non-novel stimulus after several repetitions of its occurrence. In related psychophysical investigations, Singer, Zihl, and Pöppel (1977) and Frome, MacLeod, Buck, and Williams (1981) reported that the detection threshold in humans for stimuli repeatedly flashed in the same location of the peripheral visual field increases over that of its initial presentation. Significantly, Singer *et al.* (1977) also showed that this habituation effect was eliminated when another stimulus was flashed in the symmetrical contralateral location of the retinal periphery. This suggests, as indicated by Sprague (1966), that the two contralateral hemicolliculi are in mutual and reciprocal antagonism. This result also indicates that the colliculus has an alerting or arousal function which, via the topographically organized ascending pulvinar–cortical pathway mentioned above, can enhance the cortical salience of stimuli.

Moreover, the response of transient neurones comprising the set of event detectors (Schiller and Koerner 1971) which are associated with event-triggered saccades and shifts of foveal attention can be suppressed by remote stimulation falling outside their conventionally spot-mapped receptive fields (Rizzolatti, Camarda, Grupp, and Pisa 1973, 1974; Wurtz, Richmond, and Judge 1980*b*). These purely visual suppressive effects of remote stimulation on collicular transient cells, as noted by Wurtz *et al.* (1980*b*), are not to be identified with the periphery or shift effects (McIlwain 1964; Fischer *et al.* 1975, Krüger *et al.* 1975; see Section 6.6) found in the transient cells of the retina and lateral geniculate nucleus, since these latter effects exert an *excitatory* rather than inhibitory influence on transient cell responses. Furthermore, the timing of the suppressive remote flash relative to the suppressed flash falling on the transient cell's receptive-field centre can vary; suppressive effects were obtained at flash asynchronies from 0 to about 100 ms. These long-range suppressive effects found in transient cells of the superior colliculus may be related to several psychophysical findings indicating that when two spatially separated events occur simultaneously or nearly simultaneously, they actively compete for allocation of attention and saccade-mediated foveation (Findlay 1980; Levy-Schoen 1969; Posner *et al.* 1980; Wilson and Singer 1981).

Besides these properties of transient event-detectors, let us introduce and discuss two additional collicular response properties which may be involved in the information-guided, rather than event-triggered, control of saccades and foveal attention. Recall that information-guided control arises in situations in which one serially scans or reads a stationary display. Here the location to which an upcoming saccade or shift of foveal attention is made is determined by information picked up during the ongoing fixation interval; hence that location signal must be coded in tonic- or sustained-type collicular neurones rather than the transient ones. I have already noted in Section 6.7 that sustained-type cells have been identified in the superior colliculus on the basis of anatomical and electrophysiological criteria (Cleland and Levick 1974, Fukuda and Stone 1974; Hoffmann 1973; Hoffmann and Sherman 1975; Marrocco 1978; Marrocco and Li 1977; Peck *et al.* 1980; Schiller and Malpeli 1977; Sparks, Mays, and Pollack 1977; Stein and Arigbede 1972; Stone and Keens 1980; Straschill and Schick 1977; Wässle and Illing 1980). These cells respond tonically to a stationary stimulus and are particularly evident in unanaesthetized, alert, as compared with anaesthetized and paralysed animal preparations (Peck *et al.* 1980). Many of these cells may be innervated by the 'sluggish' sustained retinotectal pathways identified by Cleland and Levick (1974); although a small proportion of the 'brisk' sustained cells, of which most project to the lateral geniculate nucleus, also seem to project to superior colliculus (Cleland and Levick 1974; Fukuda and Stone 1974; Wässle and Illing 1980).

A significant feature of other, tonically responding cells of the superior colliculus is that they signal a given spatial or environmental position rather than a fixed retinal position of the stimulus (Peck *et al.* 1980). That is to say, these cells perform spatiotopic rather than retinotopic coding of stimuli; their response, as anticipated by Robinson's (1975) model of saccadic control, is determined not only by retinal error, i.e. their receptive field location on the retina relative to the fovea, but also by the orbital position of the eyes. Recently, such cells, responding tonically to even a brief stimulus until a saccade is executed—and, therefore, presumably capable of holding or 'storing' information about both retinal error and orbital eye position—also have been identified by Mays and Sparks (1980*a*, 1981). As noted further by Peck *et al.* (1980), the existence of such cells may be related to goal-directed, spatially coded saccades, which recently have been shown to exist in monkey (Mays and Sparks 1980*b*) and man (Hallett and Lightstone 1976*a, b*), but which, on the basis of eye movement recordings following collicular stimulation have not previously been unequivocally established (L. R. Harris 1980; Guitton *et al.* 1980; Robinson and Jarvis 1974; Roucoux and Crommelinck 1976; Roucoux *et al.* 1980, 1981; Schiller and Stryker 1972; Stein *et al.* 1976; Straschill and Rieger 1973; Stryker and Schiller 1975).

The visual cortex, due to its smaller receptive fields and a more precise topographic map of the retina, very likely is used in the more finely tuned localizations of visual stimuli (Mohler and Wurtz 1977). Moreover, as noted above, it and the visual association areas may be used in the qualitative or figural analysis of stimuli. In that case, these cortical areas may furthermore be used in guiding eye movements and visual attention on the basis of not only *where* a stimulus is but also *what* it is. Thus, besides the superior colliculus, these areas may also be used in the second of the three visual settings described above, in which a static scene is sequentially scanned, via information-guided saccades, for a particular pattern or stimulus. Despite their use in determining what a stimulus might be, most cortical visual cells, as noted, are not sensitive to the behavioural significance of a stimulus (Robinson and Goldberg 1978a; Wurtz and Mohler 1976b). That is, although area 17 (striate) and area 18 (prestriate) cells can yield some response enhancement after a saccade is executed, this enhancement, unlike that found in the superior colliculus is non-selective in that it is produced by any saccades and not only goal-directed, behaviourally significant ones (Robinson *et al.* 1980; Wurtz 1976; Wurtz and Mohler 1976b). I shall show (in the next section) how this non-selective post-saccade enhancement may be related to the facilitatory influence of reticular activity, generated by a saccade, on excitability in the geniculostriate pathway. However, also noted previously, Fischer and Boch (1981a, b, c; Fischer *et al.* 1981) recently found that the response of cells in area 19 of prestriate visual cortex, which receives input from the colliculus–pulvinar complex (Diamond 1980; Diamond and Hall 1969; Jones 1974), were selectively enhanced *prior* to a saccade, that is, only when it was made toward a behaviourally significant stimulus. Hence, area 19 of visual cortex may also be used in directing visual attention to a peripheral stimulus prior to its saccade-mediated foveation.

Since the few primate collicular cells which are directionally selective (Goldberg and Wurtz 1972a; Schiller, Stryker, Cynader, and Berman 1974) respond primarily to high-velocity motion, the superior colliculus of primates probably is not used in controlling and directing ocular smooth pursuit movements which typically are characterized by a range of low angular velocities. The visual cortex, on the other hand, seems to be a candidate for such a purpose (Bridgeman 1972, 1973); its cells not only are motion sensitive but the vast majority of them also are directionally and velocity selective down to fractions of a degree per second (Robinson 1976)—two attributes required for accurately tracking stimuli moving at low and intermediate velocities.

Posterior parietal cortex and frontal eye fields The posterior parietal cortex very likely also is used in the control of smooth pursuit movements. Mountcastle and co-workers (Lynch 1980; Lynch, Mountcastle, Talbot, and Yin 1977; Mountcastle 1975, 1976; Mountcastle, Lynch, Georgopoulos, Sakata, and Acuna 1975; Mountcastle, Motter, and Anderson 1980; Yin and Mountcastle 1977) have identified neurones in the parietal lobe of alert, behaving monkey which may be used in directing eye movements and visual attention in one of three ways. 'Visual tracking neurones' respond during ocular pursuit of slowly moving stimuli but do not respond during steady fixation. Their responses are suppressed before and during a saccade superimposed on the smooth pursuit movement. Moreover, they are directionally selective to moving stimuli (Goldberg and Robinson 1977; Robinson and Goldberg 1978*b*; Robinson, Goldberg, and Stanton 1978). These neurones, like many of the visual cortical ones, therefore are most likely used in maintaining or monitoring foveation of a moving visual object. This function corresponds to the dynamic foveations characterizing the last of the three visual settings described previously.

Another neurone-type, 'saccade neurones', discharge phasically prior to visually guided saccades. They do not respond before saccadic eye movements made in the absence of a behaviourally significant stimulus. That is, their response to onset of a stimulus is enhanced only if that stimulus is the target of a saccade (Bushnell, Goldberg, and Robinson 1981; Goldberg and Robinson 1977; Robinson and Goldberg 1978*b*; Robinson *et al.* 1978). Moreover, their activity is suppressed post-saccadically during the new fixation interval. They, therefore, behave like superficial collicular and area 19 cortical neurones which also show selective, pre-saccadic response enhancement to a behaviourally significant stimulus and post-saccadic response suppression. In fact, parietal neurones are known to project to the intermediate and deep layers of the superior colliculus (Kuypers and Lawrence 1967) (which, as noted above, also respond prior to a saccade and may provide the enchancing influence via ascending projections to the superficial neurones). According to Goldberg and Robinson (1977) the parietal saccade neurones seem to integrate visual stimulus information from the environment with internally generated information about the behavioural significance of the stimulus. Moreover, Robinson and co-workers (Bushnell, Goldberg, and Robinson 1978; Bushnell *et al.* 1981; Goldberg and Robinson 1980; Robinson, Bushnell, and Goldberg 1980; Robinson *et al.* 1978) report that the enhancement effects found in parietal saccade neurones can be dissociated from eye movements. This neural facilitatory effect may have its perceptual correlate when an extrafoveal stimulus is attended to but is not the target of a saccade (Klein 1979; Posner 1980; Posner *et al.* 1978, 1980; Remington, 1980; Shulman *et al.* 1979; Zinchenko and Vergiles 1972). A third variety of parietal neurones, 'visual fixation neurones', discharge tonically while fixation of a stationary stimulus occurs, and, like visual tracking neurones, their responses are suppressed during a saccade. Their receptive fields, moreover, always contain the fovea much like the 'object-fixation' neurones described by Gross, Bender, and Gerstein (1979) in their study of inferotemporal neurones in alert monkeys.

The frontal eye fields, like the parietal lobe, also are used in the control of eye movements and visual attention (Crowne, Yeo, and Russell 1981; Guitton 1981; Latto 1978b; Leichnetz, Spencer, Hardy, and Astruc 1981; Luria *et al.* 1966; Robinson and Fuchs 1969; Schiller *et al.* 1979). The response of many visual neurones in the frontal eye fields—like those in the superficial layers of the superior colliculus, area 19 of visual cortex, and parietal lobe—are selectively enhanced when a goal-directed saccade is made to stimuli falling on their receptive fields but not elsewhere (Goldberg and Bushnell 1981a, b; Goldberg and Robinson 1977; Mohler, Goldberg, and Wurtz 1973; Wurtz and Mohler 1976b; Wurtz *et al.* 1980a). However, in neurones of the frontal eye field the enhancement occurs after (Bizzi 1968; Bizzi and Schiller 1970) as well as before saccades, whereas in parietal saccade neurones and neurones in the superficial layers of the superior colliculus it occurs about 200 ms before and maximally at the time of saccade onset (Wurtz and Mohler 1976a; Robinson *et al.* 1978). As will be seen later, this difference may be of behavioural significance, since the frontal eye fields project not only to the deep and intermediate layers (Kuypers and Lawrence 1967; Leichnetz *et al.* 1981) but also to the superficial layers (stratus opticum) (Künzle, Akert, and Wurtz 1976; Leichnetz *et al.* 1981) of the superior colliculus. Robinson and Wurtz (1976) and Robinson and Goldberg (1978a) have conjectured that this efferent projection may be responsible for the *suppression* of the response of visual cells in the superficial layers of the superior colliculus *following* saccadic eye movements; although more recent evidence (see Richmond and Wurtz 1980) does not support this hypothesis. Unpublished observations by Richmond, Wurtz and Mishkin (cited in Richmond and Wurtz 1980) revealed that this post-saccadic suppression of collicular neurones occurred even after ablation of the frontal eye fields. Hence the suppression must have its sources elsewhere.

In this connection, Tsumoto and Suzuki (1976) reported that electrical stimulation of the frontal eye fields has a *facilitatory* effect, independent of activation of the midbrain reticular formation (see below), on *sustained* neurones in the dorsal lateral geniculate nucleus and visual cortex. The latency of this effect is on the order of 50–100 ms after stimulation of the frontal eye fields. In whatever manner they are produced, these opposite effects on transient cells of the upper layers of the superior colliculus and sustained cells of the dorsal lateral geniculate nucleus could be related to the finding that activity of many of the former, collicular cells are temporally out of phase with that of the latter geniculate cells (Molotchnikoff *et al.* 1977). That is to say, when the former cells are actively responding the latter are not and vice versa.

The following scheme is proposed conjecturally to account for these relationships. The post-saccadic enhancement of frontal eye field neurones facilitates the response of sustained cells in the dorsal lateral geniculate nucleus and visual cortex so that they are prepared to process newly arriving pattern information contained in the post-saccadic fixation interval. Simultaneously, activity of transient event-detectors of the superior colliculus (and parietal saccade neurones), whose response was enhanced selectively prior to and up to the execution of the saccade, are suppressed immediately after the saccade for about 100 ms so that attention is not diverted from the currently fixated pattern information (see the second paragraph of Section 10.2.1). Only after this pattern information has been processed sufficiently by the foveal sustained channels during this 100-ms interval is attention again able to be switched to another peripheral locus. The outstanding problem of how the post-saccadic suppression of collicular cells used in event-triggered redirection of foveal attention occurs still must be addressed. I noted that Robinson and Wurtz (1976) and Robinson and Goldberg (1978a) conjectured—as it turned out, incorrectly—that the post-saccadic enhancement in frontal eye field neurones acts negatively via descending fibres onto the upper layer collicular cells which were selectively enhanced prior to a saccade or shift of foveal attention. In other words, the post-saccadic suppression was believed to occur by inhibiting the terminal superficial layer sites of the pre-saccadic enhancement effect which, as noted above (p. 313) is thought to *originate* in the pre-saccadic activity of visually non-responsive cells of the deeper layers of the superior colliculus (Goldberg and Robinson 1978; Mohler and Wurtz 1976; Wurtz and Mohler 1976a). These deep layer cells either are themselves reticular (Edwards, in press) or, in turn, receive input from among a variety of subcortical, midbrain–reticular, and cerebellar areas (Edwards *et al.* 1979; Roldán and Reinoso-Suárez 1981).

The cerebellum is another brain structure used in the generation and control of eye movements (Cohen, Goto, Shanzer, and Weis 1965; Ritchie 1976; Ron and Robinson 1973). In this regard, Kase, Miller, and Noda (1980) recently found that although responses of some Purkinje cells were enhanced in synchrony with saccade onset, others were enhanced in synchrony with saccade offset (for related findings, see also Hepp, Henn, Jaeger, and Waespe 1981). In particular, the latter set of Purkinje cells showed response enhancement which began about 40 ms prior to the end of a saccade and persisted up to about 70 ms after the saccade was completed. The total enhancement duration consequently was approximately 110 ms. According to Kase *et al.* (1980), the fact that these Purkinje cells also exhibited a high (non-enhanced) discharge rate during fixations implies that the cerebellum can exert tonic inhibition on eye movement centres and thus also assures a steady orbital eye position during inter-saccadic fixation periods. Moreover, the burst of response enhancement synchronized with the end of a saccade or, alternatively, the beginning of a new fixation may therefore play a significant role in the initiation of steady fixation. Now, let us in particular assume that cerebellar signals corollary to these bursts of Purkinje cell activity are projected to the deeper layers of the superior colliculus (Edwards *et al.* 1979), where they inhibit the visually non-responsive eye movement cells (providing pre-saccadic enhancement to upper-layer cells). This inhibition would, in turn, result in a suppression or attenuation of the post-saccadic response of the upper-layer cells, which otherwise would signal or effect a shift of attention away from the currently fixated visual display area.

Once the transient suppression terminates after a *minimal* interval of about 70–100 ms into the fixation interval, the deep-layer collicular cells *can* again be activated and via their enhancing effect on upper-layer cells can effect a pre-saccadic shift of attention to another location signalled by a new event. Note, by the way, that this minimal interval of about 70–100 ms limiting the rate of shifts of visual attention from one event to another corresponds also to the interval of purely visual inhibitory interaction between a remote stimulus and a stimulus falling on the receptive field centre of a transient event-detector in the upper layers of the superior colliculus (Rizzolatti *et al.* 1973, 1974; Wilson and Singer 1981; Wurtz *et al.* 1980*b*). Another possibility is that the corollary discharge, which post-saccadically inhibits the deep-layer cells of the superior colliculus, could also arise from reticular areas of the brain stem (Edwards *et al.* 1979) which, as will be seen below, also facilitate the response of sustained neurones in the lateral geniculate and visual cortex. In either case, the net result is that eye-movement and attention-shifting responses in subcortical oculomotor centres (as well as parietal cortex) are out of phase with eye-fixation, and attention-maintaining responses. In relation to neural process in the retinogeniculocortical pathway, each of these two responses, in turn, is respectively in phase with the pattern-inhibiting effect of transient channels activity during saccades and the pattern processing of sustained channels during the post-saccadic fixation interval.

The brain stem and midbrain reticular formation Electrical stimulation of the midbrain reticular formation is known to produce saccadic eye movements (Büttner, Büttner-Ennever, and Henn 1977; Keller 1974; Peterson, in press; Singer and Bedworth 1974). Moreover, it is also known that a corollary reticular activation occurring with eye movements modulates cortical and lateral geniculate excitability (Baker, Sanseverino, Lamarre, and Poggio 1969; Cohen, Feldman, and Diamond 1969; McIlwain 1972; Ogawa 1963; Pecci-Saacedra, Wilson and Doty 1966; Singer 1973*a*, *b*, 1977; Singer and Bedworth 1974; Singer, Tretter, and Cynader 1976; Tatton and Crapper 1972). According to Singer and co-workers' findings (Singer 1977; Singer and Bedworth 1974; Singer *et al.* 1976) the excitability of lateral geniculate relay cells and cortical neurones is generally enhanced by electrical stimulation of the reticular formation.

This enhancement of excitability in the geniculocortical pathway is thought to occur in one of several ways. Singer (1977) suggests that inhibition of the relay cells in the lateral geniculate nucleus can occur due to the activity of intrinsic interneurones or else due to an extrinsic source of activity generated in the adjacent perigeniculate nucleus which comprises part of the so-called 'non-specific' thalamic reticular nucleus. It is believed that whereas the intrinsic inhibitory loops are highly local in nature, the extrinsic ones are used in more global modifications of lateral geniculate excitability (Dubin and Cleland 1977) and are probably related to changes in an organism's state of alertness or to orienting behaviour associated with eye movements (Singer 1977). Both types of loops are known to be inhibited, in turn, by stimulation of the midbrain reticular formation. In the former case of intrinsic inhibitory loops, such stimulation also is known to have a particularly strong facilitatory effect on pyramidal cells of layer 6 of the striate cortex which, in turn, project corticofugally in a retinotopic manner to the lateral geniculate nucleus. This corticofugal activity could produce local disinhibition of geniculate relay cells.

However, besides this indirect pathway, the disinhibition could additionally be due to direct projections from the midbrain reticular formation to the lateral geniculate nucleus (Singer 1977). Stimulation of the midbrain reticular formation also is known to directly inhibit the neurones of the thalamic reticular nucleus, thus also increasing the excitability of lateral geniculate relay cells by eliminating or attenuating the extrinsic sources of global inhibition. Frizzi (1979) recently reported psychophysical findings in rhesus monkeys which demonstrated that stimulation of the midbrain reticular formation lowers the brightness detection threshold. This may be a perceptual correlate of the enhanced lateral geniculate and cortical excitability produced by reticular stimulation. Moreover, since neurones in cerebellum (Hepp, Henn, and Jaeger 1982; Suzuki, Noda, and Kase 1981) and areas of the brain stem associated with reticular activity are used in the generation not only of saccades but also of smooth pursuit movements (Eckmiller 1981; Eckmiller *et al.* 1980; Eckmiller and Mackeben, 1978, 1980; Keller 1981; Robinson 1976; Sparks and Sides 1974) such facilitation of geniculocortical neurones could also be used during ocular pursuit of a moving stimulus.

The latency of this facilitation effect is about 60–100 ms, which, incidentally, also is the latency of enhancement effects in the sustained neurones of the lateral geniculate nucleus produced by stimulation of frontal eye fields independent of reticular activation (Tsumoto and Suzuki 1976). When saccadic eye movements are used this corollary effect would, due to its latency, be maximal at and after the termination of a saccade and, as suggested by Singer and co-workers (Singer 1977; Singer and Bedworth 1974; Singer *et al.* 1976), would reset excitability of lateral geniculate and cortical sustained neurones at the beginning of each new fixation interval by counteracting the inhibition of these neurones at the beginning of and during a prior saccade. The latter inhibition may be a neural correlate of saccadic suppression which seems to have its peripheral, afferent as well as central, efferent components (Bartlett, Doty, Lee and Sakakura 1976; Breitmeyer and Valberg 1979; Breitmeyer *et al.* 1980; Brooks and Fuchs 1975; Brooks and Holden 1973; Duffy and Burchfield 1975; Jeannerod and Chouvet 1973; Judge *et al.* 1980; MacKay 1970*a, b*; Mateeff *et al.* 1976; Matin 1974*a, b*; Mitrani *et al.* 1971; Noda 1975*a, b*; Noda and Adey 1974; Riggs, Merton, and Morton 1974; Valberg and Breitmeyer 1980).

This review of some of the major central sites and mechanisms used in controlling eye movements and directing visual attention ends here. It was shown how each of these sites may be used in one or the other of the three visual settings (event-triggered saccades, information-guided saccades, and dynamic foveations) described in the introductory paragraph of the current section. The question of how the oculomotor and attentional control functions of these areas are coupled to the visual sensory and masking effects produced in those visual settings remains to be answered. To address this question, we shall attempt to show for each setting how the direction of eye movements and attention are related to the responses and interactions of sustained and transient channels in human vision. This will provide the basis for exploring the 'why?' or the functional significance and purpose of visual masking in naturalistic settings.

10.3. Visual masking and the control of eye movements and visual attention

The human brain, in its ability to process and integrate sensory information and to organize and execute response, is the most complex among primates. The evolution of the human visual system, as well as that of higher primates, can be characterized by the following main features:

1. Frontal eyes with extensive binocular overlap of two monocular visual fields contributing to stereoscopic depth perception.

2. An area of high visual acuity, the fovea, present in each eye.

3. Co-ordinated eye movements that are disjunctive in the case of vergence and conjunctive in the case of pursuit, compensatory, and saccadic movements.

4. An expansive development of occipital lobes (the primary visual areas) and temporal–parietal lobes (the association areas) contributing to visual differentiation (acuity) and abstraction (pattern recognition).

These four characteristics most likely evolved under the same set of selection pressures that guided the evolution of peripheral and cortical structures capable of differentiating and recognizing a large variety of visual objects in the environment. Physically most environmental objects and, in fact, the environment as a whole are three-dimensional structures. In evolutionary terms, the development of binocular visual systems can be considered as a solution to the problem of locating, differentiating, and recognizing objects in visual space.

An extensive binocular stereoscopic space may have been selected, at least among primates, as a trait for locomotion in an arboreal habitat or for visually directed predation (Raczkowski and Cartmill 1975). Among non-arboreal vertebrates such as the feline (Blake and Hirsch 1975) or among birds like the hawk or falcon (Fox, Lehmkuhle, and Bush 1977) stereopsis may have been selected solely for visually directed predation. For these animals as well as primates the evolution of stereopsis generally has been accompanied by the coevolution of an area centralis or a fovea permitting high-acuity vision in a restricted area of each monocular visual field (Fox, Lehmkuhle, and Westendorf 1976; Walls 1942). In either case, reasonable arguments can be made for the behavioural benefits of the development of an extensive stereoscopic field that allows an organism to identify an object and specify its location in the three-dimensional environment.

In primates, the anatomical supports of high foveal resolution are characterized by a high cone receptor density and small receptive fields in the two foveae as well as cortical magnifications of the foveal projections of both monocular fields (Cowey and Rolls 1974; Daniel and Whitteridge 1961; Hubel and Wiesel 1974*b*; Myerson 1977; Rolls and Cowey 1970; Rovamo and Virsu 1979; Talbot and Marshall 1941). The importance of central, binocular vision is attested to by the fact that strabismus or squint present or induced during cortical, neonatal periods of development produces not only loss of binocularity and binocular single vision (Hohmann and Creutzfeldt 1975; Hubel and Wiesel 1965; Maffei and Bisti 1976; Yinon 1976) but also a loss of visual acuity among sustained neural pathways representing the foveal and parafoveal region of the squint eye (Duke-Elder and Wybar 1972; Ikeda and Tremain 1979; Ikeda, Tremain, and Einon 1978; Ikeda and Wright 1976).

Berkley (1976) recently proposed an interesting hypothesis for the co-evolution of binocular single vision and high, foveal spatial resolution:

If it is assumed necessary for many animals, both predatory and nonpredatory, to have acute depth perception, then these animals should have . . . accurate eye alignment. This would permit the use of binocular parallax to detect small differences in depth (stereopsis). To achieve the accurate alignment, the regions of acute vision (fovea) may act as a pair of crosshairs for the two eyes, permitting them to be aligned with great accuracy so that small differences in retinal image position produced by targets either in front or behind plane of fixation can provide cues for depth. If this assumption is made, many animals with different lifestyles that require good depth perception also turn out to have quite outstanding visual acuity (Berkley 1976; p. 83).

Besides this possible role of precisely aligning the two monocular fields, the foveae, by virtue of their high spatial resolution, also permit a detailed scrutiny of visual patterns and objects in the environment.

Despite this dual advantage of foveal vision, according to Steinman (1975) and Robinson (1976), the existence of the foveae poses the paramount complicating problem to the oculomotor system. As the anatomical and, usually, functional locus of visual attention, the foveal region of the visual field is several orders of magnitude smaller than the entire visual field. Consequently, problems can arise in several ways. For example, they arise whenever widely separated areas of visual space must be attended to or inspected. This problem is solved by conjugate, ballistic saccades which permit a rapid change of foveation from one area of the visual field to another. Other problems occur whenever an object requiring foveal scrutiny moves over a distance larger than the foveal region or whenever an object (stationary or moving) requires foveation while the observer is in motion. The former problem is solved by conjugate, smooth ocular pursuit movements; the latter, by conjugate smooth compensatory eye movements. Both types of movements allow continued foveation of objects in a continually changing visual scene.

10.3.1. Saccades and their functional significance

Direction of visual attention   As noted above, saccades can be activated in several visual settings requiring changes of visual attention. At the risk of being repetitious, let us list two typical settings. As a first example, assume an organism is attending to a stationary object when suddenly a moving object enters its peripheral visual field or else an up-to-now stationary but extrafoveal object abruptly begins to move. Attention to either of these objects is accompanied by appropriate event-triggered saccades. As a second example, suppose that the organism is visually scanning a stationary scene, perhaps searching for a particular object in the scene. The scan is characterized by a series of fixations mediated by information-guided saccades and directed to different parts of the visual scene.

Although both settings use saccades, they are qualitatively different, and the direction of visual attention and saccades to extrafoveal stimuli is most likely and correspondingly under the control of different brain areas. Nevertheless, in either setting, a shift of attention to the extrafoveal location is known to enhance perceptual sensitivity there (Bashinski and Bacharach 1980; Crovitz and Daves 1962; Parasuraman 1979). In the former setting, while the organism is attending to a given stationary object, one can assume that visual gaze is in part monitored by parietal visual fixation neurones and pause neurones in the brain-stem areas (Keller

1981). When a behaviourally relevant object suddenly moves or appears in the periphery of the visual field, the activity of parietal fixation neurones is suppressed whereas that of saccade neurones is increased. Moreover, the abruptly appearing or moving peripheral stimulus activates transient cells in the superficial layers of the superior colliculus. The activity of these collicular cells (as well as parietal saccade neurones) is enhanced prior to the saccade directed to the peripheral stimulus. This collicular enhancement, besides amplifying the long-range inhibitory effect which the peripheral event may exert on foveal allocation of attention (Rizzolatti *et al.* 1973, 1974), in turn, is projected to circumstriate and inferotemporal cortices by way of the pulvinar complex, and there enhances the attentional saliency of pattern information falling in the region of the visual field containing the peripheral stimulus.

Since at higher levels of visual processing transient neurones have a response latency shorter by at least 50–100 ms than sustained neurones (Dow 1974), retinal transient activity projecting (directly or indirectly via visual cortex) to the superior colliculus could in this manner serve as an early-warning signal or as an alerting mechanism for switching visual attention from the fovea to the extrafoveal location signalled by the peripheral event (Ikeda and Wright 1972*a*; Wässle *et al.* 1981*a*). That is to say, the visual system is allowed an interval of at least 50–100ms to redirect its functional locus of attention to the extrafoveal location of visual space so that a preliminary pattern analysis of the newly introduced stimulus can be made by the affected peripheral sustained channels before the anatomical locus of attention, i.e. the fovea, with its high-resolution sustained channels, is directed to it for more detailed scrutiny. Psychophysical estimates of this attention-switching time interval in fact range between 50 and 100 ms as shown by LaBerge (1973), Posner (1980), and Posner and Cohen (1980). Thus, the shift of the functional locus of attention to the extrafoveal location is completed at about the time that the visual cortex begins to process in preliminary, coarse fashion the pattern information carried in the extrafoveal, cortical sustained channels. This process is followed by a saccade, or switch of the anatomical locus of attention allowing foveal sustained channels to do the fine-grain analysis of the figural aspects of the novel stimulus.

In the latter viewing setting, the direction of visual attention and saccades does not use transient channels as an early-warning device, since a static scene is being viewed. It most likely uses activation of the sustained or tonically responding cells in the superior colliculus (Mays and Sparks 1980*a*; Peck *et al.* 1980; see p.315) as well as the possible modulation of activity imposed on the pulvinar–colliculus areas by corticofugal fibres from the temporal–occipital or frontal cortex (Goldberg and Robinson 1978; Kuypers and Lawrence 1967; Leichnetz *et al.* 1981; McIlwain 1977). Consequently, the cortical areas may be used in the direction of visual attention and saccades on the basis of a coarse, preliminary extrafoveal pattern analysis, which, in conjunction with the sustained or tonically coded location information in the pulvinar/colliculus complex, inform the visual system about the extrafoveal locations potentially containing the most behaviourally significant pattern information. Once a particular one of these locations is determined, a saccade is made to that location allowing foveal sustained channels to optimally analyse the pattern information.

Saccadic suppression and enhancement   In both of the above settings, we have discussed the role of saccades merely as a means of changing visual fixations or attention abruptly and rapidly over varying spatial extents, and thus effecting a serial foveal scan of a visual scene. Let us follow up the consequences of assigning only the attention and fixation switching role of saccades. Fig. 10.2 schematizes what would occur in a fixation–saccade sequence. Assume that we are dealing, as shown in the top panel, with three 250-ms fixations separated by two 25-ms saccades. Recall that the fovea, due to its high concentration and activity of sustained channels, is specialized for analysing high spatial frequency information. Consequently, at early, e.g. retinal levels of processing visual persistence should be

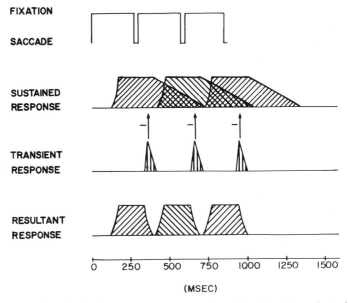

Fig. 10.2.   A hypothetical response sequence of sustained and transient channels during three 250-ms saccades (panel 1). Panel 2 illustrates response persistence of sustained channels acting as a forward mask from preceding to succeeding fixation intervals. Panel 3 shows the activation of transient channels shortly after each saccade which exert inhibition (arrows with minus sign) on the trailing, persisting sustained activity generated in prior fixation intervals. Panel 4 shows the resultant sustained channel response after the effects of transient-on-sustained inhibition have been taken into account. (From Breitmeyer (1980).)

particularly long and strong in the fovea (Breitmeyer and Halpern 1978; see Fig. 10.1). As a result, as shown in the second panel, the following situation would arise: the sustained activity generated in a preceding fixation interval would persist into the following one and interfere there with processing via forward masking by integration of uncorrelated retinotopic pattern information in the same foveal sustained channels.

Fig. 10.3, adapted from Hochberg (1978), shows what would most likely occur under these circumstances when, for example, reading a line of print requiring one, two, or three fixations. For tasks such as this and similar ones which require a sequential foveal scan of the visual scene, it is evident that visual persistence creates a problem. The problem is that at early, retinotopic levels of visual processing sustained activity generated in a prior fixation interval would mask similar activity generated in the immediately later one. Possible solutions to this problem could be to increase the duration of fixation intervals (so that the duration of the forward masking effects is short relative to that of the fixations) or to increase the duration of saccades. However, the gain produced by either of these two solutions is accompanied by a loss; namely, a significant decrease in the rate of visual information processing would ensue. Nature opted for a more efficient solution legible in the bottom line of Fig. 10.3: *normal* (i.e., sequential) *vision is iconoclastic.* By that is meant that normal, sequential vision proceeding from one fixation to another suppresses visual pattern persistence in peripheral, retinotopically organized sustained channels so that a cortical visible icon of *that* pattern information cannot be formed. However, the pattern information already residing at the cortical iconic level is not suppressed. The reasons for this will become apparent in the next section.

The suppression is performed in part by the sensory consequences of saccades occurring at the later stages of the retinogeniculostriate pathway. Saccades produce transient, rapid image motion over the retina. Consequently, they not only serve to change visual attention and fixation but also to activate short-latency transient channels, as shown in the third panel of

NORMAL VI**N**O**B**NA**L**SVI**N**O**BBAL**A**V**Î**co**a**o**c**a**s**Î**c**ONOCLASTIC   (THREE FIXATIONS)

NORMAL VISION **N**O**R**h**A**O**NU**C**BA**o**H**ICS ICONOCLASTIC   (TWO FIXATIONS)

NORMAL VISION IS ICONOCLASTIC   (ONE FIXATION)

Fig. 10.3. The perceptual masking effects of temporal integration of persisting sustained activity from preceding fixation intervals with sustained activty generated in succeeding ones when the reading of a printed sentence requires one, two, or three fixations. Here, as in panel 2 of Fig. 10.2, the effect of transient-on-sustained inhibition produced by saccades is not taken into account. (From Breitmeyer (1980).)

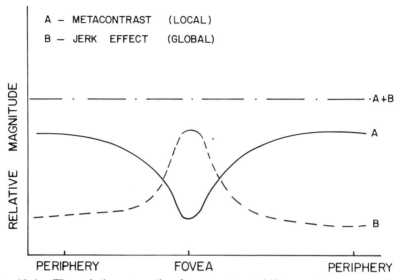

Fig. 10.4. The relative strength of metacontrast (A) as a short-range, local mechanism of saccadic suppression and the jerk effect (B) as a long-range, global mechanism of saccadic suppression as a function of retinal locus. (From Breitmeyer (1980).)

Fig. 10.2, which at lateral geniculate or cortical levels (see Section 6.7), inhibit the sustained activity that would persist from a preceding to a succeeding fixation interval (Breitmeyer and Ganz 1976; Matin 1974*b*; Singer 1977; Singer and Bedworth 1973). That is to say, transient-on-sustained inhibition provides an afferent mechanism of saccadic suppression which depends on a sudden displacement of contours across the retina.*

Let us proceed to another complication which uses the function of metacontrast or transient-on-sustained inhibition as a mechanism of saccadic suppression. At early stages of processing visual persistence is particularly strong and long in foveal viewing; and, consequently saccadic suppression ought to be optimally potent there. However, several psychophysical experiments clearly show that metacontrast typically is weaker (Alpern 1953; Bridgeman and Leff 1979; Lyon *et al.* 1981; Saunders 1977; Stewart and Purcell 1970) and of shorter spatial range (Kolers and Rosner 1960) in the fovea than in the parafovea or periphery

---

* Another form of saccadic suppression may entail what Campbell and Wurtz (1978) have called 'greyout elimination'. Campbell and Wurtz (1978) showed that greyout or blur occurred when the laboratory room was illuminated only during a saccade. However, if the room was illuminated just prior to and after the saccade, the greyout was eliminated (see also Matin *et al.* 1972). Neurophysiological correlates of this greyout elimination have recently been investigated in monkey striate cortex by Judge *et al.* (1980) and also seem to have a correlate in the cortical visually evoked potential of humans (Jeffreys 1979).

(see Section 7.4.2, Explanations 10 and 11). Hence, the complication for visual processing can be stated as follows: metacontrast, as a short-range, afferent mechanism of saccadic suppression, is too weak in the fovea.

The solution of this complicating factor may be related to the jerk effect discussed in Section 9.2. Recall that this effect produces local, foveal inhibition of sustained channels when global, peripheral excitation of transient channels occurs. It, therefore, is an additional, global, and long-range afferent mechanism of inhibiting local foveal, sustained channels. As illustrated in Fig. 10.4, the global, long-range jerk effect amplifies and compensates for a rather weak, short-range foveal metacontrast mechanism (for related views see also Steinman 1975, and MacLeod 1978).

These two afferent suppressive effects seem to be time-locked to central (efferent) suppressive mechanisms. For example, Duffy and Burchfiel (1975) found that the spontaneous activity of single neurones in striate cortex of monkey was suppressed during saccades generated in total darkness, that is, under conditions that cannot produce retinal contour displacements necessary to activate afferent mechanisms of saccadic suppression. Similar suppression has been reported in geniculostriate fibres by Bartlett *et al.* (1976). Presumably a central, corollary discharge is responsible for these inhibitory effects. These efferent suppressive effects began about 30 ms after the onset of saccadic eye movements. It should be noted that this latency corresponds approximately to the latency of transient geniculate and cortical responses after stimulation of the retina (Dow 1974; Noda 1975*a*, *b*) and to afferent suppressive effects produced by retinal contour shifts (Adey and Noda 1973). Therefore the central, efferent suppressive effects coincide temporally with those produced by afferent mechanisms.

The afferent suppressive mechanisms consequently are a part of a more complex system which, between fixations, clears the activity of peripheral sustained channels activated in a prior fixation interval that otherwise would constitute a source of intrusive noise in the cortical, iconic representation of pattern information from succeeding fixation intervals (see the lowest panel of Fig. 10.2). However, like excitation (Cleland *et al.* 1973; see Fig. 6.5), inhibition in sustained channels also persists for a relatively long duration (Winters and Hamasaki 1976; see also the results of the single–transient masking paradigm; Section 4.4.4). In fact, as noted by Creutzfeldt (1972), the time course of inhibitory neural activity generally is longer than that of neural excitation. Hence, although the transient-on-sustained inhibition produced by saccades would clear the visual system of unwanted sustained activity generated in a preceding fixation interval, we are left with the undesirable effects of a prolonged sustained channel inhibition which would extend for an appreciable duration into the succeeding fixation interval (Singer and Bedworth 1974).

Ideally the system ought to be designed in order to have strong inhibition of sustained channels just prior to and during a saccade, with a return to sustained excitation, rather than prolonged inhibition, near the beginning of the following fixation interval. In fact, this requirement prevails in the visual system. Recall that saccadic activation of both the frontal eye fields (Tsumoto and Suzuki 1976) and the midbrain reticular formation (Singer 1977, 1979; Singer and Bedworth 1974; Singer et al. 1976) facilitates or enhances the excitability of sustained channels in the lateral geniculate nucleus and striate cortex (Hunsperger and Roman 1976). The 50–70-ms latency of these facilitatory effects is such that they would be maximal at or shortly after the end of a saccade. Thus, these corollary discharges generated in the frontal eye fields and the midbrain reticular formation by saccades have the effect of counteracting, at or near the beginning of a succeeding fixation interval, the prior saccade-generated, transient-on-sustained inhibition (Jung 1972; Singer 1977; Singer and Bedworth 1974).

The account given so far of the functional significance of saccades may seem a bit complicated. Therefore, let us have an overview with the aid of some illustrative assistance. In the following three figures I shall illustrate some of the main peripheral sensory, and cortical corollary consequences of saccading to a newly attended stimulus. Recall that in a normal visual environment filled with objects, edges, and so on, saccades not only abruptly change the visual locus of fixation but also produce a sudden, transient, high-velocity displacement of contours across the retina. Let us look again at the effects of the latter contour shifts on the excitability of transient and sustained channels in the geniculostriate pathway. Noda (1975a, b) has shown that, on the one hand, a saccadic image displacement activates lateral geniculate transient neurones approximately 30 ms after its onset and for a duration of about 150 ms. On the other hand, the spontaneous activity of sustained cells is simultaneously suppressed after initiation of a saccade, and, moreover, the timecourse of this suppression is coupled to the duration of transient neural discharge. Similar results in cat and monkey striate cortex units have been reported by Noda et al. (1972), Kimura et al. (1980), and Wurtz (1976).

The reported effects of eye movements on excitability of sustained and transient channels in the retinogeniculostriate pathway have not been uniformly corroborative. For instance, several investigators (Kimura et al. 1980; Noda 1975a, b; Noda and Adey 1974b; Noda et al. 1972) report that, whereas saccades excite transient cells, they simultaneously suppress sustained cells. This is very likely due to interchannel transient-on-sustained inhibition (Noda 1975a) and is one form of saccadic suppression of pattern information. On the other hand, Bartlett et al. (1976), Noda (1975a), and Noda and Adey (1974a) report that transient cells or fibres of the geniculostriate pathway also are suppressed after a saccade is executed. However, this suppression of transient cells may have a different basis than the suppression of sustained neurones by concurrent transient-neurone activity. As noted by Noda (1975*a*), the response depression of lateral geniculate transient cells to optic tract stimulation after saccade onset paradoxically occurs during the time interval of their highest spike activity generated by the saccade onset; accordingly, the response depression of transient cells to optic tract stimulation may be due to a phasic occlusion of the optic tract impulse by coincident high-frequency spike discharges of transient cells. In other words, the 'signal' carried in the activity generated by the optic tract pulse is weak relative to the 'noise' produced by the high-frequency discharge. Hence, one may have two types of 'saccadic suppression' in the geniculostriate pathway. Type I is produced by transient-on-sustained inhibition resulting in a decreased 'signal' strength in sustained pattern-analysing channels. The other, Type II, is produced by increasing the saccade-generated internal noise level of transient channels, which effectively reduces the signal-to-noise ratio of a superimposed transient response to a brief flash occurring near or during the saccade. Since increment thresholds to brief (< 50 ms) and longer (> 50 ms) test flashes are determined primarily by transient and sustained channels, respectively (Dortmann and Spillmann 1981; Kulikowski 1977; see Section 9.2), one would expect saccadic suppression to be produced primarily by the Type I mechanism at longer test flash durations and by the Type II mechanism at briefer test flash durations.

Another source of saccadic suppression of transient responses to brief flashes may be due to central corollary activity manifesting itself as an inhibition of a significant number of transient cells (e.g., event detectors) in the superficial layers of the superior colliculus shortly after onset of a saccade (Goldberg and Wurtz 1972*a*; Richmond and Wurtz 1980; Robinson and Wurtz 1976; Wurtz 1976). Moreover, the brisk response of these transient cells to sudden stimulus onset or rapid image displacement under steady fixation is suppressed for a period corresponding to the corollary inhibition after saccade-induced image displacement (Robinson and Wurtz 1976; see especially Fig. 12). Presumably this corollary inhibition allows these transient cells to distinguish afferent or stimulus-induced events or displacements of retinal images from those produced reafferently by saccadic image displacement; and thus it may contribute to a maintenance of the stability of the visual environment during saccades (White, Post, and Leibowitz 1980) and a suppression of their attention–diversion capacity at the beginning of a new fixation interval. In any case, the corollary inhibition of such event and displacement detectors in the superior colliculus, in so far as that structure is used in detecting brief light flashes, should produce a third type of saccadic suppression. This Type III suppression is produced by a reduction or suppression of the signal strength in collicular transient cells, thus effecting a reduced signal-to-noise ratio, which in transient cells of the geniculostriate pathway is produced by an increase in internal noise (Noda 1975a). For the other transient cells in the superficial layers of the superior colliculus which are not suppressed, but rather are activated during saccades, the Type II mechanism of saccadic suppression described above would presumably be additionally used. Future psychophysical work ought to be able to differentiate between these three types of mechanisms by, among other parameters, varying the duration, size, or spatial frequency and retinal location of a test flash in order to tap selectively transient or sustained activity.

Fig. 10.5, taken from Adey and Noda (1973) shows the effects on excitability of lateral geniculate and cortical sustained neurones to electrical stimulations of the optic chiasm when a rapid retinal contour displacement of 200°/s occurs. Here, the eyes of the animal were in fact stationary and contour displacement was optically produced. In other words, we have a *simulation* of rapid image displacements which accompany real saccades without, however, the concomitant involvement of the oculomotor system and its central, corollary activity. Therefore, purely peripheral sensory consequences of rapid image displacement are involved. Note that at both levels of the visual system, and, in particular, at the striate cortex, there is a pronounced decrease in the response amplitude of neurones. The latency of this suppression is about 30 ms and its duration is approximately 150–200 ms. We can interpret this as an afferent neural correlate of saccadic suppression produced by saccade-like image displacements. During an actual saccade this suppression may also

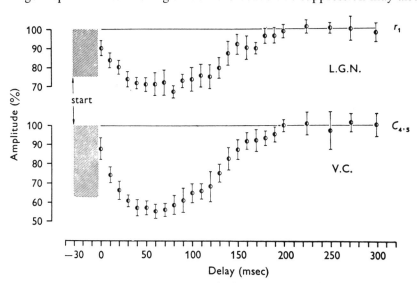

Fig. 10.5. The time course of the change in the response amplitude of neurones in the lateral geniculate nucleus (LGN) and the visual cortex (VC) produced by a rapid displacement of a grating. (From Adey and Noda (1973).)

be supplemented or reinforced by central, corollary discharges of about the same latency (Bartlett *et al.* 1976; Duffy and Burchfiel 1975; Kimura *et al.* 1980; Riggs, Merton, and Morton 1974). At any rate, since a saccade lasts approximately 50 ms, it can be seen that the saccadic suppression produced by peripheral consequences of retinal contour displacements and the temporally locked central corollary effects would by far outlast the saccade and extend appreciably into the succeeding fixation interval. It was noted above that this could pose a problem for efficient visual information processing. Optimally, one would want suppression for the interval corresponding to the duration of a saccade but not extending beyond it.

Figure 10.6 illustrates that when a saccade is actually executed the temporal extension of the suppressive effect is, in fact, not only curtailed, but, moreover, it also is replaced at the end of the saccade by a facilitatory effect. This facilitatory effect is produced by central corollary discharges accompanying activation of the oculomotor centres controlling saccades. To clarify this claim, let us look at results on cortical excitability when a saccade is made either in complete darkness (filled circles) or in the presence of a uniform, contourless Ganzfeld (open circles). In both cases, any changes in excitability cannot be due to the activation of afferent sensory mechanisms sensitive to contour displacements but, rather, must be due to central influences. Note that in both conditions cortical excitability increases. This facilitation, as mentioned above, could be produced by ascending reticulocolliculopulvinar pathways (Edwards, in press; Edwards *et al.* 1979; Perryman *et al.* 1980; Singer 1977; Singer and

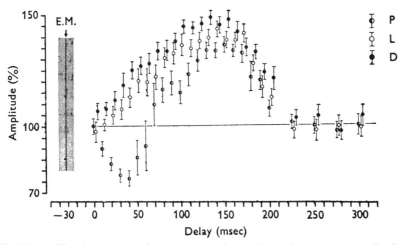

Fig. 10.6. The time course of a saccade-produced change in response amplitude of a cortical neurone when the saccade is made in the dark (filled circles), in a contourless Ganzfeld (open circles), and in a contoured or patterned visual setting (half-filled circles). (From Adey and Noda (1973).)

Bedworth 1974; Singer *et al.* 1976) or from frontal cortex (Tsumoto and Suzuki 1976).

Next let us look at the consequences of making a saccade in a patterned environment (half-filled circles). Here, an initial suppressive effect with a duration of about 70 ms is followed by a facilitatory effect with a duration of approximately 150 ms. In other words, something close to a summation of the afferent suppressive effects produced by saccade-like contour displacement shown in Fig. 10.5 and the facilitatory effects of central corollary discharges shown in Fig. 10.6 seem to prevail. The net result is that the suppression lasts for the duration of a saccade whereas facilitation occurs at the beginning and well into the following fixation interval.

Do we have any psychophysical evidence for this initial saccadic suppression followed by subsequent facilitation of sustained neurones at or near the beginning of a succeeding fixation? Fig. 10.7 shows some results on the time course of saccadic suppression reported by Volkmann, Riggs, Moore, and White (1978a). Notice that the percentage of correct identifications of a flashed target varies non-monotonically with its temporal asynchrony relative to the saccade onset. Identification perform-ance decreases to a minimum near the occurrence of the saccade, then rises

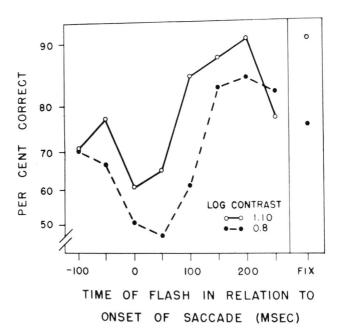

Fig. 10.7. The time course of a saccade-produced change in the visibility of a test flash. (From Volkman, Riggs, Moore, and White 1978a; reprinted by permission of the publisher from: Sanders, J., Fisher, D. and Monty, R. A. (eds.) (1980), *Eye movements and higher psychological functions.* Lawrence Erlbaum Associates, Hillsdale, N.J..)

sharply, and, at least for the lower contrast test flash, it is increased over the baseline performance obtained without saccading (shown at the right) within 100 ms and more after the new fixation has begun. This finding has been replicated and generalized to the detection of sinusoidal gratings flashed during and near saccade execution by Volkmann *et al.* (1978*b*; Fig. 5).

In summary, the following may be stated about these functional roles of saccades:

1. They activate transient channels which inhibit sustained channel activity generated in a preceding fixation interval that otherwise would mask the sustained channel activity generated during the succeeding fixation interval.

2. They activate central corollary discharges which, on the one hand, reinforce this suppression* during the saccade and, on the other hand, arising from the frontal eye fields and the midbrain reticular formation, counteract this saccadic suppression at and near the beginning of the following fixation interval, thereby preventing it from extending into that interval.

These functions are temporally orchestrated and correlated with appropriate oculomotor signals in order to produce as efficient and noiseless a sequential analysis of a pattern display as possible.

Iconic persistence and integration across saccades We are now faced with another problem posed to the visual system. Based on the prior section's material and the masking studies outlined in Sections 7.4.2 and 8.3.3, it was noted that the transient-on-sustained inhibition exerts its effect at peripheral i.e. previsible, iconic levels of processing so that sustained channel pattern activity at these levels is prevented from being transferred to the cortical, visible iconic level of integration (Breitmeyer and Ganz 1976; Michaels and Turvey 1979). The former levels of sustained channel suppression may include the lateral geniculate nucleus and layer 4 of the striate cortex (Bullier and Henry 1980; Hendrickson *et al.* 1978; Hubel and Wiesel 1972; Lund and Boothe 1975; see Section 6.7). Hence, the formation or iconic synthesis of visible patterns presumably occurs at higher cortical levels of processing.

Above, it was argued that metacontrast and the jerk effect provide afferent mechanisms of saccadic suppression of sustained pattern information. Consequently, since these mechanisms are based on transient-on-sustained inhibition, saccadic suppression must also occur prior to the level of visible, cortical icon formation. By reinspection of the lowest panel in Fig. 10.2, one can see that the preiconic, retinotopically organized

---

* See p.331 for an additional type of saccadic suppression of collicular transient and motion detectors produced by corollary activity. This suppression eliminates motion signals from retinal image displacements produced by saccadic or pursuit movements and may contribute to stability of the visual environment of such movements (White *et al.* 1980).

sustained activity present between fixation intervals is cleared by saccadic suppression. If this sequentially separated sustained activity is faithfully transferred to the visible iconic level without any further changes in its temporal course or spatial organization, one would expect the sequential analysis of an extensive pattern display to be characterized by temporal gaps between spatially uncorrelated (unstable) patterns picked up during successive fixations. However, our common experience is that, under such sequential visual inspection, perception is stable, clear, and continuous.

The problem posed here was recognized as early as the turn of the century of Dodge (1900), who regarded the apparent clear and stable perception of an object during saccadic eye movements as illusory. In contrast to this position, it will be argued below that it is not at all an illusion, but rather a consequence of central iconic persistence. If sequential perception is to be clear and continuous, one can hypothesize that the central, cortical icon itself persists across saccades and thereby produces phenomenal continuity from one fixation to another. I noted, in Section 8.3.1, Michaels and Turvey's (1979) finding that the central iconic integration of sustained pattern information persists for at least 80 ms. Since this also approximates the upper range of saccade durations (Fuchs 1976), such persistence could itself contribute to central iconic integration across saccades. However, Michaels and Turvey's (1979) investigations included the use of static, tachistoscopic vision. Therefore, it is entirely possible that their results are not applicable to more naturalistic visual behaviour characterized by saccade-fixation sequences. One reason for this possible lack of applicability is that, under static viewing, there exists a fixed one-to-one relation between spatial or environmental and retinal coordinates of visual patterns. Here, the retinotopic and spatiotopic levels of neural representations of the patterns are confounded. In contrast, under sequential vision, the retinotopic representatives of visual space change from one fixation to another yet map onto a single, stable and invariant spatiotopic representation of the visual world. As a consequence, in order to assure stable and clear vision, one would expect that central iconic integration across saccades occurs at the spatiotopic level of visual coding of the world.

However, results from several investigations of visual masking and iconic integration under eye movement conditions provide evidence supporting either a retinotopic or a spatiotopic coding of visual persistence. A masking study which had findings consistent with both types of coding has been reported by Davidson *et al.* (1973). In this study a horizontal array of five letters, e.g. S H V R Y, was flashed for 10 ms just prior to a saccade while the observer was gazing at a fixation point just below the H (see Fig. 10.8, top or first panel). Immediately after a 3° rightward saccade (second panel) to a fixation point just below the prior location of the letter R a spatially overlapping grid or else a surrounding metacontrast ring (third panel) was flashed, say, at the prior display location of the letter Y. Since, now, the letter Y and the post-saccadic mask fall one letter position to the right of the fovea, the relevant question is whether the post-saccadic mask, let us say the metacontrast ring, interacted with the letter falling on its same pre-saccadic retinal coordinates, (fourth panel), namely, the V or else with the letter falling on its same pre- and post-saccadic display coordinates, namely the Y (fifth panel). The rather interesting percept reported by the observer (sixth panel) is that, although the mask *appears in its actual display location, surrounding the letter Y*, it nevertheless *inhibits the visibility of the letter in the same pre-saccadic retinal location, namely, the V*. What is seen then resembles the following: S H   R Ⓨ.

As noted by Turvey (1977) and implied earlier in Averbach and Coriell's (1961) and Sperling's (1963) investigations of iconic persistence, *if* we identify the *target icon as the maskable correlate of stimulation*, then *per force* we must conclude that this form of visual persistence (of the letter V) is *retinotopically organized* and localized with reference to retinal

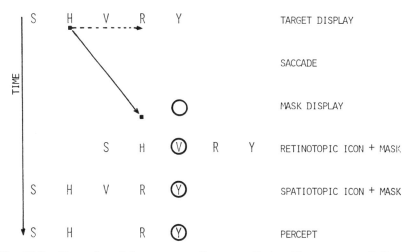

Fig. 10.8. Examples of the pre-saccadic target display (first, top panel) fixated below the second letter, H, and the post-saccadic mask display (third panel) after a saccade (second panel) is executed two letter positions to the right (fixation point below pre-saccadic display position of letter R). Note that the post-saccadic masking ring falls one letter position to the right of the fixation point at a location of the display which pre-saccadically contained the rightmost target letter (Y). The fourth panel shows the rightward shift of the (previsible) retinotopically organized target icon and of the retinotopically organized mask activity. The fifth panel shows the stable (non-shifting) spatiotopically organized target and mask icon. The sixth, bottom panel illustrates the percept of the display with the third letter, V, masked but with the masking ring itself appearing around the fifth letter, Y. Note not only that masking occurs at the pre-visible retinotopic level of pattern processing but also that visible integration occurs at the spatiotopic level.

coordinates since it, like an after image, moves with the eye. Moreover, since the letter V is masked and the masking ring is not seen surrounding the display position of this masked letter, the masking effect, presumably transient-on-sustained inhibition, must have occurred at *previsible levels* of processing of the letter V and the masking ring. However, there is no *a priori* reason or criterion for accepting only this definition of the icon. With equal legitimacy one could identify the *target icon as the correlate of stimulation which visibly or phenomenally integrates with the surrounding mask icon*. After all, such temporal integration of successive forms is one of several ways of measuring visual persistence (Coltheart 1980; DiLollo and Wilson 1978; DiLollo and Woods 1981; Eriksen and Collins 1967, 1968; Hogben and DiLollo 1974; see Section 3.2.1) and masking by common synthesis of form at a central, i.e. cortical iconic level (Michaels and Turvey 1979). At any rate, *if* we additionally accept the legitimacy of this definition of the icon, we must conclude that this latter form of visual persistence (of the letter Y and the mask ring) is *spatiotopically organized* and localized with reference to visual display or environmental coordinates. Moreover, it must be an immediate precursor of or constitute the *visible level* of processing characterized by (i) absence of transient-on-sustained inhibition and (ii) integration of sustained channel pattern information.

This conclusion is reinforced by results from several other experiments. For instance, White (1976) showed that when a target line, slightly tilted clockwise or counter-clockwise from the vertical orientation, is followed by a surrounding, vertically oriented rectangular mask while an observer is visually tracking a moving dot, optimal masking occurs when the target and rectangle-mask occupy the same perceived display coordinates whereas masking is minimal when the two stimuli occupy the same retinal coordinates. Such masking could be produced by a form of simultaneous tilt induction occurring between sustained orientation-selective mechanisms (Magnussen and Kurtenbach 1980; O'Toole 1979; see Section 9.1.1) at the spatiotopically coded iconic level, in which the slightly tilted target lines are both tilted toward the vertical by the surrounding rectangular mask. This would render the discrimination between the orientation of the two target lines more difficult. Thus, according to Turvey's (1977) criterion of identifying the icon with the maskable correlate of stimulation, we arrive at the seemingly paradoxical conclusion that the icon exists both in retinotopic and spatiotopic coordinates. But the seeming paradox rests on the presupposition that visual persistence assumes only one form. There is no reason to exclude the possibility of several forms of visual persistence or at least two of them, one of which is peripherally located and retinotopically organized whereas the other is centrally located and spatiotopically organized (see Section 3.5). Additional evidence supporting the latter, spatiotopically organized form of visual persistence derives from studies conducted by Ritter (1976), Jonides, Irwin, and Yantis (1982), and by Breitmeyer, Kropfl, and Julesz (1982).

Ritter (1976) showed that one can obtain temporal integration of brief double light flashes—one presented just prior to, the other just after a saccade—when they fall on the same environmental display coordinates but disparate retinal coordinates.

In the related study of iconic integration, Jonides *et al.* (1982) exploited the form-integration method of measuring visible persistence previously used by DiLollo and co-workers (DiLollo 1977*a,b*; DiLollo and Wilson 1978; DiLollo and Woods 1981; Hogben and DiLollo 1974). The experimental task required observers to localize an empty element position in an otherwise filled 5 × 5 (3° × 3°) element matrix, similar to the one shown in Fig. 8.9. The 24 elements were presented in two separate frames, each containing 12 randomly chosen elements which, when integrated, filled all but one of the 25 possible matrix positions. In the saccade condition, the first frame was flashed 4° to the right of fixation, for a duration adjusted to each of three observers' mean saccade latency (127, 147, and 187 ms). A blank interval of 37 ms followed, allowing each observer to execute his or her saccade. This interval was in turn followed by a 17-ms flash containing the second frame of elements.

Note that in this condition frame 1 and 2 each fell in the same environmental or display location but in disparate, peripheral and foveal, retinal locations. In the control condition the same observers viewed the fixation marker throughout the presentation of the two frame sequences. In contrast to the saccade condition, here, the two successive frames fell in separate retinal and environmental locations. Correct identification of the missing element position in the control condition varied from 4.6 to 8.4 per cent, a range only slightly above the *a priori* guessing probability of 4 per cent (1 out of 25 random locations). However, in the saccadic condition the correct performance levels varied from 53 to 62 per cent, a range more than an order of magnitude higher than the *a priori* guessing probability. These results show that iconic form integration occurred only when environmental coordinates of the two frames matched; but not when there was a mismatch of environmental coordinates. Consequently, they indicate the presence of a central spatiotopically organized icon, the existence of which, under the term of *integrative visual buffer*, was hypothesized by Rayner (1975), McConkie and Rayner (1976), and Haber (1978) in their investigations and reviews of the reading process. Unfortunately, ex-perimental results obtained subsequently by McConkie, Rayner and co-workers (McConkie and Zola 1979; Rayner 1978*a, b*, 1976; Rayner *et al.* 1978, 1980) militated against its presence in reading; and likely reasons for this prior failure to obtain confirmatory results are given by Breitmeyer (1983).

In an independent study, Breitmeyer *et al.* (1982), also reported the presence of a spatiotopically coded icon. Their procedure and rationale took the following form (see Fig. 10.9). In the first flash, lasting 200 ms, eight elements, falling within a square frame of six elements per side, were presented; in the second flash either seven or eight non-overlapping, complementary elements were flashed for 20 ms. The composite, when integrated, formed a 6 × 6 array of elements which, with an *a priori* probability of 0.5, was either completely filled or else had one internal element position left empty (as shown in Fig. 10.9). On any trial, the observer's task was simply to indicate whether an empty element position occurred. The ISI separating the two flashes was set at 40 ms, a value which, due to the long duration of the leading frame, should not yield any iconic integration under static viewing conditions. Recall—as demonstrated by DiLollo (1977*a*, *b*), DiLollo and Wilson (1978), and Hogben and DiLollo (1974) and in line with the findings of Haber and Standing (1970) and Bowen *et al.* (1974) (see Section 3.2.1)—that the visible persistence of the first flash was drastically curtailed as its duration increased above about 100 ms so that temporal integration of the two flashes was not apparent at ISIs even as low as 20 ms. A possible mechanistic explanation, based on transient-on-sustained inhibition, for this inverse-duration effect on visible persistence was discussed in Section 7.4.2, Explanation 25.

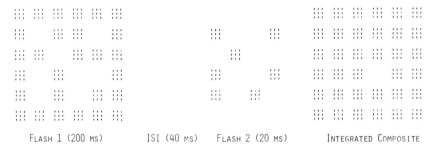

			RETINAL LOCUS		DISPLAY LOCUS	
CONDITION	SACCADE	FLASH 1	FLASH 2	FLASH 1	FLASH 2
1	NO	FOVEA	FOVEA	CENTER	CENTER
2	NO	PERIPHERY	FOVEA	OFF-CENTER	CENTER
3	YES	PERIPHERY	FOVEA	OFF-CENTER	OFF-CENTER
4	YES	FOVEA	FOVEA	CENTER	OFF-CENTER

Fig. 10.9. Typical example of successive flashes and their integrated composite employed by Breitmeyer *et al.* (1982). In the example the composite is characterized by a vacant internal element position. However, on any trial either a completely filled composite or else one with a randomly chosen vacant element position occurred with an *a priori* probability of 0.5. The experimental conditions and their respective saccade status, retinal loci, and display loci of the successive stimuli are also shown.

Four experimental conditions were investigated. In condition 1 and 2, viewing was static. The observer fixated a central dot while the two frames were flashed either to the fovea (condition 1) or else the first frame was flashed to the fovea and the second, 6.6° to the right of the fovea (condition 2). Condition 2 corresponds to the control condition in the study by Jonides *et al.* (1982). The *a priori* probability of simply guessing the presence or absence of an empty element location correctly was 0.5. Both observers employed in these two conditions yielded correct response proportion values varying between 0.47 and 0.55, none of which were significantly different from the *a priori* chance guessing probability of 0.5. These results show, in accord with those of DiLollo (1977) and DiLollo and Wilson (1978) and those of the control condition of Jonides *et al.*, that under static viewing (i) the iconic representation of the first flash did not persist over the 40 ms ISI to integrate with the iconic representation of the second flash; and (ii) no iconic form integration occurred across disparate retinal and visual-display locations.

Two other conditions also were employed by Breitmeyer *et al.* (1982). In these conditions the observer initially also fixated a central point. In condition 3, the first, 200-ms flash was presented 6.6° to the right of fixation. Moreover, the onset of the first flash served as a signal to saccade to its extrafoveal location. Since the ISI was 40 ms, the observer was allowed 240 ms to initiate and execute the saccade, a value comfortably within the range of saccade latency and execution time (Ditchburn 1973). The second flash was then presented in the same spatial location as the first one. This condition corresponded to the saccade condition in the study by Jonides *et al.* (1982). In condition 4, the first flash, again serving as a signal to execute a saccade to a marker located 6.6° to the right of fixation, was presented at the fovea, and after the saccade was executed, the second flash was presented at the location of the marker.

In condition 3, the second frame, due to the intervening saccade, fell in 6.6°, *disparate retinal locations* but on the same *visual-display locations*. The correct response proportions obtained in this condition for two observers were 0.65 and 0.77; both significantly above the 0.5 *a priori* guessing probability. In condition 4, however, where the two frames were flashed at disparate environmental coordinates but same retinal (foveal) coordinates, the correct proportions, like those in conditions 1 and 2, were not significantly different from 0.5.

Before discussing the implications of these results and those of Jonides *et al.* (1982) let us, for the moment, review other findings which dovetail rather neatly with the results and analysis provided above. Recently, Wolf, Hauske, and Lupp (1978, 1980) investigated the effect of viewing suprathreshold, parafoveally presented, pre-saccadic gratings on threshold visibility of similar gratings flashed to the fovea immediately after a saccade was directed to the prior parafoveal location. The parafoveal suprathreshold grating was extinguished at the beginning of the saccade onset so that any influence it exerted on the post-saccade, foveal grating-detection threshold would have to persist across saccades. Note that here, as in the experiments described above, the pre- and post-saccadic gratings fell at the same display location but disparate retinal locations. Wolf *et al.* (1978, 1980) found that the parafoveal pre-saccadic grating facilitated threshold visibility of foveal post-saccadic test gratings. This effect, besides being spatial frequency-dependent (Wolf *et al.* 1978, 1980), was also found to be spatial phase-specific (Wolf *et al.* 1980). That is to say, the pre-saccadic parafoveal grating had the greatest facilitatory effect on threshold visiblity of the post-saccadic foveal grating when the two gratings were phase-matched relative to display or environmental coordinates. Moreover, Wolf *et al.* (1980) reported that at the termination of the saccade, but *prior* to the presentation of the foveal post-saccadic test gratings, observers often reported a weak image of the pre-saccadic grating. Significantly, this foveal icon or image, 'seen' at the centre of gaze, was located in the same display coordinates as the pre-saccadic parafoveal or off-centre grating. These phenomenal reports indicate that 'after images' are not necessarily tied to retinal coordinates but can also be tied to visual-display coordinates. In other words, just as after images can be positive or negative (Brown 1965; Corwin *et al.* 1976), they may also vary in terms of their spatial coding, a view which reinforces the distinction between retinotopic and spatiotopic forms of visual persistence and militates against Sakitt's (1976; see Section 3.5) and Turvey's (1977) claims that the icon is of peripheral origin and, consequently, exclusively tied to retinal coordinates.

Moreover, based on these results and those reported by Jonides *et al.* (1982) and Breitmeyer *et al.* (1982), the following summary conclusions may be drawn:

1. A saccade facilitates central iconic persistence and integration, which under static viewing is drastically curtailed due to the inverse-duration effect.

2. Such facilitation occurs only for sequential stimuli separated by a saccade interval and falling at the same environmental or visual-display location but not for sequential stimuli presented under static vision at the same retinal (and environmental) location (compare the results of conditions 3 and 4 in the study by Breitmeyer *et al.* (1982).

These conclusions, in turn, indicate that a central signal corollary to the initiation and execution of a saccade enhances visible persistence only of a central, cortical icon which is coded spatiotopically. Consequently, one can claim that at this cortical, spatiotopically coded level, unlike at the more peripheral levels of visual persistence (see p.327) *normal (i.e. sequential) vision is iconoplastic* rather than iconoclastic. Moreover, the same, or perhaps a parallel, central corollary discharge must also recalibrate the retinotopic iconic representation of sustained information at the preiconic level and map it onto a spatiotopic iconic representation.* In this way, under normal viewing of a stable world, the successive, cortical, visible icon respresentations, unlike the previsible preiconic representations, are autocorrelated or template-matched and thus reinforced in spatiotopic coordinates rather than mutually masked in retinotopic coordinate space. Moreover, such a form of iconic integration would be more literal than the schematic integration across saccades suggested by Neisser (1976) or Hochberg (1978). In fact, phenomenally based reports of observers in the study by Jonides *et al.* (1982) and my own phenomenal observations (Breitmeyer *et al.* (1982)) indicate that one literally 'sees' the integration of frames across saccades when they are flashed at the same visual-display location. At any rate, via this central iconic persistence our perception of the world across saccades remains phenomenally clear, continuous, and stable.

Besides this enhancement and integration of a central, cortical icon across saccades, a similar integration ought to be used across eye blinks which, like a closing shutter, produce a momentary, optical suppression of retinal stimulation. In fact, Volkmann and co-workers (Riggs, Volkmann, and Moore 1981; Volkmann, Riggs, and Moore 1980; see also Phillips and Singer 1974) have shown that although an equivalent reduction of retinal stimulation—simulated by a reduction of visual field intensity during open-eye, static viewing—is quite noticeable as a discontinuity of perception, it is not when the reduction of retinal stimulation is produced by an eye-blink. If we assume that an eye-blink is also accompanied by persistence-enhancing, corollary or extra-retinal discharges (either centrally generated, perhaps in the cerebellum (Hepp *et al.* 1982) or arriving through rapidly conducting fibres from the extraocular eye-blink muscles) which converge onto sustained pattern channels at the cortical iconic level, continuity rather than discontinuity across eye-blinks ought to result.

A simplified neural scheme, illustrating (i) the peripheral, retinotopic process of iconic integration and its suppression followed by (ii) a cortical, spatiotopic process of iconic enhancement and integration across saccades, is presented in Fig. 10.10. (The same scheme would hold for eye-blinks except that here the mapping of retinotopic to spatiotopic representations would not change across eye-blinks.) The one-to-one mapping of sustained channel pattern information from the retinal surface to the surface of

---

* A direct neural analogy for a similar mapping of visual information from retinotopic to spatiotopic representations can be found in the *change of retinal* orientation specificity of some cortical and collicular neurones to compensate for head or body tilt in order to *maintain environmental* orientation specificity or orientation constancy (Denny and Adorjani 1972; Horn and Hill 1969; Horn, Stechler, and Hill 1972; Tomko, Barbaro, and Ali 1981; for related psychophysical and electrophysiological studies on humans, see Fiorentini, Chez, and Maffei 1972). Since the vestibular system is activated during body tilt, this plastic modification of cortical and collicular orientation responses is likely due to convergence of vestibular signals onto visual cortex and superior colliculus (Dichgans and Brandt 1978; Grüsser and Grüsser-Cornehls 1972; Henn, Young, and Finley 1974).

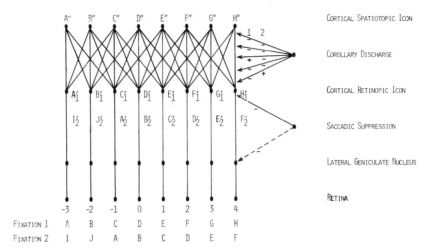

Fig. 10.10. The hypothetical transfer of retinal pattern information (indicated by capital letters at bottom) to the retinotopically coded and spatiotopically coded iconic levels of the cortical visual system. The retinotopic cortical pattern representations are shown by primed letters (subscripts correspond to the fixation as shown at bottom). The spatiotopically coded cortical pattern representations are shown by double-prime letters. Fixation 1 and 2 correspond to a right-word saccade (from top-down orientation) of two letter positions. Numbers at bottom indicate the fovea (0) and positions to the right (negative numbers) and to the left (positive numbers) of the fovea as viewed from the top-down orientation. Corollary discharges from oculomotor centres and afferent saccadic suppression are indicated by arrows pointing to their effective sites in visual pathways.

striate visual cortex (area 17) at previsible levels of processing is retinotopically organized. Beyond this level pathways diverge and converge, thus losing retinotopically coded information. Evidence for this non-topographical mapping is based on the fact that although retinotopically organized receptive fields of neurones in striate cortex are spatially restricted (Hubel and Wiesel 1962, 1968; Wurtz 1969) visual receptive fields at later stages such as parastriate cortex and inferotemporal cortex are successively larger (Barlow 1981; Gross *et al.* 1974, 1979; Vital-Durand and Blakemore 1981; Zeki 1971, 1978, 1980). Such large receptive fields which are indifferent to the exact location of adequate stimuli arise through extensive convergence and divergence of inputs from receptive fields at earlier stages.

Assume that an observer is looking straight ahead (fixation 1) and fixating the pattern D with patterns A, B, C, falling to the left of the fovea and patterns E, F, G, and H, falling to the right of the fovea. The retinal representations of these patterns are projected via retinotopically organized sustained channels to locations A'_1 through to H'_1 at area 17 of the visual cortex.

Since the oculomotor system is in dynamic balance under straight-ahead fixation, we assume that the corallary discharges originating from the oculomotor system send inhibitory influences to the convergent and divergent pathways symmetrically around the central pathways A'_1–A'' through H'_1–H'', but leave these central pathways uninhibited.*For example, pathway D'_1–D'' is left open whereas pathways B'_1–D'', C'_1–D'' and pathways E'_1–D'', F'_1–D'' are symmetrically inhibited. In this way, the retinal patterns A through to H are mapped spatially and not retinotopically onto A'' through H'', respectively. Now, assume that from the top-down orientation a saccade is made two pattern positions to the right. In this case the objects in the stable visual world corresponding to these patterns are also projected two positions to the right on the retina (fixation 2). Here pattern B falls on the fovea formerly occupied by pattern D and so on. What is desired is that retinal patterns A–F again be mapped onto spatially coded locations A''–F''.

The new retinotopic mapping of pattern information from the retina to the cortex is shown by I–I'_2 through to F–F'_2. As a first step, via saccadic suppression at (or before) the cortical preiconic level, the sustained pattern information, A–A'_1 through to H–H'_1 integrated respectively with I–I'_2 through to F–F'_2 in common peripheral, retinotopic pathways is suppressed. In this way, the cortical, retinotopic, sustained representations of I'_2 through to F'_2 occurring in fixation 2 and masked by common integration with similar persisting sustained representations of A'_1 through to H'_1

---

* In order not to overcomplicate the already complex illustration of Fig. 10.8, the oculomotor corollary discharge is given an inhibitory function. However, an alternative but logically equivalent illustration scheme would have been to include mutually lateral inhibitory processes (via interneurones) between the five divergent pathways emanating from any point at the preiconic cortical level and projecting fan-like to the iconic cortical level. The role of the corollary discharge would then be to disinhibit the appropriate pathway while leaving the inapporpriate ones mutually inhibited.

Logically equivalent *vis-à-vis* the final system response, the former one at least finds some physiological support. For instance, consistent with the inhibition scheme, Gross *et al.* (1974) noted that stimulation of the pulvinar effectively constricts the size of the activation area of inferotemporal receptive fields. Since the fibres ascending along the mid-brain reticular–collicular–pulvinar pathways terminate, among other cortical sites, in the inferotemporal cortex, signals along these pathways generated as a consequence of eye movements could, via inhibition, reduce the effective activation area of inferotemporal receptive fields. This makes the finding of spatially coded collicular cells (Mays and Sparks 1980*a*, 1981; Peck *et al.* 1980) particularly relevant since their signals may projcet to the pulvinar which in turn projects to the cortical visual areas and there effect a possible remapping of cortical retinotopically coded information onto a spatiotopic representation. In so far as the superior colliculus is used in the generation and control of saccades, such collicular cells would be consistent with Mays and Sparks' (1980*b*) and Hallett and Lightstone's (1976*a*, *b*) finding that saccades are spatially, not retinotopically coded.

Moreover, since the activity of inferotemporal neurones depends crucially on selective attention (Gross *et al.* 1979), a spatiotopic coding of information may also provide a basis for the shifts of attention which precede saccades (Posner 1980; Posner *et al.* 1978) and, therefore, are dissociable from them (Klein 1979; Posner *et al.* 1980; Shulman *et al.* 1979; Zinchenko and Vergiles 1972).

occurring in fixation 1 is prevented from being transferred to, and forming a noisy composite iconic integration at, the spatiotopically coded level.

Since, due to oblique viewing, the oculomotor system is now in an unbalanced dynamic state, its pattern of corollary inhibitory discharges is correspondingly unbalanced. For example, pathways B'_2-C'' through to B'_2-F'' are now all inhibited whereas pathway B'_2-B'' is left uninhibited. A similar recalibration of the cortical retinotopic map onto the subsequent iconic spatiotopically coded map is performed for pathways A'_2-A'', C'_2-C'', D'_2-D'', and F'_2-F''. In this way, retinal patterns A through to F are again mapped onto the spatiotopically coded positions A'' through to F'' residing at the cortical, iconic level. Of course, similar arguments hold for saccades of other directions and amplitudes. Moreover, this schematic model assumes that the oculomotor system can monitor the eyes' positions in their orbits. Such monitoring could be done via reafferent signals from the extraocular muscles or via a central programme (E. Matin 1976; L. Matin 1976; Steinbach and Smith 1981). * Extensions of this model would include such monitoring of head position and body position as well.

The above schematic model is similar to E. Matin's (1974b, 1976, 1982) dual-mechanism approach to direction constancy and dovetails neatly with recent reports by Mays and Sparks (1980b) and Hallett and Lightstone (1976a, b) that saccades themselves are spatially rather than retinotopically organized (see also Mays and Sparks 1980a; Peck et al. 1980; see p.315). In this regard, White et al. (1980) have shown that although visual displacement of a wide field surrounding an observer can produce body sway due to visuovestibular interactions (Dichgans and Brandt 1978; Dichgans et al. 1973; Grüsser and Grüsser-Cohrnels 1972; Henn et al. 1974; Young et al. 1973) similar retinal image displacements produced by saccades have little influence on body sway, a result consistent with the view that saccades are spatially organized and, thus, despite consequent retinal image displacement, preserve visual stability and spatially coded information (see also p.335, footnote).

In contrast to voluntary saccades, the rapidly alternating saccadic eye movements characterizing voluntary nystagmus do not result in the phenomenal stabilization of the environment although they do produce suppression of visual sensitivity as large as that produced by voluntary saccades of equivalent amplitude (Nagle et al. 1980). Presumably during voluntary nystagmus, no extra-retinal signals (e.g., central corollary

* Besides reafferent proprioceptive signals from the extraocular muscles, another source of reafferent information controlling or monitoring the position of the eyes seems to be purely retinal or visuosensory (Allik et al. 1981; Festinger and Holtzman 1978; Fiorentini and Maffei 1977). This source may correspond to what Gibson (1966) has termed *visual proprioception* and may consist of information on retinal error (Mays and Sparks 1980a, b, 1981) and changes produced by ambient vision during eye movements or by ambulatory vision when the entire organism is in motion. Such visual proprioception seems plausible since visual signals can modulate the activity of vestibular cells (Henn et al. 1974).

discharges) which influence the phenomenal postition of stimuli are generated, although afferent mechanisms of saccadic suppression still are active. A prediction which follows is that during voluntary nystagmus cross-saccadic integration of the central, cortical icon cannot be obtained at a spatiotopically organized level as it can be during a voluntary saccade. Hence, here, in comparison with voluntary saccades, the experimental procedures employed by Jonides *et al.* (1982) and Breitmeyer *et al.* (1982) should yield no, or significantly reduced, cross-saccadic iconic integration at the spatiotopic level.

Spatial instabilities and distortions also are common among amblyopes (Bedell and Flom 1981; Hess, Campbell, and Greenhalgh 1978; Hess and Howell 1977; Pugh 1958). With vertical test gratings at suprathreshold contrast, profound visual distortions, such as a spatial 'jumbling' of the individual bars of the grating, particularly at high and medium spatial frequencies, occur over time (Hess *et al.* 1978). This spatiotemporal instability of vision found at medium to high spatial frequencies may be related to the oculomotor instability characterizing the amblyopic eye (Hess and Howell 1977). For instance, if the oculomotor instability of an amblyopic eye is characterized by unsteady fixation and loss of control and accuracy of small saccades such as overshoot and undershoot errors (Schorr 1972), one would expect a mismatch between the magnitude of the extraretinal signal (e.g., a central corollary discharge) and the magnitude of saccades. As a consequence, the mapping of the pre-visible retinotopic icon onto the visible spatiotopically organized icon would be characterized by a mean constant error which, in turn, would manifest itself in progressively more jumbled or blended pattern vision as pattern spatial frequency increases or as pattern size decreases (Pugh 1958).

Contrast sensitivity and range changes In Section 9.2 I noted that the jerk effect reduced foveal sensitivity of the increment-threshold detecting mechanism using sustained channels but, in compensation, increased the range of luminance or contrast to which it can respond (Valberg and Breitmeyer 1980). Consequently, similar contour jerks produced by successive saccades, which, like the optimal oscillation frequencies producing the jerk effect, occur about 3–5 times per second, could also entail a loss of sensitivity but a gain in effective contrast range. It is a common experience that the reading of fine print or inspecting of fine pattern detail is accompanied by relatively few saccades and longer-duration fixations. This would enhance foveal sensitivity relative to the case where one is initially performing a coarse search for a particular pattern among many via a higher frequency of saccades and shorter-duration fixations (Antes 1974). In this latter case contrast sensitivity is sacrificed for range until the sought-for pattern is fixated when an increase in contrast sensitivity again prevails among foveal sustained channels at the expense of contrast range.

10.3.2. Smooth pursuit and compensatory eye movements

It was noted above, in the context of sustained and transient channel interactions, what the sensory consequences of saccadic shifts of visual attention and fixation are, and how the temporal sequences of these afferent processes and central corollary discharges produced by saccades are orchestrated with the timing of eye movements to produce an efficient, clear, and continuous processing of pattern information in a stable world.

However, despite this functional utility, the transient-on-sustained inhibition produced by short-range, local metacontrast mechanisms and long-range, global jerk effects actually, in turn, raises another functional complication for the visual system. Visual behaviour is characterized not only by fixation-saccade sequences but also by maintained fixation and attention on a given pattern under a variety of other dynamic viewing

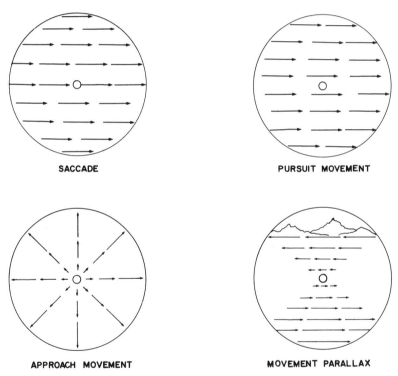

Fig. 10.11. Displacements of patterns across the visual field or retina under various types of dynamic viewing conditions. Top left: saccadic displacement. Top right: displacement during visual pursuit movement. Bottom left: displacement during approach movement. Bottom right: displacement during compensatory eye movements producing motion parallax. Note that a saccade displaces the foveal pattern (arrow attached to circle). In the other three conditions, the foveal pattern remains fixed on the fovea, whereas only extrafoveal patterns are displaced across the visual field or retina. (From Breitmeyer (1980).)

conditions as shown in Fig. 10.11. We have named such viewing dynamic foveation. One form of dynamic foveation is visual pursuit of a moving object. Presumably visual cortical mechanisms sensitive to direction and velocity of motion as well as visual tracking neurones of the parietal lobe are actively used here. Although the smoothness of ocular pursuit movement often breaks down at velocities of 30–40°/s, at which point pursuit saccades are superimposed (Brown 1972a), these smooth movements none the less can attain velocities equal to those generated by smaller amplitude saccades (Alpern 1972; Atkin 1969; Fuchs 1976). Moreover, as shown in Sections 6.6 and 9.2, these velocities are substantially greater than the threshold velocity of the neural shift effect and the psychophysical jerk effect (Breitmeyer *et al.* 1980; Fischer *et al.* 1975; Krüger 1980).

Another form of dynamic fixation entails fixating and attenting to a given object as one rapidly approaches (or recedes from) that object; or, perhaps, as in rapid sideways displacement of an observer, motion parallax may result during compensatory eye movements aimed at maintaining fixation of a stationary object. In these cases one again would expect cortical mechanisms such as parietal fixation or visual tracking neurones to be used in maintaining foveal fixation and attention. At any rate, all three of these dynamic viewing conditions or any combination thereof should activate transient channels either locally at the fovea and parafovea (visual pursuit) and/or due to velocity gradients, more globally in the periphery (visual pursuit, aproach, motion parallax). This raises the following problem: while fixating and attending to a pattern under these viewing conditions, metacontrast and/or the jerk effect should exert substantial inhibition on the foveal sustained channels activated by the fixated stimulus (see Mitrani *et al.* 1975).

That this may partly be so is indicated by the deterioration of foveal acuity (which in addition may be due to foveal slippage resulting in pursuit velocity and position errors) under dynamic conditions such as visual pursuit (Brown 1972a, b; Mackworth and Kaplan 1962; Murphy 1978).* However, note the following crucial difference between a saccadic change of fixation and attention and the three other dynamic conditions in which foveal attention is maintained on a stimulus. During a saccade not only are extrafoveal contours displaced over the retina but so is the previously

---

* In the past, reduction of visual acuity during pursuit movement typically has been attributed to two sources found when either an overshoot or undershoot error of ocular pursuit velocity relative to target velocity occurs. In either case of error, we have (i) relative motion (rather than perfect stability) of the target across the foveal or near-foveal region of the retina; and (ii) a displacement of the target from the foveal area of highest acuity. In fact, Murphy (1978) claims that visual pursuit *per se* does not produce deterioration of dynamic visual acuity. However, he employed pursuit velocities no greater than 7°/s under small display conditions not including large peripheral areas of stimulation. Hence, since the neural periphery or shift effect and the psychophysical jerk effect are below threshold at velocities lower than about 10°/s (Breitmeyer *et al.* 1980; Fischer *et al.* 1975), and since they accumulate

fixated foveal pattern itself. This is illustrated by the displacement arrow attached to the prior foveal pattern. Since sustained channels are insensitive to high-velocity saccadic image displacement (Derrington and Fuchs 1979; see Section 6.1) this rapid displacement of the foveal pattern *per se* decreases or interrupts foveal sustained activity during a saccade (for an excellent illustration of such interruption, see Fig. 1 in Noda and Adey 1974*b*).

However, in the other three conditions, the fixated, attended pattern remains more or less stationary on the retina (depending on the magnitude of retinal slippage) during eye movements whereas the extrafoveal contours or patterns are displaced over the retina. Thus, in these conditions foveal sustained activity is maintained. Moreover, since lower brain centres associated with the midbrain reticular formation are used in the control of smooth eye movements (Eckmiller 1981; Eckmiller *et al.* 1980; Eckmiller and Mackeben 1978, 1980), these centres may also send corollary discharges, like those presumably produced during saccades, and enhance the maintained foveal sustained activity. Consequently, via the reciprocal sustained-on-transient channel inhibition, which should be particularly potent at and near the fovea, sustained activity can counteract the inhibition exerted on foveal and near-foveal sustained channels by nearby transient channels activated either locally by metacontrast-like mechanisms or by the remote extrafoveal jerk effect. Accordingly, sustained-on-transient inhibition provides a mechanism whereby the fovea maintains a relatively high level of acuity, which, under dynamic viewing conditions, would otherwise deteriorate to a significantly greater extent.

10.4. Concluding remarks

Attention, if we define it as a selective pick-up of information, can at very elementary and peripheral levels of processing manifest itself as directed receptor adjustments. As such, attention must be placed in the framework of visual information processing as it occurs at relatively early, peripheral sensory levels, as well as later cognitive levels. As an illustrative case, we have used visual attention during several associated types of eye movements. The generation and monitoring of eye movements to either change or maintain foveal attention in a static or dynamic visual scene is accompanied by (i) activity corollary to (ii) the central eye movement commands and by (iii) peripheral reafferent sensory processes which are

---

strength in more extensive displays including progressively more of the retinal periphery (Derrington *et al.* 1979), one would not expect to find deterioration of dynamic visual acuity in Murphy's experiment. Similarly, Brown (1972*a, b*) used extremely large displays (200° × 34°), which, however, were uniformly illuminated. Without peripheral patterns one would, again, not expect to obtain a periphery or shift effect. Consequently, here also, remote peripheral stimulation could not have been a source of deterioration of visual acuity during pursuit movements.

consequences of image displacements over the retinae of the two eyes. All three of the components are temporally orchestrated to produce a complex but efficient means of processing visual information over time. The orchestration occurs through mutually antagonistic excitation and inhibition at peripheral and central levels.

One can think of these orchestrations as being contrapuntal. For instance, during fixation-saccade sequences, the following afferent and central, efferent events occur. The sustained-on-transient inhibition in afferent pathways during foveation parallels the excitation of fixation neurones and the simultaneous inhibition of saccade neurones in the parietal lobe or the excitation of pause neurones and the simultaneous inhibition of burst neurones in the brain stem. During a saccade, on the other hand, the excitation of saccade and burst neurones and simultaneous inhibition of fixation and pause neurones in the parietal and brain stem areas parallels the more peripheral transient-on-sustained inhibition, i.e. the saccadic suppression of persisting, retinotopic sustained pattern information. A similar contrapuntal relationship exists between activity in the superior colliculus associated with fixations, saccades, and shifts of attention and the activity of sustained and transient neurones in the geniculostriate pathway during attentive fixation intervals and saccades (see p.318). Moreover, superimposed on these two 'melodic lines' is a third one produced by central corollary discharges, which on the one hand, reinforce the transient-on-sustained inhibition during a saccade and, on the other hand, enhance the excitability of sustained neurones immediately after the saccade is executed. Furthermore, other corollary discharges map retinotopic onto spatiotopic representations of the visual world and enhance the cross-saccadic persistence and integration at the spatiotopically coded visible, cortical icon to produce a clear, stable continuous perception of the visual world.

Similar temporal couplings between movement generating signals, corollary discharges, and peripheral, reafferent sensory processes occur when the eyes are performing compensatory or pursuit movements. Here, for instance, at the central parietal level the excitation of visual pursuit neurones and the simultaneous inhibition of saccade neurones is paralleled by the local, foveal sustained-on-transient inhibition at more peripheral levels. Moreover, it is also possible that central corollary discharges enhance or reinforce sustained channel activity during these eye movements.

These descriptions are admittedly speculative and, hence, tentative. However, they are internally consistent not only with extant neurophysiological and psychophysical findings but also with the explanation of these visual mechanisms in the context of naturalistic, extralaboratory visual behaviour. None the less, more empirical work needs to be done, at the psychophysical, neurophysiological, and neuroanatomical levels, before the speculations proposed here can be verified or disconfirmed. Hence, the current chapter outlines a set of problems to which future research efforts can be directed.

Moreover, the current complementary mechanistic and purposive approach to visual masking and cognate phenomena has several other important consequences. Firstly, it lends visual masking research a means of complying with not only the criterion of ecological validity (Neisser 1976) but also the criterion of physiological and anatomical validity. That is, it provides a framework for relating mechanism and structure to behaviour. Recent attempts at providing theories of visual perception meeting the criterion of ecological validity have stressed, in addition to explaining *how* visual information processing occurs, defining or describing *what* constitutes the information in the environment being processed (Gibson 1966, 1979; Neisser 1976; Shaw and Bransford 1977; Turvey 1978). These particular calls for ecological validity and the recognition that such validity has been lacking in much of the laboratory work on human visual information processing in the last two decades (Neisser 1976) are not entirely without merit, despite their usual neglect of physiological and anatomical validity. Secondly, although such attempts often acknowledge in prolegomenous terms the existence of purposive, ecologically valid visual behaviour, these attempts, unlike the current approach, have typically failed to provide a more specific, systematic account of the relationship between purpose and mechanism of such behaviour in terms of *why* a mechanism of visual information functions as it does. It is hoped that the current dual approach can, in part, compensate for this failure and serve as a useful heuristic in guiding future research directed to an integrated understanding of the how, what, and why of visual perception.

10.5. Summary

The major topics discussed in prior chapters were given an integrative context by relating them to visual behaviour in an extralaboratory, naturalistic setting. Primary emphasis in the prior chapters was placed on theories and findings revealing *how* experimentally isolatable processes and mechanisms such as visual persistence, integration, masking, attention and so on can be characterized. In the present chapter, a broader ecological and evolutionary perspective on visual perception provided a unifying theme, a basis for discussing the relation of these component processes to each other via their composite, interdependent functional roles—or *why* they work as they do—in the service of goal-directed visual behaviour. Several major cortical and subcortical sites—among them the superior colliculus–pulvinar complex, visual cortex, posterior parietal cortex, frontal eye fields, and reticular areas—implicated in the control and direction of visual gaze and attention were discussed. It was noted that saccades, besides shifting foveal gaze, also engage local and global mechanisms of transient-on-sustained inhibition (metacontrast and jerk effect) which clear afferent, retinotopically organized pattern processing channels, especially at and near the fovea, between successive fixations. This prevents transfer of a retinotopically coded composite, i.e. a mutually masked pattern representation, of two successive fixations from being transferred to higher, spatiotopically coded, cortical levels.

The transfer of visual pattern information from separate fixations presumably is accompanied by signals corollary to those generating saccadic eye movements. Whereas one type of corollary signal enhances or amplifies the afferent suppression of pattern information during saccades, another, possibily originating from frontal eye fields or subcortical reticular areas, is temporally programmed to counteract this suppression and enhance sensitivity of afferent sustained channels immediately after a saccade is executed and a new fixation begins. The end result is a temporal segregation of sustained pattern information from successive fixations of the environment represented at cortical, retinotopically organized levels of processing.

Additional important functions of saccade-related corollary signals are (i) mapping this sequence of temporally segregated and variable retinotopic 'frames' of the visual world onto a higher level, spatiotopically organized cortical representation of the world and also (ii) enhancing the persistence of each such frame during or after its mapping so that our perceptions across saccades correspond to not only a spatially invariant and stable environment but also a temporally continuous rather than segregated one. This scheme also applies to our perception of spatiotemporal continuity across eye blinks. Here, however, the above two functions are performed by signals corollary to eye-blink rather than saccade execution. The temporal coupling of the afferent visuosensory and the motor-corollary signals also was shown to be tied to the temporal sequence of reciprocally antagonistic posterior parietal and colliculopulvinar and brain stem activities concerned, on the one hand, with allocating and maintaining visual attention and foveal gaze on a stimulus during a fixation and, on the other, with switching of visual attention and foveal gaze between fixations.

Maintaining fixation on an object while it is in motion (visual pursuit) or while it is stationary and the observer is in motion (visual approach, motion parallax during visual compensation) produces global, extrafoveal or, additionally, local foveal transient activity, which via local metacontrast or the global jerk effect can inhibit foveal pattern processing. However, here, as compared with saccades, the retinal image of the fixated object is relatively stable on the fovea, and since the fovea has the highest concentration of sustained activity responding to a stationary or slowly drifting image, it also yields the most powerful sustained-on-transient inhibition. Furthermore if, during these dynamic foveations, the response sensitivity of foveal sustained channels is enhanced (or that of foveal transient channels is attenuated) by central activity corollary to self- or eye-movements, the foveal sustained-on-transient inhibition would be especially powerful and would thus effectively counteract the reciprocal transient-on-sustained inhibition produced by image displacements of non-fixated stimuli in the visual field. Consequently, foveal sustained-on-transient inhibition, possibly reinforced by central corollary discharges, assures that, despite residual retinal image slippage, dynamic visual acuity, which otherwise would suffer from an unchecked transient-on-sustained inhibition, is maintained at as high a level as possible.

The current approach to the study of vision based on our knowledge of, on the one hand, purposive, goal-directed behaviour and, on the other, neuronatomy, neurophysiology, and psychophysics, provides a powerful heuristic tool which not only guides research of perceptual processes and mechanisms but also renders such research a measure of anatomical and physiological validity as well as ecological validity. Moreover, by incorporating several confluent and mutually interdependent explanatory levels, the approach provides a unifying theme for an integrative understanding of, otherwise, more-or-less disconnected, paradigmatically segregated areas of vision research.

11 Epilogue

Progress, far from consisting in change, depends on retentiveness . . . Those who cannot remember the past are condemned to repeat it.

George Santayana

The psychological necessity for believing in causality is that we cannot imagine a process without a purpose.

Friedrich Nietzsche

To create is to integrate.

Teilhard de Chardin

The charm of the integrative route is that you create your concepts as you go along trying to peep behind the curtain.

Ragnar Granit

I have chosen the above quotations because they not only are germane to what follows in the present chapter but also illustrate and abstract the winding conceptual threads, the warp and woof providing the integrative elements to the preceding chapters. I directed as well as followed these threads along their routes throughout the book, tied or unbound knots here and there, and arrived at what I believe to be a broad and unifying approach to the study of visual masking. To extend the metaphor: the threads took their lengthy route toward and within an integrated conceptual pattern or Gestalt—something more than a mere catalogue of historicoempirical knots. Such a Gestalt has not only its 'spatial' properties expressible in terms of the conceptual interconnectedness of theory, finding, and application but also its 'temporal' properties expressible as historical continuity and connectedness.

11.1. The historical approach

The significance of recent or extant findings, theories, and applications of visual masking can, of course, legitimately rest on several limited ahistoric criteria: how well and to what extent does a theory of visual masking explain empirical findings? What is its predictive or explanatory scope? Is it based on a sound methodology? Can it be applied to relevant problems? and so on. Despite the adequacy of this 'lateral', ahistoric context, a 'vertical' or historical dimension can significantly expand and deepen our overall, contextual understanding of visual masking (or other scientific phenomena).

George Santayana, the turn-of-the-century philosopher and poet, claimed that progress depends not only on change but also on retention; hence those who forget the past are bound to repeat it. Apropos of scientific endeavours, like the burgeoning psychological sciences, this claim is particularly valid and applicable. Many individuals, including scientists, either ignore or forget some significant parts of their heritage. In science, a new finding or theory is often reported either at an inopportune time or in some obscure or inaccessible publication or one that is written in a language with which the potentially interested scientist is not familiar. In any case, under such circumstances, the finding or theory is as good as lost to the general scientific community. This is a lamentable but unavoidable aspect of the acquisition of scientific knowledge and is not meant to impugn individual scientists. Fortunately, independent rediscovery of a finding or theory within an appropriate *Zeitgeist* often proves to be the redemptive counterpoise which rekindles interest and stimulates further empirical and theoretical investigations. Although the repetition of the past in the political and other sociocultural domains may have tragic, regressive consequences (this is what I take Santayana's claim to imply), we see, in contrast, that in the scientific domain it can have a generally stimulating and progressive effect.

Moreover, repetition in the sense of experimental replication is, among other aspects, a hallmark of science. The replicability of prior findings of which we are already aware lends credence to their empirical validity and to their applicability in formulating and testing theories meant to explain them. Consequently, from a purely methodological viewpoint, replication is a positive feature of science. Fortunately, experimental replication is not merely repetition in the sense of applying the exact same procedures and obtaining the same results—although such replicability also has its place in scientific work. In contrast, replication often occurs in the context of a more extensive scientific investigation and through the use of different, yet convergent experimental techniques. In the framework of scientific study both types of replication, repetitious and convergent, can be stated in the following reformulation of Santayana's claim: Those who know or remember the past are bound to repeat it.

What roles do discovery, rediscovery, and replication play in the development of a science? Let us attempt an answer specifically in relation to visual masking. In Chapter 1 we saw that several years before metacontrast was named as such by Stigler (1910), some of its properties already had been investigated via different but convergent techniques in a variety of lateral masking studies conducted by Sherrington (1897) and McDougall (1904a). Who then discovered metacontrast? Should we credit Sherrington or McDougall with the discovery since metacontrast is a particular form of lateral interaction, or should we credit it to Stigler for explicitly naming it and extensively studying it? I believe that all of these scientists deserve to be called discoverers of metacontrast, since all three discovered *independent methods* of investigating backward (or forward) lateral masking.

However, besides its methodological importance, metacontrast raises important theoretical issues which also must be discovered and exposed. To paraphrase Granit (1972a) a scientific discovery of *theoretical* importance breaks with a dogma, a prior intuition or an established view. Moreover, it is an importand discovery if new empirical distinctions are defined. Consequently there is an element of not only novelty but also unexpectedness in the discovery of the theoretically or empirically significant finding.

In the historical *Zeitgeist* encompassing Sherrington's, McDougall's and Stigler's experimental ventures into backward lateral masking (metacontrast as method), metacontrast as a finding presented no inexplicable fact in terms of the then-extant intuitions or established views which rested on (i) the prior known existence of lateral interaction of retinal stimuli (Mach 1865, 1866a, b, 1868) and (ii) the prior known existence of visual response persistence (Baxt 1871; Cattell 1885a, 1886; Charpentier 1880; Exner 1868; Martius 1902). Both McDougall (1904a) and Stigler (1910) explained metacontrast in terms of these extant established findings. Due to visual response persistence and possible synaptic delays, the trailing end of the persisting neural response produced by the first stimulus was thought to be laterally suppressed by the leading portion of the neural response produced by the second stimulus. Such a theory certainly was not a new discovery, rather it was simply an elaboration and juxtaposition of already discovered explanatory schemata. In Chapter 5, we saw how this basic interpretation was also incorporated in modified form in Crawford's (1947) explanation of masking by light and in Ganz's (1975) more recent model of metacontrast.

The theoretical significance of metacontrast became evident much later when it was noted that target–mask combinations in which the mask energy was equal to or even less than the target energy can yield a *type B metacontrast effect*, a finding which posed a genuine puzzle, an unexpected datum to the extant versions of the lateral suppression-visual persistence hypotheses.* Instead of the visual persistence hypothesis, some sort of multi-response, differential latency or overtake assumption seemed to be

---

* Even Stigler (1910) reported that metacontrast was maximal at positive SOAs even though the mask energy was significantly lower than the target energy. Stigler overlooked or neglected the importance of this finding, perhaps, because it would not be consistent with his theory based on (i) lateral inhibition and (ii) visual response persistence whose strength is proportional to stimulus energy.

required which, although implicit in some of McDougall's (1904*a*) and Stigler's (1926) explanations, was first made explicit in one form or another by Alpern (1953). In this sense Alpern furthered the discovery and the theoretical importance of metacontrast phenomena. Since then this basic multichannel, differential-latency assumption also has been incorporated in a variety of modified forms by subsequent investigators of metacontrast (Breitmeyer and Ganz 1976; Matin 1975; Weisstein 1968; Weisstein *et al.* 1975). The above examples illustrate how the evolution of visual masking theories proceeded from a few basic assumptions, along lines which became manifest in a variety of refined ways, as empirical observations, restructured and differentiated in the light of these same emergent theories, took on novel meanings.* Thus, the development of new and more extensively applicable explanatory schemata through which the basic assumptions are realized goes hand in hand with the discovery of novel empirical distinctions or a redefinition of prior and extant observations. Here the theoreticoempirical enterprise, though having come full cycle, continues. The replicability of the novel or redefined empirical observations is tested via repetitious or convergent techniques. Consistent and replicable, yet contingent, observations comprise the closest approximation to an objective sounding board for a theory.

Besides giving us an appreciation of the roles of discovery, rediscovery, and replication in the development of a science, a historical approach also can be a stimulating source of new and productive ideas. As a personal illustrative case, I shall use metacontrast masking. Upon reading the Gestalt psychologist Wertheimer (1912) on his extensive studies of stroboscopic motion, which, though lacking the present-day quantification of data, were rich in phenomenal descriptions, I noted a short passage stating that during stroboscopic motion—in particular phi-motion—the stimulus flashed first often appeared either entirely absent or dimmer than the second one. This led me to conjecture that a form of metacontrast masking may be involved, a fact that my collaborators and I confirmed in subsequent studies of our own (Breitmeyer and Horman 1981; Breitmeyer *et al.* 1974, 1976). Thus, through a historical approach I learned to appreciate our indebtedness to our intellectual forebears.† We learn humility but also the optimism that we too, like our forebears, can enjoy the benefits of creative research and discovery.

---

* For a more extensive and detailed discussion of the influence of theory on empirical observations see Popper (1959) and Brown (1977).

† As pointed out by an anonymous reviewer of the first version of this chapter, we should be especially indebted to the systematic and careful phenomenological analysis made in the past of perceptual experiments or demonstrations. Such descriptive data may not provide the grist for statistical or other quantitative 'number crunching'; however, fortunately they often do serve as bases for fruitful conjectures or hypotheses which do lend themselves to empirical tests satisfying quantitatively defined criteria.

11.2. Integration: creating connections between mechanistic and purposive explanations

When Aristotle introduced his category of four causes, in particular, the efficient, i.e. mechanical, and the final causes, he also introduced a controversy among scientists and philosophers of science which has remained unsettled to this day. Besides being a most eminent philosopher, Aristotle was a prominent biologist in his time. His interests included anatomy, botany, the development of animal and plant taxonomies, and the study of biological evolution. Teleological explanations based on final causes assumed prime importance, especially in regard to the last interest. Final causes, conceived to be a structure or form *towards* which intraspecies changes were *directed*, endowed these changes among members of a given species with meaning and with an equally meaningful goal- or end-state. Moreover, the world within which these processes occurred was a world of quality, charged with entelechies or various forms of animism and vitalism. Hence the science of nature was primarily an animistic, qualitative one rather than a mechanistic, quantitative one, and remained so through the Middle Ages. Renaissance science, characterized among other aspects by a rekindled interest in Pythagorean and Platonic philosophy, fostered a Neoplatonic revolution to which Kepler and Galileo made major contributions. For both of these scientists the world of nature was a world of quantity and the science of nature was likewise quantitative. As with the Pythagoreans, the book of nature, according to Galileo, was written in the language of mathematics instead of descriptive properties.

The attempted laying to the ground of Aristotle's world of quality was accompanied by another radical move. Descartes, Kepler's and Galileo's contemporary, in a what seemed to be ruthlessly objective manner, divested the biological organism of soul and of vital or animistic forces and regarded it essentially as an inanimate or dead structure working according to the then-known laws of mechanics. Outlined in Descartes' 1628 *Regulae ad directionem ingenii (Rules for the Direction of the Mind)*, this reductionist epistemology, applied to humans, was made explicit in materialist jargon a century later in LaMettrie's 1748 publication of *L'homme machine (Man as Machine)* and d'Holbach's 1770 *Systeme de la nature au des loix de monde physique et du monde moral (System of Nature or the Laws of the Physical World and the Moral World)*. We have here the first clear statements, true or false, that the laws and processes characterizing life, mind, consciousness, and so on can be *reduced* to mechanical laws and processes characterizing the material world. Accordingly, the ideal of one branch of the biological sciences was to explain the behaviour of living and, moreover, sentient organisms in terms of quantifiable, reductionist, mechanical laws without recourse to such concepts of *design, purpose* or *end*.

The postulate or presupposition upon which such an ideal epistemological programme rests can be called the postulate of objectivity. The biochemist Monod (1972) offers the following rendition*:

The cornerstone of the scientific method is the [ontological] postulate that nature is objective [and not projective; from which follows] . . . the *systematic* denial that 'true' knowledge can be got at by interpreting phenomena in terms of final causes—that is to say, of 'purpose . . . ' . . . [This canon can be dated to] the formulation by Galileo and Descartes of the principle of inertia [which] laid the groundwork not only for mechanics but for the epistemology of modern science, by abolishing Aristotelian physics and cosmology. (p. 21.)

One can see that this objectivist canon is comprised of at least one ontological and one epistemological claim; hence, there are two mutually related arguments, one methodological, the other ontological, which mechanistics and objectivists advance against teleological explanations. In this regard, Taylor (1972) offers the following arguments:

The crux of the methodological argument is the belief . . . that any explanation other than a mechanistic one leaves some questions unanswered. If we introduce purposive concepts, like the famous 'entelechies' of vitalism, then we simply close off certain questions, viz., those dealing with the mechanisms which underlie the functions specified by our entelechies. This would only be justifiable if there were nothing here to discover; however, the history of biology rather leads us to believe the opposite. Great progress has been made by attempting to discover the mechanisms underlying certain holistic functions; we have only to think of the recent breakthrough by Crick and others which has opened the mechanisms of cell-reproduction.' (p. 460.)

The crux of the methodological argument, then, is that teleological explanations based on purposive behaviour may obviate the search for mechanisms. Further on Taylor (1972) states:

The ontological objection is deeper and harder to state, and hence, naturally to criticize. It may be phrased as follows: human beings are after all physical objects; they must therefore obey the laws of physics and chemistry, those which have been found true for all physical objects. It follows that some form of reductionism must hold, that is, that higher level explanations, like the psychological, the sociological, etc., must be ultimately explicable on a more basic level in terms of physics and chemistry; on the way down this reductionist road, we would obviously pass a neurophysiological state. (p. 461.)

* Exorcising the world of its Aristotelian categories such as vitalism, purpose and so on, and returning to its pristine Democritean state, a Sisyphean Paradise, befits a purist like Monod whose metaphysical view of the world, going back to Democritus, is as austere and ascetic as his view of science. Chance and Necessity—'pure chance, nothing but chance, absolute, blind freedom' (Monod), unwittingly in league with 'bitter necessity' (S. E. Luria)—provide the superfecundity and the rarefied soil from which spring the fruit (Democritus) and the texture (Goethe) of all that is the world. Expressed this way, the scientific postulate of objectivity finds a receptive niche in that genre of literature which is a mixture of naturalism and existentialism. It is perhaps not surprising, as pointed out by Eigen and Winkler (1975), that Monod, perhaps lapsing from objective to literary thought, has elevated Chance to the very thing, namely, animist principle, which he tried so hard to exorcise from the world. A Sisyphean world, indeed!

In short, the gist of this argument, following from the ontological premise, is that we apply, at least whenever we can, reductionist rather than holistic explanations based on purpose.

Nevertheless, despite these changes in the view of nature and science, Aristotelianism prevailed in both its qualitative and animistic emphasis, particularly in the life sciences, at least through the past century. Darwin's hypotheses of *random variation* selectively exploited or filtered by *natural selection* as well as the later developments in classical and molecular *genetics*—all of which can be regarded as forms of mechanical, non-vitalistic processes—provided all that was deemed necessary to explain the evolutionary changes of species. However, the controversy between mechanistic and purposive explanations in the life sciences did not die with Aristotelian science. In psychology, the use or purposive explanations alongside mechanistic ones (Köhler 1966; Tolman 1967; Tolman and Brunswick 1935), despite the criticism of staunch objectivists (e.g., Hull 1943) has continued to the present. As we shall see, the warrant to override the objectivist criticism does not come merely from an 'irrational', psychologically intransigent clinging of the concept of purpose to that of process (see Nietzsche's quote at the beginning of this chapter), but rather because it is underwritten, for one, by reconceptualization of purposeful and goal-directed behaviour along control and information theoretic lines (Ashby 1952; Walters 1953). Hence, purposive, end-directed behaviour or—to borrow von Neumann's (1966) term—effectivities, considered from an evolutionary and cybernetic framework, can be rendered scientifically explainable and, thus, acceptable to contemporary physiologists (Granit 1977), biochemists (Bresch 1979; Eigen and Winkler 1975; Monod 1972), biologists (Mayr 1961), ethologists (Lorenz 1974, 1977), and psychologists of even such objectivist–behaviourist vintage as Skinner (1976).

However, despite exorcising the world of vitalism, animism and other Aristotelian entities, the fare attached to these current austere and ascetic canons manifests itself in several hidden premises. At times the meanings of words like *mechanism, reductionism, objectivism*, and so on are taken to be as evident and clear as the black-on-white which embodies them. For better or worse, words and the concepts which they signal are not stark, irreducible, semantic atoms. Rather, they, like scientific canons and *isms* of all sort, are illuminating by their power to attract, camouflage or conceal as well as discriminate or reveal.*

The reductionist enterprise is taken to follow (imperatively, according to Taylor (1972)) from the objectivist–materialist presupposition. Yet how—

---

* With the exception taken here of noting a specific form of conceptual confounding, I do not intend to make it my task to disentangle the meanings and forms that terms like *mechanism, objectivism, reductionism, emergentism* and so on can take. For the reader interested in the ambiguities and latitudes of their meaning when not more carefully constrained, some relevant and useful sources are Bunge (1977, 1979), Burks (1977), McMullin (1972), and Smart (1981).

even if the ontological claim that human or other living beings are after all physical objects were true—does this presupposition lead to reductionism? Certainly not via any logical warrant carried by it alone; for, here we are confronting a case of historical contingency rather than logical necessity. As noted by Collingwood (1940), by virtue of their logical efficacy, presuppositions of this sort do give rise to questions; however, they do not entail a singular answer. A case in point is that without specifically postulating ontological *reductionism* (which entails epistemic analysability), the presupposition of physicalism or materialism is as compatible with an emergentist programme (e.g., as delineated by Bunge 1977) as with reductionist ones.

Moreover, explicit in Monod's (1972) and implicit in Taylor's (1972) version of reductionism is acceptance of a Democritean or, more appropriately, and Epicurean world view—a view which, by any standard, represents a strong form of ontological reductionism. However, one can also envisage a world, in particular, a world inhabited by living and sentient creatures, in which the unit is not solely the atom, nor the molecule, nor the gene, nor the cell, nor the individual organism, but, for instance and additionally, a macro-evolutionary system or process such as the species (Gould 1982). Here, reductionism does not force atomism but rather respects several distinct ontological levels without thereby denying either mechanism or objectivity.

In view of the conflating of materialism and objectivism with a Democritean atomism, it is not surprising that programmes based on the presuppositions of field-gradient, and multi-tiered organization of reality, of which Polanyi's (1962) is a prime example, are accused of being vitalist (e.g., see Monod 1972). Specifically regarding the characterization of purpose, this conceptual conflation, when seen in *its* negativing efficacy, can be expressed as the *teleologia-cum-entelechia* premise in so far as this premise is part of the materialist–objectivist arsenal.

When, for instance, the biochemist Monod (1972) premises a *teleologia-cum-entelechia* and aptly attributes it to, and thereby justifies his critique of, particularly well-suited representatives of vitalism such as Bergson (1913) and Teilhard de Chardin (1959, 1964, 1966), one can hardly fault him. However, when Taylor (1972) attributes it generally and non-selectively or when Monod (1972) misattributes it to radical emergentists (see Bunge 1977; McMullin 1972) like Polanyi (1962) and implicitly to other pluralists like Popper (1972; Popper and Eccles 1977) and Lorenz (1977), the premise is plainly a *petitio principii*, permitting philosophic attack by proxy.

Despite Monod's (1972) well-intended but poorly executed forays on vitalism, his own objectivist programme is far from unambiguous. According to Monod, the biosphere is not deducible from or reducible to first principles; nevertheless, he proposes to *explain* an extensive set of physical, biological, psychological, social and cultural phenomena non-causally on the basis of the free play of chance (e.g., random mutation occurring without foresight) and necessity (selective pressures exerted without hindsight by the environment). With 'purpose' a mere result of blind chance and undiscerning rule of necessity one need not involve causality (see above). It is for this reason that Monod (1972) speaks of the *ultima ratio*, the basic or ultimate ground or reason for life, rather than the *ultima causa*, its basic or ultimate cause.* The claim by Monod (1972) is an ambiguously phrased epistemological and ontological one and is expressed by him as follows: 'The *ultima ratio* of *all* [italics mine] teleonomic [i.e., purposeful or goal-directed] structures and performances of living beings is thus enclosed in the sequences of residues making up polypeptide fibres.' (p. 95.) Although no form of causal, mechanist reductionism is expressed, the positing of an ultimate ground or reason (*ratio*) in terms of sequential characteristics of polypeptide fibres nonetheless supports a programme of *explanatory* reductionism for all purposive behaviour. Perhaps, more appropriately Monod (1972) should have used the term *prima ratio* to express his Democritean reductionist programme.

Dawkins (1976) adopts a similar programme. Moreover, he manages to invert the stratification and nesting of successively more complex levels of anatomical and behavioural organizations, adopted (see below) by Gould (1982), Lorenz (1977), Polanyi (1962), and Popper (1972), by claiming that all evolved higher-level systems and their function (this includes the organism *in toto*) not only depend on but also are subordinate to the biochemical, in particular, the genetic level. The biosphere, life, serves merely as a giant factory for the construction, duplication, and preservation of genetic material; all other phenomena of life *per se* are incidental; and if not deducible from, they are at least referable to this primary function or purpose. Instead of an Aristotelian teleology, we have, for want of an available term, a Dawkinsian 'arkhiology' which has replaced finalism (*telos*) with initialism (*arkhos*). The Alpha has become the Omega.

It is on the premises of this sort that biologism, physicalism and so on—the aim of which it is to explain (away) disciplines such as sociology, psychology, and anthropology in terms of a 'basically-nothing-but-ism'—

* To me, it is not clear in what sense Monod is using the term *ultima ratio*. Presumably he is sensitive to the historically documented confusion of the epistemological with the ontological senses of the terms *causa* and *ratio*. The two terms have been used interchangeably. (One can still speak of cause or reason in both the ontological or epistemological senses.) However, the semantic richness and nuances of these terms go far beyond the ontological–epistemological dimension. Historicophilosophical treatments and differential analyses of the several meanings of *causa* and *ratio* are rendered in Schopenhauer's (1957; originally published in 1813) dissertation *Über die vierfache Wurzel des Satzes vom zureichendem Grunde* (*On the Fourfold Root of the Principle of Sufficient Reason*), Collingwood's (1940) *Essay on Metaphysics* and Bunge's (1979) *Causality and Modern Science* (in particular, Chapter 9), and Burks's (1977) *Chance, Cause and Reason*.

rest. Such reductionism, as Popper (1972) noted, is a 'bad' form in the sense of being an *ad hoc* and *linguistic* reductionism since it is based more often on ideology (e.g., scientism), eponymous litanies, pleas in turn founded on principled plausibility (Smart 1981), temporary yet conquerable ignorance (Monod 1972) and allied forms of apologia, futurism and extrapolation, rather than on current and clear observation, idea and concept.

It is not the case that non-reductionists are necessarily vitalists as Monod (1972) or Taylor (1972) imply, although the converse attribution is correct without exception. Holism and globalism *per se* are not synonymous with vitalism. But when attempts are made to explain local phenomena or processes by *exclusively* global or holistic ones, the explanatory programme takes on definite vitalist characteristics. However, we shall see below that global explanations need not vitiate or otherwise oppose local ones. An epistemology in which these explanations stand appositely rather than oppositely reflects a metaphysic of mutual coexistence rather than exclusion.

Polanyi's (1962) main point, more recently iterated by Gould (1982), is that the successively more complex behavioural levels are not strictly reducible to the laws governing lower levels, in particular, to physicochemical laws; that the more complex levels are characterized by higher-order units and principles of behaviour which, though not independent of, none the less are not reducible to lower-order physicochemical units and principles. Polanyi (1962) adopts a purposive and functional approach to his understanding of *not only* animate behaviour *but also* that of inanimate machines (and presumably also to our objectified and devitalized Cartesian human *qua* machine). As a representative and summary statement of his pluralistic approach, Polanyi (1962) maintains that '*the complete knowledge of a machine as an object tells us nothing about it as a machine* . . . we identify a machine by understanding it technically; that is, by participation in its purpose and an endorsement of its operational principles. We do not exercise such participation within a physical or chemical investigation.' (p. 330.) To my knowledge nowhere does Polanyi state that higher order, operational principles or purposive behaviour characterizing a machine demand or necessitate vitalism; although he admits that non-reductionist explanations may be interpreted as vitalistic by committed mechanists and anti-pluralists (see Polanyi 1962; pp. 382–390). In a similar vein, Lorenz (1977, citing Hartmann (1964)) maintains that the ' . . . vital point is not that the differences between these levels are unbridgeable—indeed, it may be only to us that they appear unbridgeable—but that *new* laws and categories are established which, though *dependent* on those of the strata below, have their own character and assert their own autonomy.' (p. 38; italics mine.)

Thus, according to Polanyi (1962) and also Lorenz (1977), microlevel laws or principles of physics and chemistry define the conditions for a machine's success and account for its failure—be that machine animate or inanimate, sentient or non-sentient; none the less, the machine's function qua machine proceeds according to operating principles which, although dependent on the constraints imposed by its material (e.g., physicochemical or neural) state or substrate, are nevertheless autonomous and irreducible to, or undeducible from knowledge of, specific physicochemical, neural or other lower-level processes.

Even Monod (1972) embellishes his austere epistemology with, what he calls, *gratuity*. Monod (1972) begins by defining a living being as a chemical machine, whose activity, functional coherence, and architecture are governed and controlled by self-regulating, self-constructing, and self-organizing cybernetic systems. Furthermore, certain classes of biochemical processes—namely, molecular, allosteric interactions—although physiologically useful or 'rational', are none the less chemically arbitrary or 'gratuitous'. In regard to allosteric enzymes, Monod claims that gratuity can be defined, chemically speaking, as the *independence between* their *function* and the *chemical signals* controlling that function. To wit, the informational function of allosteric enzymes, although compatible with is not deducible from or reducible to the biochemical processes synthesizing them. Thus Monod would have it that gratuity, a handmaiden to 'chance', permits defining these molecules as Cartesian machines in terms of their functional and ratiomorphic properties, i.e. according to the principles governing their communication of information, independent of their physicochemical structure. Despite Monod's (1972) criticism of Polanyi's (1962) invocation of *operational* principles non-inferable from the physicochemical laws governing them *structurally*, such operating principles slip in, if you will, gratuitously in his own philosophic programme. Monod (1972) requires only that the functional, informational properties of his Cartesian molecular machines be *compatible* with, rather than deducible from, basic physicochemical laws. However, as noted, that is also the point made, among others (Lorenz 1977), by Polanyi (1962); the functional or operational principles of a machine cannot demand more than its constituents, defined by their physicochemical properties, can deliver. The conditions for the machine's successful operation or the occasions of its breakdown are specified by its physicochemical properties and, therefore, its operational principles must be compatible with the laws of physics and chemistry.

Beyond this arguable issue, Monod's (1972) example of gratuity raises another important point, namely, that of the possibility of 'multi-purpose' function of a single component of a system. Again in regard to allosteric enzymes, Monod points out that they can perform double duty as specific chemical *catalysts* and as transducers of chemical *signals* which, in turn, are required for their own biosynthesis. In other words, in service of the system of which it is a component, the enzyme molecule performs the dual chemical and informational or 'cognitive' functions; one as *chemical energizer*; the other as *information transmitter*.

Thanks to Rössler's work (cited in Lorenz 1977), Lorenz (1977) notes the importance of dual feedback of energy and information in organic evolution as follows: 'Life', he says, 'is an eminently active enterprise aimed at acquiring both a fund of energy and a stock of knowledge, the possession of one being instrumental to the acquisition of the other.' (p. 27.) In the service of an organism's (in particular, the human's) commerce with its environmental sources of energy and information, the dual functional role of enzymes as well as the more advanced machinery realized in physiological organs are required both for energy acquisition and information transmission functions.

As an apt example at the level of organ, consider the tongue. One can imagine that in its primitive state it served solely to ingest and digest food. With the additional development of vocal apparatus, its function was modified to include the vocalization and articulation of sounds for communicative purposes; and still later with the development of higher mental functions among humans it was additionally modified to specifically communicate speech or language signals. At each successively higher level, the specific lower-level function of the tongue was assimilated yet extended to accommodate a more general, higher-order function of the organism. The accommodation is a form of adaptation to changes in the internal environment of an organism which in turn responds by adapting to external environmental pressures. Now a particular organ can either adapt functionally or anatomically (Lorenz 1977). For instance, if the structure of the primitive tongue is sufficiently 'plastic' and allows for an expansion of its behavioural or functional repertoire to include vocal generation and articulation, we have a case of functional adaptation. However, if the vocal generation or articulation cannot be accommodated by such expansion due to the primitve tongue's structural 'rigidity', the form of adaptation that is required is anatomical rather than functional. If the latter is the case, then the higher-order, sound (or language)-expressive system exerts selective *internal* pressures under which the primitive tongue can be structurally and therefore, functionally modified to accommodate the demands placed on it for communication and specifically, language. In either case, the function and at times the structure of an organ depend on a more general system of which the organ is a part. The self-regulation, self-organization, and self-construction characterizing the cybernetics of biological systems presumably would be capable of generating such structural or functional accommodations.

To forego further critiques of, for instance, Monod's (1972) as well as Taylor's (1972) polemic use of the *teleologia-cum-entelechia* premise, it must be noted that one need not subscribe at all to teleology *and* entelechy, particularly the Aristotelian variety, in order to describe purposeful end-directed or goal-seeking behaviour. On this subject, Mayr (1961) says:

Pittendrigh (1958) has introduced the term *teleonomic* as a descriptive term for all end-directed systems 'not committed to Aristotelian teleology'.
this negative definition places the entire burden on the word *system* . . . It would be useful to restrict the term *teleonomic* rigidly to systems operating on the basis of a program, a code of information.(p. 1504.)

Furthermore, placing purposive behaviour in an evolutionary context, Mayr (1961) states that:

The purposive action of an individual, in so far as it is based on the properties of its genetic code, therefore is no more or less purposive than the actions of a computer that has been programmed to respond appropriately to various inputs. It is, if I may say so, a purely *mechanistic* purposiveness (p. 1504; italics mine.)

The implication of this claim is that if purposiveness *is* mechanistic, its explanation also can be made *compatible* with mechanistic ones, a claim which Monod (1972) seems to share wholeheartedly. Henceforth, to highlight this teleonomic feature and not to confuse it with the Aristotelian sense of teleology. I shall use the term 'teleonomy' instead of 'teleology', just as we differentiate 'chemistry' from its historical predecessor 'alchemy' or 'astronomy' from its predecessor 'astrology'.

Given this clarification of the relation between mechanistic and teleonomic explanations, let us take a quote from the ethologist Lorenz (1974) to illustrate—despite continued biases against—the methodological advantages of purposive thinking: 'Thus I thank the writings of the otherwise highly esteemed animal psychologist, J. A. Bierens de Haan, for bringing to my attention the unforgiveable sin committed against the spirit of natural science when one answers the question "why"? with a "so that" '(p. 7, translation mine.) In other words, the so-called infraction against natural science was to interpret the question 'why?' as a teleonomic 'what for?' and answer it with a 'so that'. Lorenz was aware of the aforementioned danger and limitations inherent in posing teleonomic questions; none the less, as his animal investigations show, he judiciously but extensively applied such thinking in his studies (Lorenz 1974, 1977), fully aware that *both* teleonomic and mechanistic analyses were methodologically useful and could be applied advantageously in a new life science.

The physiologist Ragnar Granit (1972*b*) in large part shares in Lorenz's approach when he says that:

the real issue in teleological prediction is [the following]: Since purposive thinking is concerned with integration, often with quite extensive interconnections, you can predict properties of the mechanisms by exclusion if your knowledge of the integrated totality is good enough. The more you know about a system and think about it, as an entity, the better you can say what is teleologically illogical in view of the general design of that entity. To some extent you can also make positive predictions in view of what would seem teleologically logical or, in simpler words, what would seem to make common sense (p. 406.)

This methodological advantage of purposive thinking finds its correlate (perhaps fortuitously) in the nature of goal-directed or purposive behaviour. According to Sommerhoff (1950, 1974) and Dalenroot (1979) the central aspect of goal-directed behaviour is the small amount of final variation that ensues a relatively large variety of initial conditions. An adaptive structure such as the human nervous system and in particular the brain could serve as the mechanism necessary to specify convergent boundary conditions defining the goal. Hence, such an adaptive mechanism provides a means of convergence or *directive correlation*, that is, a progressively higher correlation of initially independent environmental and system (e.g., organism) state variables.

Elsewhere, in another publication Granit (1977), like Mayr (1961), places purposiveness in the context of evolution:

In looking for principles, one should begin with some references to what is known about the evolution of beings who have to eat, reproduce, defend themselves, and communicate—things that no physicist need be concerned about. In this field of inquiry one known general principle has universal validity, the idea of natural selection producing adaptations to these challenges of the environment (p. 6.)

In this evolutionary context, Granit further illustrates his general defence of teleological thinking by commenting as follows on the specific example of industrial melanism among moths*:

Before anything was known about the genome as the substrate of heredity, an observer of the moths in their environment could invoke natural selection by itself as an explanation of *why* the light forms disappeared. Knowledge of the Mendelian gene was required to explain *how* the mechanism of the colour shift operated. On the other hand, without the initial question there would have been no real understanding of what happened (p. 8.)

In the above series of statements, Granit makes two general but important claims. One is methodological. That is to say, teleonomic and mechanistic explanations stand in apposition and not in opposition. This ' . . . means [that] the kind of interpretation in which "why" is a relevant question [stands] alongside the "how" of classic natural science.'(Granit 1977; p. 7.) Methodologically the two approaches complement and enhance each other. Moreover, Granit (1977) also implies that mechanistic *explanation* of natural phenomena alone is not sufficient in science; we also must come to *understand* them (see also Toulmin 1972), and here the teleonomic approach can be as fruitful as the mechanistic one in that it, along with the latter approach, can provide a broader, holistic and more integrated understanding of natural phenomena.

* Industrial melanism among moths was common in industrial areas relying primarily on the burning of coal as an energy source. The exhausted smoke and soot would often cover white or light surfaces, e.g. the trunks of birch trees, on which white-winged moths were camouflaged from potential predators. However, with the discolouring of these surfaces by soot and other precipitates from the exhaust, the formerly white or light colour of the moths disappeared and was replaced through selective adaptation by a darker pigment.

Let me cite one of several personal examples of the effectiveness of such integrative, apposing approaches to explanations of visual masking and cognate phenomena. In Chapter 10, p.329, I argued from a functional or purposve viewpoint that metacontrast and the inhibitory jerk effect were, respectively, local, short-range and global, long-range mechanisms of saccadic suppression of pattern information. Both effects are due to transient-on-sustained inhibition. Furthermore, since metacontrast is found to be weakest in the fovea, I reasoned that the jerk effect ought to be strongest at the fovea so as to assure an overall strong saccadic suppression of foveal pattern information; for otherwise the retinotopically coded pattern information generated in a prior fixation interval could persist and mask that of the succeeding fixation interval. The results reported by Breitmeyer and Valberg (1979) confirmed this property of the jerk effect. However, on the basis of what was known about the related neurophysiological periphery or shift effect, the opposite, wrong prediction would have been generated. We knew the following properties of the responses and retinal distributions of transient and sustained cells:

1. The strength of inhibition that transient cells exert on sustained cells is directly proportional to the transient cells excitatory activity.

2. Transient cells are characterized by a generally excitatory periphery effect.

3. Hence, at post-retinal levels of transient–sustained cell interactions, the periphery effect generally inhibits sustained cells.

4. The excitatory periphery effect in transient cells increases with retinal eccentricity whereas the responsivity of sustained cells decreases.

From these facts alone it follows quite incorrectly, as shown by the actual results, that the suppression of sustained activity by the periphery effect ought to be weakest in the fovea and increase in strength with retinal eccentricity. Hence a prediction based on global considerations of function and purpose in visual behaviour proved to be far superior to a mechanistic prediction. In a way, the mechanistic prediction, based on an algorithmically applied logic, itself was tautologous and thus added no information not already contained in properties 1–4 whereas purposive thinking served as a heuristic device correctly guiding the discovery of a new property of the human visual system.

In summary, we can make at least the following claim about mechanistic and teleonomic levels of explanation. Teleonomic or purposeful thinking (alongside mechanistic thinking) provides a useful methodological heuristic for directing research to the discovery of novel properties and mechanisms characterizing the organism. Hence, rather than obviating the search for novel mechanisms as claimed by Taylor (1972), teleonomic thinking, via its own logic, facilitates their discoveries.

Our understanding of the functional architecture of a complex system such as a biological being has been greatly enhanced in recent decades by control and information theory. Cybernetic principles, incorporated in an organism, are ideal for the expression of self-regulation, self-construction, and self-organization (Eigen and Winkler 1975; Monod 1972). Claude Bernard (1878) noted the importance of autoregulation in the maintenance of a stable *milieu interieur* where metabolic energy demands placed on the organism vary due to its own activity or to a variable *milieu exterieur*. As an example of such a homeostatic control of the organisms internal environment, Bernard demonstrated that the process of respiration, by which inhaled oxygen combines chemically with other substances in the organism to provide its energy, develops a teleonomic character through the action of the vasomotor system. Through positive and negative feedback mechanisms, the vasomotor system regulates the amount of oxygen distributed throughout the body. If the body, under stress or exercise, demands a high level of oxygen the vasomotor system reacts by vasodilatation; when the body comes to rest again the vasomotor system reacts by vasoconstriction.

These processes of positive and negative feedback provide mechanisms for an autoregulated internal equilibrium or homeostasis. Analogically we can think of this autoregulative process as being akin to a thermostat in a house which controls the temperature level. Once the level of desired temperature, like the homeostatic level in the body, is specified, the temperature control of the house is assumed by an autoregulative system. If the ambient temperature is less than the specified temperature, the heating unit is ignited by positive feedback from the thermostat. As the house heats up, it eventually reaches a temperature in excess of the specified one and through negative feedback from the ambient temperature, i.e. the house environment, the thermostat turns off the heating unit, and so on with repeated cycles of this process. The question in this autoregulative process is what acts as control and what as controlled. The switching on and off of the thermostat indeed controls the heating unit. However, by the same token the heating unit, once its effects on the environment are expressed as a sufficient rise or fall of temperature, in turn controls the off- or on-status of the thermostat's switch. Thus, one can see the thermostat as a controlling agent affecting the heating unit and vice versa. In this system the mechanisms cannot be characterized by a simple, linear, feedforward process but rather by a circular or cyclic sequence which gives rise to autoregulative processes.

This type of cyclic activity in non-biological as well as biological, autoregulated systems, as noted above, can be expressed in the form of system control language (Piaget 1971; Waddington 1975) applying, specifically, to feedforward and feedback units. In this regard, Granit (1972*b*) says:

A feedback circuit can be regarded as a teleological [i.e., teleonomic] proposition and it is generally known that the mathematics of such circuits exclude certain alternatives of response and predict others. For this reason negative feedback is sometimes regarded as the ultimate vindication of immanent teleology. This attitude is very different from mine. To me a mathematical function is the very essence of causal analysis and feedback circuits are no exception to this rule. They can obviously serve in integration but so can any physical or chemical event as such. In both cases it is a question of causal analysis in order to acquire scientific knowledge. (p. 405.)

Here we see again that we *need not* invoke final causes, entelechies, or varieties of vitalism to explain goal-directed or purposive behaviour and systems.

In our example of the thermostat–heating unit interaction, a human agent provided the initial state which then was left to autoregulative processes. In nature we, as scientists, can assume that the entire process of the development and growth of autoregulative systems in living organisms was and is controlled by evolutionary mechanisms driving and constraining phylogenetic, and ontogenetic development; one form of development— starting from the simplest inorganic chemicals and evolving to organic molecules, genes, cells, and multicellular organism—which produced the rich variety of living organisms (Bresch 1979; Dawkins 1976; Monod 1972); the other starting with the fertilized egg, to the embryo, the infant, and then to the adult organism of a species (Piaget 1971). Piaget (1971), among others, has adopted the term *epigenesis*, after Waddington (1975), for this growth and development, be it phylogenetic, embryonic, and ontogenetic. In this epigenetic process the form a species takes phylogentically or a member of a species takes ontogenetically is not only under the control of genetic programming but also under the control of environmental feedback. The two types of control are envisaged in terms of cyclic control which in conjunction with indeterminate, random processes can give rise to the variety of species and of phenotypes within a species. As Kantor (1935) puts it: 'What the organism does is a function in the mathematical sense of what it is structurally. On the other hand, what it is structurally is a function of its activities [partially under control of genetically programmed instructions] in connection with the objects and events with which it is interacting [i.e., the environment].' (p.460.) Thus, the selective pressures which the variable environment brings to bear on organisms affects their long-term development phylogenetically or their short-term development ontogenetically.

Let us take an example from the ontogenesis of an organism's visual system. In the higher mammalian visual system, certain structural elements already exist at birth. In particular, Wiesel and Hubel (1974) have demonstrated the existence of an ordered arrangement of orientation columns in visual cortex (area 17) of visually inexperienced, neonatal monkeys. The columns consist of several hundred cells arranged perpendicular to the surface of visual cortex. All cells in a given column have receptive fields oriented along a given axis; that is to say, the optimal stimulus for a given cell in a column is a black or white bar or an edge falling along the orientation of its receptive field. Within a cortical slab of about 400 by 800 μm, called a hypercolumn, all possible orientations are represented for a given region of visual space. At birth these receptive fields are somewhat plastic and can be modified through experience by increasing their orientational selectivity. Presumably the existence of these plastic structured arrays of cortical columns depends on embryonic epigenesis under the dual control of genetic programming and the embryonic environment (for one line of possible evidence for a related form of growth, see Kalil and Reh 1979). This form of plasticity may be crucial to the development of an organism, for it assures adaptive outcomes for the organism when facing a varied environment and many different object–response contingencies (Grobstein and Chow 1975).

To illustrate more specifically, Spinelli and Jensen (1979) performed the following experiment on neonatal cats. The cats were reared in a normal laboratory environment. The following shock-conditioned procedure was employed. On the one hand, when, say, the left foreleg of the cat was down it meant the presence of shock to the dorsal region of this foreleg and, say, vertical (or else horizontal) lines for one eye, e.g. the left one. On the other hand, when the same foreleg was up it meant no shock and lines of horizontal (or else vertical) orientation, respectively, for the other eye. That is to say, there was a 'safe' stimulus and an 'unsafe' one (for each eye), a type of situation that often exists in a cat's normal environment. Subsequent to the conditioning procedure, single cells were recorded from primary visual cortex and visual association cortex, presumably areas 18 and 19. Among the tested animals the physiological results mirrored their performance results obtained with the differential training in the shock-conditioning procedure. For each eye, the more easily and strongly excited cells showed a larger percentage of responses to the 'unsafe' stimulus than to the 'safe' one. Moreover, in comparison with untrained cats, who demonstrate a high degree of binocularity among cortical single cells at birth (Wiesel and Hubel 1963), Spinelli and Jensen's trained cats, in contrast, showed a significant shift toward cortical-cell monocularity; and, furthermore, these monocular cells were activated predominantly by the eye associatd with the shocked forearm. In summary, what we have here is an important demonstration of how environmental influences realized in object–subject, environment–organism contingencies can alter the functional structure of the visual system given at birth—thus, a striking example of how feedback from the environment can affect the ontogentic development of the visual system, which, in turn, is instrumental in an organism's controlling and influencing the environment. We can call this process ontogentic epigenesis, and as such it falls into the general scheme of a cyclic control existing between the epigenetic components of genetic programming and environmental feedback.

The native structures developing through phylogenetic and embryonic epigenesis can be considered the basic ground plan for an organism's visual response at birth to significant variables in the environment. That is to say, at birth an organism already has a neural representation of some of the structures and objects in the environment. Indeed, the biologist Gunther Stent (1978) has the following to say about these native structures:

these neurobiological insights into the visual system show that information about the world reaches the mind not as raw data but as highly abstract structures which are the result of a preconscious set of step-by-step transformations of the sensory input. Each transformation step involves the selective destruction of information according to a program which pre-exists in the brain. Under this program our visual perception of the world is filtered through a stage in which the input is processed in terms of straight lines, because of the manner in which the input channels coming from the primary light receptors of the retina are connected to the brain. *This fact cannot but have profound psychological consequences; evidently a geometry based on straight, parallel lines, and hence by extension, on plane surfaces, is most immediately compatible with our mental equipment.* This need not have been this way, since—at least from the neurophysiological point of view—the retinal ganglion cells might just as well have been connected to the higher cells in the visual cortex in such a way that their concentric "on" and "off" center receptive fields form‑arcs rather than straight lines. If evolution had given rise to that other circuitry, curved rather than plane surfaces would have been our primary spatial concept. *Hence neurobiology has shown what philosophical speculation led Immanuel Kant to claim 200 years ago: Euclidean geometry and its non-intersecting coplanar parallel lines is the "natural" geometry, at least for man.'* (p. 166; italics mine.)

Let us also take the following, perceptually relevant quote from Popper (1968):

we always operate with theories, some of which are even incorporated in our physiology. And a sense-organ is akin to a theory: according to evolutionist views a sense-organ is developed in an attempt to adjust ourselves to a real external world, to help us to find our way through the world. A scientific theory is an organ we develop outside our skin, while an organ is a theory we develop inside our skin. (p. 163.)

What are the consequences of these viewpoints for our conceptualization of visual perception? Basically, visual perception is a process which has evolved and has its roots in phylogeny and ontogeny. As a means of knowing the world, perception therefore can be viewed within the context of a variety of evolutionary and genetic epistemologies (Bresch 1979, Bunge 1977; Campbell 1974; Eigen and Winkler 1975; Lorenz 1941, 1977; Piaget 1971, 1978; Popper 1972; Stent 1975; Waddington 1954). To paraphrase Campbell (1966), in the functional architecture of the visual system reside implicit or tacit presumptions about the nature of the world built into the system either at various phylogenetic stages or at various ontogenetic stages. The former phylogenetically acquired functional architecture may be analogues* of Kantian synthetic (i.e., necessary or valid) *a priori* categories (Lorenz 1941, 1977; Stent 1978) that can be expressed as innate working hypotheses (Lorenz 1941, 1977) about the structure of the world. As Waddington (1954) aptly states, our cognitive–perceptual faculties, resting on these ontogenetically prior functional structures ' . . . may, in Kantian terms, not give direct contact with the thing-in-itself [noumenon], but they have been molded by things-in-themselves so as to be competent in coping with them . . . [and thus presume] a congruity between our apparatus for acquiring knowledge and the nature of the thing to be known.' (p. 880.) Specifically, in regard to vision, evolution has adopted a presumptive strategy (Campbell 1966) of creating in the nervous system an ontogenetically *a priori model, representation or schema of the world as a basis for anticipating or predicting its real properties.*

In this connection, Conant and Ashby (1970) have presented a theorem showing that any good (maximally successful and simple) regulator of a system *must* be isomorphic with the system being regulated. The construction, by the regulator, of a model of the system being regulated is therefore necessary. As an extension or corollary of the theorem, the brain, in so far as it is a successful and efficient regulator for an organism's survival of environmental challenges, *must*, in the process of learning (by its long-term adaptation to phylogenetic and its short-term adaptability to ontogenetic contingencies (Granit 1977) form a model of its environment.

To help clarify this point, let us look at some concrete examples, again from vision. Most objects and surfaces in the environment are impenetrable, and they are also opaque. In view of this, as Campbell (1974) notes, 'vision represents an opportunistic exploitation of a coincidence . . . of locomotor impenetrability with opaqueness, for a narrow band of electromagnetic waves.' (p. 414.) In other words, built into our visual system is the presumption that opaqueness corresponds to impenetrability. As another example, the world around us in its global aspect is stable and unchanging over very short intervals of time. The mapping of retinotopic onto spatiotopic icons and the latter's enhanced persistence across saccades (and eye blinks)—as noted in Chapter 10, p.343—reflects the built-in presumption that the world is stable and does not suddenly undergo deformation or disappearance during the 20–70 ms duration of a saccade. The importance of these and other built-in presumptions to our daily commerce with the visual world of objects and surfaces should be clearly apparent.

---

* They are analogues because, from a phylogenetic standpoint, they are *not* characterized by a synthetic or necessary property. The functional architecture given at birth, that is, ontogenetically *a priori*, has itself evolved and consequently changed over phylogenetic time. Therefore, in no way can these ontogenetically *a priori* structures be considered as synthetic or necessary as envisaged by a pre-Darwinian like Kant (for a similar view see Piaget 1971).

As noted previously (see the introduction of Chapter 10), in phylogenetically advanced organisms the genetic programming specifying the innate functional architecture of the visual system as well as other systems is open (Lorenz 1977; Mayr 1961; Piaget 1971). That is to say, not only is the innate functional architecture of the visual system intricately organized, but it also is a highly plastic structure subject to ontogentic, environmental influences (Blakemore and Cooper 1970; Hirsch and Spinelli 1970; Singer 1978; Spinelli and Jensen 1979). As a consequence, the innate working hypotheses residing within an organism about the general structure of the world may be modifiable to significant extents (Spinelli and Jensen 1979) during critical periods of ontogenetic development to more adequately model, schematize or represent that particular instantiation of the world known as the organism's visual ecology. Moreover, as noted by Singer (1978), the selection processes through which visual experience modifies the functional architecture of, for example, the developing visual cortex are specified by at least two factors: (i) the local pattern of retinal activity which (ii) must be adequate for the requirements of more integral sensori-motor processes found in a total behavioural setting (Spinelli and Jensen 1979). What is perhaps noteworthy here is that these specific conditions are suggestive of and consistent with Piaget's (1971) notions of accommodation and assimilation—the two main components thought to operate in ontogenetic epigenesis—during the first period, i.e. the sensori-motor period, of the cognitive development of the human infant.

The topics and issues covered in this epilogue became especially relevant in Chapter 10, where we discussed visual masking mechanisms in a functional context of evolution and purposive, goal-directed visual behaviour conducted in a natural, extralaboratory setting. Thus in addition to satisfying the criterion of physiological validity, of *how* a mechanism functions, this integrated approach to visual masking also satisfies the criterion of ecological validity (Neisser 1976), of *why* a mechanism functions as it does in an organism's ecological niche. The latter criterion's emphasis on naturalistic, extralaboratory visual behaviour was anticipated, as indicated in Section 1.1, by Ebbecke's (1920) caution that the visual laboratory provides an unnatural, impoverished visual setting, which in the rigorously controlled quest of isolated 'elemental' processes and mechanisms can produce several distracting, artefactual effects that may contaminate our understanding of whatever visual processes one wishes to investigate.

By implication other aspects of human behaviour besides the visual ones are amenable to an approach integrating lower and higher level explanations. Since visual perception is only one of several component functions of the human organism, or more specifically, of its central nervous system, the brain and the human within which it serves must be placed into an ever-widening, inclusive world. Here, Toulmin (1972) aptly summarizes, in somewhat more general ways, the integration of neural, psychophysical, and purposive levels of explanation adopted in the present monograph:

The larger organizations in which the brain serves embrace not only the individual human being by himself, but also his natural environment, his culture and beliefs, and the whole mode of life to which the demands of his situation, and the accidents of his upbringing jointly introduce him. So the proper tasks of the *neurosciences* will be on their way to completion only when their own results are seen to dovetail in with a sufficently rich amount of all the complexities of *human ecology and culture*.' (p. 422.)

References

Adelson, E.H. (1978). Iconic storage: The role of rods. *Science N.Y.* **201**, 544–6.
—— (1979). Visual persistence without rods. *Percept. Psychophys.* **26**, 245–6.
—— and Jonides, J. (1980). The psychophysics of iconic storage. *J. exp. psychol.: human percept. Perf.* **6**, 486–93.
Adey, W.R. and Noda, H. (1973). Influence of eye movements on geniculostriate excitability in the cat. *J. physiol., Lond.* **235**, 805–21.
Albano, J.E. and Wurtz, R.H. (1981). The role of primate superior colliculus, pretectum and posterior-medial thalamus in visually guided eye movements. In *Progress in oculomotor research* (ed. A. F. Fuchs and W. Becker) pp. 153–60. Elsevier/North-Holland, Amsterdam.
Alexander, K.R. (1974). Sensitization by annular surrounds: sensitization and the contrast-flash effect. *Vis. Res.* **14**, 623–31.
Allik, J., Rauk, M., and Luuk, A. (1981). Control and sense of eye movement behind closed lids. *Perception* **10**, 39–51.
Allport, D.A. (1968). Phenomenal simultaneity and the perceptual moment hypothesis. *Br. J. Psychol.* **59**, 395–406.
—— (1970). Temporal summation and phenomenal simultaneity: Experiments with the radius display. *Q.J. exp. Psychol.* **22**, 686–701.
Alpern, M. (1952). Metacontrast: Historical introduction. *Am. J. Optometry* **29**, 631–46.
—— (1953). Metacontrast. *J. opt. soc. Am.* **43**, 648–57.
—— (1963). Simultaneous brightness contrast for flashes of light of different durations. *Invest. Ophthalmol.* **2**, 47–54.
—— (1965). Rod-cone independence in the after-flash effect. *J. physiol., Lond.* **176**, 462–72.
—— (1972). Eye movements. In *Handbook of sensory physiology*, Vol. 7/4, *Visual psychophysics* (ed. D. Jameson and L. M. Hurvich) pp. 303–30. Springer Verlag, New York.
—— and Barr, L. (1962). Durations of the after-images of brief light flashes and the theory of the Broca and Sulzer effect. *J. opt. soc. Am.* **52**, 219–21.
—— and Faris, J. (1956). Luminance–duration relationship in the electrical response of the human eye during dark adaptation. *J. opt. soc. Am.* **46**, 845–50.
—— and Rushton, W.A.H. (1965). The specifcity of the cone interaction in the after-flash effect. J. physiol., Lond. **176**, 473–82.
—— —— and Torii, S. (1970a). The size of rod signals. *J. physiol., Lond.* **206**, 193–208.
—— —— —— (1970b). The attenuation of rod signals by backgrounds. *J. physiol., Lond.* **206**, 209–27.
—— —— —— (1970c). The attenuation of rod signals by bleachings. *J. physiol., Lond.* **207**, 449–61.
—— —— —— (1970d). Signals from cones. *J. physciol., Lond.* **207**, 463–75.
Alwitt, L.F. (1981). Two neural mechanisms related to modes of selective attention. *J. exp. psychol.: hum. percept. Perf.* **7**, 324–32.
Andrews, D.P. and Hammond, P. (1970). Mesopic increment threshold spectral sensitivity of single optic tract fibres in the cat: Cone–rod interaction. *J. physiol., Lond.* **209**, 65–81.
Anstis, S.M. (1970). Phi movement as a subtraction process. *Vis. Res.* **10**, 1411–30.

—— and Atkinson, J. (1967). Distortions in moving figures viewed through a stationary slit. *Am. J. Psychol.* **80**, 572–85.

—— and Moulden, B.P. (1970). After effects of seen movement: Evidence for peripheral and central components. *Q. J. exp. Psychol.* **22**, 222–9.

Antes, J.R. (1974). The time course of picture viewing. *J. exp. Psychol.* **103**, 62–70.

Arand, D. and Dember, W.N. (1978). The effect of luminance on metacontrast with internally contoured targets. *Bull. psychonomic Soc.* **11**, 57–9.

Ashby, W. (1952). *Design for a brain.* Chapman and Hall, London.

Atkin, A. (1969). Shifting fixation to another pursuit target: Selective and anticipatory control of ocular pursuit initiation. *Exp. Neurol.* **23**, 157–73.

Aubert, H. (1865). *Physiologie der Netzhaut.* Morgenstern, Breslau, Germany.

Averbach, E. and Coriell, A.S. (1961). Short-term memory in vision. *Bell systems tech. J.* **40**, 309–28.

Baade, W. (1917*a*) Selbstbeobachtungen and Introvokation. *Zeitschr. psychol. physiol. Sinnesorgane* **79**, 68–96.

—— (1917*b*). Experimentelle Untersuchungen zur darstellenden Psychologie des Wahrnehmungsprozesses. *Zeitschr. psychol. physiol. Sinnesorgane* **79**, 97–127.

Bachmann, T. and Allik, J. (1976). Integration and interruption in the masking of form by form. *Perception* **5**, 79–97.

Badcock, D. and Lovegrove, W. (1981). The effects of contrast, stimulus duration and spatial frequency on visual persistence in normal and specifically disabled readers. *J. exp. psychol.: human percept. Perf.* **7**, 495–505.

Baker, F.H., Sanseverino, E.R., Lamarre, Y., and Poggio, G.F. (1969). Excitatory responses of geniculate neurons of the cat. *J. Neurophysiol.* **32**, 916–29.

Baker, H.D. (1953). The instantaneous threshold and early dark adaptation. *J. opt. soc. Am.* **43**, 789–803.

—— (1955). Some direct comparisons between light and dark adaptation. *J. opt. soc. Am.* **45**, 839–44.

—— (1963). Initial stages of light and dark adaptation. *J. opt. soc. Am.* **53**, 98–103.

—— (1973). Area effects and the rapid threshold decrease in early dark adaptation. *J. opt. soc. Am.* **63**, 749–54.

Banks, W.P. and Barber, G. (1977). Color information in iconic memory. *Psychol. Rev.* **84**, 536–46.

—— —— (1980). Normal iconic memory for stimuli invisible to rods. *Percept. Psychophys.* **27**, 581–4.

Barlow, H.B. (1957). Increment thresholds at low intensities considered as signal/noise discriminations. *J. physiol., Lond.* **136**, 469–88.

—— (1958). Temporal and spatial summation in human vision at different background intensities. *J. physiol., Lond.* **141, 337–50.**

—— (1981). Critical limiting factors in the design of the eye and visual cortex. *Proc. R. soc., Lond.* **212B**, 1–34.

—— and Brindley, G.S. (1963). Inter-ocular transfer of movement aftereffects during pressure blinding of the stimulated eye. *Nature, Lond.* **200**, 1347.

—— Derrington, A.M., Harris, L.R., and Lennie, P. (1977). The effects of remote retinal stimulation on the responses of cat retinal ganglion cells. *J. physiol, Lond.* **269**, 177–94.

—— Fitzhugh, R., and Kuffler, S.W. (1957*a*). Dark adaptation, absolute threshold and Purkinje shift in single units of the cat's retina. *J. physiol., Lond.* **137**, 327–37.

—— —— —— (1957*b*). Change of organization of the receptive fields of the cat's retina during dark adaptation. *J. physiol., Lond.* **137**, 338–54.

—— and Sparrock, J.M.B. (1964). The role of after-images in dark adaptation. *Science N.Y.* **144**, 1309–14.

Baron, J. and Thurston, I. (1973). An analysis of the word superiority effect. *Cog. Psychol.* **4**, 207–28.

Baroncz, Z. (1911). Versuch über den sogenannten Metakontrast. *Pfl. arch. gesamten Physiol.* **140**, 491–507.

Barris, M.C. and Frumkes, T.E. (1978). Rod–cone interaction in human scotopic vision—IV. Cones stimulated by contrast flashes influence rod threshold. *Vis. Res.* **18**, 801–8.

Barry, S.H. and Dick, O. (1972). On the 'recovery' of masked targets. *Percept. Psychophys.* **12**, 117–20.

Bartlett, J.R., Doty, R.W., Lee, B.B., and Sakakura, H. (1976). Influence of eye movements on geniculostriate excitability in normal monkeys. *Exp. brain Res.* **25**, 487–509.

Bartlett, N.R. (1965). Thresholds as dependent on some energy relations and characteristics of the subject. In *Vision and visual perception* (ed. C. H. Graham) pp. 154–84. John Wiley and Sons, New York.

Bartley, S.H. (1938). A central mechanism in brightness discrimination. *Proc. soc. exp. biol. Med.* **38**, 535–6.

Bashinski, H.S. and Bacharach, V.R. (1980). Enhancement of perceptual sensitivity as the result of selectively attending to spatial locations. *Percept. Psychophys.* **28**, 241–8.

Battersby, W.S., Oesterreich, R.E., and Sturr, J.F. (1964). Neural limitations of visual excitability. VII. Nonhomonymous retrochiasmal interactions. *Am. J. Physiol.* **206**, 1181–8.

—— and Wagman, I.H. (1962). Neural limitations of visual excitability. IV. Spatial determinants of retrochiasmal interaction. *Am. J. Physiol.* **203**, 359–65.

Baumgardt, E. (1972). Threshold quantal problems. In *Handbook of sensory physiology*, Vol. VII/4. *Visual psychophysics* (ed. D. Jameson and L. M. Hurvich) pp. 29–55. Springer Verlag, New York.

—— and Segal, J. (1942). Facilitation et inhibition parametres de la fonction visuelle. *L'Annee Psychologique* **43–44**, 54–102.

Baxt, N. (1871). Ueber die Zeit, welche nöthig ist, damit ein Gesichtseindruck zum Bewusstsein kommt und über die Grösse (Extension) der bewussten Wahrnehmung bei einem Gesichtseinrucke von gegebener Dauer. *Arch. gesammte Psychol.* **4**, 325–36.

Beck, J. (1966). Effect of orientation and shape similarity on perceptual grouping. *Percept. Psychophys.* **1**, 300–2.

—— and Ambler, B. (1973). The effects of concentrated and distributed attention on peripheral acuity. *Percept. Psychophys.* **14**, 225–30.

Bedell, H.E. and Flom, M.C. (1981). Monocular spatial distortion in strabismic amblyopia. *Invest. ophthalmol. vis. Sci.* **20**, 263–8.

Benevento, L.A. and Yoshida, K. (1981). The afferent and efferent organization of lateral geniculo-prestriate pathways in the macaque monkey. *J. comp. Neurol.* **203**, 455–74.

Benevento, L.A., Creutzfeldt, O.D., and Kuhnt, U. (1972). Significance of intracortical inhibition in visual cortex. *Nature, Lond.* **238**, 124–6.

—— and Fallon, J.H. (1975). The ascending projections of the superior colliculus in the rhesus monkey (*Macaca mulatta*). *J. comp. Neurol.* **160**, 339–62.

—— and Rezak, M. (1976). The cortical projections of the inferior pulvinar and adjacent lateral pulvinar in the rhesus monkey (*Macaca mulatta*): An autoradiographic study. *Brian Res.* **108**, 1–24.

Berbaum, K., Weisstein, N., and Harris, C. (1975). A vertex-superiority effect. *Bull. psychonomic Soc.* **6**, 418.

Berger, C. (1954). Illumination of surrounding field and flicker fusion frequency with foveal images of different sizes. *Acta physiol. Scand.* **30**, 161–70.

Bergson, H. (1913). *Creative evolution*. Holt, Rinehart and Winston, New York.

Berkley, M.A. (1976). Visual acuity. In *Evolution of brain and behavior in vertebrates* (ed. R. M. Masterton, M. E. Bitterman, C. B. G. Campbell, and N. Hotton) pp. 73–88. Lawrence Erlbaum and Associates, Hillsdale, N.J.

Bernard, C. (1878). Definition de la vie. *Science experimentale*, Paris.

Bernhard, C.G. (1940). Time correlations in man of electrophysiological and sensory phenomena following light stimuli. *Acta physiol. Scand.* **1**, Suppl. 1, 52–94.

Bernstein, I. (1978). Metacontrast as a contextual effect. Paper presented at *The Symposium on Models and Mechanisms of Visual Masking; annual meeting of the American Psychological Association, Toronto, September*.

—— Amundson, V.E., and Schurman, D.L. (1973). Metacontrast inferred from reaction time and verbal report: Replication and comments on the Fehrer-Biederman experiment. *J. exp. Psychol.* **100**, 195–201.

Bernstein, I.H., Fisicaro, S.A., and Fox, J.A. (1976). Metacontrast suppression and criterion content: A discriminant function analysis. *Percept. Psychophys.* **20**, 198–204.

—— Proctor, J.D., Proctor, R.W., and Schurman, D.L. (1973). Metacontrast and brightness discrimination. *Percept. Psychophys.* **14**, 293–7.

Berson, D.M. and McIlwain, J.T. (1982). Retinal Y-cell activation of deep-layer cells in superior colliculus of the cat. *J. Neurophysiol.* **47**, 700–14.

Bevan, W., Jonides, J., and Collyer, S.C. (1970). Chromatic relationships in metacontrast suppression. *Psychonomic Sci.* **19**, 367–8.

Bisti, S., Clement, R., Maffei, L., and Mecacci, L. (1977). Spatial frequency and orientation tuning curves of visual neurons in the cat: Effects of mean luminance. *Exp. Brain Res.* **27**, 335–45.

Bizzi, E. (1969). Discharge of frontal eye field neurons during saccadic and following eyemovements in unanesthetized monkeys. *Exp. brain Res.* **6**, 69–80.

—— Maffei, L. and Piccolino, M. (1972). Variations in the visual response of the superior colliculus in relation to body roll. *Science, N.Y.* **175**, 456–7.

—— and Schiller, P.H. (1970). Single unit activity in the frontal eye fields of unanesthetized monkeys during eye and head movement. *Exp. brain Res.* **10**, 151–8.

Blake, R. and Fox, R. (1973). The psychophysical inquiry into binocular summation. *Percept. Psychophys.* **14**, 161–85.

—— —— (1974). Adaptation to invisible gratings and the site of binocular rivalry suppression. *Nature, Lond.* **249**, 488–90.

—— and Hirsch, H.V.B. (1975).Deficits in binocular depth perception in cats after alternating monocular deprivation. *Science, N.Y.* **190**, 1114–6.

Blakemore, C. and Campbell, F.W. (1969). On the existence of neurones in the human visual system selectively sensitive to orientation and size of retinal image. *J. physiol., Lond.* **203**, 237–60.

—— Carpenter, R.H.S., and Georgeson, M.A. (1970). Lateral inhibition between orientation detectors in human vision. *Nature, Lond.* **228**, 37–9.

—— and Cooper, G. (1970). Development of the brain depends on the visual environment. *Nature, Lond.* **228**, 477–8.

—— and Eggers, H.M. (1978). Animal models for human visual development. In *Frontiers in visual science* (ed. S. J. Coole and E. Smith III) pp. 651–9. Springer, New York.

—— and Hague, B. (1972). Evidence for disparity detecting neurones in the human visual system. *J. physiol., Lond.* **225**, 437–55.

—— and Julesz, B. (1971). Stereoscopic depth aftereffect produced without monocular cues. *Science, N.Y.* **171**, 286–8.

—— and Tobin, E.A. (1972). Lateral inhibition between orientation detectors in the cat's visual cortex. *Exp. brain Res.* **15**, 439–40.

Blanchard, J. (1918). The brightness sensibility of the retina. *Phys. Rev.* **11**, Series 2, 81–99.

Blick, D.W. and MacLeod, D.I.A. (1978). Rod threshold: Influence of neighboring cones. *Vis. Res.* **18**, 1611–16.

Bloch, A.M. (1885). Experience sur la vision. *Comptes Rendus de Seances de la Societe de Biologie, Paris* **37**, 493–5.

Bodis-Wollner, I. and Hendley, C.D. (1977). Relation of evoked potentials to pattern and local luminance detectors in the human visual system. In *Visual evoked potentials in man: New developments* (ed. J. E. Desmedt) pp. 197–207. Clarendon Press, Oxford.

—— —— (1979). On the separability of two mechanisms involved in the detection of grating patterns in humans. *J. physiol., Lond.* **291**, 251–63.

Bolz, J., Rosner, G. and Wässle, H. (1982). Response latency of brisk-sustained (X) and brisk-transient (Y) cells in the cat retina. *J. physiol., Lond.* **328**, 171–90.

Bonnett, C. (1977). Visual motion detection models: Features and frequency filters. *Perception* **6**, 491–500.

Bouman, M.A. (1955). On foveal and peripheral interaction in binocular vision. *Optica Acta* **1**, 177–83.

Bowen, R.W. (1981). Latencies for chromatic and achromatic visual mechanisms. *Vis. Res.* **21**, 1457–66.

—— Pokorny, J., and Cacciato, D. (1977). Metacontrast masking depends on luminance transients. *Vis. Res.* **17**, 971–5.

—— Pola, J., and Matin, L. (1974). Visual persistence effects of flash luminance, duration, and energy. *Vis. Res.* **14**, 295–303.

Bowling, A. and Lovegrove, W. (1980). The effect of stimulus duration on the persistence of gratings. *Percept. Psychophys.* **27**, 574–8.

—— —— (1981). Two components to visible persistence: Effects of orientation and contrast. *Vis. Res.* **21**, 1241–51.

—— —— and Mapperson, B. (1979). The effect of spatial frequency and contrast on visual persistence. *Perception* **8**, 529–39.

Boycott, B.B. and Wässle, H. (1974). The morphological types of ganglion cells in the domestic cat's retina. *J. physiol., Lond.* **240**, 397–419.

Boynton, R.M. (1958). On-response in the human visual system as inferred from psychophysical studies of rapid-adaptation. *AMA Archiv. Ophthalmol.* **60**, 800–10.

—— (1961). Some temporal factors in vision. In *Sensory communication* (ed. W. A. Rosenblith) pp. 739–56. John Wiley and Sons, New York.

—— (1969). Temporal summation during backward visual masking. In *Information processing in the nervous system*, (ed. K. N. Leibovic) pp. 167–76. Springer Verlag, New York.

—— (1972). Discrimination of homogeneous double pulses of light. In *Handbook of sensory physiology*, Vol. VII/4, *Visual psychophysics* (ed. D. Jameson and L. J. Hurvich) pp. 202–32. Springer Verlag, New York.

—— and Kandel, G. (1957). On responses in the human visual system as a function of adaptation level. *J. opt. soc. Am.* **47**, 275–86.

—— and Miller, N.D. (1963). Visual performance under conditions of transient adaptation. *Illuminating Engineer* **58**, 541–50.

—— and Siegfried, J.B. (1962). Psychophysical estimates of on-responses to brief light flashes. *J. opt. soc. Am.* **52**, 720–1.

—— and Triedman, M.H. (1953). A psychophysical and electrophysiological study of light adaptation. *J. exp. Psychol.* **46**, 125–34.

—— and Whitten, D.N. (1970). Visual adaptation in monkey cones: Recordings of late receptor potentials. *Science, N.Y.* **170**, 1423–6.

Braddick, O. (1973). The masking of apparent motion in random-dot patterns. *Vis. Res.* **13**, 355–69.

Breitmeyer, B.G. (1972). *The relation between the detection of size and velocity in human vision.* Ph.D. thesis, Stanford University.

—— (1973). A relationship between the detection of size, rate, orientation and direction in the human visual system. *Vis. Res.* **13**, 41–58.

—— (1975). Simple reaction time as a measure of the temporal response properties of transient and sustained channels. *Vis. Res.* **15**, 1411–12.

—— (1978a). Disinhibition of metacontrast masking of Vernier acuity targets: Sustained channels inhibit transient channels. *Vis. Res.* **18**, 1401–05.

—— (1978b). Metacontrast masking as a function of mask energy. *Bull. psychonomic Soc.* **12**, 50–2.

—— (1978c). Metacontrast with black and white stimuli: Evidence for inhibition of *on* and *off* sustained activity by either *on* or *off* transient activity. *Vis. Res.* **18**, 1443–8.

—— (1980). Unmasking visual masking: A look at the 'why' behind the veil of 'how'. *Psychol. Rev.* **87**, 52–69.

—— (1983). Sensory masking, persistence and enhancement in visual exploration and reading. In *Eye movements in reading: perceptual and language processes* (ed. K. Rayner) pp. 3–30. Academic Press, New York.

—— Battaglia, F., and Weber, C. (1976). U-shaped backward contour masking during stroboscopic motion. *J. P. exp. psychol.: hum. percept. Perf.* **2**, 167–73.

—— and Ganz, L. (1976). Implications of sustained and transient channels for theories of visual pattern masking, saccadic suppression, and information processing. *Psychol. Rev.* **83**, 1–36.

—— —— (1977). Temporal studies with flashed gratings: Inferences about human transient and sustained channels. *Vis. Res.* **17**, 861–5.

—— and Halpern, M. (1978). Visual persistence depends on spatial frequency and retinal locus. Paper presented at *The annual meeting of the Psychonomic Society, San Antonio, Texas, November.*

—— and Horman, K. (1981). On the role of stroboscopic motion in metacontrast. *Bull. psychonomic Soc.* **17**, 29–32.

—— and Julesz, B. (1975). The role of on and off transients in determining the psychological spatial frequency response. *Vis. Res.* **15**, 411–15.

—— Julesz, B., and Kropfl, W. (1975). Dynamic random-dot stereograms reveal up-down anisotropy and left-right isotropy between cortical hemifields. *Science, N.Y.* **187**, 269–70.

—— and Kersey M. (1981). Backward masking by pattern stimulus offset. *J. exp. psychol.: hum. percept. Perf.* **7**, 972–7.

—— Kropfl, W., and Julesz, B. (1982). The existence and role of retinotopic and spatiotopic forms of visual persistence. *Acta Psychol.* **52**, 175–96.

—— Levi, D.M., and Harwerth, R.S. (1981a). Flicker-masking in spatial vision. *Vis. Res.* **21**, 1377–85.

—— Love, R., and Wepman, B. (1974). Contour suppression during stroboscopic motion and metacontrast. *Vis. Res.* **14**, 1451–6.

—— and Rudd, M.E. (1981). A single-transient masking paradigm. *Percept. Psychophys.* **30**, 604–6.

—— Rudd, M., and Dunn, K. (1981*b*). Spatial and temporal parameters of metacontrast disinhibition. *J. exp. psychol.: hum. percept. Perf.* **7**, 770–9.

—— and Valberg, A. (1979). Local foveal, inhibitory effects of global, peripheral excitation. *Science, N.Y.* **203**, 463–5.

—— Valberg, A., Kurtenbach, W., and Neumeyer, C. (1980). The lateral effect of oscillation of peripheral luminance gratings on the foveal increment threshold. *Vis. Res.* **20**, 799–805.

Bresch, C. (1979). *Zwischenstufe Leben: Evolution ohne Ziehl?* Fischer Taschenbuch Verlag, Frankfurt, W. Germany.

Bridgeman, B. (1971). Metacontrast and lateral inhibition. *Psychol. Rev.* **78**, 528–39.

—— (1972). Visual receptive fields sensitive to absolute and relative motion during tracking. *Science, N.Y.* **178**, 1106–8.

—— (1973). Receptive fields in single cells of monkey visual cortex during visual tracking. *Int. J. Neurosci.* **6**, 6–17.

—— (1977). A correlational model applied to metacontrast: Reply to Weisstein, Ozog, and Szoc. *Bull. psychonomic Soc.* **10**, 85–8.

—— (1978). Distributed sensory coding applied to simulations of iconic storage and metacontrast. *Bull. math. Bio.* **40**, 605–23.

—— and Leff, S. (1979). Interaction of stimulus size and retinal eccentricity in metacontrast masking. *J. exp. psychol.: hum. percept. Perf.* **5**, 101–9.

Briggs, G.G. and Kinsbourne, M. (1972). Visual persistence as measured by reaction time. *Q. J. exp. Psychol.* **24**, 318–25.

Brindley, G.S. (1962). Two new properties of foveal after-images and a photochemical hypothesis to explain them. *J. physiol., Lond.* **164**, 168–79.

Broadbent, D.E. (1977). The hidden preattentive process. *Am. Psychol.* **32**, 109–18.

Broca, A. and Sulzer, D. (1902). La sensation lumineuse en fonction du temps. *Journal de Physiologie et de Pathologie Generale* **4**, 632–40.

Brooks, B. and Jung, R. (1973). Neuronal physiology of the visual cortex. In *Handbook of sensory physiology,* Vol. 7, Part 3, *Central processing of visual information* (ed. R. Jung), pp. 325–440. Springer, New York.

Brooks, B.A. and Fuchs, A.F. (1975). Influence of stimulus parameters on visual sensitivity during saccadic eye movement. *Vis. Res.* **15**, 1389–98.

—— and Holden, A.L. (1973). Suppression of visual signals by rapid image displacement in the pigeon retina: A possible mechanism for 'saccadic' suppression. *Vis. Res.* **13**, 1387–90.

—— and Impelman, D.M. (1981). Suppressive effects of a peripheral grating displacement during saccadic eye movements and during fixation. *Exp. Brain Res.* **42**, 489–92.

—— —— and Lum, J.R. (1980). Influence of background luminance on visual sensitivity during saccadic eye movements. *Exp. Brain Res.* **40**, 322–9.

Brown, B. (1972*a*). Dynamic visual acuity, eye movements and peripheral acuity for moving targets. *Vis. Res.* **12**, 305–21.

—— (1972*b*). The effect of target contrast variation on dynamic visual acuity and eye movements. *Vis. Res.* **12**, 1213–24.

Brown, D.L. and Salinger, W.L. (1975). Loss of X-cells in lateral geniculate nucleus with monocular paralysis: Neural plasticity in adult cats. *Science, N.Y.* **189**, 1011–12.

Brown, H.I. (1977). *Perception, theory and commitment.* University of Chicago Press.

Brown, J.L. (1965). Afterimages. In *Vision and visual perception* (ed. C. H. Graham) pp. 479–503. John Wiley and Sons, New York.

Brown, K.T., Wanatabe, K., and Murakami, M. (1965). The early and late receptor potentials of monkey cones and rods. *Cold Springs Harbor symp. quant. Biol.* **30**, 457–82.

Brussell, E.M., Adkins, J., and Stober, S.R. (1977). Increment thresholds, luminance, and brightness. *J. opt. soc. Am.* **67**, 1354–6.

—— Stober, S.R., and Favreau, O.E. (1978). Contrast reversal in backward masking. *Vis. Res.* **18**, 225–7.

Buck, S.L., Peeples, D.R., and Makous, W. (1979). Spatial patterns of rod–cone interaction. *Vis. Res.* **19**, 775–82.

—— and Norton, T.T. (1977). Receptive field properties of X-, Y-, and intermediate cells in the cat lateral geniculate nucleus. *Brain Res.* **121**, 151–6.

Bullier, J.H. and Henry, G.S. (1979). Neural path taken by afferent streams in the striate cortex of cat. *J. Neurophysiol.* **42**, 1264–70.

—— —— (1980). Ordinal position and afferent input of neurons in monkey striate cortex. *J. comp. Neurol.* **193**, 913–35.

Bunge, M. (1979). *Causality and modern science.* Dover, New York.

—— (1977). Emergence and the mind. *Neuroscience* **2**, 501–9.

Bunt, A.H., Hendrickson, A.E., Lund, J.S., Lund, R.D., and Fuchs, A.F. (1975). Monkey retinal ganglion cells: Morphometric analysis and tracing of axonal projections, with a consideration to the peroxidase technique. *J. comp. Neurol.* **164**, 265–86.

Burbeck, C.A. (1981). Criterion-free pattern and flicker thresholds. *J. opt. soc. Am.* **71**, 1343–50.

—— and Kelly, D.H. (1981). Contrast gain measurements and the transient/ sustained dichotomy. *J. opt. soc. Am.* **71**, 1335–42.

Burchard, S. and Lawson, R.B. (1973). A U-shaped detection function for backward masking of similar contours. *J. exp. Psychol.* **99**, 35–41.

Burian, H.M. (1969). Pathophysiologic basis of amblyopia and of its treatment. *Am J. Ophthalmol.* **67**, 1–12.

Burks, A.W. (1977). *Chance, cause and reason.* University of Chicago Press.

Bushnell, M.C., Goldberg, M.E., and Robinson, D.L. (1981). Behavioral enhancement of visual responses in monkey cerebral cortex. I. Modulation in posterior parietal cortex related to selective visual attention. *J. Neurophysiol.* **46**, 755–72.

Bushnell, M.C., Goldberg, M.E., and Robinson, D.L. (1978). Dissociation of movement and attention: Neuronal correlates in posterior parietal cortex. *Neurosci. Abstr.* **4**, 621.

Butler, T.W., King-Smith, P.E., Moore, R.K., and Riggs, L.A. (1976). Visual sensitivity to retinal image motion. *J. physiol., Lond.* **263**, 170–1P.

Butter, C.M., Weinstein, C., Bender, D.B., and Gross, C.G. (1978). Localization and detection of visual stimuli following superior colliculus lesion in rhesus monkey. *Brain Res.* **156**, 33–49.

Büttner, U., Büttner-Ennever, J.A., and Henn, V. (1977). Vertical eye movement related unit activity in the rostral mesencephalic reticular formation of the alert monkey. *Brain Res.* **130**, 239–52.

—— Grüsser, O.J., and Schwanz, E. (1975). The effect of area and intensity on the response of cat retinal ganglion cells to brief light flashes. *Exp. Brain Res.* **23**, 259–78.

Campbell, D.T. (1966). Pattern matching as an essential of distal knowing. In *The psychology of Egon Brunswick* (ed. K. R. Hammond) pp. 81–106. Holt, Rinehart and Winston, New York.

—— (1974). Evolutionary epistemology. In *The philosophy of Karl Popper* (ed. P. A. Schillp) pp. 413–463. Open Court, LaSalle, Illinois.

Campbell, F.W., Cleland, B.G., Cooper. G.F., and Enroth-Cugell, C. (1968). The angular selectivity of visual cortical cells to moving gratings. *J. physiol., Lond.* **198**, 237–50.

—— and Kulikowski, J.J. (1966). Orientation selectivity of the human visual system. *J. physiol., Lond.* **187**, 437–45.

—— and Wurtz, R.H. (1978). Saccadic omission: Why do we not see a greyout during a saccadic eye movement. *Vis. Res.* **18**, 1297–1303.

Carpenter, P.A. and Ganz. L. (1972). An attentional mechanism in the analysis of spatial frequency. *Percept. Psychophys.* **12**, 57–60.

Cattell, J.McK. (1885*a*). The inertia of the eye and brain. *Brain* **8**, 295–312.

—— (1885*b*). The influence of the intensity of the stimulus on the length of the reaction time. *Brain* **8**, 511–15.

—— (1886). Ueber die Trägheit der Netzhaut und des Sehcentrums. *Philosophische Studien* **3**, 94–127.

Chalupa, L.M., Anchel, H., and Lindsley, D.B. (1973). Effects of cryogenic blocking of pulvinar upon visually evoked responses in the cortex of cat. *Exp. Neurol.* **39**, 112–22.

—— Coyle, R.S., and Lindsley, D.B. (1976). Effect of pulvinar lesion on visual pattern discrimination in monkeys. *J. Neurophysiol.* **39**, 354–69.

Charpentier, A. (1890). Recherches sur la persistance des impressions retiniennes et sur la excitations lumineuses de courte duree. *Archives D'Ophthalmologie* **10**, 108–35.

Citron, M.C., Emerson, R.C., and Ide, L.S. (1981). Spatial and temporal receptive-field analysis of the cat's geniculocortical pathway. *Vis. Res.* **21**, 385–96.

Cleland, B.G., Dubin, M.W., and Levick, W.R. (1971). Sustained and transient neurones in the cat's retina and lateral geniculate nucleus. *J. physiol., Lond.* **217**, 473–96.

—— Harding, T.H., and Tulunay-Keesey, U. (1979). Visual resolution and receptive field size: Examination of two kinds of cat retinal ganglion cell. *Science, N.Y.* **205**, 1015–17.

—— and Levick, W.R. (1974). Brisk and sluggish concentrically organized ganglion cells in the cat's retina. *J. physiol., Lond.* **240**, 421–56.

—— Levick, W.R., and Sanderson, K.J. (1973). Properties of sustained and transient ganglion cells in the cat retina. *J. physiol., Lond.* **228**, 649–80.

—— —— and Wässle, H. (1975). Physiological identification of a morphological class of cat retinal ganglion cells. *J. physiol., Lond.* **248**, 151–71.

Coenen, A.M.L. and Eijkman, E.G.J. (1972). Cat optic tract and geniculate unit responses corresponding to human visual masking effects. *Exp. brain Res.* **15**, 441–51.

Cohen, B., Feldman, H., and Diamond, S.P. (1969). Effects of eye movement, brain-stem stimulation, and alertness on transmission through lateral geniculate body of monkey. *J. Neurophysiol.* **32**, 583–94.

—— Goto, K., Shanzer, S., and Weis, A.H. (1965). Eye movements induced by electric stimulation of the cerebellum in the alert cat. *Exp. neurol.* **13**, 145–62.

Cohen, H.I., Winters, R.W., and Hamasaki, D.I. (1980). Response of X and Y cat retinal ganglion cells to moving stimuli. *Exp. brain Res.* **38**, 299–303.

Collingwood, R.G. (1940). *Essay on metaphysics*. Oxford University Press, London.

—— (1956). *The idea of history*. Oxford University Press, London.

Coltheart. M. (1980). Iconic memory and visible persistence. *Percept. Psychophys.* **27**, 183–228.

Conant, R.C. and Ashby, W.R. (1970). Every good regulator of a system must be a model of that system. *Int. J. systems Sci.* **1**, 89–97.

Cooper, H.M., Kennedy, H., Magnin, M., and Vital-Durand, F. (1979). Thalamic projections to area 17 in a prosimian primate, *Microcebus murinus*. *J. comp. Neurol.* **187**, 145–68.

Corfield, R., Frosdick, J.P., and Campbell, F.W. (1978). Grey-out elimination: The roles of spatial waveform, frequency and phase. *Vis. Res.* **18**, 1305–11.

Corwin, T.R., Volpe, L.C., and Tyler, C.W. (1976). Images and after-images of sinusoidal gratings. *Vis. Res.* **16**, 345–9.

Cowey, A. and Rolls, F.T. (1974). Human cortical magnification factor and its relation to visual acuity. *Exp. brain Res.* **21**, 447–54.

Cox, S.I. and Dember, W.N. (1970). Backward masking of visual targets with internal contours. *Psychonomic Sci.* **19**, 255–6.

—— —— (1972). U-shaped metacontrast functions with a detection task. *J. Exp. Psychol.* **95**, 327–33.

Crawford, B.H. (1940). The effect of field size and pattern on the change of visual sensitivity with time. *Proc. R. soc., Lond.* **129B**, 94–106.

—— (1947). Visual adaptation in relation to brief conditioning stimuli. *Proc. R. soc., Lond.* **134B**, 283–302.

Creutzfeldt, O.D. (1972). Some neurophysiological considerations concerning 'memory'. In *Memory and transfer of information* (ed. H. P. Zippel) pp. 293–302. Plenum Publishing, New York.

—— and Heggelund, P. (1975). Neural plasticity in visual cortex of adult cats after exposure to visual patterns. *Science, N.Y.* **188**, 1025–7.

—— Kuhnt, U., and Benevento, L.A. (1974). An intracellular analysis of visual cortical neurones to moving stimuli: Responses in a co-operative neuronal network. *Exp. brain Res.* **21**, 251–74.

—— Lee, B.B., and Elepfandt, A. (1979). A quantitative study of chromatic organization and receptive fields in the lateral geniculate body of the rhesus monkey. *Exp. brain Res.* **35**, 527–45.

Crommelinck, M., Roucoux, A., and Meulders, M. (1977). Eye movements evoked by stimulation of lateral posterior nucleus and pulvinar in the alert cat. *Brain Res.* **124**, 361–6.

Crook, M.N. (1937). Visual discrimination of movement. *J. Psychol.* **3**, 541–58.

Crovitz, H.F. and Daves, W. (1962). Tendencies to eye movements and perceptual accuracy. *J. exp. Psychol.* **63**, 495–8.

Crowne, D.P., Yeo, C.H., and Russell, I.S. (1981). The effects of unilateral frontal eye field lesions in the monkey: Visual-motor guidance and avoidance behavior. *Behav. brain Res.* **2**, 165–7.

Cynader, M. and Berman, N. (1972). Receptive-field organization of monkey superior colliculus. *J. Neurophysiol.* **35**, 187–201.

Dalenroot, G.J. (1979). Deterministic versus teleological explanation in general system's theory. Paper presented at *The Sixth International Congress of Logic, Methodology, and Philosophy of Science, Hannover, Federal Republic of Germany, August.*

Daniel, P.M. and Whitteridge, D. (1961). The representation of the visual field on the cerebral cortex in monkeys. *J. physiol., Lond.* **159**, 203–21.

Davidson, M.L., Fox, M.J., and Dick. A.O. (1973). Effects of eye movements on backward masking and perceived location. *Percept. Psychophys.* **14**, 110–16.

Dawkins, R. (1976). *The selfish gene*. Oxford University Press, New York.

De Bruyn, E.J., Wise, V.L., and Casagranda, V.A. (1980). The size and topographic arrangement of retinal ganglion cells in the galago. *Vis. Res.* **20**, 315–37.

Del Pezzo, E.M. and Hoffman, H.S. (1980). Attentional factors in the inhibition of a reflex by a visual stimulus. *Science, N.Y.* **210**, 673–4.

Dember, W.N., Mathews, W.D., and Stefl. M. (1973). Backward masking and enhancement of multisegmented visual targets. *Bull. psychonomic Soc.* **1**, 45–7.

—— and Purcell, D.G. (1967). Recovery of masked visual targets by inhibition of the masking stimulus. *Science, N.Y.* **157**, 1335–6.

—— Schwartz, M., and Kocak, M. (1978). Substantial recovery of a masked visual target and its theoretical interpretation. *Bull. psychonomic Soc.* **11**, 285–7.

—— and Stefl, M. (1972). Backward enhancement? *Science, N.Y.* **175**, 93–5.

—— Stefl, M., and Kao, K.C. (1974). Backward masking of gratings varying in spatial frequency. *Bull. psychonomic Soc.* **3**, 439–41.

Demkiw, P. and Michaels, C. (1976). Motion information in iconic memory. *Acta Psychol.* **40**, 257–64.

De Monasterio, F.M. (1978*a*). Properties of concentrically organized X and Y ganglion cells of macaque retina. *J. Neurophysiol.* **41**, 1394–1417.

—— (1978*b*). Center and surround mechanisms of opponent-colour X and Y ganglion cells of retina of macaques. *J. Neurophysiol.* **41**, 1418–34.

—— and Gouras, P. (1975). Functional properties of ganglion cells of the rhesus monkey retina. *J. physiol. Lond.* **251**, 167–95.

——, Gouras, P., and Tolhurst, D.J. (1976). Spatial summation, response pattern and conduction velocity of ganglion cells of the rhesus monkey retina. *Vis. Res.* **16**, 674–8.

—— and Schein, S.J. (1980). Proton-like spectral sensititivity of foveal Y ganglion cells of the retinal of macaque monkeys. *J. physiol. Lond.* **299**, 385–96.

Denney, D. and Adorjani, C. (1972). Orientation specificity of visual cortical neurons after head tilt. *Exp. brain Res.* **14**, 312–17.

Derrington, A.M. and Fuchs, A.F. (1978). Spatial and temporal frequency tuning in cat lateral geniculate nucleus. *J. physiol., Lond.* **282**, 44–5P.

—— —— (1979). Spatial and temporal properties of X and Y cells in the cat lateral geniculate nucleus. *J. physiol., Lond.* **293**, 347–64.

—— —— (1979). Effects of visual deprivation on cat LGN neurones. *J. physiol., Lond.* **300**, 61–2P.

—— —— (1981). Spatial and temporal properties of cat geniculate neurones after prolonged deprivation. *J. physiol., Lond.* **314**, 107–20.

—— and Henning, G.B. (1981). Pattern discrimination with flickering stimuli. *Vis. Res.* **21**, 597–602.

—— Lennie, P., and Wright, M.J. (1979). The mechanism of peripherally evoked responses in retinal ganglion cells. *J. physiol., Lond.* **289**, 299–310.

Diamond, I.T. (1980). Changing views of the organization and evolution of the visual pathway. In *Proceedings of the First Institute of Neurological Sciences, Symposium in Neurobiology, Changing Concepts of the Nervous System.*

—— and Hall, W.C. (1969). Evolution of neocortex. *Science, N.Y.* **164**, 251–62.

Dichgans, J. and Brandt, T., (1978). Visual-vestibular interaction: Effects of self-motion perception and postural control. In *Handbook of sensory physiology*, Vol. 8, *Perception* (ed. R. Held, H. W. Leibowitz, and H.-L. Teuber) pp. 755–804. Springer, Berlin.

—— Schmidt, C.L., and Graf, W. (1973). Visual input improves the speedometer function of the vestibular nuclei in the goldfish. *Exp. brain Res.* **18**, 319–22.

Dick, A.O. (1974). Iconic memory and its relation to perceptual processing and other memory mechanisms. *Percept. Psychophys.* **16**, 575–96.

—— and Dick S.O. (1969). An analysis of hierarchical processing in visual perception. *Can. J. Psychol.* **23**, 203–11.

Didner, R. and Sperling, G. (1980), Perceptual delay: A consequence of metacontrast and apparent motion. *J. exp. psychol: hum. percept. Perf.* **6**, 235–43.

DiLollo, V. (1977a). On the spatio-temporal interactions of brief visual displays. In *Studies in perception* (ed. R. H. Day and G. V. Stanley) pp. 39–55. University of Western Australia Press, Perth.

—— (1977b). Temporal characteristics of iconic memory. *Nature, Lond.* **267**, 241–3.

—— (1980). Temporal integration in vision. *J. exp. psychol: General* **109**, 75–97.

—— (1981). Hemispheric symmetry in duration of visible persistence. *Percept. Psychophys.* **29**, 21–5.

—— and Wilson, A.E. (1978). Iconic persistence and perceptual moment as determinants of temporal integration in vision. *Vis. Res.* **18**, 1607–10.

—— and Woods, E. (1981). Duration of visible persistence in relation to range of spatial frequencies. *J. Exp. psychol: hum. percept. Perf.* **7**, 754–69.

Ditchburn, R.W. (1973). *Eye-movements and perception*. Clarendon Press, Oxford.

Dixon, N.F. and Hammond, E.J. (1972). The attenuation of visual persistence. *Br. J. Psychol.* **63**, 243–54.

Dodge, R. (1900). Visual perception during eye movement. *Psychol. Rev.* **7**, 454–65.

Doerflein, R.S. and Dick, A.O. (1973). The effect of an eye movement on iconic memory and perceived location. Paper presented at *The annual meeting of the Psychonomic Society, Boston, November*.

Donagan, A. (1966). The Popper–Hempel theory reconsidered. In *Philosophical analysis and history* (ed. W. H. Dray) pp. 127–59. Harper and Row, New York.

Donchin, E. (1967). Retroactive visual masking: Effects of test flash duration on the masking interval. *Vis. Res.* **7**, 79–87.

Dortmann, U. and Spillmann, L. (1979). Shift-induced changes of foveal increment thresholds: The role of test flash delay. Paper presented at *The European Conference on Visual Perception, Noordwijkerhout, Netherlands, October*.

—— —— (1981). Facilitation and inhibition in the jerk effect depend upon test flash duration and delay. *Vis. Res.* **21**, 1783–91.

Dow, B.M. (1974). Functional classes of cells and their laminar distribution in monkey visual cortex. *J. Neurophysiol.* **37**, 927–46.

Dowling, J.E. (1963). Neural and photochemical mechanisms of visual adaptation in the rat. *J. gen. Physiol.* **46**, 1287–1301.

—— (1967). The site of visual adaptation. *Science, N.Y.* **155**, 273–9.

Dreher, B., Fukuda, Y., and Rodieck, R.W. (1976). Identification, classification and anatomical segregation of cells with X-like and Y-like properties in the lateral geniculate nucleus of old-world primates. *J. physiol., Lond.* **258**, 433–52.

—— Leventhal, A.G., and Hale, P.T. (1980). Geniculate input to cat visual cortex: A comparison of area 19 with areas 17 and 18. *J. Neurophysiol.* **44**, 804–26.

Dubin, M.W. and Cleland, B.G. (1977). Organization of visual inputs to interneurons of lateral geniculate nucleus of the cat. *J. Neurophysiol.* **40**, 410–27.

Du Croz, J.J. and Rushton, W.A.H. (1963). Cone dark-adaptation curves. *J. physiol., Lond.* **168**, 52P.

Duffy, F.H. and Burchfiel, J.L. (1975). Eye movement related inhibition of primate visual neurons. *Brain Res.* **89**, 121–32.

Duke-Elder, S. and Wybar, K. (1972). *System of ophthalmology*, Vol. 1, *Ocular motility and strabismus* (ed. S. Duke-Elder). Henry Kimpton, London.

Ebbecke, U. (1920). Über das Augenblicksehen. *Pfl. arch. gesamte Physiol.* **185**, 181–95.

Eckhorn, R. and Pöpel, B. (1981). Responses of cat retinal ganglion cells to the random motion of a spot stimulus. *Vis. Res.* **21**, 435–43.

Eckmiller, R. (1981). A model of the neural network controlling foveal pursuit eye movements. In *Progress of oculomotor research* (ed. A. F. Fuchs and W. Becker) pp. 541–550. Elsevier/North-Holland, Amsterdam.

—— Blair, S.M., and Westheimer, G. (1980). Fine structure of saccade bursts in macaque pontine nucleus. *Brain Res.* **181**, 460–4.

—— and Mackeben, M. (1978). Pursuit eye movements and their neural control in the monkey. *Pfl. arch. gesamten Physiol.* **377**, 15–23.

—— and Mackeben, M. (1980). Pre-motor single unit activity in the monkey brain stem correlated with eye velocity during pursuit. *Brain Res.* **184**, 210–14.

Edwards, S.B. (in press). The deep cell layers of the superior colliculus: Their reticular characteristics and structural organization. In *The reticular formation revisited* (ed. A. Hobson and M. Brazier). Raven Press, New York.

—— Ginsburgh, C.L., Henkel, C.K., and Stein, B.E. (1979). Sources of subcortical projections to the superior colliculus in the cat. *J. comp. Neurol.* **184**, 309–30.

Efron, R. (1967). The duration of the present. *Ann. N.Y. Acad. Sci.* **138**, 713–29.

—— (1970*a*). The relationship between the duration of a stimulus and the duration of a perception. *Neuropsychologia* **8**, 37–55.

—— (1970*b*). The minimum duration of a perception. *Neuropsychologia* **8**, 57–63.

—— (1970*c*). Effect of stimulus duration on perceptual onset and offset latency. *Percept. Psychophys.* **8**, 231–4.

—— and Lee, D.N. (1971). The visual persistence of a moving stroboscopically illuminated object. *Am. J. Psychol.* **84**, 365–75.

Egeth, H. and Gilmore, G. (1973). Perceptibility of the letters in words and nonwords with complete control for redundancy. *Bull. psychonomic Soc.* **2**, 329.

Eigen, M. and Winkler, R. (1975). *Das Spiel: Naturgesetze steuern den Zufall.* R. Piper and Co. Verlag, Munich.

Ellis, D. and Dember, W.N. (1971). Backward masking of visual targets with internal contours: A replication. *Psychonomic Sci.* **22**, 91–2.

Engel, G.R. (1970). An investigation of visual responses to brief stereoscopic stimuli. *Q. J. exp. Psychol.* **22**, 148–60.

Enoch, J.M., Sunga, R.N., and Bachmann, E. (1970*a*). A static perimetric techique believed to test receptive field properties. I. Extension of Westheimer's experiments on spatial interaction. *Am. J. Ophthalmol.* **70**, 113–26.

—— —— —— (1970*b*). A static perimetric technique believed to test receptive field properties. II. Adaptation of the method to the quantitative perimeter. *Am. J. Ophthalmol.* **70**, 126–37.

Enroth-Cugell, C., Hertz, B.G., and Lennie, P. (1977). Convergence of rod and cone signals in the cat's retina. *J. physiol., Lond.* **269**, 297–318.

—— and Robson. J.G. (1966). The contrast sensitivity of retinal ganglion cells of the cat. *J. physiol., Lond.* **187**, 517–52.

—— and Shapley, R.M. (1973*a*). Adaptation and dynamics of cat retinal ganglion cells. *J. physiol., Lond.* **233**, 271–309.

—— —— (1973*b*). Flux, not retinal illumination, is what cat retinal ganglion cells really care about. *J. physiol., Lond.* **233**, 311–26.

Erdmann, B. and Dodge, R. (1898). *Psychologische Untersuchungen über das Lesen auf experimenteller Grundlage.* Max Niemeyer, Halle.

Eriksen, C.W. (1966). Temporal luminance summation effects in backward and forward masking. *Percept. Psychophys.* **1**, 87–92.

—— Becker, B.A. and Hoffman, J.E. (1970). Safari to masking land: A hunt for the elusive U. *Percept. Psychophys.* **8**, 245–50.

—— and Colegate, R.L. (1970). Identification of forms at brief durations when seen in apparent motion. *J. exp. psychol.* **84**, 137–40.

—— and Collins, J.F. (1967). Some temporal characteristics of visual pattern perception. *J. exp. Psychol.* **74**, 476–84.

—— —— (1968). Sensory traces versus the psychological moment in the temporal organization of form. *J. exp. Psychol.* **77**, 376–82.

—— and Eriksen, B.A. (1972). Visual backward masking as measured by voice reaction time. *Percept. Psychophys.* **12**, 5–8.

—— and Hoffman, M. (1963). Form recognition at brief duration as a function of adapting field and interval between stimulations. *J. exp. Psychol.* **66**, 485–99.

—— and Lappin, J.S. (1964). Luminance summation–contrast reduction as a basis for certain forward and backward masking effects. *Psychonomic Sci.* **1**, 313–14.

Erismann, Th. (1935). Die Empfindungszeit. *Arch. gesammte Psychol.* **93**, 453–519.

Erwin, D.E. (1976). Further evidence for two components in visual persistence. *J. exp. psychol.: hum. percept. Perf.* **2**, 191–209.

—— and Nebes, R.D. (1976). Right hemispheric involvement in the functional properties of visual persistence. Paper presented at *the annual meeting of the Eastern Psychological Association, New York, April.*

Exner, S. (1868). Über die zu einer Gesichtswahrnehmung nöthige Zeit. *Wiener Sitzungsbericht der mathematisch—naturwissenschaftlichen Classe der kaiserlichen Akademie der Wissenschaften* **58**, Part 2, 601–32.

—— (1875). Experimentelle Untersuchung der einfachsten psychischen Processe. *Pfl. arch. gesamten Physiol.* **11**, 403–32.

—— (1888). Ueber optische Bewegungsempfindungen. *Biologisches Zentralblatt (Leipzig)* **8**, 437–48.

—— (1898). Studien auf dem Grenzgebiet des lokalisierten Sehens. *Pfl. arch. gesamten Physiol.* **73**, 117–71.

Fain, G.L. (1975). Interactions of rod and cone signals in the mudpuppy retina. *J. physiol., Lond.* **252** 735–69.

—— and Dowling, J.E. (1973). Intracellular recordings from single rods and cones in the mudpuppy retina. *Science, N.Y.* **180**, 1178–81.

Famiglietti, E.V., Jr., and Kolb. H. (1976). Structural basis for ON- and OFF-center responses in retinal ganglion cells. *Science, N.Y.* **194**, 193–5.

Fechner, G.T., (1840a). Ueber die subjektiven Nachbilder und Nebenbilder. I. *Poggendorf Annalen der Physik und Chemie* **50**, 193–221.

—— (1840b). Ueber die subjektiven Nachbilder und Nebenbilder. II. *Poggendorf Annalen der Physik und Chemie* **50**, 427–70.

Fehmi, L.G., Adkins, J.W., and Lindsley, D.B. (1969). Electrophysiogical correlates of visual perceptual masking in monkeys. *Exp. brain Res.* **7**, 299–316.

Fehrer, E. (1966). Effect of stimulus similarity on retroactive masking. *J. exp. Psychol.* **71**, 612–15.

—— and Biederman, I. (1962). A comparison of reaction and verbal report in the detection of masked stimuli. *J. exp. Psychol.* **64**, 126–30.

—— and Raab. D. (1962). Reaction time to stimuli masked by metacontrast. *J. exp. psychol.* **63**, 143–7.

—— and Smith, E. (1962). Effects of luminance ratio on masking. *Percept. Motor Skills* **14**, 243–53.

Feinbloom, W. (1938). A quantitative study of visual afterimage. *Arch. Psychol.* **33**, Nov., No. 233.

Felsten, G. and Wasserman, G.W. (1978). Masking by light in *Limulus* receptors. *J. comp. physiol. Psychol.* **92**, 778–84.

—— —— (1979). Masking-induced sensitivity changes in Limulus photoreceptors. *Vis. Res.* **19**, 943–5.

—— —— (1980). Visual masking: Mechanisms and theories. *Psychol. Bull.* **88**, 329–53.

Ferry, E.S. (1892). Persistence of vision. *Am. J. Sci.* **44**, Series 3, 192–207.

Ferster, D. and LeVay, S. (1978). The axonal arborizations of lateral geniculate neurons in the striate cortex of cat. *J. comp. Neurol.* **182**, 932–44.

Festinger, L. and Holtzmann, J.D. (1978). Retinal image smear as a source of information about magnitude of eye movements. *J. exp. psychol.: hum. percept. Perf.* **4**, 573–85.

Findlay, J.M. (1980). The visual stimulus for saccadic eye movements in human observers. *Perception* **9**, 7–21.

Finkel. D.L. (1973). A developmental comparison of processing of two types of visual information. *J. exp. child Psychol.* **16**, 250–66.

—— and Smythe, L. (1973). Short-term storage of spatial information. *Dev. Psychol.* **9**, 424–8.

Finlay, B. S., Schiller, P.H., and Volman, S.F. (1976). Quantitative studies of single-cell properties in monkey striate cortex. IV. Cortico-tectal cells. *J. Neurophysiol.* **39**, 1352–61.

Fiorentini, A., Chez, C., and Maffei, L. (1972). Physiological correlates of adaptation to a rotated visual field. *J. physiol., Lond.* **227**, 313–22.

—— and Maffei, L. (1977). Instability of the eye in the dark and proprioception. *Nature, Lond.* **269**, 330–1.

Fischer, B., Barth, R., and Sternheim, C.E. (1978). Interaction of receptive field responses and shift-effect in cat retinal and geniculate neurons. *Exp. brain Res.* **31**, 235–48.

—— and Boch, R. (1981*a*). Activity of neurons in area 19 preceding visually guided eye movements of trained rhesus macaques. In *Progress in oculomotor research* (ed. A. F. Fuchs and W. Becker) pp. 211–14. Elsevier/North-Holland, Amsterdam.

—— —— (1981*b*). Selection of visual targets activates prelunate cortical cells in trained rhesus monkey. *Exp. brain Res.* **41**, 431–3.

—— —— (1981*c*). Enhanced activation of neurons in prelunate cortex before visually guided saccades of trained rhesus monkey. *Exp. brain Res.* **44**, 129–37.

—— —— and Bach, M. (1981). Stimulus *vs.* eye movements: Comparison of neural activity in the striate and prelunate visual cortes (A17 and A19) of trained rhesus monkey. *Exp. brain Res.* **43**, 69–77.

—— and Krüger, J. (1974). The shift-effect in the cat's lateral geniculate nucleus. *Exp. brain Res.* **21**, 225–7.

—— Krüger, J., and Droll, W. (1975). Quantitative aspects of the shift-effect in cat retinal ganglion cells. *Brain Res.* **82**, 391–403.

Fisicaro, S.A., Bernstein, I.H., and Narkiewicz, P. (1977). Apparent movement and metacontrast suppression. *Percept. Psychophys.* **22**, 517–25.

Flaherty, T.B. and Matteson, H.H. (1971). Comparison of two measures of metacontrast. *J. opt. soc. Am.* **61**, 828–30.

Foster, D.H. (1976). Rod–cone interaction in the after-flash effect. *Vis. Res.* **16**, 393–6.

—— (1977). Rod- and cone-mediated interactions in the fine-grain movement illusion. *Vis. Res.* **17**, 123–7.

—— (1978). Action of red-sensitive colour mechanism on blue-sensitive colour mechanism in visual masking. *Optica Acta* **25**, 1001–4.

—— (1979). Interactions between blue- and red-sensitive colour mechanisms in metacontrast masking. *Vis. Res.* **19**, 921–31.

—— and Mason, R.J. (1977). Interaction between rod and cone systems in dichoptic visual masking. *Neurosci. Lett.* **4**, 39–42.

Fotta, M. E. (1979). Disinhibition of type B metacontrast and paracontrast. Paper present at *The annual meeting of the Midwestern Psychological Association, Chicago, May*.

Fox. R. (1978). Visual masking. In *Handbook of sensory physiology*, Vol. 8, *Perception* (ed. R. Held, H. Leibowitz, and H. L. Teuber) pp. 621–53. Springer Verlag, New York.

—— and Lehmkuhle, S. (1977). Iconic memory in stereo space: Seeing without storing. Paper presented at *The annual meeting of the Psychonomic Society, Washington, D.C., November*.

—— —— and Bush R.C. (1977). Stereopsis in the falcon. *Science, N.Y.* **197**, 79–81.

—— —— and Westendorf, D.H. (1976). Falcon visual acuity. *Science, N.Y.* **192**, 263–5.

Friedlander, M.J., Lin, C.-S., and Sherman, S.M. (1979). Structure of physiologically identified X and Y cells in the cat's lateral geniculate nucleus. *Science, N.Y.* **204**, 1114–17.

—— —— Stanford, L.R., and Sherman S.M. (1981). Morphology of functionally identified neurons in lateral geniculate nucleus of the cat. *J. Neurophysiol.* **46**, 80–129.

Frizzi. T.J. (1979). Midbrain reticular stimulation and brightness detection. *Vis. Res.* **19**, 123–30.

Fröhlich, F.W. (1921). Untersuchungen über periodische Nachbilder. *Zeitschr. Sinnesphysiol.* **52**, 60–8.

—— (1922*a*). Über den Einfluss der Hell-und Dunkeladaptation auf den Verlauf der periodischen Nachbilder. *Zeitschr. Sinnesphysiol.* **53**, 79–107.

—— (1922*b*). Über die Abhängigkeit der periodischen Nachbilder von der Dauer der Belichtung. *Zeitschr. Sinnesphysiol.* **53**, 108–21.

—— (1923). Über die Messung der Empfindungszeit. *Zeitschr. Sinnesphysiol.* **54**, 58–78.

—— (1929). *Die Empfindungszeit*. Fischer Verlag, Jena.

Frome, F.S., MacLeod, D.I.A., Buck, S.L., and Williams, D. R. (1981). Large loss of visual sensitivity to flashed peripheral targets. *Vis. Res.* **21**, 1323–8.

Frost, D., and Pöppel, E. (1976). Different programming modes of human saccadic eye movements as a function of stimulus eccentricity: Indications of functional subdivisions of the visual field. *Biol. Cybernet.* **23**, 39–48.

Frumkes, T.E., Sekuler, M.D., Barris, M.C., Reiss, E.H., and Chalupa, L.M. (1973). Rod–cone interaction in human scotopic vision—I. Temporal analysis. *Vis. Res.* **13**, 1269–82.

—— and Temme, L.A. (1977). Rod–cone interaction in human scotopic vision—II. Cones influence rod increment thresholds. *Vis. Res.* **17**, 673–9.

Fry. G.A. (1934). Depression of the activity aroused by a flash of light by applying a second flash immediately afterwards to adjacent areas of the retina. *Am. J. Physiol.* **108**, 701–7.

—— (1936). Color sensations produced by intermittent white light and the three-component theory of color vision. *Am. J. Physiol.* **48**, 464–9.

—— and Alpern. M. (1946). Theoretical implications of the response of a photoreceptor to a flash of light. *Am. J. Optometry* **23**, 509–25.

—— —— (1953). The effect of a peripheral glare source upon the apparent brightness of an object. *J. opt. soc. Am.* **43**, 189–95.

—— and Bartley, S.H. (1936). The effect of steady illumination of one part of the retina upon the critical flicker frequency in another. *J. exp. Psychol.* **19**, 351–6.

Fuchs, A.F. (1976). The neurophysiology of saccades. In *Eye movements and psychological processes* (ed. R. A. Monty and J. W. Senders) pp. 39–53. Lawrence Erlbaum Associates, Hillsdale, N.J.

Fukuda, Y. (1971). Receptive field organization of the cat optic nerve fibers with special reference to conduction velocity. *Vis. Res.* **11**, 209–26.

—— (1973). Differentiation of principal cells of the rat lateral geniculate body into two groups; fast and slow. *Exp. brain Res.* **17**, 242–60.

—— and Saito, H.-A. (1971). The relationship between response characteristics to flicker stimulation and receptive field organization in the cat's optic nerve fibers. *Vis. Res.* **11**, 227–40.

—— and Stone, J. (1974). Retinal distribution and central projections of Y-, X- and W-cells of the cat's retina. *J. Neurophysiol.* **37**, 749–72.

—— —— (1976). Evidence of differential inhibitory influences of X- and Y-type relay cells in the cat's lateral geniculate nucleus. *Brain Res.* **113**, 188–96.

—— and Sugitani, M. (1974). Cortical projections of two types of principal cells of the rat lateral geniculate body. *Brain Res.* **67**, 157–61.

—— —— and Iwama. K. (1973). Flash-evoked responses of two types of principal cells in the rat lateral geniculate body. *Brain Res.* **57**, 208–12.

—— Sumitoma, I., Sugitani, M., and Iwama, K. (1979). Receptive field properties of cells in the dorsal part of the albino rat's lateral geniculate nucleus. *Jap. J. Physiol.* **29**, 283–307.

Ganz, L. (1975). Temporal factors in visual perception. In *Handbook of perception*, Vol. 5, (ed. E. C. Carterette and M. P. Friedman) pp. 169–231. Academic Press, New York.

Garey, L.J. and Blakemore, C. (1977). Monocular deprivation: Morphological effects on different classes of neurons in the lateral geniculate nucleus. *Science, N.Y.* **195**, 414–16.

Geldard, F.A. (1932). Foveal sensitivity as influenced by peripheral stimulation. *J. gen. Psychol.* **7**, 185–9.

—— (1934). Flicker relations within the fovea. *J. opt. soc. Am.* **24**, 299–302.

Gibson. J.J. (1966) *The senses considered as perceptual systems*. Houghton Mifflin, Boston.

—— (1977). The theory of affordances. In *Perceiving, acting, and knowing* (ed. R. Shaw and J. Bransford) pp. 67–82. Lawrence Erlbaum Associates, Hillsdale, N.J.

—— (1979). *The Ecological Approch to Visual Perception*. Houghton Mifflin, Boston.

Gielen, C.C.A.M., van Gisbergen, J.A.M., and Vendrik. A.J.H. (1981). Characterization of spatial and temporal properties of monkey LGN Y-cells. *Biol. Cybernet.* **40**, 157–70.

Gilinksy, A.S. (1967). Masking of contour-detectors in the human visual system. *Psychonomic Sci.* **8**, 395–6.

—— (1968). Orientation-specific effects of patterns of adapting light on visual acuity. *J. opt. soc. Am.* **58**, 13–18.

—— and Mayo, T.H. (1971). Inhibitory effects of orientational adaption. *J. opt. soc. Am.* **61**, 1710–14.

Ginsburg, A.P. (1976). The perception of visual form: A two-dimensional filter analysis. Paper presented at *The Fourth Symposium on Sensory Systems Physiology: Information Processing in Visual Systems, Pavlov Institute of Physiology of the USSR Academy of Sciences, Leningrad, November.*

—— (1978). *Visual information processing based on spatial filters constrained by biological data.* Technical Report AMRL–TR–78–129. Aerospace Medical Research Laboratory, Wright Patteron Air Force Base, Ohio, USA.

—— Carl, J.W., Kabrisky, M., Hall, F.C., and Gill. R.A. (1972). Psychological aspects of a model for the classification of visual images. Paper presented at *The International Congress of Cybernetics and Systems, Oxford University, England, August.*

Glass, R.A. and Sternheim, C.E. (1973). Visual sensitivity in the presence of alternating monochromatic fields of light. *Vis. Res.* **13**, 689–99.

Goldberg, M.E. and Bushnell, M.C. (1981*a*). Behavioral enhancement of visual responses in monkey cerebral cortex. II. Modulation in frontal eye fields specifically related to saccades. *J. Neurophysiol.* **46**, 773–87.

Goldberg, M.E. and Bushnell, M.C. (1981*b*). Role of frontal eye fields in visually guided saccades. In *Progress in oculomotor research* (ed. A. F. Fuchs and W. Becker) pp. 185–92. Elsevier/North-Holland, Amsterdam.

—— and Robinson, D.L. (1977). Visual mechanisms underlying gaze: Function of the cerebral cortex. In *Developments in neuroscience*, Vol. 1, *Control of gaze by brain stem neurons* (ed. R. A. Baker and A. Berthoz) pp. 445–451. Elsevier/North-Holland Biomedical Press, Amsterdam.

—— —— (1978). Visual system: Superior colliculus. In *Handbook of behavioral neurobiology* (ed. R. B. Masterton) pp. 119–64. Plenum Publishing, New York.

—— —— (1980). The significance of enhanced visual responses in posterior parietal cortex. *Behav. brain Sci.* **3**, 503–5.

—— and Wurtz, R.H. (1972*a*). Activity of superior colliculus cells in monkey. I. Visual receptive fields of single neurons. *J. Neurophysiol.* **35**, 542–59.

—— —— (1972*b*). Activity of superior colliculus in behaving monkey. II. Effect of attention on neuronal responses. *J. Neurophysiol.* **35**, 560–75.

Gordon, J. and Graham, N. (1973). Early light and dark adaptation in frog on–off retinal ganglion cells. *Vis. Res.* **13**, 647–59.

Gould, S.J. (1982). Darwinism and the expansion of evolutionary theory. *Science, N.Y.* **216**, 380–7.

Gouras, P. (1968). Identification of cone mechanisms in monkey ganglion cells. *J. physiol., Lond.* **199**, 533–47.

—— (1969). Antidromic responses of orthodromically identified ganglion cells in monkey retina. *J. physiol., Lond.* **204**, 407–19.

—— and Link, K. (1966). Rod and cone interaction in dark-adapted monkey ganglion cells. *J. physiol., Lond.* **184**, 499–510.

—— and Zrenner, E. (1981). Color coding in the primate retina. *Vis. Res.* **21**, 1591–8.

Graham, C.H. and Kemp, E.H. (1938). Brightness discrimination as a function of the duration of the increment in intensity. *J. gen. Physiol.* **21**, 635–50.

Graham, C. H. (1965). Some fundamental data. In *Vision and visual perception* (ed. C. H. Graham) pp. 68–80. John Wiley and Sons, New York.

—— and Granit, R. (1931). Comparative studies of peripheral and central retina, VI. *Am. J. Physiol.* **98**, 664–73.

Granit, R. (1972*a*). Discovery and understanding. *Annu. rev. Physiol.* **34**, 1–12.

—— (1972*b*). In defense of teleology. In *Brain and human behavior* (ed. A. G. Karczmar and J. C. Eccles) pp. 400–8. Springer, New York.

—— (1977). *The purposive brain.* MIT Press, Cambridge. Ma.

Green, M. (1981*a*). Psychophysical relationships among mechanisms sensitive to pattern, motion and flicker. *Vis. Res.* **21**, 971–83.

—— (1981*b*). Spatial frequency effects in masking by light. *Vis. Res.* **21**, 861–6.

Greenspoon, T.S. and Eriksen, C.W. (1968). Interocular non-independence. *Percept. Psychophys.* **3**, 93–6.

Grobstein, P. and Chow, K.L. (1975). Receptive field development and individual experience. *Science, N.Y.* **190**, 352–8.

Gross, C.G., Bender, D.B., and Gerstein, G.L. (1979). Activity of inferior temporal neurons in behaving monkeys. *Neuropsychologia* **17**, 215–29.

—— —— and Roch-Miranda, C.E. (1974). Inferotermporal cortex: A single unit analysis. In *The neurosciences third study program* (ed. F. O. Schmitt and F. G. Worden) pp. 229–238. MIT Press, Cambridge. Ma.

Grossberg, M. (1970). Frequencies and latencies in detecting two-flash stimuli. *Percept. Psychophys.* **7**, 377–80.

Growney, R. (1976). The function of contour in metacontrast. *Vis. Res.* **16**, 253–61.

—— and Weisstein, N. (1972). Spatial characteristics of metacontrast. *J. opt. soc. Am.* **62**, 690–6.

—— —— and Cox, S.I. (1977). Metacontrast as a function of spatial separation with narrow line targets and masks. *Vis. Res.* **17**, 1205–10.

Grüsser, O.-J. (1960). Receptor abhängige Potentiale der Katzenretina und ihre Reaktionen auf Flimmerlicht. *Pfl. arch. gesamten Physiol.* **221**, 511–25.

—— and Grüsser-Cornehls, U. (1972). Interaction of vestibular and visual inputs in the visual system. In *Progress in brain research*, Vol. 37, *Basic aspects of central vestibular mechanisms* (ed. A. Brodal and O. Pompeiano) pp. 573–583. Elsevier/North-Holland Biomedical Press, Amsterdam.

Guitton, D. (1981). On the participation of the feline 'frontal eye field' in the control of eye and head movements. In *Progress in oculomotor research* (ed. A. F. Fuchs and W. Becker) pp. 193–201. Elsevier/North Holland, Amsterdam.

—— Crommelinck, M., and Roucoux, A. (1980). Stimulation of the superior colliculus in the alert cat. I. Eye movements and neck EMG activity evoked when the head is restrained. *Exp. brain Res.* **39**, 63–73.

Haber, R.N. (1968). Perceptual reports *vs.* recognition responses: A reply to Parks. *Percept. Psychophys.* **4**, 374.

—— (1969). Repetition, visual persistence, visual noise and information processing. In *Information processing in the nervous system* (ed. D. N. Leibovic) pp. 121–140. Springer Verlag, New York.

—— (1978). Visual perception. *Annu. rev. Psychol.* **29**, 31–59.

—— and Nathanson, L.S. (1968). Post-retinal storage? Some further observations on Parks' camel as seen through the eye of a needle. *Percept. Psychophys.* **3**, 349–55.

—— and Standing, L. (1969). Direct measures of short-term visual storage. *Q.J. exp. Psychol.* **21**, 43–54.

—— —— (1970). Direct estimates of the apparent duration of a flash. *Can. J. Psychol.* **24**, 216–29.

Hale, P.T., Sefton, A.J., and Dreher, B. (1979). A correlation of receptive field properties with conduction velocity of cells in the rat's retino-geniculo-cortical pathway. *Exp. brain Res.* **35**, 425–42.

Hallett, P.E. (1969). Rod increment thresholds on steady and flashed backgrounds. *J. physiol., Lond.* **202**, 344–77.

Hallett, P.E. and Lightstone, A.D. (1976a). Saccadic eye movements towards stimuli triggered by prior saccades. *Vis. Res.* **16**, 99–106.

—— —— (1976b). Saccadic eye movements to flashed targets. *Vis. Res.* **16**, 107–114.

Hammond, P. (1968). Spectral properties of dark-adapted retinal ganglion cells in the plaice (*Pleuronectes platessa*, L.). *J. physiol., Lond.* **195**, 535–56.

—— (1971). Chromatic sensitivity and spatial organization of cat visual cortical cells: Cone–rod interaction. *J. physiol., Lond.* **213**, 475–94.

—— (1972). Chromatic sensitivity and spatial organization of LGN neurone receptive fields in cat: Cone–rod interaction. *J. physiol., Lond.* **225**, 391–413.

—— (1974). Cat retinal ganglion cells: Size and shape of receptive field centers. *J. physiol., Lond.* **242**, 99–118.

—— (1975). Receptive field mechanisms of sustained and transient retinal ganglion cells in the cat. *Exp. brain Res.* **23**, 113–28.

Harris, C.S. (1980). *Visual coding and adaptability.* Lawrence Erlbaum Associates, Hillsdale, N.J.

Harris, L.R. (1980). The superior colliculus and movements of the head and eyes in cats. *J. physiol., Lond.* **300**, 367–91.

Harris, M.G. (1980). Velocity specificity of the flicker to pattern sensitivity ratio in human vision. *Vis. Res.* **20**, 687–91.

Harrison, K. and Fox, R. (1966). Replication of reaction time to stimuli masked by metacontrast. *J. exp. Psychol.* **71**, 162–3.

Harter, M.R. (1967). Excitability cycles and cortical scanning: A review of two hypotheses of central intermittency in perception. *Psychol. Bull.* **68**, 47–58.

—— and White, C.R. (1967). Perceived number and evoked potential. *Science, N.Y.* **156**, 406–8.

Hartline, H.K. and Ratliff, F. (1957). Inhibitory interaction of receptor units in the eye of *Limulus*. *J. gen. Physiol.* **40**, 357–76.

—— —— (1958). Spatial summation of inhibitory influences in the eye of *Limulus*, and the mutual interaction of receptor units. *J. gen. Physiol.* **41**, 1049–66.

Hartmann, E., Lachenmayr, B., and Brettel, H. (1979). The peripheral critical flicker frequency. *Vis. Res.* **19**, 1019–23.

Hartmann, N. (1964). *Der Aufbau der realen Welt.* Walter de Gruyter, Berlin.

Harvey, A.R. (1980). A physiological analysis of subcortical and commisural projections of area 17 and 18 of the cat. *J. physiol., Lond.* **302**, 507–34.

Harwerth, R.S. and Levi, D.M. (1978). Reaction time as a measure of suprathreshold grating detection. *Vis. Res.* **18**, 1579–86.

—— Boltz, R.L., and Smith, E.L. (1980). Psychophysical evidence for sustained and transient channels in monkey visual system. *Vis. Res.* **20**, 15–22.

Hearty, P.J. and Mewhort, D.J.K. (1975). Spatial localization in sequential letter displays. *Can. J. Psychol.* **29**, 348–59.

Hebb, D.O. (1949). *The organization of behavior.* John Wiley and Sons, New York.

Heinemann, E.G. (1955). Simultaneous brightness induction as a function of inducing- and test-field luminances. *J. exp. Psychol.* **50**, 89–96.

Hellige, J.B., Walsh, D.A., Lawrence, V.S., and Cox, P.J. (1977). The importance of figural relationships between target and mask. *Percept. Psychophys.* **21**, 285–6.

—— —— —— and Prasse, M. (1979). Figural relationship effects and mechanisms of visual masking. *J. exp psychol.: human percept. Perf.* **5**, 88–100.

Helmholtz, H.V. (1866). *Handbuch der physiologischen Optik* (lst ed.). Voss, Leipzig. (trans. J. P. C. Southall (1962). *Handbook of Physiological Optics* (3rd ed.). Dover, New York).

Hempel, C.G. (1966). Explanations in science and history. In *Philosophical analysis and history* (ed. W. H. Dray) pp. 95–126. Harper and Row, New York.

Hendrickson, A.E., Wilson, J.R., and Ogren, M.P. (1978). The neuroanatomical organization of pathways between the dorsal lateral geniculate nucleus and visual cortex in old world and new world primates. *J. comp. Neurol.* **182**, 123–36.

Henn, V., Young, L.R., and Finley, C. (1974). Vestibular nucleus units in alert monkeys are also influenced by moving visual fields. *Brain Res.* **71**, 144–9.

Henry, G.H., Bishop, P.O., Tupper, R.M., and Dreher, B. (1973). Orientation specificity and response variability of cells in the striate cortex. *Vis. Res.* **13**, 1771–9.

—— Harvey, A.R., and Lund, J.S. (1979). The afferent connections and laminar distribution of cells in the cat striate cortex. *J. comp. Neurol.* **187**, 725–44.

Henson, D.B. (1979). Investigation into corrective saccadic eye movements for refixation amplitudes of 10 degrees and below. *Vis. Res.* **19**, 57–61.

Hepp, K., Henn, V., and Jaeger, J. (1982). Eye movement related neurons in the cerebellar nuclei of the alert monkey. *Exp. brain Res.* **45**, 253–64.

Hepp, K., Henn, V., Jaeger, J., and Waespe, W. (1981). Oculomotor pathways through the cerebellum. In *Progress in oculomotor research* (ed. A. F. Fuchs and W. Becker) pp. 97–105. Elsevier/North-Holland, Amsterdam.

Hermann, L. (1870). Eine Erscheinung simultanen Kontrastes. *Pfl. arch. gesamten Physiol.* **3**, 13–15.

Hernandez, L.L. and Lefton, L.A. (1977). Metacontrast as measured under signal detection model. *Perception* **6**, 695–702.

Hering, E. (1872). Zuhr Lehre vom Lichsinn. Ueber successive Lichtinduction. *Wiener Sitzungsberichte der mathematisch-naturwissenschaftliche Classe der kaiserlichen Akademie der Wissenschaft* **66**, Part 3, 5–24.

—— (1878). *Zuhr Lehre vom Lichtsinn*. Carl Gerold's Sohn, Vienna.

Herrick, R.M. (1974). Foveal light-detection thresholds with two temporally spaced flashes: A review. *Percept. Psychophys.* **15**, 361–7.

Hess, R.F., Campbell, F.W., and Greenhalgh, T. (1978). On the nature of the neural abnormality in human amblyopia; neural aberrations and visual sensitivity loss. *Pfl. arch. gesamten Physiol.* **377**, 201–7.

—— and Howell, E.R. (1977). The threshold contrast sensitivity function in strabismic amblyopia: Evidence for a two-type classification. *Vis. Res.* **17**, 1049–55.

Hickey, T.L. and Guillery, R.W. (1981). A study of Golgi preparations from the human lateral geniculate nucleus. *J. comp. Neurol.* **200**, 545–77.

—— Winters, R.W., and Pollack, J.G. (1972). Receptive field center and surround interactions in single cat retinal ganglion cells. *Brain Res.* **43**, 250–3.

—— —— —— (1973). Centre-surround interactions in two types of on-centre retinal ganglion cells in the cat. *Vis. Res.* **13**, 1511–26.

Hirsch, H.V.B. and Spinelli, D.M. (1970). Visual experience modifies distribution of horizontally and vertically oriented receptive fields in cat. *Science, N.Y.* **168**, 869–71.

Hochberg, J. (1968). In the mind's eye. In *Contemporary theory and research in visual perception* (ed. R. N. Haber) pp. 309–331. Holt, Rhinehart and Winston, New York.

—— (1978). *Perception*. Prentice-Hall, Englewood Cliffs, N.J.

Hochstein, S. (1979). Visual cell X/Y classification: Characteristics and correlations. In *Developmental neurobiology of vision* (ed. R. D. Freedman) pp. 185–94. Plenum Press, New York.

—— and Shapley, R.M. (1976a). Quantitative analysis of retinal ganglion cell classifications. *J. physiol., Lond.* **262**, 237–64.

—— —— (1976b). Linear and nonlinear spatial sub-units in Y cat retinal ganglion cells. *J. physiol., Lond.* **262**, 365–84.

Hoffman, J. (1980). Interaction between global and local levels of form. *J. exp. psychol.: hum. percept. Perf.* **6**, 222–34.

Hoffman, K.-P. (1973). Conduction velocity in pathways from retina to superior colliculus in the cat: A correlation with receptive-field properties. *J. Neurophysiol.* **36**, 409–24.

—— and Sherman, S.M. (1975). Effects of early binocular deprivation of visual input to cat superior colliculus. *J. Neurophysiol.* **38**, 1049–59.

—— and Stone, J. (1971). Conduction velocity of afferents to cat visual cortex: A correlation with cortical receptive field properties. *Brain Res.* **32**, 460–6.

—— —— and Sherman, S.M. (1972). Relay of receptive field properties in the dorsal lateral geniculate nucleus of the cat. *J. Neurophysiol.* **35**, 518–31.

Hogben, J.H. and DiLollo, V. (1974). Perceptual integration and perceptual segregation of brief visual stimuli. *Vis. Res.* **14**, 1059–69.

Hohmann, A. and Creutzfeldt, O.D. (1975). Squint and the development of binocularity in humans. *Nature, Lond.* **254**, 613–14.

Holland, H.C. (1963). 'Visual masking' and the effects of stimulant and depressant drugs. In *Experiments with drugs* (ed. H. J. Eysenck) pp. 69–106. Pergamon Press, Oxford.

Holtzworth, R.J. and Doherty, M.E. (1971). Visual masking by light offset. *Percept. Psychophys.* **10**, 327–30.

—— —— (1974). Visual masking by light offset: An experiment in reply to Hogben and DiLollo. *J. exp. Psychol.* **103**, 815–16.

Hood, D.C. (1973). The effects of edge sharpness and exposure duration on detection threshold. *Vis. Res.* **13**, 759–66.

—— and Grover, B.G. (1974). Temporal summation of light by a vertebrate visual receptor. *Science, N.Y.* **184**, 1003–5.

Horn, G. and Hill, R.M. (1969). Modifications of receptive fields of cells in the visual cortex occurring spontaneously and associated with bodily tilt. *Nature, Lond.* **221**, 186–8.

—— Stechler, A.B., and Hill, R.M. (1972). Receptive fields of units in the visual cortex of the cat in the presence and absence of body tilt. *Exp. brain Res.* **15**, 113–32.

Houlihan, K. and Sekuler, R.W. (1968). Contour interactions in visual masking. *J. exp. Psychol.* **77**, 281–5.

Hubel, D.H. and Wiesel, T.N. (1965). Binocular interaction in striate cortex of kittens reared with artificial squint. *J. Neurophysiol.* **28**, 1041–59.

Hubel, D.H. and Wiesel, T.N. (1962). Receptive fields, binocular interaction and functional architecture in the cat's visual cortex. *J. physiol., Lond.* **160**, 106–54.

—— —— (1968). Receptive fields and functional architecture of monkey striate cortex. *J. physiol., Lond.* **195**, 215–43.

—— —— (1972). Laminar and columnar distribution of geniculo-cortical fibers in the macaque monkey. *J. comp. Neurol.* **146**, 421–50.

—— —— (1974*a*). Sequence regularity and geometry of orientation columns in the monkey striate cortex. *J. comp. Neurol.* **158**, 267–94.

—— —— (1974*b*). Uniformity of monkey striate cortex: A parallel relationship between field size, scatter, and magnification factor. *J. comp. Neurol.* **158**, 295–306.

—— —— (1977). Functional architecture of macaque monkey visual cortex. *Proc. R. soc., Lond.* **198B**, 1–59.

—— —— and LeVay, S. (1977). Plasticity of ocular dominance columns in monkey striate cortex. *Phil. transact. R. soc., Lond.* **278B**, 377–409.

Hull, C.L. (1943). *Principles of behavior.* Appleton-Century-Crofts, New York.

Humphrey, N.K. (1974). Vision in a monkey without striate cortex: A case study. *Perception* **3**, 241–55.

Hunsperger, R.W. and Roman, D. (1976). The integrative role of the intralaminar system of the thalamus in visual orientation and perception in the cat. *Exp. brain Res.* **25**, 231–46.

Ikeda, H. (1979). Physiological basis of amblyopia. *Trends in Neurosci.* **2**, 209–12.

—— (1980). Visual acuity, its development and amblyopia. *J. R. soc. med.* **73**, 546–55.

—— Jacobson, S.G., Plant, G., and Tremain, K.E. (1976). Behavioral, neurophysiological and morphological evidence for a nasal visual field loss in cats reared with monocular convergent squint. *J. physiol., Lond.* **260**, 48–9P.

—— and Tremain, K.E. (1969). Amblyopia occurs in retinal ganglion cells in cats reared with convergent squint without alternating fixation. *Exp. brain Res.* **35**, 559–82.

—— Tremain, K.E., and Einon, G. (1978). Loss of spatial resolution of lateral geniculate nucleus neurones in kittens raised with convergent squint produced at different stages in development. *Exp. brain Res.* **31**, 207–20.

—— and Wright, M.J. (1972a). Receptive field organization of 'sustained' and 'transient' retinal ganglion cells which subserve different functional roles. *J. physiol., Lond.* **227**, 769–800.

—— —— (1972b). Differential effects of refractive errors and receptive field organization of central and peripheral ganglion cells. *Vis. Res.* **12**, 1465–76.

—— —— (1972c). Functional organization of the periphery effect in retinal ganglion cells. *Vis. Res.* **12**, 1875–9.

—— —— (1974). Evidence for 'sustained' and 'transient' neurones in the cat's visual cortex. *Vis. Res.* **14**, 133–6.

—— —— (1975a). Spatial and temporal properties of 'sustained' and 'transient' neurones in area 17 of the cat's visual cortex. *Exp. brain Res.* **22**, 363–83.

—— —— (1975b). Retinotopic distribution, visual latency and orientation tuning of 'sustained' and 'transient' cortical neurones in area 17 of the cat. *Exp brain Res.* **22**, 385–98.

—— —— (1975c). Amblyopic cells in the lateral geniculate nucleus in kittens raised with surgically produced squint. *J. physiol., Lond.* **256**, 41–2P.

—— —— (1976). Properties of LGN cells in kittens reared with convergent squint: A neurophysiological demonstration of amblyopia. *Exp. brain Res.* **25**, 63–77.

Ikeda, M. (1965). Temporal summation of positive and negative flashes in the visual system. *J. opt. soc. Am.* **55**, 1527–34.

—— and Boynton, R.M. (1965). Negative flashes, positive flashes, and flicker examined by increment threshold technique. *J. opt. soc. Am.* **5**, 560–6.

—— and Uchikawa, K. (1978). Integrating time for visual pattern perception and a comparison with the tactile mode. *Vis. Res.* **18**, 1565–71.

Jacobson, J.A. and Rhinelander, G. (1978). Geometric and semantic similarity in visual masking. *J. exp. psychol.: hum. percept. Perf.* **4**, 224–31.

Jakiela, H.G. (1978). Periphery effect of cat retinal ganglion cells re-examined. Poster displayed at *The annual meeting of the Association for Research in Vision and Opthalmology, Sarasota, Florida, April.*

—— (1975). Movement of background patterns reduces sensitivity of cat retinal ganglion cells. Paper presented at *The annual meeting of the Association for Research in Vision and Opthalmology, Sarasota, Florida, April.*

—— and Enroth-Cugell, C. (1976). Adaptation and dynamics in X-cells and Y-cells of the cat retina. *Exp. brain Res.* **24**, 335–42.

Jeannerod, M. and Chouvet, G. (1973). Saccadic displacements of the retinal image: Effects on the visual system in the cat. *Vis. Res.* **13**, 161–9.

Jeffreys, D.A. (1979). Contour specific potentials evoked by saccadic image displacements. *J. physiol., Lond.* **298**, 24P.

Johnston, J.C. and McClelland, J.L. (1974). Perception of letters in words: Seek not and ye shall find. *Science, N.Y.* **184**, 1192–3.

Jones, E.G. (1974). The anatomy of extrageniculate visual mechanisms. In *The neurosciences third study program* (ed. F. O. Schmitt and F. G. Worden) pp. 215–227. MIT Press, Cambridge, Ma.

Jones, R. and Keck, M.J. (1978). Visual evoked response as a function of grating spatial frequency. *Invest. opthalmol. vis. Sci.* **17**, 652–9.

Jonides, J., Irwin, D.E., and Yantis, S. (1982). Integrating visual information from successive fixations. *Science, N.Y.* **215**, 192–4.

Judge, S.J., Wurtz, R.H., and Richmond, B.J. (1980). Vision during saccadic eye movements. I. Visual interactions in the striate cortex. *J. Neurophysiol.* **43**, 1133–55.

Julesz, B. (1971). *Foundations of cyclopean perception*. University of Chicago Press.

—— Breitmeyer, B., and Kropfl, W. (1976). Binocular-disparity-dependent upper-lower hemifield anisotropy and left-right isotropy as revealed by dynamic random-dot stereograms. *Perception* **5**, 129–41.

—— and Chiarucci, E. (1973). Short-term memory for stroboscopic motion perception. *Perception* **2**, 294–60.

—— Gilbert, E.N., Shepp, L.A., and Frisch, H.L. (1973). Inability of humans to discriminate between visual textures that agree in second order statistics-revisited. *Perception* **2**, 391–405.

Jung, R. (1972). Neurophysiological and psychophysical correlates in vision research. In *Brain and human behavior* (ed. A. G. Karczmar and J. C. Eccles) pp. 209–258. Springer, Berlin.

—— (1973). Visual perception and neurophysiology. In *Handbook of sensory physiology*, Vol. 7/3A, *Central processing of visual information* (ed. R. Jung) pp. 1–152. Springer, Berlin.

Kaas, J.H., Huerta, M.F., Weber, J.T., and Harting, J.K. (1978). Patterns of retinal terminations and laminar organization of the lateral geniculate nucleus of primates. *J. comp. Neurol.* **182**, 517–54.

Kahneman, D. (1964). Temporal summation in an acuity task at different energy levels—a study of the determinants of summation. *Vis. Res.* **4**, 557–66.

—— (1965). Exposure duration and effective figure-ground contrast. *Q. J. exp. Psychol.* **17**, 308–14.

—— (1966). Time–intensity reciprocity in acuity as a function of luminance and figure-ground contrast. *Vis. Res.* **6**, 207–15.

—— (1967*a*). Time–intensity reciprocity under various conditions of adaptation and backward masking. *J. exp. Psychol.* **71**, 543–9.

—— (1967*b*). An onset–onset law for one case of apparent motion and metacontrast. *Percept. Psychophys.* **2**, 577–84.

—— (1968). Method, findings, and theory in studies of visual masking. *Psychol. Bull.* **70**, 404–25.

—— and Norman, J. (1964). The time–intensity relation in visual perception as a function of the observer's task. *J. exp. Psychol.* **68**, 215–20.

—— —— and Kubovy, M. (1967). Critical duration for the resolution of form: Centrally or periphally determined? *J. exp. Psychol.* **73**, 323–7.

—— and Wolman, R.E. (1970). Stroboscopic motion: Effects of duration and interval. *Percept. Psychophys.* **8**, 161–4.

Kalil, K. and Reh, T. (1979). Regrowth of severed axons in the neonatal central nervous system: Establishment of normal connections. *Science, N.Y.* **205**, 1158–61.

—— and Worden, I. (1978). Cytoplasmic laminated bodies in the lateral geniculate nucleus of normal and dark reared cats. *J. comp. Neurol.* **178**, 469–86.

Kantor, J.R. (1935). The evolution of mind. *Pyschol. Rev.* **42**, 455–65.

Kao, K.C. and Dember, W.N. (1973). Effect of size of ring on backward masking of a disk by a ring. *Bull. psychonomic Soc.* **2**, 15–17.

Kaplan, E. and Shapley, R.M. (1982). X and Y cells in the lateral geniculate nucleus of macaque monkeys. *J. physiol., Lond.* **330**, 125–43.

Kase, M., Miller, D.C., and Noda, H. (1980). Discharges of Purkinje cells and mossy fibres in the cerebellar vermis of the monkey during saccadic eye movements and fixation. *J. physiol., Lond.* **300**, 539–55.

Katayama, S. and Aizawa, T. (1956). The mechanisms of spatial induction in the retina. *Tohoku J. exp. Med.* **64**, 179–87.

Keller, E.L. (1974). Participation of medial pontine reticular formation in eye movement generation in monkey. *J. Neurophysiol.* **37**, 316–32.

—— (1981). Brain stem mechanisms in saccadic control. In *Progress in oculomotor research* (ed. A. F. Fuchs and W. Becker) pp. 57–2. Elsevier/ North–Holland, Amsterdam.

Kelly, D.H. (1961). Visual responses to time-dependent stimuli. I. Amplitude sensitivity measurements. *J. opt. soc. Am.* **51**, 422–9.

—— (1971*a*). Theory of flicker and transient reponses. I. Uniform fields. *J. opt. soc. Am.* **61**, 537–46.

—— (1971*b*). Theory of flicker and transient responses. II. Counterphase gratings. *J. opt. soc. Am.* **61**, 632–40.

—— (1972). Adaptation effects on spatio-temporal sine-wave thresholds. *Vis. Res.* **12**, 89–101.

—— (1973). Lateral inhibition in human color mechanisms. *J. physiol., Lond.* **228**, 55–72.

Kerr, J.L. (1971). Visual resolution in the periphery. *Percept. Psychophys.* **9**, 375–8.

Keys, W. and Robinson, D.L. (1979). Eye movement-dependent enhancement of visual responses in the pulvinar nucleus of the monkey. *Soc. neurosci. Abstr.* **5**, 791.

Kimura M., Komatsu, Y., and Toyama, K. (1980). Differential responses of 'simple' and 'complex' cells of cat's striate cortex during saccadic eye movements. *Vis. Res.* **20**, 553–6.

Kinchla, R. and Wolfe, J. (1979). The order of visual processing: 'Top-down', 'bottom-up' or 'middle-out'. *Percept. Psychophys.* **25**, 225–31.

King-Smith, P.E. and Carden, D. (1976). Luminance and opponent-color contributions to visual detection and adaptation and to temporal and spatial integration. *J. opt. soc. Am.* **66**, 709–17.

—— and Kulikowski, J.J. (1973). Spatial arrangement of flicker and pattern detectors for a fine line. *J. physiol., Lond.* **234**, 33–5P.

—— —— (1975). Pattern and flicker detection analysed by subthreshold summation. *J. physiol., Lond.* **249**, 519–48.

—— Moore, R., and Riggs, L.A. (1976). A new approach to the frequency response of human vision. *J. physiol., Lond.* **257**, 36–7P.

Kinnucan, M.T. and Friden, T.P. (1981). Visual form integration and discontinuity detection. *J. exp. psychol.: hum. percept. Perf.* **7**, 948–53.

Kinsbourne, M. (1978). *Assymetrical function of the brain.* Cambridge University Press.

—— and Warrington, E.K. (1962*a*). The effect of an aftercoming random pattern on the perception of brief visual stimuli. *Q. J. exp. Psychol.* **14**, 223–34.

—— —— (1962*b*). Further studies on the visual masking of brief visual stimuli by a random pattern. *Q. J. exp. Psychol.* **14**, 235–45.

Kirby, A.W. (1979). The effect of strychnine, bicuculline, and picrotoxin on X and Y cells in the cat retina. *J. gen. Physiol.* **74**, 71–84.

—— and Enroth-Cugell, C. (1976). The involvement of gamma-aminobutyric acid in the organization of cat retinal ganglion cell receptive fields. *J. gen. Physiol.* **68**, 465–84.

Kirk, D.L., Cleland, B.G., Wässle, H., and Levick, W.R. (1975). Axonal conduction latencies of cat retinal ganglion cells in central and peripheral retina. *Exp. brain Res.* **23**, 85–90.

Klein, R. (1979). Does oculomotor readiness mediate cognitive control of visual attention? In *Attention and performance VIII* (ed. R. S. Nickerson) pp. 259–276. Lawrence Erlbaum Associates, Hillsdale, N.J.

Köhler, W. (1966). *The place of value in a world of facts.* Liveright, New York.

Kolb, H. (1974). The connections between horizontal cells and photoreceptors in the retina of the cat: Electron microscopy of Golgi preparations. *J. comp. Neurol.* **155**, 1–14.

Kolers, P. (1962). Intensity and contour effects in visual masking. *Vis. Res.* **2**, 277–94.

—— (1963). Some differences between real and apparent visual movement. *Vis. Res.* **3**, 191–206.

—— (1968). Some psychological aspects of pattern recognition. In *Recognizing patterns* (ed. P. A. Kolers and M. Eden) pp. 4–61. MIT Press, Cambridge, Ma.

—— and Rosner, B.S. (1960). On visual masking (metacontrast): Dichoptic observations. *Am. J. Psychol.* **73**, 2–21.

—— and von Grünau, M.W. (1976). Shape and color in apparent motion. *Vis. Res.* **16**, 329–35.

—— —— (1977). Fixation and attention in apparent motion. *Q. J. exp. Psychol.* **29**, 389–95.

Kratz, K.E., Sherman, S.M., and Kalil, R. (1979). Lateral geniculate nucleus in dark-reared cats: Loss of Y cells without changes in cell size. *Science, N.Y.* **203**, 1353–5.

—— Webb, S.V., and Sherman, S.M. (1978). Electrophysiological classification of X- and Y- cells in the cat's lateral geniculate nucleus. *Vis. Res.* **18**, 489–92.

Krauskopf, J. and Mollon, J.D. (1971). The independence of the temporal integration properties of individual chromatic mechanisms in the human eye. *J. physiol., Lond.* **219**, 611–23.

Kriegman, D.H. and Biederman, I. (1980). How many letters in Bidwell's ghost? An investigation into the upper limits of full report from a brief visual stimulus. *Percept. Psychophys.* **28**, 82–4.

Kristofferson, A.B. (1967a). Attention and psychophysical time. *Acta Psychol.* **27**, 93–100.

—— (1967b). Successiveness discrimination as a two-state quantal process. *Science, N.Y.* **158**, 1337–9.

—— Galloway, J., and Hanson, R.G. (1979). Complete recovery of a masked visual target. *Bull. psychonomic Soc.* **13**, 5–6.

Krüger, J. (1977a). Stimulus dependent colour specificity of monkey lateral geniculate neurons. *Exp. brain Res.* **30**, 297–311.

—— (1977b). The shift-effect in the lateral geniculate body of the rhesus monkey. *Exp. brain Res.* **29**, 387–92.

—— (1980). The shift-effect enhances X- and suppresses Y-type response characteristics of cat retinal ganglion cells. *Brain Res.* **201**, 71–84.

—— and Fischer, B. (1973). Strong periphery effect in cat retinal ganglion cells. Excitatory responses in On- and Off-center neurons to single grid displacements. *Exp. brain Res.* **18**, 316–18.

—— —— and Barth, R. (1975). The shift-effect in retinal ganglion cells of the rhesus monkey. *Exp. brain Res.* **23**, 443–6.

Kuffler, S.W. (1953). Discharge patterns and functional organization of mammalian retina. *J. Neurophysiol.* **16**, 37–68.

Kuhn, T.S. (1957). *The Copernican revolution*. Harvard University Press, Cambridge, Ma.

—— (1962). *The structure of scientific revolution*. University of Chicago Press.

Kulikowski, J.J. (1973). *Limiting conditions of visual perception*. English translation of original manuscript entitled *Warunki graniczne percepcji wzrokowej*. (1969). Prace Instytutu Automatyki PAN No. 77, Warsaw.

—— (1974). Human averaged occipital potentials evoked by pattern and movement. *J. physiol., Lond.* **242**, 70–1P.

—— (1977). Visual evoked potentials as a measure of visibility. In *Visual evoked potentials in man: new developments* (ed. J. E. Desmedt) pp. 168–83. Clarendon Press, Oxford.

—— Bishop, P.O., and Kato, H. (1979). Sustained and transient responses by cat striate cells to stationary flashing light and dark bars. *Brain Res.* **170**, 362–7.

—— and Tolhurst, D.J. (1973). Psychophysical evidence for sustained and transient detectors in human vision. *J. physiol., Lond.* **232**, 149–62.

Kunkel, A. (1874). Über die Abhängigkeit der Farbempfindung von der Zeit. *Pfl. arch. gesamten Physiol.* 197–200.

Künzle, H., Akert, K., and Wurtz, R.H. (1976). Projection of area 8 (frontal eye field) to superior colliculus in the monkey. An autoradiographic study. *Brain Res.* **117**, 487–92.

Kuypers, H.G.J.M. and Lawrence, D.G. (1967). Cortical projections to the red nucleus of the brain stem in the rhesus monkey. *Brain Res.* **4**, 151–88.

LaBerge, D. (1973). Identification of the time to switch attention: A test of a serial and parallel model of attention. In *Attention and performance*, Vol. 4, (ed. S. Kornblum) pp. 71–85. Academic Press, New York.

Latch, M. and Lennie, P. (1977). Rod–cone interaction in light adaptation. *J. physiol., Lond.* **269**, 517–34.

Latour, P.L. (1967). Evidence of internal clocks in the human operator. *Acta Psychol.* **27**, 341–8.

Latto, R. (1978*a*). The effects of bilateral frontal eye-field posterior parietal and superior collicular lesions on visual search in the rhesus monkey. *Brain Res.* **146**, 35–50.

—— (1978*b*). The effects of bilateral frontal eye-field lesions on the learning of visual search tasks by rhesus monkey. *Brain Res.* **147**, 370–6.

Lee, B.B., Elepfandt, A., and Virsu, V. (1981). Phase of responses to moving sinusoidal gratings in cells of cat retina and lateral geniculate nucleus. *J. Neurophysiol.* **45**, 807–17.

Lefton, L.A. (1970). Metacontrast: Further evidence for monotonic functions. *Psychonomic Sci.* **21**, 85–7.

—— (1973). Metracontrast: A review. *Percept. Psychophys.* **13**, 161–71.

—— (1974). Internal contours, intercontour distance, and interstimulus intervals: The complex interaction in metacontrast. *J. exp. Psychol.* **103**, 891–5.

—— and Griffin, J.R. (1976). Metacontrast with internal contours: More evidence for monotonic functions. *Bull. psychonomic Soc.* **7**, 29–32.

—— and Hernandez, L.L. (1977). Metacontrast: Internal contours and different dependent variables. *Bull. psychonomic Soc.* **9**, 427–30.

—— and Newman, Y. (1976). Metacontrast and paracontrast: both photopic and scotopic luminance levels yield monotones. *Bull. psychonomic Soc.* **8**, 435–8.

Legge, G.E. (1978). Sustained and transient mechanisms in human vision: Temporal and spatial properties. *Vis. Res.* **18**, 69–81.

Lehmkuhle, S. and Fox, R. (1980). Effect of depth separation on metacontrast masking. *J. exp. psychol.: hum. percept. Perf.* **6**, 605–21.

—— Kratz, K.E., Mangel, S.C. and Sherman S.M. (1978). An effect of early monocular lid suture upon the development of X-cells in the cat's lateral geniculate nucleus. *Brain Res.* **157**, 346–50.

—— —— —— —— (1980*a*). Spatial and temporal sensitivity of X- and Y-cells in dorsal lateral geniculate nucleus of the cat. *J. Neurophysiol.* **43**, 520–41.

—— —— —— —— (1980*b*). Effects of early monocular lid suture on spatial and temporal sensitivity of neurons in dorsal lateral geniculate nucleus of the cat. *J. Neurophysiol.* **43**, 542–56.

Leichnetz, G.R., Spencer, R.F., Hardy, S.G.P., and Astruc, J. (1981). An anatomical view of the role of cerebral cortex in eye movement. Projections from prefrontal cortex to the superior colliculus and oculomotor complex in the monkey. In *Progress in oculomotor research* (ed. A. F. Fuchs and W. Becker), pp. 177–84. Elsevier/North-Holland, Amsterdam.

Lennie, P. (1980*a*). Parallel visual pathways: A review. *Vis. Res.* **20**, 561–94.

—— (1980*b*). Perceptual signs of parallel pathways. *Phil. trans. R. soc., Lond.* **290B**, 23–37.

—— and Perry, V.H. (1981). Spatial contrast sensitivity of cells in the lateral geniculate nucleus of the rat. *J. physiol., Lond.* **315**, 69–79.

LeVay, S. and Ferster, D. (1977). Relay cell classes in the lateral geniculate nucleus of the cat and the effects of visual deprivation. *J. comp. Neurol.* **172**, 563–84.

—— and Gilbert, C.D. (1976). Laminar patterns of geniculocortical projections in the cat. *Brain Res.* **113**, 1–19.

Leventhal, A.G. (1979). Evidence that the different classes of relay cells in the cat's lateral geniculate nucleus terminate in different layers of the striate cortex. *Exp. brain Res.* **37**, 359–72.

—— and Hirsch, H.V.B. (1978). Receptive-field properties of neurons in different laminae of visual cortex of the cat. *J. Neurophysiol.* **41**, 948–62.

—— Rodieck, R. W., and Dreher, B. (1981). Retinal ganglion cell classes in the old world monkey: Morphology and central projections. *Science, N.Y.* **213**, 1139–42.

Levi, D.M. and Harwerth, R.S. (1977). Spatio-temporal interactions in anisometropic and strabismic amblyopia. *Invest. ophthalmol. vis. Sci.* **16**, 90–5.

—— —— (1978*a*). A sensory mechanism for amblyopia: Electrophysiological studies. *Am. J. opt. physiol. Optics* **55**, 163–71.

—— —— (1978*b*). Contrast evoked potentials in strabismic and anisometropic amblyopia. *Invest. ophthalmol. vis. Sci.* **17**, 571–5.

—— —— (1980). Contrast sensitivity in amblyopia due to stimulus deprivation. *Br. J. Ophthalmol.* **64**, 15–20.

—— —— and Manny, R. (1979). Suprathreshold spatial frequency detection and binocular interaction in strabismic and anisometropic amblyopia. *Invest. ophthalmol. vis. Sci.* **18**, 714–25.

Levick, W.R., Oyster, C.W., and Davis, D.L. (1965). Evidence that McIlwain's periphery effect is not a stray light artifact. *J. Neurophysiol.* **28**, 555–9.

—— and Zacks, J.L. (1970). Responses of cat retinal ganglion cells to brief flashes of light. *J. physiol., Lond.* **206**, 677–700.

Levine, R., Didner, R., and Tobenkin, N. (1967). Backward masking as a function of interstimulus distance. *Psychonomic. Sci.* **9**, 185–6.

Levinson, J.Z. and Frome, F.S. (1979). Perception of size of one object among many. *Science, N.Y.* **206**, 1245–6.

Levy-Schoen, A. (1969). Détermination et latence de la réponse oculomotrice à deux stimulus simultanés ou successifs selon leur excentricité relative. *L'Annee Psychologique* **69**, 373–92.

Lewis, J.H., Dunlap, W.P. and Matteson, H.H. (1972). Perceptual latency as a function of stimulus onset and offset and retinal location. *Vis. Res.* **12**, 1725–31.

Lie, I. (1980). Visual detection and resolution as a function of retinal locus. *Vis. Res.* **20**, 967–74.

—— (1981). Visual detection and resolution as a function of adaptation and glare. *Vis. Res.* **21**, 1793–7.

Lin, C.S. and Kaas, J.H. (1979). The inferior pulvinar complex in owl monkey: Architectonic subdivisions and patterns of input from the superior colliculus and subdivisions of the visual cortex. *J. comp. Neurol.* **187**, 655–78.

—— —— (1980). Projections from the medial nucleus of the inferior pulvinar complex to the middle temporal area of the visual cortex. *Neuroscience* **5**, 2219–28.

Lindsley, D.B., Fehmi, L.G., and Adkins, J.W. (1967). Visually evoked potentials during perceptual masking in man and monkey. *Electroenceph. clin. Neurophysiol.* **23**, 79.

Lipkin, B.S. (1962). Monocular flicker discrimination as a function of the luminance and area of contralateral steady light. I. Luminance. *J. opt. soc. Am.* **52**, 1287–1300.

Lohmann, W. (1906). Über Helladaptation. *Zeitschr. Sinnesphysiol.* **41**, 290–311.

Long, G.M. (1979). Comment on Hawkins and Shulman's Type I and Type II visual persistence. *Percept. Psychophys.* **26**, 412–14.

—— (1980). Iconic memory: A review and critique of the study of short-term visual storage. *Psychol. Bull.* **88**, 785–820.

—— and Beaton, R.J. (1982). The case for peripheral persistence: Effects of target and background luminance on a partial-report task. *J. exp. psychol.: hum. percept. Perf.* **8**, 383–91.

—— and Gildea, T.J. (1981). Latency for the perceived offset of brief target gratings. *Vis. Res.* **21**, 1395–9.

—— and Sakitt, B. (1980*a*). Target duration effects on iconic memory: The confounding role of changing stimulus dimensions. *Q. J. exp. Psychol.* **32**, 269–85.

—— —— (1980*b*). The retinal basis of iconic memory: Eriksen and Collins revisited. *Am. J. Psychol.* **92**, 195–206.

—— —— (1981). Differences between flicker and non-flicker persistence tasks: The effects of luminance and the number of cycles in the grating patterm. *Vis. Res.* **21**, 1387–93.

Long, N.R. and Gribben, J.A. (1971). The recovery of a visually masked target. *Percept. Psychophys.* **10**, 197–200.

—— and Over, R. (1973). Stereoscopic depth aftereffects with random-dot patterns. *Vis. Res.* **13**, 1283–7.

—— and Scheirlinck, J.G.M. (1981). Spatial disinhibition of orientation analysers. *Percept. Psychophys.* **29**, 212–16.

Lorenz, K. (1941). Kants Lehre vom apriorischen im Lichte gegenwartiger Biologie. *Blätter Dtsche. Philosophie* **15**, 94–125.

—— (1974). *Vom Weltbild des Verhaltensforscher*. Deutscher Taschenbuch Verlag, Munich.

—— (1977). *Behind the mirror*. Methuen, London.

Lovegrove, W., Bowling, A., and Gannon. S. (1981). Orientation specificity in the visual persistence for gratings and checker boards. *Vis. Res.* **21**, 1239–40.

—— Heddle, M., and Slaghuis, W. (1980). Reading disability: Spatial frequency specific deficits in visual information store. *Neuropsychologia* **18**, 111–15.

Lund, J.S. and Boothe, R.G. (1975). Interlaminar connections and pyramidal neuron organization in the visual cortex, area 17, of the macaque monkey. *J. comp. Neurol.* **159**, 305–34.

—— Lund, R.D., Hendrickson, A.E., Bunt, A.H., and Fuchs, A.F. (1975). The origin of efferent pathways from the primary visual cortex, area 17, of the macaque monkey as shown by retrograde transport of horseradish peroxidase. *J. comp. Neurol.* **164**, 287–304.

Lupp, U. (1977). Differences in processing high and low spatial frequencies in the human visual system. *Proc. soc. photogr. sci. Eng.*, October 24–25.

—— Hauske, G., and Wolf, W. (1976). Perceptual latency to sinusoidal gratings. *Vis. Res.* **16**, 969–72.

—— —— —— (1978). Different systems for the visual detection of high and low spatial frequencies. *Photogr. Sci. Eng.* **22**, 80–4.

Luria, A.R. (1959). Disorders of 'simultaneous perception' in a case of bilateral occipito-parietal brain injury. *Brain* **82**, 437–49.

—— Karpov, B.A., and Yarbuss, A.L. (1966). Disturbances of active visual perception with lesions of the frontal lobes. *Cortex* **2**, 202–12.

Luria, S.E. (1973). *Life—The unfinished experiment*. Souvenir Press, London.

Lynch, J.C. (1980). The functional organization of posterior parietal association cortex. *Behav. brain Sci.* **2**, 485–99.

—— Mountcastle, V.B., Talbot, W.H., and Yin, T.C.T. (1977). Parietal lobe mechanisms for directed visual attention. *J. Neurophysiol.* **40**, 362–89.

Lyon, J.E., Matteson, H.H., and Maras, M.S. (1981). Metacontrast in the fovea. *Vis. Res.* **21**, 297–9.

Mach, E. (1865). Über die Wirkung der räumlichen Vertheilung des Lichtreizes auf die Netzhaut. *Wiener Sitzungsberichte der mathematisch-naturwissenschaftlichen Classe der kaiserlichen Akademie der Wissenschaft* **52**, Part 2, 303–22.

—— (1866a). Über den physiologischen Effect räumlich vertheilter Lichtreize (Zweite Abhandlung). *Wiener Sitzungsberichte der mathematisch-naturwissenschaftlichen Classe der kaiserlichen Akademie der Wissenschaft* **54**, Part 2, 131–44.

—— (1866b). *Über die physiologische Wirkung räumlich vertheilter Lichtreize (Dritte Abhandlung). Wiener Sitzungsberichte der mathematisch-naturwissenschaftlichen Classe der kaiserlichen Akademie der Wissenschaft* **52**, Part 2, 393–408.

—— (1868). Über die physiologische Wirkung räumlich vertheilter Lichtreize (Vierte Abhandlung). *Wiener Sitzungsberichte der mathematisch-naturwissenschaftlichen Classe der kaiserlichen Akademie der Wissenschaft* **57**, Part 2, 11–19.

McClelland, J.L. (1978). Perception and masking of wholes and parts. *J. exp. psychol.: hum. percept. Perform.* **4**, 210–23.

McCloskey, M. and Watkins, M.J. (1978). The seeing-more-than-is-there phenomenon: Implications for the locus of iconic storage. *J. exp. psychol.: hum. percept. Perf.* **4**, 553–64.

McConkie, G.W. and Rayner, K. (1976). Identifying the span of the effective stimulus in reading: Literature review and theories of reading. In *Theoretical models and processes in reading* (ed. H. Singer and R. B. Rudell) pp. 137–162. International Reading Association, Newark, Delaware.

—— and Zola, D. (1979). Is visual information integrated across successive fixations in reading? *Percept. Psychophys.* **25**, 221–4.

McDougall, W. (1904a). The sensations excited by a single momentary stimulation of the eye. *Br. J. Psychol.* **1**, 78–113.

—— (1904*b*). The variations of the intensity of visual sensation with the duration of the stimulus. *Br. J. Psychol.* **1**, 151–89.

McFadden, D. and Gummerman, K. (1973). Monoptic and dichoptic metacontrast across the vertical meridian. *Vis. Res.* **13**, 185–96.

McIlwain, J.T. (1964). Receptive fields of optic tract axons and lateral geniculate cells. Peripheral extent and barbiturate sensitivity. *J. Neurophysiol.* **27**, 1154–73.

—— (1977). Topographic organization and convergence in corticotectal projections from areas 17, 18, and 19 in the cat. *J. Neurophysiol.* **40**, 189–98.

—— (1966). Some evidence concerning the physiological basis of the periphery effect in the cat's retina. *Exp. brain Res.* **1**, 265–71.

—— (1972). Nonretinal influences in the lateral geniculate nucleus. *Invest. ophthalmol. vis. Sci.* **5**, 311–22.

—— (1973*a*). Topographic relationships in projection from striate cortex to superior colliculus of the cat. *J. Neurophysiol.* **36**, 690–701.

—— (1973*b*). Retinotopic fidelity of striate cortex-superior colliculus interactions in the cat. *J. Neurophysiol.* **36**, 702–10.

—— (1975). Visual receptive fields and their images in superior colliculus of the cat. *J. Neurophysiol.* **38**, 219–30.

—— and Lufkin, R.B. (1976). Distribution of direct Y-cell inputs to the cat's superior colliculus: Are there spatial gradients? *Brain Res.* **103**, 133–8.

Mackavey, W.R., Bartley, S.H., and Casella, C. (1962). Disinhibition in the human visual system. *J. opt. soc. Am.* **52**, 85–8.

MacKay, D.M. (1970*a*). Elevation of visual threshold by displacement of retinal image. *Nature, Lond.* **225**, 90–2.

—— (1970*b*). Interocular transfer of suppressive effects of retinal image displacement. *Nature, Lond.* **225**, 872–3.

McKee, S.P. and Westheimer, G. (1970). Specificity of cone mechanisms in lateral interaction. *J. physiol., Lond.* **206**, 117–28.

MacLeod, D.I.A. (1978). Visual sensitivity. *Ann. rev Psychol.* **29**, 613–45.

McMullin, E. (1972). What difference does mind make? In *Brain and human behavior* (ed. A. G. Karczmar and J. C. Eccles) pp. 423–47. Springer, New York.

Mackworth, N. and Kaplan, I.T. (1962). Visual acuity when eyes are pursuing moving targets. *Science, N.Y.* **136**, 387–8.

Maffei, L. and Bisti, S. (1976). Binocular interaction in strabismic kittens deprived of vision. *Science, N.Y.* **191**, 579–80.

—— and Campbell, F.W. (1970). Neurophysiological localization of the vertical and horizontal coordinates in man. *Science, N.Y.* **167**, 386–7.

—— Cervetto, L., and Fiorentini, A. (1970). Transfer characteristics of excitation and inhibition in cat retinal ganglion cells. *J. Neurophysiol.* **33**, 276–84.

—— and Fiorentini, A. (1972). Process of synthesis in visual perception. *Nature, Lond.* **240**, 479–82.

—— and Fiorentini, A. (1976). Monocular deprivation in kittens impairs the spatial resolution of geniculate neurones. *Nature, Lond.* **264**, 754–5.

—— —— and Bisti, S. (1973). Neural correlate of perceptual adaptation to gratings. *Science, N.Y.* **182**, 1036–8.

Magnussen, S. and Kurtenbach, W. (1980). Adapting to two orientations: Disinhibition in a visual aftereffect. *Science, N.Y.* **207**, 908–9.

Makous, W. and Peeples, D. (1979). Rod–cone interaction: Reconciliation with Flamant and Stiles. *Vis. Res.* **19**, 695–8.

Markoff, J.I. and Sturr, J.F. (1971). Spatial and luminance determinants of the increment threshold under monoptic and dichoptic viewing. *J. opt. soc. Am.* **61**, 1530–7.

Marr, D. (1978). Representing visual information. In *Computer vision systems (ed. A. Hanson and E. Riseman) pp. 61–80. Academic Press, New York.*

—— (1982). *Vision.* W. H. Freeman and Co., San Francisco.

—— and Poggio, T. (1979). A computational theory of human stereo vision. *Proc. R. soc., Lond.* **204B**, 301–28.

Marriott, F.H.C. (1962). Colour vision: The two-colour threshold technique of Stiles. In *The eye*, Vol. 2, *The visual process* (ed. H. Davson) pp. 251–72. Academic Press, New York.

Marrocco, R.T. (1976). Sustained and transient cells in monkey lateral geniculate nucleus: Conduction velocities and response properties. *J. Neurophysiol.* **39**, 340–53.

—— (1978). Conduction velocities and afferent input to superior colliculus in normal and decorticate monkeys. *Brain Res.* **140**, 155–8.

—— and Brown, J.B. (1975). Correlation of receptive field properties of monkey LGN cells with the conduction velocity of retinal afferent input. *Brian Res* **92**, 137–44.

—— and Li, R. (1977). Monkey superior colliculus: Properties of single cells and their afferent inputs. J. Neurophysiol. **40**, 844–60.

—— McClurkin, J.W., and Young, R.A. (1982). Spatial summation and conduction latency classification of cells of the lateral geniculate nucleus of macaques. *J. Neurosci.* **2**, 1275–91.

Martin, M. (1979). Local and global processing: The role of sparsity, *Mem. Cog.* **7**, 476–84.

Martius, G. (1902). Ueber die Dauer der Lichtempfindungen. *Beiträge Psychol. Physiol., Leipzig* **1**, 275–366.

Marzi, C.A., Di Stefano, M., Tassinari, G., and Crea, F. (1979). Iconic storage in the two hemispheres. *J. exp. psychol.: hum. percept. Perf.* **5**, 31–41.

Mateeff, S., Yakimoff, N., and Mitrani, L. (1976). Some characteristics of the visual masking by moving contours. *Vis. Res.* **16**, 484–92.

Matin, E. (1974a). Light adaptation and the dynamics of induced tilt. *Vis. Res.* **14**, 255–65.

—— (1974b). Saccadic suppression: A review and analysis. *Psychol. Bull.* **81**, 899–917.

—— (1975). The two-transient (masking) paradigm. *Psychol. Rev.* **82**, 451–61.

—— (1976). Saccadic suppression and the stable world. In *Eye movements and psychological processes* (ed. R. A. Monty and J. W. Senders) pp. 113–119. Lawrence Erlbaum Associates, Hillsdale, N.J.

—— (1982). Saccadic suppression and the dual mechanism of direction constancy. *Vis. Res.* **22**, 335–6.

—— Clymer, A., and Matin, L. (1972). Metacontrast and saccadic suppression. *Science, N.Y.* **178**, 179–82.

Matin, L. (1976). A possible hybrid mechanism for modication of visual direction associated with eye-movements—the paralyzed eye experiment reconsidered. *Perception* **5**, 233–9.

Matsumura, M. (1976a). Visual responses to brief flashes of different temporal stimulus wave forms. *Tohoku Psychologica Folia* **34**, 95–102.

—— (1976b). Visual masking by luminance increment and decrement: Effects of rise time and decay time. *Tohoku Psychologica Folia* **35**, 104–14.

—— (1977). Visual masking to luminance increment and decrement of temporal ramp stimuli with different rise and decay time. *Tohoku Psychologica Folia* **36**, 111–19.

Matteson, H.H. (1969). Effects of surround size and luminance on metacontrast. *J. opt. soc. Am.* **59**, 1461–8.

Matthews, M.L. (1971). Spatial and temporal factors in masking by edges and disks. *Percept. Psychophys.* **9**, 15–22.

May, J.G., Grannis, S.W., and Porter, R.J., Jr. (1980). The 'lag effect' in dichoptic viewing. *Brain Lang.* **11**, 19–29.

Mayr, E. (1961). Cause and effect in biology. *Science, N.Y.* **134**, 1501–6.

Mays, L.E. and Sparks, D.L. (1980*a*). Dissociation of visual and saccade-related responses in superior colliculus neurons. *J. Neurophysiol.* **43**, 207–32.

—— —— (1980*b*). Saccades are spatially, not retinotopically, coded. *Science, N.Y.* **208**, 1163–5.

—— —— (1981). The localization of saccade targets using a combination of retinal and eye position information. In *Progress in oculomotor research* (ed. A. F. Fuchs and W. Becker) pp. 39–47. Elsevier/North-Holland, Amsterdam.

Mayzner, M.S. (1970). The disinhibited effect in sequential masking, *Psychonomic Sci.* **20**, 218–19.

—— and Tresselt, M.E. (1970). Visual information processing with sequential inputs: A general model for sequential blanking, displacement, and overprinting phenomena. *Ann. N.Y. acad. Sci* **169**, 599–618.

Merikle, P.M. (1977). On the nature of metacontrast with complex targets and masks. *J. exp. psychol.: hum. percept Perf.* **3**, 607–21.

—— (1980). Selective metacontrast. *Can. J. Psychol.* **34**, 196–9.

Mewhort, D.J.K. and Campbell, A.J. (1978). Processing spatial information and the selective-masking effect. *Percept. Psychophys.* **24**, 93–101.

—— Hearty, P.J., and Powell, J.E. (1978). A note on sequential blanking. *Percept. Psychophys.* **23**, 132–6.

Meyer, G.E. (1977). The effect of color-specific adaptation on the perceived duration of gratings. *Vis. Res.* **17**, 51–6.

—— Jackson, W.E., and Yang, C. (1979). Spatial frequency, orientation and color: Interocular effects of adaptation on the perceived duration of gratings. *Vis. Res.* **19**, 1197–1201.

—— Lawson, R., and Cohen, W. (1975). The effects of orientation-specific adaptation on the duration of short-term visual storage. *Vis. Res.* **15**, 569–72.

—— and Maguire, W.M. (1977). Spatial frequency and the mediation of short-term visual storage. *Science, N.Y.* **198**, 524–5.

—— —— (1981). Effects of spatial-frequency specific adaptation and duration on visual persistence. *J. exp. psychol.: hum. percept. Perf.* **7**, 151–6.

Michaels, C.F. and Turvey, M.T. (1973). Hemiretinae and nonmonotonic masking functions with overlapping stimuli. *Bull. psychonomic Soc.* **2**, 163–4.

—— —— (1979). Central sources of visual masking: Indexing structures supporting seeing at a single, brief glance. *Psychol. Res.* **41**, 1–61.

Miller, J. (1981*a*). Global precedence in attention and decision. *J. exp. psychol.: hum. percept. Perf.* **7**, 1161–74.

—— (1981*b*). Global precedence: Information availability or use? Reply to Novan. *J. exp. psychol.: hum. percept. Perf.* **7**, 1183–5.

Minkowski, M. (1913). Experimentelle Untersuchungen über die Beziehung der Grosshirnrinde und der Netzhaut zu den primären optischen Zentern, besonders zum Corpus geniculatum externum. *Arbeitsblatt des hirnanatomischen Institut Zurichs* **7**, 255–62.

—— (1920*a*). Über den Verlauf, die Endigung und die zentrale Repräsentation von gekreuzten und ungekreuzten Sehnervenfasern bei einigen Säugetieren and beim Menschen. *Schweizerische arch. neurol. Psychol.* **6**, 201–52.

—— (1920*b*). Über den Verlauf, die Endigung und die zentrale Repräsentation von gekreuzten und ungekreuzten Sehnerven fasern bei einigen Säugetieren and beim Menschen. *Schweizerische arch. neurol. Psychol.* **7**, 268–303.

Mitchell, D.E. and Baker, A.G. (1973). Stereoscopic after-effects: Evidence for disparity-specific neurones in the human system. *Vis. Res.* **13**, 2273–88.

Mitov, D., Vassilev, A., and Manahilov, V. (1981). Transient and sustained masking. *Percept. Psychophys.* **30**, 205–10.

Mitrani, L., Mateeff, S., and Yakimoff, N. (1971). Is saccadic suppression really saccadic? *Vis. Res.* **11**, 1157–61.

—— Radil-Weiss, T., Yakimoff, N., Mateef, St., and Božkov, V. (1975). Deterioration of vision due to contour shifts over the retina during eye movements. *Vis. Res.* **15**,

Mohler, C.W., Goldberg, M.E., and Wurtz, R.H. (1973). Visual receptive fields of frontal eye field neurons. *Brain Res.* **16**, 385–409.

—— and Wurtz, R.H. (1976). Organization of monkey superior colliculus: Intermediate layer cells discharging before eye movements. *J. Neurophysiol.* **39**, 722–44.

—— —— (1977). Role of striate cortex and superior colliculus in visual guidance of saccadic eye movements in monkeys. *J. Neurophysiol.* **40**, 74–94.

Molotchnikoff, S., Lachapelle, P., and L'Archeveque, P. (1977). Alternating activity between neurons of lateral geniculate nucleus and superior colliculus of rabbit. *Experientia* **33**, 232–4.

Mollon, J.D. and Krauskopf, J. (1973). Reaction time as a measure of the temporal response properties of individual colour mechanisms. *Vis. Res.* **13**, 27–40.

Monahan, J.S. and Steronko, R.J. (1977). Stimulus luminance and dichoptic pattern masking. *Vis. Res.* **17**, 385–90.

Monjé, J. (1927). Die Empfindungszeitmessung mit der Methode des Löschreizes. *Zeitschr. Biol.* **87**, 23–40.

—— (1931). Über die gegenseitige Beeinflussung der Empfindungen bei binokularem Sehen. *Zeitschr. Biol.* **91**, 387–98.

Monod, J. (1972). *Chance and necessity.* Vintage Books, New York.

Moors, J. and Vendrick, A.J.H. (1979*a*). Responses of single units in the monkey superior colliculus to stationary flashing stimuli. *Exp. brain Res.* **35**, 333–47.

—— —— (1979*b*). Responses of single units in the monkey superior colliculus to moving stimuli. *Exp. brain Res.* **35**, 349–69.

Motokawa, K., Iwama, K., and Ebe, M. (1954). Velocities of spreading induction in human and mammalian retina. *Tohoku J. ex. Med.* **59**, 11–22.

Mountcastle, V.B. (1975). The view from within: Pathways to the study of perception. *John Hopkins Med. J.* **136**, 109–31.

—— (1976). The world around us: Neural command functions for selective attention. In F. O. Schmitt Lecture in Neuroscience for 1975, *Neurosci. prog. res. Bull.* **14**, Suppl: 1–47.

—— (1978). An organizing principle for cerebral function: The unit module and the distributed system. In *The mindful brain* (by G. M. Edelman and V. B. Mountcastle) pp. 7–50. MIT Press, Cambridge, Ma.

—— Lynch, J.C., Georgopoulos, A., Sakata, H., and Acuna, C. (1975). Posterior parietal association cortex of the monkey: Command functions for operations within extrapersonal space. *J. Neurophysiol.* **38**, 871–908.

—— Motter, B.C., and Andersen, R.A. (1980). Some further observations of the functional properties of neurons in the parietal lobe of the waking monkey. *Behav. brain Sci.* **3**, 520–3.

Movshon, J.A., Thompson, I.D., and Tolburst, D.J. (1978*a*). Spatial summation in the receptive fields of simple cells in the cat's striate cortex. *J. physiol., Lond.* **283**, 53–77.

—— —— —— (1978*b*). Receptive field organization of complex cells in the cat's striate cortex. *J. physiol., Lond.* **283**, 79–99.

—— —— —— (1978c). Spatial and temproal contrast sensitivity of neurones in area 17 and 18 of the cat's visual cortex. *J. physiol., Lond.* **283**, 101–20.

Mowbray, G.H. and Durr, L.B. (1964). Visual masking. *Nature, Lond.* **201**, 277–8.

Müller, J. (1834). *Handbuch der Physiologie des Menschen*. Hölscher, Coblenz.

Murphy, B.J. (1978). Pattern thresholds for moving and stationary gratings during smooth eye movement. *Vis. Res.* **18**, 521–30.

Myerson, J. (1977). Magnification in striate cortex and retinal ganglion cell layer of owl monkey: A quantitative comparison. *Science, N.Y.* **198**, 855–7.

Nachmias, J. (1967). Effect of exposure duration on visual contrast sensitivity with square-wave gratings. *J. opt. soc. Am.* **57**, 421–7.

Nagamata, H. (1954). A contribution to the knowledge of afterimages. *Acta soc. ophthalmol. Jap.* **58**, 719–22.

Nagle, M., Bridgeman, B., and Stark. L. (1980). Voluntary nystagmus, saccadic suppression, and stablization of the visual world. *Vis. Res.* **20**, 717–21.

Nagano, T. (1980). Temporal sensitivity of the human visual system to sinusoidal gratings. *J. opt. soc. Am.* **70**, 711–16.

Navon, D. (1977). Forest before trees: The precedence of global features in visual perception. *Cog. Psychol.* **9**, 353–83.

—— (1981). Do attention and decision follow perception? Comment on Miller. *J. exp. psychol.: hum. percept. Perf.* **7**, 1175–82.

—— and Purcell D. (1981). Does integration produce masking or protect from it? *Perception* **10**, 71–83.

Neff, W.S. (1936). A critical investigation of visual apprehension of movement. *Am. J. Psychol.* **48**, 1–42.

Neisser, U. (1967). *Cognitive psychology*. Appleton-Century-Crofts, New York.

—— (1976) *Cognition and reality*. W. H. Freeman, San Francisco.

Nelson, J.I. and Frost, B.J. (1978). Orientation-selective inhibition from beyond the classic visual receptive field. *Brain Res.* **139**, 359–66.

Newark, J. and Mayzner, M.S. (1973). Sequential blanking effects for two interleaved words. *Bull. psychonomic Soc.* **2**, 74–6.

Nillson, T.H., Richmond, C.F., and Nelson, T.M. (1975). Flicker adaptation shows evidence of many visual channels selectively sensitive to temporal frequency. *Vis. Res.* **15**, 621–4.

Noda, H. (1975a). Depression in the excitability of relay cells of lateral geniculate nucleus following saccadic eye movements in the cat. *J. physiol., Lond.* **249**, 87–102.

—— (1975b). Discharges of relay cells in lateral geniculate nucleus of cat during spontaneous eye movements in light and darkness. *J. physiol., Lond.* **250**, 579–95.

—— and Adey, W.R. (1974a). Excitability changes in cat lateral geniculate cells during saccadic eye movements. *Science N.Y.* **183**, 543–5.

—— —— (1974b). Retinal ganglion cells of the cat transfer information on saccadic movement and quick target motion. *Brain Res.* **70**, 340–5.

—— Freeman, R.B., Jr., and Creutzfeldt, O.D. (1972). Neuronal correlates of eye movements in the visual cortex of the cat. *Science N.Y.* **175**, 661–4.

Normann, R.A., and Werblin, F.S. (1974). Control of retinal sensitivity: I. Light and dark adapation of vertebrate rods and cones. *J. gen. Physiol.* **63**, 37–61.

Norton, T.T., Casagrande, V.A., and Sherman, S.M. (1977). Loss of Y-cells in the lateral geniculate nucleus of monocularly deprived tree shrews. *Science, N.Y.* **197**, 784–6.

Nunokawa, S. (1973). Effects of background illumination on the receptive field organization of single cortical cells in area 18 of the immobilized cat. *Jap. J. Physiol.* **23**, 13–23.

Ogawa, T. (1963). Midbrain reticular influences upon single neurons in lateral geniculate nucleus. *Science, N.Y.* **139**, 343–4.

Olsen, B.T., Seim, T., and Valberg, A. (1982). Remote pattern reversal reduces the proximal negative response of the goldfish retina. *J. physiol. Lond.* **323**, 463–72.

Onley, J.W. and Boynton, R.M. (1962). Visual responses to equally bright stimuli of unequal luminance. *J. opt. soc. Am.* **52**, 934–40.

O'Toole, B.I. (1979). The tilt illusion: Length and luminance changes of induction line and third (disinhibiting) line. *Percept. Psychophys.* **25**, 487–96.

Owens, W.G. (1972). Spatio-temporal integration in the human peripheral retina. *Vis. Res.* **12**, 1011–26.

Oyama, T. (1970). The visually perceived velocity as a function of aperture size, stripe size, luminance and motion direction. *Jap. psychol. Res.* **12**, 163–71.

Oyster, C.W. and Takahashi, E.S. (1975). Responses of rabbit superior colliculus neurons to repeated visual stimuli. *J. Neurophysiol.* **38**, 301–12.

Palmer, L.A. and Rosenquist, A.C. (1974). Visual receptive fields of single striate cortical units projecting to the superior colliculus in the cat. *Brain Res.* **67**, 27–42.

Pantle, A. (1971). Flicker adaptation—I.: Effect on visual sensitivity to temporal fluctuations of light intensity. *Vis. Res.* **11**, 943–52.

—— and Picciano, L. (1976). A multistable movement display: Evidence for two separate motion systems in human vision. *Science, N.Y.* **193**, 500–2.

Parasuraman, R. (1979). Memory load and event rate control sensitivity decrements in sustained attention. *Science. N.Y.* **205**, 924–7.

Parker, D.M. (1980). Simple reaction times to onset, offset and contrast reversal of sinusoidal grating stimuli. *Percept. Psychophys.* **28**, 365–8.

—— and Salzen, E.A. (1977a). Latency changes in the human visual evoked response to sinusoidal gratings. *Vis. Res.* **17**, 1201–04.

—— —— (1977b). The spatial selectivity of early and late waves within the human visual evoked response. *Perception* **6**, 85–95.

—— —— (1982). Evoked potentials and reaction times to the offset and contrast reversal of sinusoidal gratings. *Vis. Res.* **22**, 205–7.

Parks. T.E. (1965). Post-retinal visual storage. *Am. J. Psychol.* **78**, 145–7.

—— (1970). A control for ocular tracking in the demonstration of post-retinal visual storage. *Am. J. Psychol.* **83**, 442–4.

—— (1968). Further comments on the evidence for post-retinal storage. *Percept. Psychophys.* **4**, 373.

Pease, V.P. and Sticht, T.G. (1965). Reaction time as a function of onset and offset stimulation of the fovea and periphery. *Percept. Motor Skills* **20**, 549–54.

Pecci-Saacreda, J., Wilson, P.D., and Doty, R.W. (1966). Presynaptic inhibition in primate lateral geniculate nucleus. *Natures, Lond.* **210**, 740–2.

Peck, C.K., Schlag-Rey, M., and Schlag, J. (1980). Visuo-oculomotor properties of cells in the superior colliculus of the alert cat. *J. comp. Neurol.* **194**, 97–116.

Peichl, L. and Wässle, H. (1979). Size, scatter and coverage of ganglion cell receptive field centres in the cat retina. *J. physiol., Lond.* **291**, 117–41.

Pentney, R.P. and Cotter, J.R. (1978). Structural and functional aspects of the superior colliculus in primates. In *Sensory systems of primates* (ed. C. R. Noback) pp. 109–34. Plenum Press, New York.

Perryman, K.M., Lindsley, D.F., and Lindsley, D.B. (1980). Pulvinar neuron responses to spontaneous and trained eye movements and to light flashes in squirrel monkeys. *Electroenceph. clin. Neurophysiol.* **49**, 152–61.

Peterson, B.W. (in press). Participation of pontomedullary reticular neurons in specific motor activity. In *The reticular formation revisited* (ed. A. Hobson and M. Brazier). Raven Press, New York.

Petrén, K. (1893). Untersuchungen über den Lichtsinn. *Skandinavische arch. Physiol.* **4**, 421–47.

Petry, S. (1978). Perceptual changes during metacontrast. *Vis. Res.* **18**, 1337–41.

—— Grigonis, A., and Reichert, B. (1979). Decrease in metacontrast masking following adaptation to flicker. *Perception* **8**, 541–7.

Phillips, W.A. (1974). On the distinction between sensory storage and short-term visual memory. *Percept. Psychophys.* **16**, 283–90.

—— and Singer, W. (1974). Function and interaction of on and off transients in vision: I.: Psychophysics. *Exp. brain Res.* **19**, 493–506.

Piaget. J. (1971). *Biology and knowledge.* The University of Chicago Press.

—— (1978). *Behavior and evolution.* Pantheon Books, New York.

Piéron, H. (1935). Les processes du metacontraste. *J. Psychologie* **32**, 5–24.

Piper, H. (1903). Über Dunkeladaptation. *Zeitschr. psychol. physiol. Sinnesorgane* **31**, 169–214.

Pittendrigh, C.S. (1958). Adaptation, natural selection, and behavior. In *Behavior and evolution* (ed. A. Roe and G.G. Simpson) pp. 390–416. Yale University Press, New Haven, Ct.

Plateau, J. (1834). Über das Phänomen der zufälligen Farben. *Poggendorf Annalen der Physik und Chemie* **32**, 543–54.

Poggio, G.F., Baker, F.H., Lamarre, Y., and Sanseverino, E.R. (1969). Afferent inhibition at input to visual cortex of the cat. *J. Neurophysiol.* **32**, 892–915.

Polanyi, M. (1962). *Personal knowledge.* University of Chicago Press.

Pollack, I. (1973). Interaction effects in successive visual displays: An extension of the Eriksen–Collins paradigm. *Percept. Psychophys.* **13**, 367–73.

Pollack, J.G. and Winters, R.W. (1978). A comparison of the strength of lateral inhibition in X and Y cells in the cat retina. *Brain Res.* **143**, 538–43.

Pomerantz, J.R. and Garner, W.R. (1973). Stimulus configuration in selective attention tasks. *Percept. Psychophys.* **14**, 565–9.

Pöppel, E. (1970). Excitability cycles in central intermittency. *Psychologische Forschung* **34**, 1–9.

—— Held, R., and Frost, D. (1973). Residual function after brain wounds involving the central pathways in man. *Nature, Lond.* **243**, 295–6.

Popper, K.R. (1959). *The logic of scientific discovery.* Harper and Row, New York.

—— (1968). Is there an epistemological problem of perception? In *The problem of inductive knowledge* (ed. I. Lakatos and A. Musgrave) pp. 163–4. North–Holland, Amsterdam.

—— (1972). *Objective knowledge.* Oxford University Press.

—— and Eccles, J.C. (1977). *The self and its brain.* Singer, New York.

Posner, M. (1978). *Chronometric exploration of mind.* Lawrence Erlbaum Associates, Hillsdale. N.J.

—— (1980). Orienting of attention. *Q.J. exp. Psychol.* **32**, 3–25.

—— and Cohen, Y. (1980). Attention and the control of movements. In *Tutorials in motor behavior* (ed. G. E. Stelmach and J. Requin) pp. 243–58. Elsevier/North-Holland, Amsterdam.

—— Nissen, M.J., and Snyder, D.R. (1978). Relationships between attention shifts and saccadic eye movements. Paper presented at *The annual meeting of the Psychonomic Society, San Antonio, Texas, November.*

—— Snyder, C.R.R., and Davidson, B.J. (1980). Attention and the detection of signals. *J. exp. psychol.: General* **109**, 160–74.

Pugh, M. (1958). Visual distortion in amblyopia. *B. J. Ophthalmol.* **42**, 449–60.

Pulos, E., Raymond, J.E. and Makous, W. (1980). Transient sensitization by a contast flash. *Vis. Res.* **20**, 281–8.

Purcell, D.G. and Dember, W.N. (1968). The relation of phenomenal brightness reversal and re-reversal to backward masking and recovery. *Percept. Psychophys.* **3**, 290–2.

—— and Stewart, A.L. (1970). U-shaped backward masking functions with nonmetacontrast paradigms. *Psychonomic Sci.* **21**, 361–3.

—— —— (1975). Visual masking by a patterned stimulus and recovery of observer performance. *Bull. psychonomic Soc.* **6**, 457–60.

—— —— and Brunner, R.L. (1974). Metacontrast target detection under light and dark adaptation. *Bull. psychonomic Soc.* **3**, 199–201.

—— —— Davis, J., Huntermark, J., Robbins, S., Rowland, P., and Salley, K. (1975). U-shaped masking functions under backward masking by pattern mask. *Bull. psychonomic Soc.* **5**, 498–500.

—— —— and Hochberg, E.P. (1982). Recovery and nonmonotone masking effects. *Vision Res.* **22**, 1087–96.

Purkinje, J.E. (1819). *Beiträge zur Kenntnis des Sehens in subjektiver Hinsicht.* J. G. Calve, Prague.

Raczkowski, D. and Cartmill, M. (1975). Primate evolution: Were traits selected for arboreal locomotion or visually directed predation? *Science, N.Y.* **187**, 455–6.

Ransom-Hogg, A. and Spillmann, L. (1980). Perceptive field size in fovea and periphery of the light- and dark-adapted retina. *Vis. Res.* **20**, 221–8.

Rashbass, C. (1970). The visibility of transient changes of luminance. *J. physiol., Lond.* **210**, 165–86.

Ratliff, F. (1965). *Mach bands: quantitative studies on neural networks in the retina.* Holden-Day, San Francisco.

Rayner, K. (1975). The perceptual span and peripheral cues in reading. *Cog. Psychol.* **7**, 65–81.

—— (1978*a*). Eye movements in reading and information processing. *Psychol. Bull.* **85**, 618–60.

—— (1978*b*). Foveal and parafoveal cues in reading. In *Attention and performance*, vol. 7 (ed. J. Requin) pp. 149–62. Lawrence Erlbaum and Associates, Hillsdale, N.J.

—— McConkie, G.W., and Ehrlich, S. (1978). Eye movements and integrating information across fixations. *J. exp. psychol.: hum. percept. Perf.* **4**, 529–44.

—— —— and Zola, D. (1980). Integrating information across eye movements. *Cog. Psychol.* **12**, 206–26.

Reeves, A. (1981). Metacontrast in hue substitution. *Vis. Res.* **21**, 907–12.

—— (1982). Metacontrast U-shaped functions derive from two monotonic processes. *Perception* **11**, 415–26.

Regan, D. (1970). Evoked potentials and psychophysical correlates of changes in stimulus colour and intensity. *Vis. Res.* **10**, 163–78.

Reicher, G.M. (1969). Perceptual recognition as a function of meaningfulness of stimulus material. *J. exp. Psychol.* **81**, 275–80.

Rentschler, I. and Hilz, R. (1976). Evidence for disinhibition in line detectors. *Vis. Res.* **16**, 1299–1302.

Remington, R.W. (1980). Attention and saccadic eye movements. *J. exp. psychol.: hum. percept. Perf.* **6**, 726–44.

Rezak, M. and Benevento, L.A. (1979). A comparison of the organization of the projections of the dorsal lateral geniculate nucleus, the inferior pulvinar and adjacent lateral pulvinar to primary visual cortex (area 17) in the macaque monkey. *Brain Res.* **167**, 19–40.

Richmond, B.J. and Wurtz, R.H. (1980). Vision during saccadic eye movements II. A corollary discharge to monkey superior colliculus. *J. Neurophysiol.* **43**, 1156–67.

Riggs, L.A. (1940). Recovery from the discharge of an impulse in a single visual receptor unit. *J. Cell. Comp. Physiol.* **15**, 273–83.

—— and Graham, C.H. (1940). Some aspects of light adaptation in a single photoreceptor unit. *J. Cell. Comp. Physiol.* **16**, 15–23.

—— Merton, P.A., and Morton, H.B. (1974). Suppression of visual phosphenes during saccadic eye movements. *Vis. Res.* **14**, 997–1010.

—— Volkmann, F.C., and Moore, R.K. (1981). Suppression of the black-out due to blinks. *Vis. Res.* **21**, 1075–9.

—— and Wooten, B. (1972). Electrical measures and psychophysical data on human vision. In *Handbook of sensory physiology*, Vol. VII/4. *Visual psychophysics* (ed. D. Jameson and L. H. Hurvich) pp. 690–731, Springer Verlag, New York.

Rijsdijk, J.P., Kroon, J.N. and von der Weldt, G.J. (1980). Contrast sensitivity as a function of position on the retina. *Vis. Res.* **20**, 235–41.

Ritchie, L. (1976). Effects of cerebellar lesions on saccadic eye movements. *J. Neurophysiol.* **39**, 1246–56.

Ritter, M. (1976). Evidence for visual persistence during saccadic eye movements. *Physiol. Res.* **39**, 67–85.

Rizzolatti, G., Camarda, R., Grupp, L.A. and Pisa, M. (1973). Inhibition of visual responses of single units in the cat superior colliculus by the introduction of a second visual stimulus. *Brain Res.* **61**, 390–4.

—— —— —— —— (1974). Inhibitory effect of remote visual stimuli on visual responses of cat superior colliculus: Spatial and temporal factors. *J. Neurophysiol.* **37**, 1262–74.

Robinson, D.A. (1975). Oculomotor control Signals. In *Basic mechanisms of ocular motility and their clinical implications* (ed. G. Lennerstrand and P. Bach-y-Rita) pp. 337–74. Pergamon Press, Oxford.

—— (1976). The physiology of pursuit eye movements. In *Eye movements and psychological processes* (ed. A. Monty and J. W. Senders) pp. 19–31. Lawrence Erlbaum Associates, Hillsdale, N.J.

—— and Fuchs, A. (1969). Eye movements evoked by stimulation of frontal eye fields. *J. Neurophysiol.* **32**, 637–48.

Robinson, D.L., Baizer, J.S., and Dow, B.M. (1980). Behavioral enhancement of visual responses of prestriate neurons of the rhesus monkey. *Invest. ophthalmol. vis. Sci.* **19**, 1120–3.

—— Bushnell, M.C., and Goldberg, M.E. (1980). Role of posterior parietal cortex in selective visual attention. In *Progress in oculomotor research* (ed. A. F. Fuchs and W. Becker) pp. 203–10.

—— and Goldberg, M.E. (1977). Visual mechanisms underlying gaze: Function of the superior colliculus. In *Control of gaze by brain stem neurons*, Vol. 1, *Developments in neuroscience* (ed. R. Baker and A. Berthoz) pp. 445–51. Elsevier/North-Holland, Amsterdam.

—— —— (1978*a*). The visual substrate of eye movements. In *Eye movements and the higher psychological functions* (ed. J. W. Senders, D. F. Fisher, and R.D. Monty) pp. 3–14. Lawrence Erlbaum Associates, Hillsdale, N.J.

—— —— (1978*b*). Sensory and behavioral properties of neurons in posterior parietal cortex of the awake, trained monkey. *Proc. fed. Am. soc. exp. Biol.* **37**, 2258–61.

—— —— and Stanton, G.B. (1978). Parietal association cortex in the primate: Sensory mechanisms and behavioral modulations. *J. Neurophysiol.* **41**, 910–32.

—— and Jarvis, C.D. (1974). Superior colliculus neurons studied during head and eye movements of the behaving monkey. *J. Neurophysiol.* **37**, 533–40.

—— and Wurtz, R.H. (1976). Use of an extraretinal signal by monkey superior colliculus to distinguish real from self-induced movement. *J. Neurophysiol.* **39**, 852–70.

Robinson, D.N. (1966). Disinhibition of visually masked stimuli. *Science, N.Y.* **154**, 157–8.

—— (1968). Visual disinhibition with binocular and interocular presentations. *J. opt. soc. Am.* **58**, 254–7.

—— (1971). Backward masking, disinhibition, and hypothesized neural networks. *Percept. Psychophys.* **10**, 33–5.

Rodieck, R.W. and Rushton, W.A.H. (1976). Cancellation of rod signals by cones, and cone signals by rods in the cat retina. *J. physiol., Lond.* **254**, 775–85.

Rogowitz, B.E. (1983). Spatial/temporal interactions: Backward and forward metacontrast masking with sine-wave gratings. *Vis. Res.* **23**, 1057–73.

Rohrbaugh, J.W. and Eriksen, C.W. (1975). Reaction time measurements of temporal integration and organization of form. *Percept. Psychophys.* **17**, 53–8.

Roldán, M. and Reinoso-Suárez, F. (1981). Cerebellar projections to the superior colliculus in the cat. *J. Neuroscience* **1**, 827–34.

Rolls, E.T. and Cowey, A. (1970). Topography of the retina and striate cortex and its relationship to visual acuity in rhesus monkeys and squirrel monkeys. *Exp. brain Res.* **10**, 298–310.

Ron, S. and Robinson, D.A. (1973). Eye movements evoked by cerebellar stimulation of alert monkey. *J. Neurophysiol.* **36**, 1004–22.

Ronderos, A.G., Matteson, H.H., and Marx, M.S. (in press). Effects of spatial frequency, duration and luminance on flash and pattern thresholds. *Vis. Res.*

Rosenquist, A.C. and Palmer, L.A. (1971). Visual receptive field properties of cells of the superior colliculus after cortical lesions in the cat. *Exp. Neurol.* **33**, 629–52.

Ross, J. and Hogben, J.H. (1974). Short-term memory in stereopsis. *Vis. Res.* **14**, 1195–1201.

Roucoux, A. and Crommelinck, M. (1976). Eye movements evoked by superior colliculus stimulation in the alert cat. *Brain Res.* **106**, 349–63.

—— —— and Guitton, D. (1981). The role of superior colliculus in the generation of gaze shift. In *Progress in oculomotor research* (ed. A. F. Fuchs and W. Becker) pp. 129–35. Elsevier/North–Holland, Amsterdam.

—— Guitton, D., and Crommelinck, M. (1980). Stimulation of the superior colliculus in the alert cat. II. Eye and head movements evoked when the head is unrestrained. *Exp. brain Res.* **39**, 75–85.

Roufs, J.A. (1972). Dynamic properties of vision. I. Experimental relationships between flicker and flash thresholds. *Vis. Res.* **12**, 261–78.

Rovamo, J. and Virsu, V. (1979). An estimation of the human cortical magnification factor. *Exper. brain Res.* **37**, 495–510.

Rowe, M.H. and Stone, J. (1977). Naming of neurones. *Brain, behav. Evolution* **14**, 185–216.

Rubin, E. (1929). Kritisches und Experimentelles zur 'Empfindungszeit' Fröhlichs. *Psychologische Forschung* **13**, 101–12.

Rushton, W.A.H. (1963). Increment threshold and dark adaptation. *J. opt. soc. Am.* **53**, 104–9.

—— (1965). Bleached rhodopsin and visual adaptation. *J. physiol., Lond.* **181**, 645–55.

—— and Westheimer, G. (1962). The effect upon the rod threshold of bleaching neighboring rods. *J. physiol., Lond.* **164**, 318–29.

Saito, H. (1981). The effects of strychnine and bicuculline on the responses of X- and Y- cells of the isolated eye-cup preparation of the cat. *Brain Res.* **212**, 243–8.

Sakitt, B. (1975). Locus of short-term visual storage. *Science, N.Y.* **190**, 1318–19.

—— (1976). Iconic memory. *Psychol. Rev.* **83**, 257–76.

—— and Long, G.M. (1978). Relative rod and cone contributions in iconic storage. *Percept. Psychophys.* **23**, 527–36.

—— —— (1979a). Cones determine subjective offset of a stimulus but rods determine total persistence. *Vis. Res.* **19**, 1439–41.

—— —— (1979b). Spare the rod and spoil the icon. *J. exp. psychol.: hum. percept. Perf.* **5**, 19–30.

Salinger, W.L. and Lindsley, D.B. (1973). Patterns of unit activity in optic tract of cat during suppression–recovery effect: Relationship to high intensity effect. *Vis. Res.* **13**, 2121–7.

—— Schwartz, M.A., and Wilkerson, P.R. (1977). Selective loss of lateral geniculate cells in the adult cat after monocular paralysis. *Brain Res.* **125**, 257–63.

Sandberg, M.A., Berson, E.L., and Effron, M.H. (1981). Rod–cone interaction in the distal human retina. *Science, N.Y.* **212**, 829–31.

Sasaki, H., Saito, Y., Bear, D.M., and Ervin, F.R. (1971). Quantitative variation in striate receptive fields of cats as a function of light and dark adaptation. *Exp. brain Res.* **13**, 273–93.

Saunders, J. (1977). Foveal and spatial properties of brightness metacontrast. *Vis. Res.* **17**, 375–8.

Scharf, B. and Lefton, L.A. (1970). Backward and forward masking as a function of stimulus and task parameters. *J. exp. Psychol.* **84**, 331–8.

—— Zamansky, H.S., and Brightbill, R.F. (1966). Word recognition with masking. *Percept. Psychophys.* **1**, 110–12.

Scheerer, E. (1973). Integration, interruption and processing rate in visual backward masking. *Psychologische Forschung* **36**, 71–93.

Schiller, P.H. (1965a). Backward masking for letters. *Percept. Motor Skills* **20**, 47–50.

—— (1965b). Monoptic and dichoptic visual masking by patterns and flashes. *J. exp. Psychol.* **69**, 193–9.

—— (1966). Forward and backward masking as a function of relative overlap and intensity of test and masking stimuli. *Percept. Psychophys.* **1**, 161–4.

—— (1968). Single unit analysis of backward visual masking and metacontrast in the cat lateral geniculate nucleus. *Vis. Res.* **8**, 855–66.

—— and Chorover, S.L. (1966). Metacontrast: Its relation to evoked potentials. *Science, N.Y.* **153**, 1398–1400.

—— Finlay, B.S., and Volman, S.F. (1976). Quantitative studies of single-cell properties in monkey striate cortex. I. Spatiotemporal organization of receptive fields. *J. Neurophysiol.* **39**, 1288–1319.

—— and Greenfield, A. (1969). Visual masking and the recovery phenomenon. *Percept. Psychophys.* **6**, 182–4.

—— and Koerner, F. (1971). Discharge characteristics of single units in superior colliculus of alert rhesus monkey. *J. Neurophysiol.* **35**, 920–36.

—— and Malpeli, J. (1977). Properties and tectal projections of monkey retinal ganglion cells. *J. Neurophysiol.* **40**, 428–45.

—— —— (1978). Functional specificity of lateral geniculate nucleus laminae of the rhesus monkey. *J. Neurophysiol.* **41**, 788–97.

—— and Smith, M.C. (1965). A comparison of forward and backward masking. *Psychonomic Sci.* **3**, 77–8.

—— —— (1966). Detection in metacontrast. *J. exp. Psychol.* **71**, 32–9.

—— —— (1968). Monoptic and dichoptic metacontrast. *Percept. Psychophys.* **3**, 237–9.

—— and Stryker, M. (1972). Single-unit recoding and stimulation in superior colliculus of the alert rhesus monkey. *J. Neurophysiol.* **35**, 915–24.

—— Stryker, M., Cynader, M., and Berman, N. (1974). Response characteristics of single cells in the monkey superior colliculus following ablation or cooling of visual cortex. *J. Neurophysiol.* **37**, 181–94.

—— True, S.D., and Conway, J.L. (1979). Effects of frontal eye field and superior colliculus ablations on eye movements. *Science, N.Y.* **206**, 590–2.

—— and Wiener, M. (1963). Monoptic and dichoptic visual masking. *J. exp. Psychol.* **66**, 386–93.

Schmidt, M.J., Cosgrove, M.P., and Brown, D.R. (1972). Stablized images: Functional relationships among populations of orientation specific mechanisms in the human visual system. *Percept. Psychophys.* **11**, 187–90.

Schneider, G.E. (1969). Two visual systems. *Science, N.Y.* **163**, 895–902.

Schober, H.A.W. and Hilz, R. (1965). Contrast sensitivity of the human eye for square-wave gratings. *J. opt. soc. Am.* **55**, 1086–91.

Schopenhauer, A. (1957). *Über die vierfache Wurzel des Satzes vom zureichenden Grunde*. Felix Meiner Verlag, Hamburg.

Schorr, C.M. (1972). *Oculomotor and neurosensory analysis of amblyopia*. Ph.D. thesis, University of California, Berkeley.

Schouten, J.F. and Ornstein, L.S. (1939). Measurement of direct and indirect adaptation by means of a binocular method. *J. opt. soc. Am.* **29**, 168–82.

Schultz, E.W. and Eriksen, C.W. (1977). Do noise masks terminate target processing? *Mem. Cog.* **5**, 90–6.

Schulz, A.J. (1908). Untersuchungen über die Wirkung gleicher Reize auf die Auffasung bei momentoner Exposition. *Zeitschr. Psychol.* **52**, 238–96.

Schumann, F. (1899). Sitzungsberichte des Psychologischen Vereins zu Berlin. *Zeitschr. Psychol.* **1**, 96–100.

Schurman, D.L. and Eriksen, C.W. (1969). Summation and interaction of successive masking stimuli in visual perception. *Am. J. Psychol.* **82**, 320–32.

Sekuler, R. (1965). Spatial and temporal determinants of visual backward masking. *J. exp. Psychol.* **70**, 401–6.

—— (1973). Review of *Handbook of sensory physiology*, Vol. 7/4, *Visual psychophysics* (ed. D. Jameson and L. M. Hurvich). *Am. J. Psychol.* **86**, 876–86.

Servière, J., Miceli, D., and Galifret, Y. (1977). A psychophysical study of the visual perception of 'instantaneous' and 'durable'. *Vis. Res.* **17**, 57–63.

Shapley, R. and Hochstein, S. (1975). Visual spatial summation in two classes of geniculate cells. *Nature, Lond.* **256**, 411–13.

—— and So, Y.T. (1980). Is there an effect of monocular deprivation on the proportions of X and Y cells in the cat lateral geniculate nucleus? *Exp. brain Res.* **39**, 41–8.

—— and Victor, J.D. (1978). The effect of contrast on the transfer properties of cat retinal ganglion cells. *J. physiol., Lond.* **285**, 275–98.

Sharpe, C.R. (1972). A perceptual correlate McIlwain's periphery effect. *Vis. Res.* **12**, 519–20.

Shaw, R. and Bransford, J. (1977). *Perceiving, acting, and knowing*. Lawrence Erlbaum Associates, Hillsdale, N.J.

Sherman, S.M., Hoffman, K.-P., and Stone, J. (1972). Loss of a specific cell type from dorsal lateral geniculate nucleus in visually deprived cats. *J. Neurophysiol.* **35**, 532–41.

—— Norton, T.T., and Casagrande, V.A. (1975). X- and Y-cells in the dorsal lateral geniculate nucleus of the tree shrew (*Tupaia glis*). *Brain Res.* **93**, 152–7.

—— Wilson, J.R., Kaas, J.H., and Webb, S.V. (1976). X- and Y-cells in the dorsal lateral geniculate nucleus of the owl monkey (*Aotus trivirgatus*). *Science, N.Y.* **192**, 475–7.

Sherrick, M.F., Keating, J.K., and Dember, W.N. (1974). Metacontrast with black and white stimuli. *Can. J. Psychol.* **28**, 438–45.

Sherrington, C.S. (1897). On reciprocal action in the retina as studied by means of some rotating discs. *J. physiol., Lond.* **21**, 33–54.

—— (1906). *Integrative action of the nervous system.* Yale University Press, New Haven, Ct.

Shipley, W.C., Kenney, F.A., and King, M.E. (1945). Beta apparent movement under binocular, monocular, and interocular stimulation. *Am. J. Psychol.* **58**, 545–9.

Shulman, G.L., Remington, R.W., and McLean, J.P. (1979). Moving attention through visual space. *J. exp. psychol.: hum. percept. Perf.* **5**, 522–6.

Simon, L.G. (1974). Color specific effects in metacontrast masking. Paper presented at *The annual meeting of the Association for Research in Vision and Ophthalmology, Sarasota, Florida, April.*

Singer, W. (1973*a*). The effect of mesencephalic reticular stimulation on intracellular potentials of cat lateral geniculate neurons. *Brain Res.* **61**, 35–54.

—— (1973*b*). Brain stem stimulation and the hypothesis of presynaptic inhibition in cat lateral geniculate nucleus. *Brain Res.* **61**, 55–68.

—— (1976). Temporal aspects of subcortical contrast processing. *Neuronal mechanisms in vis. percept. Neurosci. res. prog. Bull.* **15**, 358–69.

—— (1977). Control of thalamic transmission by corticofugal and ascending reticular pathways in the visual system. *Physiol. Rev.* **57**, 386–420.

—— (1978). Requirements for experience dependent changes in the circuitry of cat visual cortex. *Arch. Italiennes de Biologie* **116**, 393–401.

—— (1979). Central-core control of visual cortex functions. In *The neurosciences 4th study program* pp. 1093–1110. MIT Press, Cambridge, Ma.

—— and Bedworth, N. (1973). Inhibitory interaction between X and Y units in cat lateral geniculate nucleus. *Brain Res.* **49**, 491–307.

—— —— (1974). Correlation between the effects of brain stem stimulation and saccadic eye movements on transmission in the cat lateral geniculate nucleus. *Brain Res.* **72**, 185–202.

—— and Creutzfeldt, O.D. (1970). Reciprocal lateral inhibition in on- and off-center neurons in the lateral geniculate body of cat. *Exp. brain Res.* **10**, 311–30.

—— Tretter, F., and Cynader, M. (1975). Organization of cat striate cortex: A correlation of receptive-field properties with afferent and efferent connections. *J. Neurophysiol.* **38**, 1080–98.

—— —— —— (1976). The effect of reticular stimulation on spontaneous and evoked activity in the cat visual cortex. *Brain Res.* **102**, 71–90.

—— Zihl, J., and Pöppel, E. (1977). Subcortical control of visual thresholds in humans: Evidence for modality specific and retinotopically organized mechanisms of selective attention. *Exp. brain Res.* **29**, 173–90.

Sireteanu, R. and Hoffman, K.-P. (1979). Relative frequency and visual resolution of X- and Y-cells in the LGN of normal and monocularly deprived cats: Interlaminar differences. *Exp. brain Res.* **34**, 591–603.

Skagestadt, P. (1975). *Making sense of history.* Universitetsforlaget, Oslo.

Skinner, B.F. (1976). *About behaviorism.* Vintage Books, New York.

Smart, J.J.C. (1981). Physicalism and emergence. *Neuroscience* **6**, 109–13.

Smith, E.E. and Haviland, S.E. (1972). Why words are perceived more accurately than nonwords: Inference versus unitization. *J. exp. Psychol.* **92**, 59–64.

Smith, M.C. and Schiller, P.H. (1966). Forward and backward masking: A comparison. *Can. J. Psychol.* **20**, 191–7.

Smith, R.A. and Richards, W. (1969). Propagation velocity of lateral interaction in the human visual system. *J. opt. soc. Am.* **59**, 1469–72.

Smythe, L. and Finkel, D.L. (1974). Masking of spatial and identity information from geometric forms by a visual noise field. *Can. J. Psychol.* **28**, 399–408.

Sommerhoff, G. (1950). *Analytical biology*. Oxford University Press.

—— (1974). *Logic of the living brain*. John Wiley and Sons, London.

Sparks, D.L. and Mays, L.E. (1980). Movement fields of saccade-related burst neurons in the monkey superior colliculus. *Brain Res.* **190**, 39–50.

—— —— (1981). The role of the monkey superior colliculus in the control of saccadic eye movements: A current perspective. In *Progress in oculomotor research* (ed. A. F. Fuchs and W. Becker) pp. 137–44. Elsevier/North-Holland, Amsterdam.

—— —— and Pollack, J.G. (1977). Saccade-related unit activity in the monkey superior colliculus. In *Control of gaze by brain stem neurones* (ed. R. Baker and H. Berthoz) pp. 437–44. Elsevier/North-Holland, Amsterdam.

—— and Sides, J.P. (1974). Brain stem unit activity related to horizontal eye-movements occurring during visual tracking. *Brain Res.* **77**, 320–4.

Spehlmann, R. (1965). The averaged electrical responses to diffuse and to patterned light in the human. *Electroenceph. clin. Neurophysiol.* **19**, 560–9.

Spencer, T.J. (1969). Some effects of different masking stimuli on iconic storage. *J. exp. Psychol.* **81**, 132–40.

—— and Shuntich, R. (1970). Evidence for an interruption theory of backward masking. *J. exp. Psychol.* **85**, 198–203.

Sperling, G. (1960). The information available in brief visual presentations. *Psychol. Monogr.* **74**, Whole No. 498, 1–29.

—— (1963). A model for visual memory tasks. *Hum. Factors* **5**, 19–31.

—— (1964). What visual masking can tell us about temporal factors in perception. *Proceedings of the seventeenth international congress of psychology* (Washington, D.C.) pp. 199–200. North-Holland, Amsterdam.

—— (1965). Temporal and spatial visual masking. I. Masking by impulse flashes. *J. opt. soc. Am.* **55**, 541–59.

—— (1967). Successive approximations to a model for short-term memory. *Acta Psychol.* **27**, 285–92.

—— Budiansky, J., Spivak, J.G., and Johnson, M.C. (1971). Extremely rapid visual search: The maximum rate of scanning letters for the presence of a numeral. *Science, N.Y.* **174**, 307–11.

—— and Joliffe, C.L. (1965). Intensity–time relationship at threshold for spectral stimuli in human vision. *J. opt. soc. Am.* **55**, 191–9.

—— and Melchner, M.J. (1978). The Attention Operating Characteristic: Examples from visual search. *Science, N.Y.* **202**, 315–18.

Spinelli, D.N. and Jensen, F.E. (1979). Plasticity: The mirror of experience. *Science, N.Y.* **203**, 75–8.

Sprague, J.M. (1966). Interaction of cortex and superior colliculus in mediation of visually guided behavior. *Science, N.Y.* **153**, 1544–7.

Stainton, W.H. (1928). The phenomenon of Broca and Sulzer in foveal vision. *J. opt. soc. Am.* **16**, 26–39.

Stein, B.E. and Arigbede, M.O. (1972). A parametric study of movement detection properties of neurons in cat's superior colliculus. *Brain Res.* **45**, 437–54.

—— Goldberg, S.J., and Clamann, H.P. (1976). The control of eye movements by the superior colliculus in the alert cat. *Brain Res.* **118**, 469–74.

Steinbach, M.J. and Smith, D.R. (1981). Spatial localization after strabismus surgery: Evidence for inflow. *Science, N.Y.* **213**, 1407–9.

Steinberg, R.H. (1969). Rod–cone interaction in S-potentials from the cat retina. *Vis. Res.* **9**, 1331–44.

—— and Schmidt, R. (1970). Identification of horizontal cells as S-potential generators in the cat retina by intracellular dye injection. *Vis. Res.* **10**, 817–20.

Steinman, R.M. (1975). Oculomotor effects in vision. In *Basic mechanisms of ocular motility and their clinical implications* (ed. G. Lennerstrand and P. Bach-Y-Rita) pp. 395–415. Pergamon Press, Oxford.

Stent, G. (1975). Limits to the scientific understanding of man. *Science, N.Y.* **187**, 1052–7.

—— (1978). *Paradoxes of progress.* W. H. Freeman, San Francisco.

Sterling, P. and Wickelgren, B.G. (1969). Visual receptive fields in the superior colliculus of the cat. *J. Neurophysiol.* **32**, 1–15.

Sternheim, C.E. and Cavonius, C.R. (1972). Sensitivity of the human ERG and VECP to sinusoidally modulated light. *Vis. Res.* **12**, 1685–95.

Stewart, A.L. and Purcell, D.G. (1970). U-shaped masking functions in visual backward masking: Effects of target configuration and retinal position. *Percept. Psychophys.* **7**, 253–6.

—— —— (1974). Visual backward masking by a flash of light: A study of U-shaped detection functions. *J. exp. Psychol.* **103**, 553–66.

—— —— and Dember, W.N. (1968). Masking and recovery of target brightness. *Proceedings of the 76th annual convention of the American Psychological Association* pp. 109–10.

Stigler, R. (1908). Über die Unterschiedsschwelle im aufsteigenden Teile einer Lichtempfindung. *Pfl. arch. gesamte Physiol.* **123**, 163–223.

—— (1910). Chronophotische Studien über den Umgebungskontrast. *Pfl. arch. gesamte Physiol.* **135**, 365–435.

—— (1913). Metacontrast (Demonstration), IX Congress International de Physiologie, Gröningen. *Arch. Int. Physiologie* **14**, 78.

—— (1926). Die Untersuchung des zeitlichen Verlaufes der optischen Erregung mittels des Metakontrastes. In *Handbuch der Biologischen Arbeitsmethoden*, Part 6, Whole No. 6 (ed. E. Aberhalden) pp. 949–68. Urban and Schwarzenberg, Berlin.

Stiles, W.S. (1939). The directional sensitivity of the retina and the spectral sensitivities of the rods and cones. *Proc. R. soc., Lond.* **127B**, 64–105.

—— (1949). Increment thresholds and the mechanisms of colour vision. *Doc. Ophthalmol.* **3**, 138–63.

—— (1959). Color vision; the approach through increment threshold sensitivity. *Proc. nat. acad. sci., Washington* **45**, 100–14.

Stober, R.S., Brussel, E.M., and Komoda, M.K. (1978). Differential effects of metacontrast on target brightness and clarity. *Bull. psychonomic Soc.* **12**, 433–6.

Stone, J. (1978). The number and distribution of ganglion cells in the cat's retina. *J. comp. Neurol.* **180**, 753–72.

—— and Dreher, B. (1973). Projection of X- and Y-cells of the cat's lateral geniculate nucleus to areas 17 and 18 of visual cortex. *J. Neurophysiol.* **36**, 551–67.

—— and Fukuda, Y. (1974). Properties of cat retinal ganglion cells: A comparison of W-cells with X- and Y-cells. *J. Neurophysiol.* **37**, 722–48.

—— and Keens, J. (1980). Distribution of small and medium-sized ganglion cells in the cat's retina. *J. comp. Neurol.* **192**, 235–46.

—— Leventhal, A., Watson, C.R.R., Keens, J., and Clarke, R. (1980). Gradients between nasal and temporal areas of the cat retina in the properties of retinal ganglion cells. *J. comp. Neurol.* **192**, 219–33.

Stoper, A.E. and Banffy, S. (1977). Relation of split apparent motion to metacontrast. *J. exp. psychol.: hum. percept. Perf.* **3**, 258–77.

Straschill, M. and Rieger, P. (1973). Eye movements evoked by focal stimulation of the cat's superior colliculus. *Brain Res.* **59**, 211–27.

—— and Schick, F. (1977). Discharges of superior colliculus neurons during head and eye movements of the alert cat. *Exp. brain Res.* **27**, 131–41.

—— and Takahashi, H. (1981). Changes of EEG and single unit activity in the human pulvinar associated with saccadic gaze shifts and fixation. In *Progress in oculomotor research* (ed. A. F. Fuchs and W. Becker) pp. 225–31. Elsevier/North–Holland, Amsterdam.

Stroud, J.M. (1956). The fine structure of psychological time. In *Information theory in psychology* (ed. H. Quastler) pp. 174–207. Free Press, Glencoe, Il.

Stryker, M.P. and Schiller, P.H. (1975). Eye and head movements evoked by electrical stimulation of monkey superior colliculus. *Exp. brain Res.* **23**, 102–12.

Sturr, J.F. and Frumkes, T.E. (1968). Spatial factors in masking with black and white targets. *Percept. Psychophys.* **4**, 282–4.

—— —— and Veneruso, D.M. (1965). Spatial determinants of visual masking: Effects of mask size and retinal position. *Psychonomic Sci.* **3**, 327–8.

Sukale-Wolf, S. (1971). *Prediction of the metacontrast phenomenon from simultaneous brightness contrast*. Ph.D. thesis, Stanford University.

Suzuki, D.A., Noda, H., and Kase, M. (1981). Visual and pursuit eye movement-related activity in posterior vermis of monkey cerebellum. *J. Neurophysiol.* **46**, 1120–39.

Swanson, J., Ledlow, A., and Kinsbourne, M. (1978). Lateral asymmetries revealed by simple reaction time. In *Asymmetrical functions of the brain* (ed. M. Kinsbourne) pp. 274–91. Cambridge University Press.

Szoc, R. (1973). *Metacontrast with stereoscopically displayed stimuli*. Master's Thesis, University of California, Santa Barbara.

Talbot, S.A. and Marshall, W.H. (1941). Physiological studies on neural mechanisms of visual localization and discrimination. *Am. J. Ophthalmol.* **24**, 1244–63.

Tartaglione, A., Goff, D.P., and Benton, A.L. (1975). Reaction time to square-wave gratings as a function of spatial frequency, complexity and contrast. *Brain Res.* **100**, 111–20.

Tatton, W.G. and Crapper, D.R. (1972). Central tegmental alteration of cat lateral geniculate activity. *Brain Res.* **47**, 371–87.

Taylor, C. (1972). Conditions for a mechanistic theory of behavior. In *Brain and human behavior* (ed. A. Karczmar and J. C. Eccles) pp. 449–65. Springer Verlag, New York.

Taylor, G.A. and Chabot, R.J. (1978). Differential backward masking of words and letters by masks of varying orthographic structure. *Mem. Cog.* **6**, 629–35.

Teilhard de Chardin, P. (1959). *The phenomenon of man*. Harper and Row, New York.

—— (1964). *The future of man*. Harper and Row, New York.

—— (1966). *Man's place in nature*. Harper and Row, New York.

Teller, D.Y. (1971). Sensitization by annular surrounds: Temporal (masking) properties. *Vis. Res.* **11**, 1325–35.

—— Matthews, C., Phillips, W.D., and Alexander, K. (1971). Sensitization by annular surrounds: Sensitization and masking. *Vis. Res.* **11**, 1445–58.

Temme, L.A. and Frumkes, T.E. (1977). Rod–cone interaction in human scotopic vision—III: Rods influence cone increment thresholds. *Vis. Res.* **17**, 681–5.

Tenkink, E. and Werner, H.H. (1981). The intervals at which homogeneous flashes recover masked targets. *Percept. Psychophys.* **30**, 129–32.

Thomas, G.J. (1954). The effect on critical flicker frequency of interocular differences and phase relations of flashes of light. *Am. J. Psychol.* **67**, 632–46.

Thompson, J.H. (1966). What happens to the stimulus in backward masking? *J. exp. Psychol.* **71**, 580–6.

Thorson, J., Lange, G.D., and Biederman-Thorson, M. (1969). Objective measure of the dynamics of a visual movement illusion. *Science, N.Y.* **164**, 1087–8.

Tigerstedt, R. and Bergqvist, J. (1883). Zur Kenntniss der Apperceptionsdauer zusammengesetzter Gesichtsvorstellungen. *Zeitschr. Biologie* **19**, 5–44.

Toch, H.H. (1956). The perceptual elaboration of stroboscopic presentations. *Am. J. Psychol.* **69**, 345–58.

Todd. J.T. and Van Gelder, P. (1979). Implications of a transient–sustained dichotomy for the measurement of human performance. *J. exp. psychol.: hum. percept. Perf.* **5**, 625–38.

Tolhurst, D.J. (1973). Separate channels for the analysis of the shape and movement of a moving visual stimulus. *J. physiol., Lond.* **231**, 385–402.

—— (1975). Reaction times in the detection of gratings by human observers: A probabilistic mechanism. *Vis. Res.* **15**, 1143–9.

Tolman, E.C. (1967). *Purposive behavior in animals and men.* Appleton-Century-Crofts, New York.

—— and Brunswick, E. (1935). The organism and the causal texture of the environment. *Psychol. Rev.* **42**, 43–77.

Tomko, D.L., Barbaro, N.M., and Ali, F.N. (1981). Effect of body tilt on receptive field orientation of simple visual cortical neurons in unanesthetized cats. *Exp. brain Res.* **43**, 309–14.

Toulmin, S. (1972). The mentality of man's brain. In *Brain and human behavior* (ed. A. Karczmar and J. C. Eccles) pp. 409–22. Springer, New York.

Treisman, A., Russell, R., and Green, J. (1975). Brief visual storage of shape and movement. In *Attention and performance V.* (ed. P. M. A. Robbitt and S. Dornic), pp. 699–721. Academic Press, London.

Treisman, M. (1963). Temporal discrimination and the indifference interval: Implications for a model of the 'internal clock'. *Psychol. Monogr.* **77**, Whole No. 576, 1–31.

Tresselt, M.E., Mayzner, M.S., Schoenberg, K.M., and Waxman, J. (1970). A study of sequential blanking and overprinting combined. *Percept. Psychophys.* **8**, 261–4.

Tretter, F., Cynader, M., and Singer, W. (1975). Cat parastriate cortex: A primary or secondary visual area? *J. Neurophysiol.* **38**, 1099–1113.

Trevarthen, C.B. (1968). Two mechanisms of vision in primates. *Psychologische Forschung* **31**, 299–337.

—— (1978). Manipulative strategies of baboons and origins of cerebral asymmetry. In *Asymmetrical function of the brain* (ed. M. Kinsbourne) pp. 329–91. Cambridge University Press.

Trojanowski, J.Q. and Jacobson, S. (1974). Medial pulvinar afferents to frontal eyefields in rhesus monkey demonstrated by horseradish peroxidase. *Brain Res.* **80**, 395–411.

Tsumoto, T. and Suzuki, D.A. (1976). Effects of frontal eye field stimulation upon activities of the lateral geniculate body of the cat. *Exp. brain Res.* **25**, 291–306.

—— (1978). Inhibitory and excitatory binocular convergence to visual cortical neurons of the cat. *Brain Res* **159**, 85–97.

Tulunay-Keesey, U. (1972). Flicker and pattern detection: A comparison of thresholds. *J. opt. soc. Am.* **62**, 446–8.

—— and Bennis, B.J. (1979). Effects of stimulus onset and image motion on contrast sensitivity. *Vis. Res.* **19**, 767–74.

Turvey, M.T. (1973). On peripheral and central processes in vision: Inferences from an information-processing analysis of masking with patterned stimuli. *Psychol. Rev.* **80**, 1–52.

—— (1977). Contrasting orientations to the theory of visual information processing. *Psychol. Rev.* **84**, 67–88.

—— (1978). Visual processing and short-term memory. In *Handbook of learning and cognitive processes*, Vol. 5, *Human information processing* (ed. W. K. Estes) pp. 91–142. Lawrence Erlbaum Association, Hillsdale, N.J.

—— Michaels, C.F., and Kewley-Port, D. (1974). Visual Storage or visual making? An analysis of the 'retroactive contour enhancement' effect. *Q. J. exp. Psychol.* **26**, 72–81.

Tytla, M.E. and McAdie, P.J. (1981). Metacontrast masking in amblyopia. Paper presented at *The annual meeting of the Association for Research in Vision and Ophthalmology, Sarasota, Florida, May*.

Ueno, T. (1977). Temporal characteristics of the human visual system as revealed by reaction time to double pulses of light. *Vis. Res.* **17**, 591–6.

Uttal, W.R. (19700. On the physiological basis of masking with dotted visual noise. *Percept. Psychophys.* **7**, 321–7.

—— (1971). The psychobiologically silly season-or-what happens when neurophysiological data become psychological theories. *J. gen. Psychol.* **84**, 151–66.

—— (1973). *The psychobiology of sensory coding*. Harper and Row, New York.

—— (1981). *A taxonomy of visual processes*. Lawrence Erlbaum Associates, Hillsdale. N.J.

Valberg, A. (1974). Lateral interactions between large retinal stimuli and symmetric receptive fields. *Physica Norvegica* **7**, 227–35.

—— and Breitmeyer, B. (1980). The lateral effect of oscillation of peripheral luminance gratings: Test of various hypothesis. *Vis. Res.* **20**, 789–98.

Valberg, A., Olsen, B.T., and Marthinsen, S. (1981). Peripheral contrast reversal inhibits visually evoked potentials in the fovea. *Vis. Res.* **21**, 947–50.

Van der Meer, H.C. (1976). The effects of adaptation to stereoscopic depth and to uniocular image magnification on the duration of short-term visual storage. *Acta Psychol.* **40**, 311–23.

Van der Wildt, G.J. and Vrolijk, P.C. (1981). Propagation of inhibition. *Vis. Res.* **21**, 1765–71.

Vassilev, A. and Mitov, D. (1976). Perception time and spatial frequency. *Vis. Res.* **16**, 89–92.

—— and Strashimirov, D. (1979). On the latency of human visually evoked response to sinusoidal gratings. *Vis. Res.* **19**, 843–5.

Vaughn, H.G., Jr. and Silverstein, L. (1968). Metacontrast and evoked potentials: A reappraisal. *Science, N.Y.* **160**, 207–8.

Ventura, J. (1980). Foveal metacontrast: I. Criterion content and practice effects. *J. exp. psychol.: human percept. Perf.* **6**, 473–85.

Verhoeff, F.H. (1940). Phiphenomenon and anomalous projection. *Arch. Ophthalmol.* **24** 247–51.

Vernoy, M.W. (1976). Masking by pattern in random-dot sterograms. *Vis. Res.* **16**, 1183–4.

Vierordt, K. (1868). *Der Zeitsinn*. Tübingen Universität. Tübingen, W. Germany.

Virsu, V., Lee, B.B., and Creutzfeldt, O.D. (1977). Dark adaptation and receptive field organization of cells in the cat lateral geniculate nucleus. *Exp. brain Res.* **27**, 35–50.

Vital-Durand, F., and Blakemore, C. (1981). Visual cortex of an anthropoid ape. *Nature, Lond.* **291**, 588–90.

Volkmann, F.C., Riggs, L.A., and Moore, R.K. (1980). Eye blinks and visual suppression. *Science, N.Y.* **207**, 900–2.

—— —— —— and White, K.D. (1978a). Central and peripheral determinants of saccadic suppression. In *Eye movements and the higher psychological functions* (ed. J. W. Senders, D. F. Fisher, and R. D. Monty) pp. 35–54. Lawrence Erlbaum Associates, Hillside, N.J.

—— —— —— (1978b). Contrast sensitivity during saccadic eye movements. *Vis. Res.* **18**, 1193–9.

von Grünau, M.W. (1976). The 'fluttering heart' and spatio-temporal characteristics of color processing—III. Interactions between the systems of the rods and the long-wavelength cones. *Vis. Res.* **16**, 397–401.

—— (1978a). Interaction between sustained and transient channels: Form inhibits motion in the human visual system. *Vis. Res* **18**, 197–201.

—— (1978b). Dissociation and interaction of form and motion information in the human visual system. *Vis. Res.* **18**, 1485–9.

—— (1979). Form information is necessary for the perception of motion. *Vis. Res.* **19**, 839–41.

—— (1981). The origin of pattern information of an apparently moving object during stroboscopic motion. *Percept. Psychophys.* **30**, 357–61.

von Neumann, J. (1966). *Theory of self-reproducing automata.* (ed. A. W. Burks). University of Illinois Press, Urbana, Ill.

Waddington, C.H. (1954). Evolution and epistemology. *Nature, Lond.* **173**, 880.

—— (1975). *The evolution of an evolutionist.* Cornell University Press, Ithaca, New York.

Wade, N.J. (1974). Some perceptual effects generated by rotating gratings. *Perception* **3**, 169–84.

Wald, G. (1961). Retinal chemistry and the physiology of vision. In *Visual Problems of Color.* Symposium, Vol. 1, Teddington, England, National Physical Laboratory, pp. 15–67. Chemical Publishing, New York.

Walley, R.E. and Weiden, T.F. (1973). Lateral inhibition and cognitive masking: A neurophysiological theory of attention. *Psychol. Rev.* **80**, 284–302.

Walls. G. (1942). *The vertebrate eye.* Cranbook Institute, Bloomfield Hills, Ill.

Walters, W. (1953). *The living brain.* Norton and Co., New York.

Wanatabe, J. and Tasaki, K. (1980). Shift-effect in the rabbit retinal ganglion cells. *Brain Res.* **181**, 198–201.

Wässle, H., Peichl, L., and Boycott, B.B. (1981a). Morphology and topography of on- and off-alpha cells in the cat retina. *Proc. R. soc., Lond.* **212B**, 157–75.

—— Boycott, B.B., and Illing, R.-B. (1981b). Morphology and mosaic of on- and off-beta cells in the cat retina and some functional characteristics. *Proc. R. soc., Lond.* **212B**, 177–90.

—— and Illing, R.-B. (1980). The retinal projection to the superior colliculus in the cat: A quantitative study with HRP. *J. comp. Neurol.* **190**, 333–56.

—— Levick, W.R., and Cleland, B.G. (1975). The distribution of the alpha type of ganglion cells in the cat's retina. *J. comp. Neurol.* **159**, 419–38.

Watson, A.B. and Nachmias, J. (1977). Patterns of temporal interaction in the detection of gratings. *Vis. Res.* **17**, 893–902.

Webb, S.V. and Kaas, J.H. (1976). The sizes and distribution of ganglion cells in the retina of the owl monkey, *Aotus trivirgatus. Vis. Res.* **16**, 1247–54.

Weisstein, N. (1968). A Rashevsky–Landahl neural net: Stimulation of metacontrast. *Psychol. Rev.* **75**, 494–521.

—— (1971). W-shaped and U-shaped functions obtained for monoptic and dichoptic disk-disk masking. *Percept. Psychophys.* **9**, 275–8.

—— (1972). Metacontrast. In *Handbook of sensory physiology*, Vol. 7/4, *Visual psychophysics* (ed. D. Jameson and L. M. Hurvich) pp. 233–72. Springer Verlag, New York.

—— and Growney, R. (1969). Apparent movement and metacontrast: A note on Kahneman's formulation. *Percept. Psychophys.* **5**, 321–8.

—— and Haber, R.N. (1965). A U-shaped backward masking function. *Psychonomic Sci.* **2**, 75–6.

—— and Harris, C.S. (1974). Visual detection of line segments: An object-superiority effect. *Science, N.Y.* **186**, 752–5.

—— Harris, C.S., and Ruddy, M. (1973). An object superiority effect. *Bull. psychonomic Soc.* **2**, 324.

—— Jurkens, T., and Onderisin, T. (1970). Effect of forced-choice *vs.* magnitude-estimation measures on the waveform of metacontrast functions. *J. opt. soc. Am.* **60**, 978–80.

—— and Maguire, W. (1978). Computing the next step: Psychophysical measures of representation and interpretation. In *Computer vision systems* (ed. A. R. Hanson and E. M. Riseman) pp. 243–60. Academic Press, New York.

—— Ozog, G., and Szoc, R. (1975). A comparison and elaboration of two models of metacontrast. *Psychol. Rev.* **82**, 325–43.

Werblin, F.S. (1971). Adaptation in the vertebrate retina: Intracellular recording in *Necturus. J. Neurophysiol.* **34**, 357–65.

—— (1974). Control of retinal sensitivity. II. Lateral interactions at the outer plexiform layer. *J. gen. Physiol.* **63**, 62–87.

—— and Copenhagen, D.R. (1974). Control of retinal sensitivity. III. Lateral interactions at the inner plexiform layer. *J. gen. Physiol.* **63**, 88–110.

—— and Dowling, J.E. (1969). Organization of the retina of the mudpuppy, *Necturus maculosus. J. Neurophysiol.* **32**, 339–55.

Werner, H. (1935). Studies on contour: I. Qualitative analysis. *Am. J. Psychol.* **47**, 40–64.

—— (1940). Studies on contour strobostereoscopic phenomena. *Am. J. Psychol.* **53**, 418–22.

Wertheim, T. (1894). Über die indirekte Sehschärfe. *Zeitschr. Psychol. Sinnesphysiol.* **7**, 172–87.

Wertheimer, M. (1912). Experimentelle Studien über das Sehen von Bewegung. *Zeitschr. Psychol.* **61**, 161–265.

—— (1958). *Principles of perceptual organization*. Van Norstrand, Princeton, N.J.

Wesson, M.D. and Loop, M.S. (1982). Temporal contrast sensitivity in amblyopia. *Invest. ophthalmol. vis. Sci.* **22**, 98–102.

Westheimer, G. (1965). Spatial interaction in the human retina during scotopic vision. *J. physiol., Lond.* **181**, 881–94.

—— (1967). Spatial interaction in human cone vision. *J. physiol., Lond.* **190**, 139–54.

—— (1968). Bleached rhodopsin and retinal interaction. *J. physiol., Lond.* **195**, 97–106.

—— (1970). Rod–cone independence for sensitizing interaction in the human retina. *J. physiol., Lond.* **206**, 109–16.

—— and Hauske, G. (1975). Temporal and spatial interference with Vernier acuity. *Vis. Res.* **15**, 1137–41.

Wheeler, D.D. (1970). Processes of word recognition. *Cog. Psychol.* **1**, 59–85.

White, C.T. (1963). Temporal numerosity and the psychological unit of duration. *Psychol. Monogr.* **77**, Whole No. 575, 1–37.

White, C.W. (1976). Visual masking during pursuit eye movement. *J. exp. psychol.: hum. percept. Perf.* **2**, 469–78.

White, K.D., Post, R.B., and Leibowitz, H. (1980). Saccadic eye movements and body sway. *Science, N.Y.* **209**, 621–3.

White, T.W., Irvin, G.E., and Williams, M.C. (1980). Asymmetry in the brightness and darkness Broca–Sulzer effects. *Vis. Res.* **20**, 723–6.

—— and Rinalducci, E.J. (1981). Sensation length and equal brightness. *Vis. Res.* **21**, 603–5.

Whitten, D.N. and Brown, K.T. (1973*a*). The time courses of late receptor potentials from monkey cones and rods. *Vis. Res.* **13**, 107–35.

—— —— (1973*b*). Photopic suppression of monkey's rod receptor potential, apparently by a cone-initated lateral inhibition. *Vis. Res.* **13**, 1629–58.

—— —— (1973*c*). Slowed decay of the monkey's cone receptor potential by intense stimuli, and protection from this effect by light adaptation. *Vis. Res.* **13**, 1659–67.

Wiener, N. (1948). *Cybernetics*. John Wiley and Sons, New York.

Wiesel, T.N. and Hubel, D.H. (1963). Single cell responses in striate cortex of kittens deprived of vision in one eye. *J. Neurophysiol.* **26**, 1003–17.

—— —— (1966). Spatial and chromatic interactions in the lateral geniculate body of the rhesus monkey. *J. Neurophysiol.* **29**, 1115–56.

—— —— (1974). Ordered arrangement of orientation columns in monkeys lacking visual experience. *J. comp. Neurol.* **158**, 307–18.

Williams, A., and Weisstein, N. (1978). Line segments are perceived better in a coherent context than alone: An object-line effect in visual perception. *Mem. Cog.* **6**, 85–90.

Williams, M.C. (1980). *Fast and slow responses to configurational factors in 'object-superiority' stimuli.* Ph.D. thesis, Department of Psychology, State University of New York at Buffalo, Buffalo, New York.

—— and Weisstein, N. (1980*a*). Apparent depth and connectedness produce spatial frequency specific effects on metacontrast. Paper presented at *The annual meeting of the Association for Research in Vision and Ophthalmology, Orlando, Florida, May.*

—— —— (1980*b*). Perceptual grouping produces spatial frequency specific effects on metacontrast. Paper presented at *The annual meeting of the Association for Research in Vision and Ophthalmology, Orlando, Florida, May.*

—— —— (1981). Spatial frequency response and perceived depth in the time-course of object superiority. *Vis. Res.* **21**, 631–46.

Williamson, S.J., Kaufman, L., and Brenner, D. (1977). Magnetic fields of the human brain. *Naval Res. Rev.* **30**, 1–18.

—— —— —— (1978). Latency of the neuromagnetic response of the human visual cortex. *Vis. Res.* **18**, 107–10.

Wilson, H.R. (1980). Spatiotemporal characterization of a transient mechanism in the human visual system. *Vis. Res.* **20**, 443–52.

—— Phillips, G., Rentschler, I., and Hilz, R. (1979). Spatial probability summation and disinhibitiion in psychophysically measured line-spread functions. *Vis. Res.* **19**, 593–98.

Wilson, J.R. and Sherman, S.M. (1976). Receptive-field characteristics of neurons in cat striate cortex: Changes with visual field eccentricity. *J. Neurophysiol.* **39**, 512–33.

Wilson, J.T.L. (1981). Visual persistence at both onset and offset of stimulation. *Percept. Psychophys.* **30**, 353–6.

—— and Singer, W. (1981). Simultaneous visual events show a long-range spatial interaction. *Percept. Psychophys.* **30**, 107–13.

Wilson, M.E. and Toyne, M.J. (1970). Retino-tectal and cortico-tectal projections in *Macaca mulatta*. *Brain Res.* **24**, 395–406.

Wilson, P.D., Rowe, M.H., and Stone, J. (1976). Properties of relay cells in cat's lateral geniculate nucleus: A comparison of W-cells with X- and Y-cells. *J. Neurophysiol.* **39**, 1193–1209.

Winters, R.W. and Hamasaki, D.I. (1975). Peripheral inhibition in sustained and transient on-center ganglion cells in cat retina. *Experentia* **31**, 305–6.

—— —— (1976). Temporal characteristics of peripheral inhibition of sustained and transient ganglion cells in cat retina. *Vis. Res.* **16**, 37–45.

Winterkorn, J.M.S., Shapley, R., and Kaplan, E. (1981). The effect of monocular paralysis on the lateral geniculate nucleus of the cat. *Exp. brain Res.* **42**, 117–21.

Wolf, W., Hauske, G., and Lupp. U. (1978). How presaccadic gratings modify postsaccadic modulation transfer function. *Vis. Res.* **18**, 1173–9.

—— —— —— (1980). Interaction of pre- and post-saccadic patterns having the same coordinates in space. *Vis. Res.* **20**, 117–25.

Woodworth, R.S. (1938). *Experimental psychology*. Holt, Rhinehart and Winston, New York.

Wundt, W. (1899). Zur Kritik tachistoskopischer Versuche I. *Philosophische Studien* **15**, 287–317.

—— (1900). Zur Kritik tachistoskopischer Versuche II. *Philosophsiche Studien* **16**, 61–81.

Wurtz, R.H. (1969). Visual receptive fields of striate cortex neurons in awake monkeys. *J. Neurophysiol.* **32**, 727–42.

—— (1976). Extraretinal influences on the primate visual system. In *Eye movements and psychological processes* (ed. R. A. Monty and J. W. Senders) pp. 231–44. Lawrence Erlbaum Associates, Hillsdale, N.J.

—— and Albano, J.E. (1980). Visual-motor function of the primate superior colliculus. *Ann. rev. Neurosci.* **3**, 189–226.

—— and Goldberg, M.E. (1971). Superior colliculus responses relates to eye movements in awake monkeys. *Science, N.Y.* **171**, 82–4.

—— —— (1972). Activity of superior colliculus in behaving monkey. Effects of lesions on eye movements. *J. Neurophysiol.* **35**, 587–96.

—— —— and Robinson, D.L. (1980*a*). Behavioral modulation of visual responses in the monkey: Stimulus selection for attention and movement. In *Progress in psychobiology and physiological psychology*. **9**, 43–83.

—— and Mohler, C.W. (1974). Selection of visual targets for the initiation of saccadic eye movements. *Brain Res.* **71**, 209–14.

—— —— (1976*a*). Organization of monkey superior colliculus: Enhanced visual response of superficial layer cells. *J. Neurophysiol.* **39**, 745–65.

—— —— (1976*b*). Enhancement of visual responses in the monkey striate cortex and frontal eye fields. *J. Neurophysiol.* **39**, 766–72.

—— Richmond, B.J., and Judge, S.J. (1980*b*). Vision during saccadic eye movements. III. Visual interactions in monkey superior colliculus. *J. Neurophysiol.* **43**, 1168–81.

Yarbus, A.L. (1967). *Eye movements and vision*. Plenum, New York.

Yellott, J. I., Jr. and Wandell, B.A. (1976). Color properties of the contrast flash effect: Monoptic *vs.* dichoptic comparisons. *Vis. Res.* **16**, 1275–80.

Yin, T.C.T. and Mountcastle. V.B. (1977). Visual input to the visuomotor mechanisms the monkey's parietal lobe. *Science, N.Y.* **197**, 1381–3.

Yinon, U. (1976). Age dependence of the effect of squint on cells in kitten's visual cortex. *Exp. brain. Res.* **260**, 151–7.

Young, L., Dichgans, J., Murphy, R., and Brandt, Th. (1973). Interaction of optokinetic and vestibular stimuli in motion perception. *Acta Otolaryngolica* **76**, 24–31.

Yukie, M. and Iwai, E. (1981). Direct projection from the dorsal lateral geniculate nucleus to the prestriate cortex in macaque monkeys. *J. comp. Neurol.* **201**, 81–97.

Zacks, J.L. (1975). Changes in response of X and Y type cat retinal ganglion cells produced by changes in background illumination. Paper presented at *The annual meeting of the Association for Research in Vision and Ophthalmology, Sarasota, Florida, April.*

Zeki, S.M. (1971). Cortical projections from two separate areas in the monkey. *Brain Res.* **34**, 19–35.

—— (1978). The cortical projections of the foveal striate cortex in the rhesus monkey. *J. physiol., Lond.* **277**, 227–44.

—— (1980). A direct projection from area V1 to area V3A of rhesus monkey visual cortex. *Proc. R. soc., Lond.* **207B**, 499–506.

Zetlan, S.R., Spear, P.D., and Geisert, E.E. (1981). The role of cortico-geniculate projections in the loss of Y-cells in monocularly deprived cats. *Vis. Res.* **21**, 1035–9.

Zinchenko, V.P. and Vergiles, N.Y. (1972). *Formation of visual images.* Consultants Bureau, New York.

Zöllner, F. (1862). Über eine neue Art anorthoskopischer Zerrbilder. *Annalen der Physik und Chemie* **27**, 477–84.

Author index

Page numbers in italics refer to the list of references

431

Subject index